Lockhart & Wiseman's Crop Husbandry Including Grassland

Related titles:

Cereal grains for the food and beverage industries
(ISBN 978-0-85709-413-1)

Environmental assessment and management in the food industry
(ISBN 978-1-84569-552-1)

Food processing technology, 3rd edition
(ISBN 978-1-84569-216-2)

Woodhead Publishing Series in Food Science, Technology and Nutrition:
Number 277

Lockhart & Wiseman's Crop Husbandry Including Grassland

Ninth Edition

H. J. S. Finch, A. M. Samuel and G. P. F. Lane

AMSTERDAM • BOSTON • CAMBRIDGE • HEIDELBERG • LONDON
NEW YORK • OXFORD • PARIS • SAN DIEGO
SAN FRANCISCO • SINGAPORE • SYDNEY • TOKYO
Woodhead Publishing is an imprint of Elsevier

WP
WOODHEAD
PUBLISHING

Woodhead Publishing is an imprint of Elsevier
80 High Street, Sawston, Cambridge, CB22 3HJ, UK
225 Wyman Street, Waltham, MA 02451, USA
Langford Lane, Kidlington, OX5 1GB, UK

Notice
No responsibility is assumed by the publisher for any injury and/or damage to persons or
property as a matter of products liability, negligence or otherwise, or from any use or
operation of any methods, products, instructions or ideas contained in the material herein.
Because of rapid advances in the medical sciences, in particular, independent verification of
diagnoses and drug dosages should be made.

British Library Cataloguing-in-Publication Data
A catalogue record for this book is available from the British Library

Library of Congress Control Number: 2014937164

ISBN 978-1-78242-371-3 (print)
ISBN 978-1-78242-392-8 (online)

For information on all Woodhead Publishing publications
visit our website at http://store.elsevier.com/

Typeset by RefineCatch Limited, Bungay, Suffolk

Printed and bound in the United Kingdom

Working together
to grow libraries in
developing countries

www.elsevier.com • www.bookaid.org

This book is dedicated to the memory of Jim Lockhart

Contents

Woodhead Publishing Series in Food Science, Technology and Nutrition

Foreword

First published in 1966, *Lockhart & Wiseman's Crop Husbandry Including Grassland* has now established itself as the standard source of information relating to modern crop production technology and management for students and farmers. Not only is it one of the most popular reference sources for students in colleges and universities studying temperate agriculture, it is a text much liked by practitioners engaged in commercial crop production, as well as by others also employed in the crop-based industries serving farming. In my view, each new edition confirms and extends the reputation of this seminal book, and the Ninth Edition is no exception. It includes many updates and revisions, as it should, in response to the many changes in agricultural practice since its last publication in 2002. The content and coverage continues to expand, and rightly so. Living in a period of considerable and complex change, it is good to see this standard textbook keeping up with the times.

I have had the privilege of knowing both original authors at the Royal Agricultural College (as was). Jim Lockhart (who sadly passed away in 2012) was a very enthusiastic potatoes and cereals man, with a love of practical farming, and this shone through his teaching to many hundreds of agriculture students at Cirencester during his long career. Tony Wiseman, now retired, was no less a significant influence on the careers of many future farmers and land agents during his own long teaching career. Tony was a particularly keen grassland and forage crop enthusiast, and this still comes through the book today as it did in his teaching at the College over many years. *Crop Husbandry* is testimony to their commitment to agricultural education, and their 'baby' has since been in the safe hands of my close colleagues in writing the most recent editions. Steve, Alison and Gerry clearly share the same enthusiasm for the subject as the original authors, and this shines through in the new Ninth Edition.

As a 'Crops' man myself, I continue to check various important issues on occasion in *Crop Husbandry* and the Ninth Edition is welcome. Important basic information needed to ensure crops can continue to be successfully grown is as important now, if not more so with increasingly more mouths to feed, as it has ever been. The Ninth Edition of *Crop Husbandry* should certainly help to meet this need in our constant struggle to boost crop performance and future profitability, and in a much more sustainable manner.

Professor Paul Davies
Vice Principal
Royal Agricultural University, UK

Acknowledgements

The authors are grateful to the following for their contributions to this book:

- Dr Keith Chaney, Principal Lecturer, Harper Adams University, UK (author of Chapter 9 *Sustainable crop management*).
- Dr Jim Monaghan, Principal Lecturer, Harper Adams University, UK (author of Chapter 17 *Fresh produce crops*).
- Dr Helen Moreton, Senior Lecturer in Animal Science, Royal Agricultural University, UK (co-author of Chapter 21 *Grazing management* and Chapter 22 *Conservation of grass and forage crops*).

Introduction

Jim Lockhart and Tony Wiseman published their first edition of *Crop Husbandry* in 1966. At the time there were many changes taking place in arable production and yields were increasing rapidly. Concerns in crop production were very different from those found today. Now, over 47 years later, we have produced the ninth edition. It is fitting with the passing away of Jim Lockhart to look at how crop production changed during his lifetime.

Cropping

Until the 1980s, barley was the most commonly grown cereal crop (Fig. 0.1). Since then, winter wheat has become the major cereal crop in the UK, while oats continue to be a minor crop. There are many reasons for the shift in popularity of the cereal crops but the main ones are increased yields, markets and most importantly profitability. In the first edition, spring barley was considered the highest yielding barley crop and some farmers were growing continuous spring barley; this is not the case now. With the increase in understanding of diseases such as BYDV in winter barley there were many years when winter crops were more popular than spring crops and in the 1980s many farmers on poorer soils were growing continuous winter barley. Subsequent problems with barley yellow mosaic virus as well as some poor harvests brought this trend to an end and now the popularity has swung back more in favour of spring barley. Rotation was also not considered to be so important with the introduction of more pesticide options and inorganic fertilisers. This has also changed again as we struggle with problems such as pesticide resistance and increases in fertiliser prices.

Of the other arable crops, there have also been some changes (Fig. 0.2). Most significant is the increase in area now grown of oilseed rape. Once the UK joined the Common Market in 1973 (now European Union – EU) there was financial

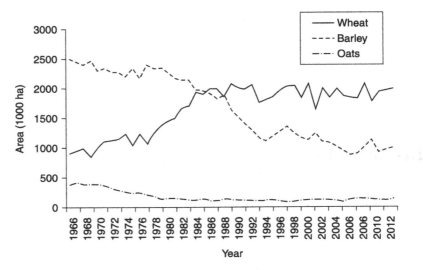

Fig. 0.1   UK cereal areas 1966–2012 (1000 ha), Defra.

support available for crop production in the Common Agricultural Policy (CAP), either in the form of production support or income support. One of these supports was for oil production from oilseed rape. This made the crop more viable economically. There were also major developments in breeding making oilseed rape better for oil production and increasing the market for rape meal. Now oilseed rape is the most commonly grown arable break crop. This had certainly not been envisaged back in the 1960s. Other break crops such as field beans, peas and linseed have fluctuated in areas grown mainly depending on support payments and prices.

Other important changes in arable cropping affected sugar beet and potatoes. Sugar beet is regulated in the EU, though we could grow more. Over the years there have been major reductions in the number of factories and this has restricted areas where the crop is grown. The reduction in area grown of sugar beet helped increase the area of oilseed rape as growers looked for suitable alternative break crops. The area grown of potatoes used to be governed by a UK quota system but after joining the EU this quota system was scrapped. Potatoes have now become a very highly mechanised operation, with limited direct farm sales and more being sold on contract to packers for supermarkets and processors; this has led to fewer growers.

The area of field vegetables has also declined despite new technologies, meaning more crops can be produced on a smaller area. The UK has continued to increase the percentage of imported vegetables as some countries are able to produce vegetables at different times of the year from the UK and sometimes more cheaply.

The other major change in cropping has been in area grown of forage crops, including forage maize. Again in the 1960s virtually no maize was grown in the

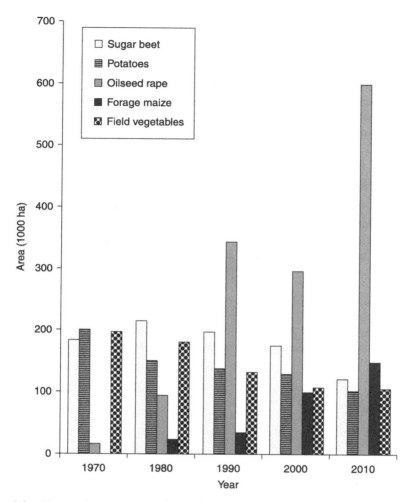

Fig. 0.2   Changes in area grown of sugar beet, potatoes, oilseed rape, forage maize and field vegetables, in England, 1970–2010 (1000 ha), Defra.

UK, though there was quite an area of other forage crops such as kale, mangolds, turnips and swedes. The development of new maize varieties that are able to cope with the UK climate, as well as the cost of importing livestock feeds, led to farmers trying to grow more of their own livestock requirements. Livestock systems in the 1960s relied more on grazed crops as fewer animals were housed during the winter months. With the increase in winter housing and silage rather than hay based systems, other forage crops have declined in importance. Grassland areas have also changed; permanent pasture has remained fairly static but there has been a large decline in area of temporary grass (under 5 years old). In the

1960s ley farming (rotating crops with short term leys) was important and, according to the first edition of this book, 'mixed farming was found on most lowland farms'. Subsequently, many farms have become much more specialised with fewer mixed farms, especially in eastern England. In the future, with changes in EU support payments, it may be that farmers have to look again at having a broader rotation.

One other area that has changed since the 1960s is the importance of industrial crops, so a new chapter on these crops has been added in this edition.

Crop yields

Many crop yields have significantly increased since the 1960s. Wheat and barley yielded similarly in 1966 at around 4 tonnes/hectare (t/ha). Yields increased steadily right up to the late 1990s with most improvement in the yield of wheat. Since the late 1990s yields have plateaued (Fig. 0.3) despite improvements in varieties and crop husbandry and a small increase in atmospheric carbon dioxide.

In root crops (Fig. 0.4) there has been a steady increase in potato yields, but it is sugar beet that is the real success story with very good yield improvements, especially over the last 30 years. These yield increases can be attributed to both improvements in crop husbandry as well as in plant breeding.

Fig. 0.3   UK cereal yields from 1966–2012 (t/ha), Defra.

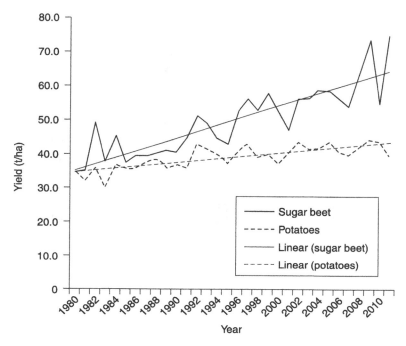

Fig. 0.4   UK sugar beet and potato yields 1980–2011, Defra.

Crop inputs

Crop inputs have increased dramatically since 1966 when, for example, in wheat N rates were only 75 kg/ha with 60 kg/ha P_2O_5 and K_2O, a seed dressing was commonly used plus one or two herbicides and occasionally a plant growth regulator was applied. Some inputs were even supported by government grants, such as the liming subsidy.

Fertiliser applications

Application rates of fertiliser N was one of the main changing inputs over the years (Fig. 0.5). N rates increased right up to the 1990s at the same time as crop yields were increasing. Since then nitrogen rates have stayed fairly constant in arable crops. In 1966 ammonium nitrate was only just being developed and used; previously the main nitrogen source other than organic manures was ammonium sulphate. It was not until the 1970s that prilled ammonium nitrate was readily available, which led to increased ease of application. In grassland, N inputs peaked in the 1990s but have since declined.

One nutrient that was barely mentioned in the first edition was sulphur. At the time there were enough atmospheric depositions of S so that none had to be

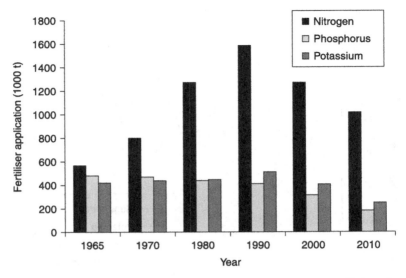

Fig. 0.5   Total amount of N, phosphorus (P$_2$O$_5$) and potassium (K$_2$O) applied to all crops and grassland in the UK, Defra.

applied to any crop (though the main bag N fertiliser then did contain some S). With the reduction of air pollution many parts of the UK now suffer from S deficiencies in a range of crops and applications of fertiliser sulphur have increased.

The first fertiliser manual (RB209) with recommendations for agricultural and horticultural crops was published by MAFF in 1973. This manual has been edited several times and is now the industry standard for fertiliser recommendations. Optimum recommended rates have changed according to trial results as well as fertiliser price, use of organic manures and legislation.

Crop protection chemicals

The use of agrochemicals is an area of crop husbandry that has changed tremendously over the years (Fig. 0.6). In the first edition, other than seed dressings there were virtually no fungicides available, herbicides were limited as were insecticides. On average in 1974 there were only two treatments applied per crop compared with over 10 now. In some crops, such as sugar beet, hand hoeing within the crop row was commonly undertaken; now a very detailed herbicide programme has been developed. Many new chemicals, especially fungicides, were developed between the 1970s and 2000. The new agrochemicals that were developed were safer environmentally and were also applied in much lower quantities (g rather than kg of active ingredients). So in practice, though more treatments are applied, lower amounts of active ingredient are applied per ha (Fig. 0.7). There has been an increase in the amount of legislation surrounding

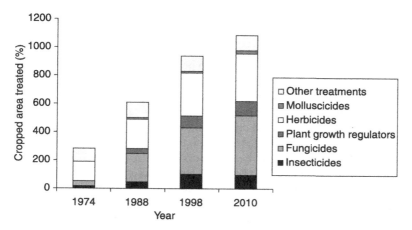

Fig. 0.6 Pesticide-treated area of crops as a percentage of cropped land (adapted from pesticide usage survey results, Defra). Note: Data were for England and Wales in 1974, for Great Britain in 1988 and 1998, and for the UK in 2010.

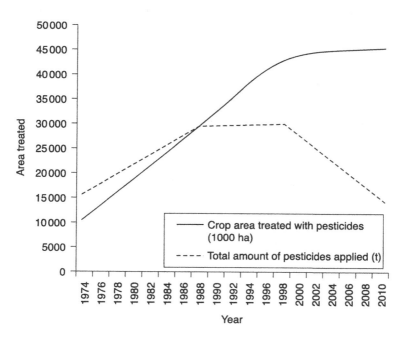

Fig. 0.7 Changes in area treated with pesticides (1000 ha) and total amount of pesticides applied (t) (Pesticide Usage Survey). Note: Data were for England and Wales in 1974, for Great Britain in 1988 and 1998, and for the UK in 2010.

pesticide use to ensure safety to operators as well as to the environment and food safety. Knowledge of disease epidemiology and pest life cycles has also improved and decision support programmes have been developed.

In the first edition there was nothing on pesticide resistance. There had been a few cases of resistance to the organochlorine insecticides but it was not considered an important issue. Now, with 40 plus years of routine use of pesticides, resistance is a common problem in many groups of agrochemicals. It is partly because of resistance issues and the limited production of new chemicals that many farmers have had to take a more integrated approach to weed, pest and disease control, so a specific chapter was added in the 8th edition.

As the development of new chemicals has become very expensive there has been a very limited number of new chemicals coming onto the market and many companies are now looking at biological control methods. In the first edition, biological control methods were considered to have little application in this country; how things have changed. It is now standard practice in glasshouse production systems to use biological control methods and it is even beginning to be looked at for use in the field.

With the increase in complexity of crop protection, a training and regulatory scheme for agrochemicals (BASIS) was established in 1978; this meant that the growing number of crop advisers could be trained in weed, pest and disease identification, lifecycles and control as well as in safe storage and applications of pesticides. (A similar fertiliser scheme was developed in 1993 (FACTS).)

Varieties and plant breeding
Variety selection has always been very important in crop production. Quite a percentage of the yield increases up to the 1990s was attributed to improvements in crop varieties. In the first edition, as now, we recommend readers to consult the recommended variety lists. What is different now is that the majority of plant breeding companies are independent. In the 1960s and 1970s the UK had its own state funded plant breeding stations including the plant breeding institute (PBI) which was very important in producing some of the most commonly grown wheat, barley and potato varieties of the time. In 1987 it was privatised. Since then private plant breeding companies have continued to develop with the use of biotechnology, production of hybrid varieties and increase in knowledge of crop gene sequencing. GM crops are still not allowed to be grown in this country but it will be interesting to see what happens in the near future.

Crop husbandry
During the 1970s and 1980s there was an increased amount of research work into crop physiology, especially for cereals and potatoes. This led to a better understanding of the interaction between seed rates, sowing dates and varieties and the effect of other inputs, especially nitrogen on crop growth and yield. In the first edition, all the different crops' husbandry was lumped together; subsequent

editions have split many of the crops into different chapters as the amount of crop husbandry knowledge increased.

Sugar beet is one of the crops whose husbandry has changed most since 1966. Monogerm pelleted (insecticide treated) seed, as predicted, is now the norm and has enabled the crop to be precision drilled, removing the need to hand single. With improvements in varieties it is now possible to drill the crop earlier, which has also helped to maximise yields. Crop protection programmes are now much more sophisticated, especially for weed control, with development of the repeated low dose programme. An understanding of soil structure and compaction led to the development of controlled wheeling systems and measures to reduce incidence of soil erosion.

Crop mechanisation and storage

There have been some very major changes in how crops are harvested and stored compared with 1966. At the time of writing the first edition, 30% of the sugar beet crop was still harvested by hand. Potato harvesters were only just being developed and the majority of potato crops were still hand lifted. Potatoes were often stored in clamps outside, whereas now many farmers have temperature controlled, insulated stores. After joining the common market there were some good capital grant schemes available (e.g. FHDS) which encouraged participating farmers to expand, especially in relation to building silage pits, grain and potato stores.

There have been major changes in the size and capacity of machinery. For example, in the 1960s fertiliser spreaders only had a spread of 9 m compared with up to 36 m now. Tramlines were initially not used in cereal crops as there were very few passes across most crops; as inputs increased and farmers required more precision they started drilling tramlines. Now some farmers are using remote sensing technology and variable rate applications of fertiliser in order to target and optimise fertiliser applications. These precision farming techniques are discussed in the new precision farming chapter added in this edition.

Seed bed cultivations have varied across the years. In the 1980s many farmers were direct drilling after burning the crop stubble. Once burning was banned many people went back to plough-based systems, but with increased fuel prices farmers are looking again at minimum tillage systems.

In horticulture there have been major developments in crop mechanisation. Many crops are now grown from plants raised in modules; planters can be tractor-less using precision farming techniques; and the use of crop covers has enabled growers to extend the crop season. Several harvesters and field rigs have been developed for harvesting and packing in the fields. Crops are often cooled as soon as they are harvested and are then refrigerated during storage and transportation in order to maintain quality before arriving on the supermarket shelves (chill chain). With the reduction in approved pesticides, many growers have to take a much more integrated approach as far as crop protection is concerned.

Support and legislation

During the 1960s and 1970s farmers were encouraged to increase production, either by UK support or subsequently the CAP. The CAP was very successful in increasing crop production across the EU. Unfortunately this led to mountains of some commodities such as wheat which had to be stored in large intervention stores. Subsequently, set-aside was introduced in 1992 to take some arable land out of production. The need for set-aside was removed in 2008 mainly due to a reduction of the grain mountain and also a fall in world supplies. Over the years there have been a number of EU directives that have affected the way we farm, including the nitrates directive and water framework directive. Some of these directives have been introduced to limit agricultural pollution and soil erosion and to ensure the sustainability of farming systems. In this current edition we have included a chapter on sustainable crop management.

As well as the various EU directives impacting on arable production, once we joined the EU it also led to the adoption of metric units. The first edition of crop husbandry was all written using imperial units; now of course everything is metricated, though in practice many farmers still use both systems!

Environmental issues

Environmental issues were not discussed or considered a problem in the 1960s. It was only during the 1970s and 1980s that many people became concerned about loss of habitats and biodiversity, soil erosion and pollution. Birds are a good indicator of what is happening to farmland biodiversity as they have a wide variety of habitat and food requirements and are fairly easy to monitor. Since 1970 farmland bird populations have fallen so that in 2011 numbers were only 60% of the base numbers in 1970. Some specialist bird species have been more seriously affected. The decline in numbers was greatest during the 1970s and 1980s. The reasons are many, including loss of mixed farms, more autumn cropping and production of silage not hay.

In 1985 the first funded Environmentally Sensitive Area (ESA) was started in England and this was the beginning of a number of funded farm conservation schemes. Since then there have been various CAP-supported agri-environmental schemes introduced. In England about two-thirds of the farmed land is covered by one of the agri-environment schemes. The current CAP single payment scheme also has a number of environmental management requirements – cross-compliance.

In 1966 virtually no one talked about organic farming – it was only after some of these environmental issues were raised, plus financial support, that farmers started converting to this farming system. Jim Lockhart was very much against this type of farming as he had grown up in a time when farmers were trying to reduce the amount of manual work carried out on farms! It was quite a struggle to persuade him to let us put a chapter in this book. We think he would have been quite surprised about the current area of organically grown crops in the UK.

Research

In the 1960s and 1970s most agricultural research and advice was provided free by the government (Ministry of Agriculture Fisheries and Food – MAFF and the Agricultural Development and Advisory Service – ADAS). The UK was a net importer of food and the government's target was to promote crop production. This all changed after the 1980s when there were 'Wheat and other food mountains', and the government started reducing the amount of support for agricultural research. Organisations such as the weed research organisation were closed. Later ADAS was privatised and the government pushed for the agricultural industry itself to support research through crop levies. Currently, the Agriculture and Horticulture Development Board (AHDB) is the statutory levy board in the UK, and farmers, growers and others in the supply chain are charged a small levy to fund much needed, independent, research and development work.

In 1979 Jim Lockhart was involved in getting one of the new national cereal events to be held in the Cotswolds – Barley '79'. These cereal events are now one of the most important annual arable shows in the country. After Barley '79' a group of progressive Cotswold farmers, with support from lecturers and advisers including Jim, started their own private cereal trials organisation – the Cotswold Cereal Centre (CCC). We are sure he would never have imagined how this private organisation would grow! With the loss of government funding for near market research, trials work had to be more farmer funded – exactly what the CCC was doing. Over the years it has grown to have centres in different parts of the country. It joined with Morley (trials and crop advisory group) to form The Arable Group (TAG) and subsequently combined with NIAB to form the national independent, agronomy research and information organisation NIAB TAG.

Education

There has been a decline in universities offering agricultural courses over the years; both Wye (Imperial) and Seale-Hayne (Plymouth University) colleges have been the most recent to close. More of the national institutions (such as the Royal Agricultural University and Harper Adams University) are providing the training, as well as county colleges. There are now a wider range of courses and levels of courses available, but student numbers have fluctuated widely over the years; numbers have been very dependent on the profitability of farming. With farming becoming more specialised and advanced mechanically and technologically, together with all the legislation that farmers need to follow it is more important than ever that future farmers have a good agricultural education – so there is still a requirement for this book!

Sources of further information and advice

Further reading

Anon, *Wild Bird Populations in the UK 1970–2011, National Statistics*, Defra, 2012.

Knight S, Knightley S, Bingham I, Hoad S, Lang B *et al.*, *HGCA Project Report 502, Desk study to evaluate contributory causes of the current 'yield plateau' in wheat and oilseed rape*, HGCA 2012.

Sly J M A, *Pesticide Usage Survey Report 8: Review of usage of pesticides in agriculture and horticulture in England and Wales 1965–1974*, MAFF, 1977.

Garthwaite D G, Thomas M R, *Pesticide Usage Survey Report 159, Arable farm crops in Great Britain*, MAFF, 1999.

Garthwaite D G, Barker I, Parrish G, Smith L, Chippendale C and Pietravalle S, *Pesticide Usage Survey Report 235, Arable crops in the UK 2010*, Defra, 2011.

Wilkins R J, *Grassland in the Twentieth Century*, IGER innovations, 2000.

Websites

www.defra.gov.uk

www.bto.org

www.hgca.com

www.statistics.gov.uk

www.fera.defra.gov.uk

www.ahdb.org.uk

www.niab.com

Part I

Principles of crop production

1

Plants

DOI: 10.1533/9781782423928.1.3

Abstract: This chapter describes the biology of plants, the most important organisms on the planet. It covers plant physiology and important biochemical processes such as photosynthesis and respiration. It describes the grouping of plants depending on their life cycle. It discusses the structures which are of most importance to crop production including seeds, roots and leaves. The chapter stresses the importance of the leguminous plants and their role in improving soil fertility. Finally it describes the requirements of plants for growth and development, and how these are controlled by plant hormones.

Key words: physiology, photosynthesis, water, nutrients, plant structure, plant hormones.

1.1   Introduction

Plants are living organisms consisting of many specialised individual cells. They differ from animals in many ways and a very important difference is that they can build up valuable organic substances from simple materials such as carbon dioxide and water. The most important part of this building process, called photosynthesis, is the production of carbohydrates such as sugars, starch and cellulose, along with oxygen. They have rigid cell walls enclosing a semi-permeable cell membrane which allows the passage of water through it (osmosis). They have specialised organs such as roots, stems and leaves, are mostly immobile and are primary producers of food in most land-based ecosystems.

1.2   Plant physiology

1.2.1   Photosynthesis

In photosynthesis a blue/green substance called chlorophyll A and a yellow/green substance called chlorophyll B use light energy (normally sunlight but sometimes artificial) to change carbon dioxide and water into sugars (carbohydrates) and oxygen in the green parts of the plant. The amount of photosynthesis per day which takes place is limited by the duration and intensity of sunlight, and the ability of the green parts of a plant to capture it. The amount of carbon dioxide available can also be a limiting factor. Shortage of water, low temperatures and leaf disease or damage can reduce photosynthesis, as can shading by other plants, e.g. by weeds in a crop. The cells that contain chlorophyll also have orange/yellow pigments such as xanthophyll and carotene, and brown pigments called phaeophytins which absorb different wavelengths of light than the chlorophylls. Crop plants can only build up chlorophyll A and B in the light, and so any leaves that develop in the dark are yellow and cannot efficiently produce carbohydrates. The yellowing of leaves (chlorosis) can also be caused by disease attack, nutrient deficiency or natural senescence (dying off).

Oxygen is released back into the atmosphere during photosynthesis and the process may be set out as follows:

(a) The light reaction (light dependent)
This takes place in the thylakoid membranes inside the 'chloroplast', an organelle found inside the cells of green tissue. Light provides energy for the chlorophyll molecule that releases electrons. These split water into oxygen and hydrogen.

The chemical reaction of this stage is:

$$2H_2O \rightarrow 2H_2 + O_2 \hspace{2cm} [1.1]$$

The hydrogen then moves into the next stage:

(b) The dark reaction (light independent)
This takes place in the watery stroma of the chloroplast. Here the hydrogen is combined with carbon dioxide by the Calvin Cycle to give carbohydrate and water:

$$2H_2 + CO_2 \rightarrow CH_2O + H_2O \hspace{2cm} [1.2]$$

The carbohydrates are simple sugars, which can be moved through the vascular system of the plant in solution to wherever they are needed. This process not only provides the basis for all food production but it also supplies the oxygen which animals and plants need for respiration. The simple carbohydrates, such as glucose, may be built up to form starch for storage purposes or as cellulose for building cell walls. Fats and oils (lipids) are formed from carbohydrates by a process of esterification which produces mostly triglycerides. These are usually found in seeds and are a form of concentrated energy. Protein material, which is an essential part of all living cells, is made from carbohydrates and nitrogen compounds and also frequently contains sulphur. These form amino acids which are held together in proteins by peptide bonds.

Most plants consist of roots, stems, leaves and reproductive parts and need a medium in which to grow. These media could be soil, compost, water where plants are grown hydroponically or even air, where the bare roots are sprayed with a fine mist of nutrients and water (aeroponics). In soil the roots spread through the spaces between the particles and anchor the plant. The amount of root growth can

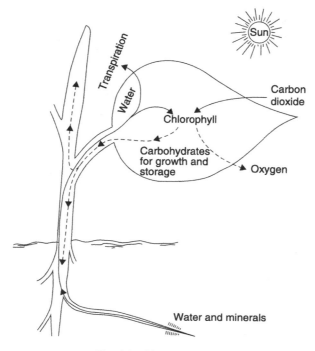

Fig. 1.1   Photosynthesis.

be phenomenal. For example, in a single plant of wheat the root system may extend to many miles.

The leaves, with their broad surfaces, are the main parts of the plant where photosynthesis occurs (Fig. 1.1). A very important feature of the leaf structure is the presence of large numbers of tiny pores (stomata) on the surface of the leaf (Fig. 1.2). There are usually thousands of stomata per square centimetre of leaf surface. Each pore (stoma) is oval-shaped and surrounded by two guard cells. The carbon dioxide used in photosynthesis diffuses into the leaf through the stomata. Most of the water vapour leaving the plant, as well as the oxygen from photosynthesis, diffuses out through the stomata.

1.2.2   Transpiration

The evaporation of water from plants is called transpiration. It mainly occurs through the stomata and has a cooling effect on the leaf cells. Water in the cells of the leaf can pass into the pore spaces in the leaf and then out through the stomata as water vapour (Fig. 1.3).

The rate of transpiration varies considerably. It is greatest when the plant is well supplied with water and the air outside the leaf is warm and dry. When the guard cells are turgid (full of water) the stomata are open. When the plant is under

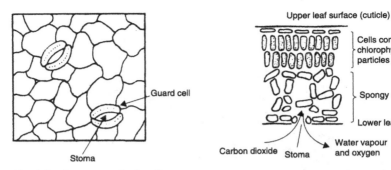

Fig. 1.2   Stomata on leaf surface.

Fig. 1.3   Cross-section of green leaf showing gaseous movements during daylight.

drought stress the guard cells lose water and the stoma closes, slowing down the loss of water vapour (transpiration) from the plant. It also slows down the rate of photosynthesis. The stoma also close in very cold weather, e.g. 0 °C. Transpiration is also retarded if the humidity of the atmosphere is high because there is only a very small water vapour gradient between the inside of the leaf and the outside atmosphere. The stomata guard cells close (and so transpiration ceases) during darkness. This is because photosynthesis ceases and water is lost from the guard cells when some of the sugars present change to starch.

1.2.3   Respiration

Plants breathe, like animals, i.e. they take in oxygen that combines with organic foodstuffs and releases energy, carbon dioxide and water. Plants are likely to be checked in growth if the roots are deprived of oxygen for respiration, which might occur in a waterlogged soil. Respiration appears, superficially, to be the reverse of photosynthesis, with carbohydrates combining with oxygen to give carbon dioxide and water with a release of energy. However, it is much more complicated with a very different metabolic process taking place.

There are two main processes. The first is glycolysis where simple sugars are split to release energy and to form pyruvic acid, water and a carrier molecule. The second stage is the 'Citric Acid' or 'Krebs' cycle where the pyruvic acid is converted to citric acid, which cycles within the system through intermediate molecules releasing energy in the form of ATP (adenosine triphosphate), carbon dioxide and water. The energy-rich ATP can then be moved around the plant to provide it with the energy it requires for metabolism, growth and development. The processes take place in the cell mitochondria, organelles which are also involved in cell cycling and cell growth.

1.2.4   Conduction

The conductive flow of water through the plant takes place in the xylem tissue that runs in bundles along the length of the root and stem and into the organs of the plant. The xylem or wood vessels, which carry the water and mineral salts from the roots to the leaves, are tubes made from dead cells called tracheids and vessel elements. The cross walls of these cells are no longer present and the longitudinal walls are thickened with lignum to form wood. These tubes help to strengthen the stem.

1.2.5   Translocation

The movement of food materials through the plant is known as translocation. The phloem tubes carry organic material through the plant, e.g. sugars and amino acids from the leaves to storage parts or growing points. These vessels are chains of living cells, not lignified, and with cross walls which are perforated. They are sometimes referred to as sieve elements which, along with companion cells providing energy and proteins, form the sieve tube members. These form a branched system throughout the plant and transport assimilates from sites of production (sometimes called sources) to sites of demand or consumption (sinks). The actual mechanism of transport is still not clear.

In the stem of dicotyledons, the xylem and phloem tubes are usually found in a ring near the outside of the stem with phloem on the outside and xylem on the inside, separated by a cambium layer. In monocotyledons the vascular bundles are more randomly arranged within the stem. In the root, the xylem and phloem tubes form separate bundles and are found near the centre.

1.2.6   Uptake of water

Water is taken into the plant from the soil. This occurs mainly through the root hairs near the root tip. As the root grows the hairs are constantly replaced. A few centimetres back from the root tip the hairs disappear. Their function is to increase the surface area available for absorption of water. There are thousands of root tips on a single healthy crop plant (Fig. 1.4).

The absorption of water into the plant in this way is due to suction pull, which starts in the leaves. As water transpires (evaporates) from the cells in the leaf, more water is drawn from the xylem tubes which extend from the leaves to the root tips. In these tubes the water is stretched like a taut wire. This is possible because the molecules of water are held together very firmly when in narrow tubes by the bonds between the hydrogen and oxygen atoms (cohesion-tension). The pull of this water in the xylem tubes of the root is transferred through the root cells to the root hairs and so water is absorbed into the roots and up to the leaves. In general, the greater the rate of transpiration, the greater the amount of water taken into the plant.

However, this 'transpiration pull' cannot completely explain water movement within plants. There appears to be a positive pressure exerted by roots which forces water up into the stem, and the very narrow vascular tubes act as a kind of wick (capillarity) which helps to hold water inside them.

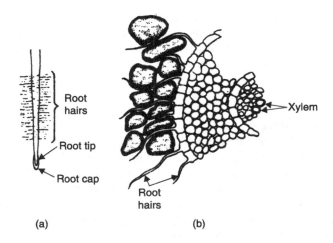

Fig. 1.4   (a) Section of root tip and root hair region. (b) Cross-section of root showing the root hairs as tube-like elongations of the surface cells in contact with soil particles.

The rate of absorption is slowed down by:

1. Shortage of water in the soil.
2. Lack of oxygen for root respiration (e.g. in waterlogged soils).
3. A high concentration of salts in the soil water near the roots.

Normally, the concentration of the soil solution does not interfere with water absorption. High soil water concentration can occur in salty soils and near bands of fertiliser. Too much fertiliser near developing seedlings may damage germination and subsequent emergence by restricting the uptake of water.

1.2.7   Osmosis
Much of the water movement into and from cell to cell in plants is due to osmosis. This is a process in which a solvent, such as water, will flow through a semi-permeable membrane (e.g. a cell wall) from a weak solution to a more concentrated one. The cell wall only allows the water to pass through, as the molecules in solution are too big. The force exerted by such a flow is called the osmotic pressure. In plants the normal movement of the water is from the soil solution into the cell. However, if the concentration of a solution outside the cell is greater than that inside, there is a loss of water from the cell, and its content contracts; this is called plasmolysis.

1.2.8   Uptake of nutrients
The absorption of chemical substances (nutrients) into the root cells is partly due to a diffusion process but it is mainly due to the ability of the cells near the root tips to accumulate such nutrients. Nutrients are taken into the root in the form of

charged ions through the root hairs, along with water. The water and solutes move through the cells into the inner ring of xylem. They are prevented from leaking back into the soil between the cell walls by a waxy layer of cells called the Casparian Strip. In this way an electrochemical gradient is produced which allows the flow of nutrient ions into the plant from the soil solution.

1.3   Plant groups

There are many ways of classifying plant groups but, from an agricultural and horticultural point of view, a useful way is to divide them into annuals, biennials and perennials according to their total length of life.

1.3.1   Annuals

Typical examples are wheat, barley and oats that complete their life history in one growing season, i.e. starting from the seed, in one year they develop roots, stem and leaves and then produce an ear which flowers and sets seed before dying.

1.3.2   Biennials

These plants grow for two years. They spend the first year in producing roots, stem and leaves, and the following year in producing the flowering stem and seeds, after which they die. Sugar beet, swedes and turnips are typical biennials, although the grower treats these crops as annuals, exploiting their life cycle by harvesting them at the end of the first year when most of the foodstuff is stored up in the root, and before the plant moves on to produce the seed head. Those plants that do behave as annuals and throw a seed head in their first year are called 'bolters'.

1.3.3   Perennials

They live for more than two years and, once fully developed, they usually produce seeds each year. Many of the grasses and forage legumes are perennials, as are many of the horticultural fruit crops such as raspberries or apples, and some energy crops such as willow and Miscanthus.

1.4   Structure of the seed

Plants are also classified as dicotyledons and monocotyledons according to the structure of the seed.

1.4.1   Dicotyledon

A good example of a dicotyledon seed is the field bean. If its pod is opened when nearly ripe it will be seen that each seed is attached to the inside of the pod by a short

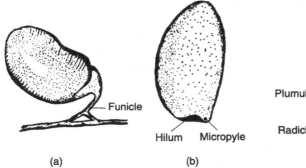

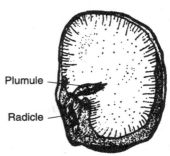

Fig. 1.5   (a) Bean seed attached to the inside of the pod by the funicle. (b) Bean seed showing the hilum and micropyle.

Fig. 1.6   Bean seed with one cotyledon removed.

stalk called the funicle. All the nourishment that the developing seed requires passes through the funicle from the bean plant. When the seed is ripe and has separated from the pod, a black scar, known as the hilum, can be seen where the funicle was attached. Near one end of the hilum is a minute hole called the micropyle (Fig. 1.5).

If a bean is soaked in water the seed coat can be removed easily and all that is left is largely made up of the embryo (germ). This consists of two seed leaves (cotyledons) which contain the food for the young seedling. Lying between the two cotyledons is the radicle (which eventually forms the primary root) and a continuation of the radicle at the other end, the plumule (Fig. 1.6). This develops into the young shoot and is the first bud of the plant.

1.4.2   Monocotyledon

This important class includes all the cereals and grasses. The wheat grain is a typical example. It is not a true seed (it should be called a single-seeded fruit). The seed completely fills the whole grain, being almost united with the inside wall of the grain or fruit. This fruit wall is made up of many different layers which are separated on milling into varying degrees of fineness, e.g. bran and pollards which are valuable livestock feed.

Most of the interior of the grain is taken up by the floury endosperm. The embryo occupies the small raised area at the base. The scutellum, a shield-like structure, separates the embryo from the endosperm. Attached to the base of the scutellum are the five roots of the embryo, one primary and two pairs of secondary rootlets. The roots are enclosed by a sheath called the coleorhiza while the shoot is enclosed by the coleoptile. The position of the radicle and the plumule can be seen in Fig. 1.7. The scutellum can be regarded as the cotyledon of the seed. There is only one cotyledon present and so wheat is a monocotyledon.

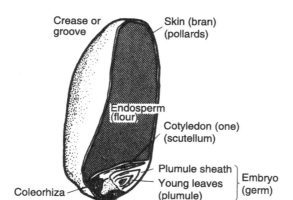

Fig. 1.7   Wheat seed cut in half at the crease.

1.4.3   Germination of the bean – the dicotyledon

Given suitable conditions for germination, i.e. water, heat and air, the seed coat of the dormant, but living, seed splits near the micropyle and the radicle begins to grow downwards through this split to form the main, or primary, root from which lateral branches will soon develop (Fig. 1.8). When the root is firmly held in the soil, the plumule starts to grow by pushing its way out of the same opening in the seed coat. As it grows upwards its tip is bent to protect it from injury in passing through the soil, but it straightens out on reaching the surface, and leaves develop very quickly from the plumular shoot.

In the field bean the cotyledons remain underground, gradually giving up their stored food materials to the developing plant. This is called hypogeal germination (Fig. 1.9a). However, in the French bean and in many other dicotyledon seeds, the cotyledons are brought above ground with the plumule. This is epigeal germination (Fig. 1.9b).

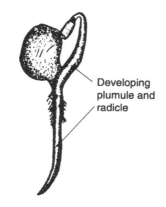

Fig. 1.8   Germination of bean seed.

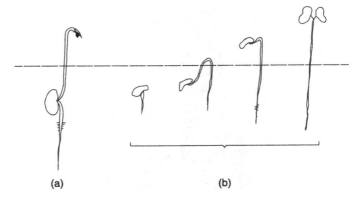

Fig. 1.9   (a) Hypogeal germination; (b) epigeal germination.

1.4.4   Germination of wheat – the monocotyledon

When the grain germinates the coleorhiza expands and splits open the seed coat and, at the same time, the roots break through the coleorhiza and enter the soil (Fig. 1.10). The primary root is soon formed, supported by the two pairs of secondary rootlets, but this root system (the seminal roots) is only temporary and is soon replaced by adventitious roots (Fig. 1.11 and 1.12). As the first root system is being formed at the base of the stem, so the plumule starts to grow upwards, and its first leaf, the coleoptile, appears above the ground as a single pale tube-like structure. From a slit in the top of the coleoptile there appears the first true leaf followed by others, the younger leaves growing from the older leaves (Fig. 1.13).

As the wheat embryo grows the floury endosperm is used up by the developing roots and plumule. The scutellum has the important function of changing the endosperm into digestible food for the growing parts, converting starch into simple, mobile sugars. In the field bean, the cotyledons provide the food for the early nutrition of the plant, whilst the wheat grain is dependent upon the endosperm and scutellum. In both cases it is not until the plumule has reached the light, and turned green, that the plant can begin to be independent. This point is important in relation to the depth at which seeds should be sown. Small seeds such as brassicas, clovers and many of the grasses must be, as far as possible, shallow sown. Their food reserves will be exhausted before the shoot reaches the surface if sown too deep. Larger seeds such as beans and peas can and should be sown deeper.

When the leaves of the plant begin to manufacture food by photosynthesis, and when the primary root has established itself sufficiently well to absorb nutrients from the soil, the plant can develop independently, provided there is sufficient moisture and air present in the soil, and conditions above ground are suitable for growth.

The main differences between the two groups of plants can be summarised in Table 1.1.

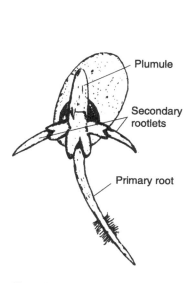

Fig. 1.10   Germination of the wheat grain.

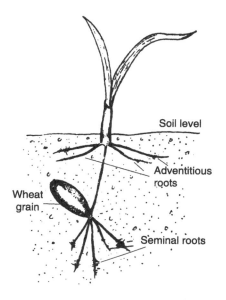

Fig. 1.11   Developing wheat plant.

Fig. 1.12   Adventitious root system.

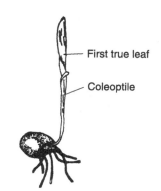

Fig. 1.13   Seedling wheat plant.

1.5   Plant structure

The plant can be divided into four parts: the root system, stem, leaf and flower.

1.5.1   The root system

The root system is concerned with the parts of the plant growing in the soil; there are two main types:

Table 1.1   Summary of the main differences between dicotyledons and monocotyledons

Dicotyledons	Monocotyledons
The embryo has two seed leaves (cotyledons).	The embryo has one seed leaf.
A primary root system is developed and persists.	A primary root system is developed, but is replaced by an adventitious root system.
Usually broad-leaved plants, e.g. legumes and sugar beet.	Usually narrow-leaved plants, e.g. cereals and grasses.

These two great groups of flowering plants can be further divided in the following way:

Families or orders: Legumes

Genus: Clovers of the legume family

Species: Red clover

Cultivar or variety: Late red clover

The tap root or primary system

This is made up of the primary root called the tap root with lateral secondary roots branching out from it and, from these, tertiary roots possibly developing obliquely to form, in some cases, a very extensive system of roots (Fig. 1.14). The root of the bean plant is a good example of a tap root system. If this is split it will be seen that there is a slightly darker central woody core, the skeleton of the root, which helps to anchor the plant and transport foodstuffs. The lateral secondary roots arise from this central core (Fig. 1.15). Carrots and other true root crops such as

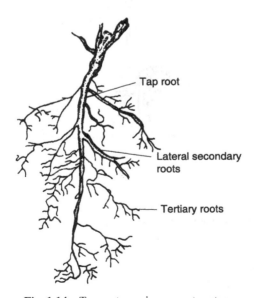

Fig. 1.14   Tap root or primary root system.

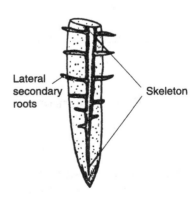

Fig. 1.15   Tap root of the bean plant.

sugar beet and mangels have very well developed tap roots. These biennials store food in their roots during the first year of growth to be used in the following year for the production of the flowering shoot and seeds. However, they are normally harvested after one season and the roots are used as food for humans and stock.

The adventitious root system
This is found on all grasses and cereals, and it is the main root system of most monocotyledons. The primary root is soon replaced by adventitious roots, which arise from the base of the stem (Fig. 1.12). These roots can, in fact, develop from any part of the stem, and they are found on some dicotyledons as well (but not as the main root system), e.g. underground stems of the potato.

Root hairs (Fig. 1.4) are very small white, hair-like structures that are found near the tips of all roots. As the root grows, the hairs on the older parts die off and others develop on the younger parts of the root. They play a very important part in the nutrition and water uptake of a plant.

1.5.2   The stem
The second part of the flowering plant is the shoot that normally grows upright above the ground. It is made up of a main stem and branches. Stems are either soft (herbaceous) or hard (woody) and in UK agriculture it is only the soft and green herbaceous stems which are of any importance. These usually die back every year. All stems start life as buds and the increase in length takes place at the tip of the shoot called the terminal bud. If a Brussels sprout is cut longitudinally and examined it will be seen that the young leaves arise from the bud axis. This axis is made up of different types of cell tissue, which is continually making new cells and thus growing (Fig. 1.16).

Stems are usually jointed with each joint forming a node. The part between two nodes is called the internode. At the nodes the stem is usually solid and thicker, and this swelling is caused by the storing up of material at the base of the leaf (Fig. 1.17).

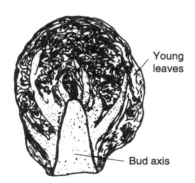

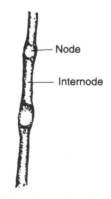

Fig. 1.16   Longitudinal section of a Brussels sprout.

Fig. 1.17   Jointed stem.

The bud consists of closely packed leaves arising from a number of nodes. It is, in fact, a condensed portion of the stem that develops by a lengthening of the internodes.

Axillary buds are formed in the angle between the stem and leaf stalk. These buds, which are similar to the terminal bud, develop to form lateral branches, leaves and flowers.

Modified stems

1. A stolon is a stem that grows along the ground surface.
   Adventitious roots are produced at the nodes, and buds on the runner can develop into upright shoots, and separate plants can be formed, e.g. strawberry plants (Fig. 1.18).
2. A rhizome is similar to a stolon but grows under the surface of the ground, e.g. common couch (Fig. 1.18).
3. A tuber is really a modified stolon. The ends of the stolons swell to form tubers. The tuber is therefore a swollen stem. The potato is a well-known example with 'eyes' (buds) which develop shoots when the potato tuber is planted. Potato tubers will also turn green by producing chlorophyll when exposed to light; this is another characteristic of stems.
4. A tendril is found on certain legumes, such as the pea. The terminal leaflet is modified as in the diagram (Fig. 1.19). This is useful for climbing purposes to support the plant.

Corms and suckers are other examples of modified stems.

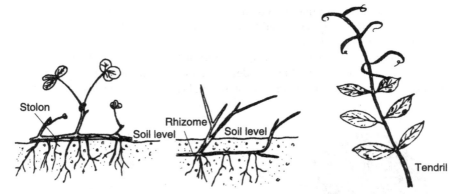

Fig. 1.18   Modified clover and couch grass stems.   **Fig. 1.19**   Modified pea stems.

1.5.3   The leaf

Leaves in all cases arise from buds. They are extremely important organs, being not only responsible for the manufacture of sugar and starch from the atmosphere for the growing parts of the plant, but they are also the organs through which transpiration of water takes place.

A typical leaf of a dicotyledon consists of three main parts:

1. The blade.
2. The stalk or petiole.
3. The basal sheath connecting the leaf to the stem. This may be modified (as with legumes) into a pair of wing-like stipules (Fig. 1.20a).

The blade is the most obvious part of the leaf and it is made up of a network of veins.

There are two main types of dicotyledonous leaves:

1. Those with a prominent central midrib, from which lateral veins branch off on either side. These side veins branch into smaller and smaller ones (Fig. 1.20a).
2. Those with no single midrib, but several main ribs spread out from the top of the leaf stalk; between these the finer veins spread out as before, e.g. horse-chestnut leaf (Fig. 1.20b).

The veins are the essential supply lines for the process of photosynthesis. They consist of two main parts: the xylem for bringing the required raw material up the leaf and the phloem that carries the finished product away from the leaf.

Leaves can show great variation in shape and type of margin, as in Fig. 1.20. They can also be divided into two broad classes as follows:

1. Simple leaves. The blade consists of one continuous piece (Fig. 1.20a).
2. Compound leaves. Simple leaves may become deeply lobed and when the division between the lobes reaches the midrib it becomes a compound leaf, and the separate parts of the blade are called the leaflets (Fig. 1.20b).

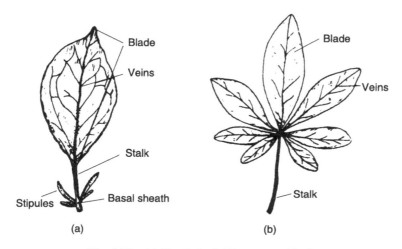

Fig. 1.20   (a) Simple leaf; (b) compound leaf.

The blade surface may be smooth (glabrous) or hairy, according to variety. This is important in legumes because it can affect their palatability. Surfaces of leaves can also be glaucous, i.e. covered with a bloom of wax. This wax coating gives the leaf some protection from disease or pest attack.

Monocotyledonous leaves are dealt with in Part IV, Chapter 19 (Grassland).

Modified leaves

1. Cotyledons or seed leaves are usually of a very simple form.
2. Scales are normally rather thin, yellowish to brown membranous leaf structures, very variable in size and form. On woody stems they are present as bud scales which protect the bud; they are also found on rhizomes such as common couch.
3. Leaf tendrils. The terminal leaflet on the stem can be modified into thin thread-like structures, e.g. the pea plant (see Fig. 1.19).

Other examples of modified leaves are leafspines and bracts.

1.5.4   The flower

In the centre of the flower is the axis that is simply the continuation of the flower stalk. It is known as the receptacle and on it are arranged four kinds of organ:

1. The lowermost is a ring of green leaves called the calyx, made up of individual sepals. The sepals offer the flower protection while still at the bud stage, when they are still closed.
2. Immediately above the calyx is a ring of petals known as the corolla. The number of petals on a flower can differ widely between species. Petals are usually brightly coloured and their function is to attract insects for pollination. At the base of the petal are modified structures called nectaries. These, as their name suggests, produce the sweet nectar that acts as a reward for a visiting insect.
3. Above the corolla are the stamens, again arranged in a ring. They are similar in appearance to an ordinary match, the swollen tip called the anther sitting on top of the filament. When ripe the anther bursts open to release the pollen grains. The filaments can be of varying sizes, either keeping the anther inside the flower or allowing it to hang outside (useful in wind pollinated plants).
4. The highest position on the receptacle is occupied by the pistil which consists of one or more small green bottle-shaped bodies called carpels. These are made up of three parts: the stigma, style and the ovary (containing ovules). It is within the ovary that the future seeds are produced (Fig. 1.21 and 1.22).

Most flowers are more complicated in appearance than the above, but basically they consist of these four main parts.

1.5.5   The formation of seeds

Pollination precedes fertilisation, which is the union of the male and female reproductive cells. When pollination takes place the pollen grain is transferred

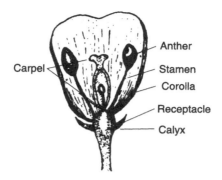

Fig. 1.21   Longitudinal section of a simple flower.

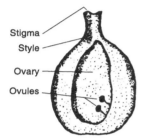

Fig. 1.22   Carpel detail.

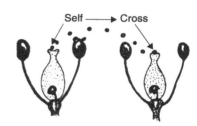

Fig. 1.23   Self- and cross-pollination.

from the anther to the stigma. This may be self-pollination where the pollen is transferred from the anther to the stigma of the same flower, or cross-pollination when it is carried to a different flower, most often on a different plant (Fig. 1.23). The vectors of pollen transfer can also vary. Some flowers are insect pollinated (entomophilous) and are usually scented and brightly coloured. The pollen itself is usually sticky or oily. Other flowers are wind pollinated (anemophilous) and do not need to be brightly coloured. They produce huge amounts of pollen (most is lost) and the pollen grains are smooth, light and small. The flowers of wind pollinated plants are often unisexual with a predominance of males. Stamens and stigmas often hang outside the flowers, and the stigmas are often feathery to give them a better chance of trapping a pollen grain as it blows past.

With fertilisation the pollen grain grows a pollen tube down the style of the carpel. This takes place very quickly and the tube grows around the ovary sac and enters through the micropyle. Three nucleii travel down the tube, one tube nucleus and two male nucleii. The tube nucleus disintegrates once the tube has reached the embryo sac, one male nucleus fuses with the egg cell in the ovule and the other joins with a second nucleus to form the primary endosperm nucleus. This double fertilisation is unique to flowering plants. The ovule itself goes on to form the seed. The ovary also changes after fertilisation to form the fruit, as distinct from the seed.

With the grasses and cereals there is only one seed formed in the fruit and, being so closely united with the inside wall of the ovary, it cannot easily be separated from it. The one-seeded fruit is called a grain.

1.5.6   The inflorescence

Special branches of the plant are modified to bear the flowers, and they form the inflorescence. There are two main types of inflorescence:

1. Where the branches bearing the flowers continue to grow, so that the youngest flowers are nearest the apex and the oldest farthest away – an indeterminate inflorescence (Fig. 1.24a). A well-known example of this inflorescence is the spike found in many species of grasses.
2. Where the main stem is terminated by a single flower and ceases to grow in length; any further growth takes place by lateral branches, and they eventually terminate in a single flower and growth is stopped – a determinate inflorescence (Fig. 1.24b), e.g. linseed.

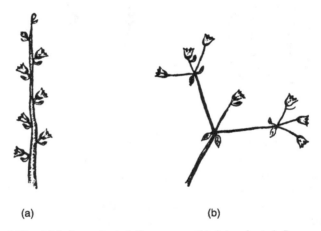

(a)   (b)

Fig. 1.24   (a) Indeterminate inflorescence; (b) determinate inflorescence.

There are many variations of these two main types of inflorescence. For example some plants, such as sunflowers, have a head called a capitulum, others are called umbels, e.g. carrots and parsley; while oats and many grass species have panicles.

1.6   Plant requirements

To grow satisfactorily a plant needs warmth, light, water, carbon dioxide, oxygen and other chemical elements which it can obtain from the soil.

1.6.1 Warmth

Most crop plants in this country start growing when the average daily temperature reaches 6 °C. Growth is best between 16 °C and 27 °C. These temperatures apply to thermometer readings taken in the shade about 1.5 m above ground. Crops grown in hotter countries usually have higher temperature requirements. Several plants require specific soil temperatures before they can successfully germinate, e.g. maize. Plants require a certain number of 'heat units' before they can successfully complete their development. This explains the difficulty of trying to grow subtropical or Mediterranean plants in the UK, e.g. durum wheat, grain maize or soya. Climate change may make it easier and more successful to grow these crops but the change may not necessarily result in 'global warming' effects and rises in average temperatures in many temperate countries.

Cold, frosty conditions may seriously damage plant growth. Crop plants differ in their ability to withstand extreme cold. For example, winter rye and wheat can stand colder conditions better than winter oats. Potato plants and stored tubers are easily damaged by frost. Sugar beet plants may go to seed (bolt) if there are frosts after germination producing a vernalisation effect (see below). Frost in December and January may destroy crops left in the ground. Cold may, however, be beneficial. A period of cold temperature is necessary to allow some autumn-sown crops to change from vegetative to sexual development. This is called 'vernalisation'. Some seeds may have to undergo cold conditions before they will germinate. This is known as 'stratification'. Cold temperatures can also help to reduce the incidence of weeds, pests and disease in crops.

1.6.2 Light

Without light, flowering plants cannot produce carbohydrates and will soon die. The amount of photosynthesis that takes place daily in a plant is partly due to the length of daylight, and partly to the intensity of the sunlight. Bright sunlight is very important where there is dense plant growth. The periods of daylight and darkness will vary according to the distance from the equator and also from season to season. This affects the flowering and seeding of crop plants and is another of the limiting factors in introducing new crops into a country.

1.6.3 Water

Water is an essential part of all plant cells and it is also required in extravagant amounts for the process of transpiration. Water carries nutrients from the soil into and through the plant; it also carries the products of photosynthesis from the leaves to wherever they are needed. It is also used to give the plant rigidity (turgor), is used in many of the biochemical pathways inside the plant, and is also used to regulate the temperature of the plant. Plants take up about 200 tonnes of water for every tonne of dry matter produced.

1.6.4   Carbon dioxide

Plants need carbon dioxide for photosynthesis. This is taken into the leaves through the stomata, and the amount which can go in is affected by the rate of transpiration. Another limiting factor is the small amount (0.03%) of carbon dioxide in the atmosphere. The percentage can increase just above the surface of soils rich in organic matter, where soil bacteria are active and releasing carbon dioxide. This is possibly one of the reasons why crops grow better on such soils. In protected crops, in greenhouses and polytunnels, the levels of carbon dioxide can be artificially raised by the grower to improve production. Tropical plants commonly have a different carbon utilisation pathway which makes them more efficient at using CO_2 and therefore can survive amongst thick tropical vegetation where the CO_2 supply is limited. They are often known as C4 plants, whereas temperate species are called C3 plants.

1.6.5   Oxygen

Plants need oxygen all the time, not just through the night, as is a popular belief. Seedlings, before they emerge and begin photosynthesis rely entirely on respiration of the seed reserves for energy and require oxygen in the seedbed. Oxygen is crucial for root growth and survival.

1.6.6   Chemical elements required by plants

Many chemical elements are needed by the plant in order that it may live and flourish. Most soils supply the majority of these nutrients and in farming practice it is only with regard to nitrogen, phosphorus, potassium and magnesium (the major elements) that there is any widespread necessity to supplement the natural supplies from the soil. Other major elements may be required in certain situations. Calcium is an essential plant food but, as lime, it is regarded more as a soil conditioner. Sodium is highly desirable for crops such as sugar beet and fodder beet (which have maritime ancestors), when it can replace some or all of the potassium requirements. Sulphur, in areas away from industry, could be needed for the grass crop when it is cut more than once in the year, and for certain brassicas such as oilseed rape.

The main plant foods are discussed more fully in Chapter 4 (fertilisers and manures) and see also Table 1.2.

Those elements required only in small amounts by the plant are known as the minor or trace elements. They are, nevertheless, essential and a shortage or lock-up in the soil, especially of boron and manganese (often as a result of liming or high organic matter), will cause deficiency symptons in particular crops. Other trace elements include magnesium, chlorine, iron, molybdenum and zinc, but these rarely cause trouble on most farm soils (see Section 4.2 'Trace elements', page 69).

Table 1.2   Sources and uses of major nutrients

Major nutrients	Use	Source
Carbon (C) Hydrogen (H) Oxygen (O)	Used in making carbohydrates	The air and water
Nitrogen (N)	Very important for building proteins	Organic matter (including farmyard manure FYM); rainfall Nitrogen-fixing soil micro-organisms Nitrogen fertilisers such as ammonium and nitrate compounds and urea
Phosphorus (P) (phosphate)	Essential for cell division and many chemical reactions	Small amounts from the mineral and organic matter in the soil Mainly from phosphate fertilisers, e.g. superphosphate, ground rock phosphate, basic slag and compounds, and residues of previous fertiliser applications
Potassium (K) (potash)	Helps with formation of carbohydrates and proteins Regulates water in and through the plant	Small amounts from mineral and organic matter in the soil Potash fertilisers, e.g. muriate sulphate of potash
Calcium (Ca)	Essential for development of growth tissues, e.g. root tips	Usually enough in the soil Applied as chalk or limestone to neutralise acidity
Magnesium (Mg)	A necessary part of chlorophyll	If soil is deficient, may be added as magnesium limestone or magnesium sulphate, also FYM
Sulphur (S)	Part of many proteins and some oils	Usually sufficient in the soil Atmospheric sulphur absorbed by the soil and plant. Added in some fertilisers (e.g. sulphate of ammonia and superphosphate)

1.7   Legumes and nitrogen fixation

Legumes are plants that have a number of interesting features, such as:

- A special type of fruit called a legume, which splits along both sides to release its seeds, e.g. pea pod.
- Nodules on the roots containing special types of bacteria (rhizobia) which can 'fix' (convert) nitrogen from the air into nitrogen compounds. These bacteria enter the plant through the root hairs from the surrounding soil.

This fixation of nitrogen is of considerable agricultural importance. Many of our farm crops are legumes, for example peas, beans, vetches, lupins, clovers, lucerne (alfalfa), sainfoin and trefoil. The bacteria obtain carbohydrates from the plant and in return they supply nitrogen as ammonium compounds which are released into the soil or directly into the host plant in the form of soluble nitrates. In the soil, the ammonium part of the compound is changed to nitrate and taken up by neighbouring plants, e.g. by grasses in a grass and clover sward, or by the following crop, e.g. wheat after clover or beans. The amount of nitrogen that can be fixed by legume bacteria varies widely – estimates of 50–450 kg/ha of nitrogen have been made. Some of the reasons for variations are:

- *The type of plant.* Some crop plants fix more nitrogen than others, e.g. lucerne and clovers (especially if grazed) are usually better than peas and beans.
- *The conditions in the soil.* The bacteria usually work best in soils that favour the growth of the plant on which they live. A good supply of calcium and phosphate in the soil is usually beneficial, although lupins grow well on acid soils.
- *The strains of bacteria present.* The majority of soils in any particular country contain the strains of bacteria required for most of the leguminous crops which are grown. However, if a legume is introduced, e.g. lucerne or alfalfa into the UK, and it has not been grown in the field within the previous three years, it is necessary to inoculate the seed, i.e. treat it with an inoculum containing *Rhizobium meliloti* before sowing to encourage effective nodulation.

1.8   The control of plant growth and development

1.8.1   Growth substances

In order for plants to grow and develop in an orderly manner there needs to be some kind of regulation of cell division, elongation and differentiation. As the plant has no nervous system, this regulation is entirely chemical in nature. The chemicals involved are called 'growth substances' and there are five main types:

1. Auxins – these substances mostly control cell enlargement and differentiation. Most auxins are made at the apex of stems and in young leaves.
2. Gibberellins – also involved in cell enlargement, but also play a key role in the breaking of dormancy in seeds and the mobilisation of seed reserves by the production of the enzyme, alpha-amylase.
3. Cytokinins – associated with cell division in the presence of auxins. Found in large amounts in fruits and seeds where they are associated with embryo growth.
4. Abscisic acid – a growth inhibitor associated with dormant buds and some seeds, and the closure of stomata.
5. Ethene (ethylene) – a gas which controls fruit ripening and senescence. Controls fruit and leaf drop from some trees.

One of the ways that plants react to external stimuli is by movement. Roots, shoots, leaves and flowers can all move either towards or away from these stimuli. This movement is called 'tropism'.

1.8.2   Phototropism

This is the movement in response to light. Shoots will be positively phototropic and move towards light. Auxins play a major part in this movement by moving to the dark side of the shoot, encouraging cell growth and thus bending the shoot towards the light source. This is a vital response if maximum sunlight interception is to take place by the plant so that this can be converted to yield.

1.8.3   Geotropism

This is the response to gravity. Shoots are negatively geotropic and therefore grow upwards while roots are positively geotropic and grow downwards. This is particularly important in the germinating and emerging seed and means that the grower does not have to ensure that the seed is planted the right way up! Other organs of the plant have an intermediate geotropism, e.g. leaves and side branches.

1.8.4   Hydrotropism

Roots are positively hydrotropic and will grow towards water.

1.8.5   Thigmotropism

This is a very specialised type related to touch. Tendrils of peas are positively thigmotropic allowing them to wrap around each other and any supporting objects. Similarly, hop stems will climb up strings in the hop garden by wrapping themselves around for support.

1.8.6   The effect of light

As well as movement caused by differentiation in cell enlargement and division, the plant's development is also controlled in part by external stimuli. By far the most important one is light, although temperature does have a part to play. Light affects plants in several ways. Without light plants become 'etiolated' – they lack chlorophyll and are therefore yellow rather than green and are fragile and collapse easily. The intensity and spectrum of the light has an effect on production, on breaking dormancy in some seeds and in the regulation of flowering.

Probably the most important aspect of light is its duration. The response of a plant to day length is called 'photoperiodism'. In fact it is the length of the dark period that is critical. The biggest effect is on flowering, although fruit and seed production, dormancy and leaf fall are also affected. Plants can be categorised into three groups when considering the differences in photoperiod requirements for flowering:

Short day (long night) plants
Flowering is induced by dark periods *longer* than a critical length. The length can vary according to species.

Long day (short night) plants
Flowering is induced by dark periods *shorter* than a critical length. Many spring crops, e.g. spring wheat and spring barley, are long day plants with the move into summer, and the subsequent shortening of the nights, stimulating them to move from vegetative growth to sexual development.

Day neutral plants
Flowering is independent of photoperiod. There are a large number of plants in this category. All this becomes much more important the further away you are from the equator, with the seasons very strongly associated with varying day lengths.

The knowledge of how each species is triggered into flowering is particularly important in the horticultural cut flower or house plant business. Growers need to synchronise flowering with important dates in the calendar, such as Mother's Day or Christmas. This ensures that their stock will be in perfect condition for the many bouquets that are bought at these times. They can manipulate flowering in the glasshouse or polytunnel by artificially adjusting lengths of daylight using high intensity lamps and exploiting the plant's photoperiodic response.

1.9   Sources of further information and advice

Barnes C, Poole N and Poore N, *Plant Science in Action*, Hodder Arnold, 1994.
Rost N Y *et al.*, *Botany: A Brief Introduction to Plant Biology*, John Wiley, 1984.
Taylor D J, Green N P O, Stout G W and Soper R (eds), *Biological Science 1 and 2* (3rd edn), Cambridge, 1997.

2

Climate and weather

DOI: 10.1533/9781782423928.1.27

Abstract: This chapter discusses the importance of climate and weather to crop yields and quality. It describes how climate is affected by local factors such as altitude and goes on to explain the importance of rainfall and temperature in particular. It discusses the influence of weather and climate on diseases and soil factors such as organic matter decomposition. Finally, the chapter discusses climate change, the possible future scenarios and how agriculture may have contributed.

Key words: rainfall, temperature, frost, greenhouse gases, carbon.

2.1   Introduction

Climate has an important influence on the types of crops which can be grown in the United Kingdom. It may be defined as a seasonal average of weather conditions. Weather is the state of the atmosphere at any one time. It is the combined effect of such conditions as heat or cold, wetness or dryness, wind or calm, clearness or cloudiness, pressure and the electric state of the air. It has a major influence on the type, yield and quality of the crops grown.

The climate of the United Kingdom is mainly influenced by:

- Its distance from the equator (50–60 °N latitude).
- The warm Gulf Stream which flows along the western coasts.
- The prevailing south-westerly winds.
- The numerous 'lows' or 'depressions' which cross from west to east and bring most of the rainfall.
- The distribution of highland and lowland; most of the hilly and mountainous areas are on the west side or run from north to south through the middle of the country.
- Its nearness to the continent of Europe, from where hot winds in summer and very cold winds in winter can affect the weather in the southern and eastern areas.

- The jet stream which consists of very strong winds 5–10 miles above the earth, which move weather systems around the planet and can reach speeds of up to 200 mph (Met Office, 2013).

Local variations are caused by altitude, aspect and slope.

Altitude: Height above sea level can affect climate in many ways. The temperature drops about 0.5 °C for every 90 m rise above sea level. Every 15 m rise in height can shorten the growing season by two days (one in spring and one in autumn) and it may check the rate of growth during the year. High land is more likely to be buffeted by strong winds and will receive more rain from the moisture-laden prevailing winds, which are cooled as they rise upwards.

Aspect: The direction in which land faces can affect the amount of sunshine absorbed by the soil and thus the soil temperature. In this country the temperature of north-facing slopes may be 1 °C lower than on similar slopes facing south.

Slope: When air cools it becomes heavier and will move down a slope and force warmer air upwards. This is why frost often occurs on the lowest ground on clear still nights whereas the upper slopes may remain free of frost. 'Frost pockets' occur where cold air collects in hollows or alongside obstructing banks, walls, hedges, etc. (Fig. 2.1). Frost-susceptible crops such as early potatoes, maize and fruit should not be grown in such places.

Crop production can be regarded as essentially the conversion of solar radiation, water and soil nutrients into useful end-products. The influences of various aspects of weather and climate on potential crop production can be numerous and varied. For instance, sunshine and rainfall could be considered as primary controls of production because of their immediate influence on the rate of crop metabolism, nutrient uptake, turgidity, biochemical processes within the plant and crop structure, as well as their effect on soil and air temperature. In addition there will be an effect from topography, geographical position and proximity to bodies of water. Other climatic elements such as frosts, wind and humidity will also have an

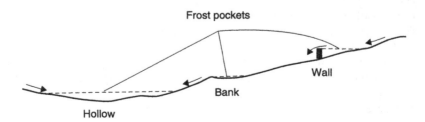

Fig. 2.1   Frost pockets formed as cold air flows down a slope.

effect on production because of their influence on such things as crop disease, pest distribution and direct physical damage.

2.2   Solar radiation and rainfall

Solar radiation in itself is important because of its role in photosynthesis. In the absence of any other restrictions, crop production of harvestable yield would be determined by the interception of sunlight. The intensity of solar radiation changes with season, latitude, aspect, slope and the amount of cloud cover and pollution. More light falls onto the south of England than in Scotland and the difference between the UK and, say, the south of France is even more marked. Yields of sugar beet in the Paris basin are often 25–30% higher than in East Anglia, whilst the yield of apples in Provence can be 100% higher than in the West Midlands. Other factors may be involved but a large part of the difference is due to the intensity and the duration of sunlight. The efficiency of use of this energy by crops is not good, with a maximum potential of about 8%, but, in commercial field crops, it is nearer 1%. The *interception* of light will vary according to ground cover: early in the season much of the sunlight will fall onto bare ground because of the small size of the plants; and late in the season, with senescence and subsequent leaf loss, the interception of energy will again fall. During the crop's 'grand growth stage' interception of light will be near 100%.

Water supply is vital for crop growth. Without enough water the yield and quality of crops are compromised in several ways. Cell division and cell growth are the first processes to be affected. Eventually photosynthesis will slow down due to the closing of the stomata. Leaves will wilt and their ability to intercept sunlight will be adversely affected and it will be more difficult for the plant to take up nutrients in solution from the soil. The soil water status underneath a crop will depend on the difference between rainfall and evapotranspiration, and there is far more variability in rainfall than there is in evaporation. Growers must therefore be able to compensate for this variability by draining land and storing water when in excess and applying it by irrigation when in deficit.

In the UK rain comes mainly from the moist south-westerly winds and from the many 'depressions' which cross from west to east. Western areas receive much more rain than eastern areas; this is partly due to the west to east movement of the rain-bearing air and also because most of the high land is along the western side of the country. As the moisture-laden air rises over this high ground it is cooled, its ability to hold moisture decreases and this moisture is deposited as rain. On the leeward (sheltered) side of the mountain the air is drier and warmer. This is sometimes known as a 'rain shadow' effect.

The average annual rainfall on lowland areas in the west is about 900 mm (35 in) and in the east is about 600 mm (24 in). Larger differences can be seen by comparing eastern lowlands with western highlands – average annual rainfall for parts of Cambridgeshire is 500 mm (20 in) whilst in the highest, western facing districts of the Lake District it is 5000 mm (200 in)!

2.3   Air and soil temperature

The optimum temperature for the metabolic processes within a temperate plant, especially photosynthesis, is about 25 °C. Therefore, in most seasons and especially during autumn and early spring, plants will not be metabolising at their optimum rate in the UK. Small increases or decreases in temperature from year to year will not have a significant effect on production, but it has been shown that, in very hot summers, e.g. those of 1975 and 1976, the photosynthetic rate and the duration of the Green Area Index (GAI) of crop plants was significantly decreased. These summers were characterised by temperatures above 25 °C. The summer of 2012 was one of the wettest and coldest on record with a mean temperature of 13.9 °C. This also had a detrimental effect on yields of crops.

Temperature has a large effect on the growing season, with growth of most crops beginning when temperatures rise above 5–6 °C and ceasing in the autumn when temperatures drop below this level. This can give growing seasons of as little as 180 days in the higher areas of Scotland and more than 350 days in the south west of England. These differences are very important when considering the potential for grass growth on livestock farms. Very high temperatures running up to harvest can also increase the rate of senescence in cash crops and hence reductions of harvestable yield.

Soil temperatures are more relevant to plant growth during the early part of the growing season. Germination is at its fastest at soil temperatures between 15 and 25 °C but for our autumn- and spring-sown crops these conditions are unlikely to be found. Winter wheat will germinate at temperatures as low as 3 °C as long as the soil is not waterlogged and it is not uncommon to find soil temperatures of 10–15 °C at 5 cm depth when most drilling occurs in September and early October. The gradual increase of mean UK temperatures due to global warming is discussed later in this chapter. Temperature changes are mainly due to:

- The seasonal changes in length of day and intensity of sunlight.
- The source of the wind, e.g. whether it is a mild south-westerly, or whether it is cold polar air from the north or from the Continent in winter, and hot, dry wind from the south in summer. Wind direction depends mainly on the position of areas of high and low pressure in the immediate vicinity of the British Isles, although local winds, e.g. offshore breezes, Fohn effect winds (warm) and katabatic winds (cold) from high ground, can occur independently of the cyclonic system.
- Local variations in altitude and aspect.
- Cloud cover during the night. Night temperatures are usually higher when there is cloud cover which prevents too much heat escaping into the upper atmosphere.

The soil temperature may also be affected by colour – dark soils absorb more heat than light-coloured soils. Also, damp soils can absorb more heat than dry soils. The presence of stones can also increase mean soil temperatures by acting as heat reserves in a similar way to storage radiators in buildings. The average January air temperature in lowland areas along the western side of the country is about 6 °C and about 4 °C along the eastern side. The average July temperature in lowland areas in the southern counties is 17 °C and 13 °C in the north of Scotland. Temperature can also have an effect on the agronomy of the crop. For example it can affect the rate of breakdown of residual herbicides applied to the soil, it can affect the efficiency of use of plant growth regulators and it can affect the volatilisation of ammonia from applied urea. High temperatures can also mean that plants with 'soft' new growth are more susceptible to scorch from agrochemicals applied onto them.

2.4   Other aspects of climate and weather

Advection is the horizontal transfer of heat. The effects of advection are most noticeable in the farming areas close to the western coasts of the British Isles. The Gulf Stream and the prevailing south-westerly winds sweeping up over the Atlantic mean that air temperatures over the sea to the south-west of the UK are often 10–15 °C higher than those on land. The effects of this are noticeable over most of the UK but are at their most intense over west-facing coastal areas. These are often known as early cropping areas, where potatoes and vegetable crops can be sown during winter and harvested in late spring, e.g. Scilly Isles, Pembrokeshire, Cornwall. Often more than one crop per year can be achieved as there is virtually year-round plant growth and little or no chance of frosts.

The likely incident of frosts in an area will have a large effect on the types of crops grown. Most areas of England and Wales will be subject to periods of night frosts during the winter months. Autumn-sown crops are bred to exhibit a certain level of frost hardiness and will survive most winters unscathed. In areas of Scotland where temperatures are lower and may never exceed 0 °C, even during the day, the crops are mostly spring-sown and this is certainly the case in the huge wheat growing areas of the US and Canada. High value crops such as soft fruit are most at risk from night frosts occurring in the spring when the flower buds are emerging. This can be partially overcome by the use of shelter, artificial protection with mulches or the use of spray irrigation. Other weather phenomena which can affect crop production are gales, floods and hailstorms.

The epidemiology and life-cycles of many diseases are heavily influenced by weather. Potato blight is a classic example with the disease requiring very specific humidity and temperature conditions to spread within the crop. These conditions can be monitored and predicted, and an early warning system can be used by farmers to help to control the disease effectively with fungicides. Many of the most serious fungal diseases on mainstream crops, such as cereals and oilseed rape, are favoured by wet, windy and humid conditions, whereas in a dry spring and summer, disease levels will usually be much lower. The introduction of a philosophy of integrated crop management means that much more use of weather data to try and predict disease incidence and spread will need to be used in the future. This can also apply to crop pests.

Plants growing in a given climate have a potential production which is modified (often limited) by the soil in which they are grown and by individual weather conditions experienced during the growing season. A very important effect of climate on soil is that of the soil condition. The soil processes which can be affected by climate include:

- Littering and humification – the accumulation of raw organic matter and its subsequent conversion to humus. The breakdown of organic matter is quicker in higher temperatures because of the increased biological activity of the soil.
- Decalcification – the removal of calcium carbonate from soil profiles by, among other things, leaching. This will be worse in light soils combined with high rainfall.
- Gleying – the reduction of iron under anaerobic conditions with mottling of ferric concretions. Found in heavy soils with poor drainage characteristics in areas where rainfall amounts mean long periods of waterlogging of the soil.
- Salinisation and alkalinisation – the build-up of salt in the soil and the accumulation of sodium (Na^+) ions on the exchange sites. Usually associated with the use of saline water for irrigation but can occur naturally where high temperatures give high evaporation from the soil surface. Salts are moved to the surface by capillary action and there is not enough rain to wash them away from the rooting zone. The overuse of water for irrigation in some areas has also led to the ingress of sea water into aquifers used for irrigation.

Perhaps the most important interaction between climate and soil is that of soil type and rainfall. At the one extreme this interaction can allow a crop to reach its full potential production and at the other extreme it can severely limit this potential, to the point where near crop failure can occur. Two identical soil types in terms of structure and texture can have very different characteristics depending on where they are located. Take, for example, a sandy loam situated in South Devon and a similar sandy loam in Suffolk. The total rainfall in Devon could be as high as 1000 mm whereas in Suffolk this would be nearer 500 mm. The soil in Devon could well be under permanent grassland because of the difficulty in finding available work days for cultivations in autumn and spring. If it was cultivated, leaching of nitrogen and calcium would be a major concern but water supply to crops would be adequate through the summer and justification for an irrigation system would be difficult. The Suffolk soil would probably be under arable or horticultural production. There would be plenty of available work days but irrigation would be vital for high value crops so that yield potential and quality could be maintained.

2.5   Climate change

The climate history of the earth is one of dramatic change cycles over millions of years. Many periods of ice ages alternating with periods of warmer climates have helped to shape the landscapes that are seen today and have determined the course of rivers, the shape of the oceans and the types of soil deposition, especially in temperate areas of the world. There are changes in the weather from year to year and there are decades where the global temperatures are higher or lower than average. There are extremes of weather and climate, often bringing with them disaster and fatality – hurricanes, floods, tidal waves, snowstorms, heat waves and droughts. But these are the phenomena caused by the natural 'pulse' of the planet's weather and climate systems; the inevitable result of the chaos existing in the atmospheric circulation and the ocean's currents.

One of the most important events influencing the world's climates is El Niño. Every three to five years a large area of warmer water appears in the Pacific Ocean off the coast of South America. This area lasts for one to three years. The physical effects of El Niño are twofold. Firstly the evaporation from the warmer ocean increases, changing the dynamics of the atmospheric circulation across the globe, and secondly a slow moving wave (called a Rossby wave) moves out from the epicentre of the event. The resulting tidal surges cause problems for low-lying areas and prevent ground water from draining effectively into the oceans.

The other natural phenomenon which affects, in particular, northern Europe is the position of the jet stream, a flow of air about 11 km above the Earth's surface at each of the poles. This travels from west to east and, depending on how far south the northern hemisphere stream sits, will bring colder or warmer seasons from year to year. In 2012 the jet stream was sitting very low and brought cold, polar air in the winter and wet, windy weather during the summer months. This resulted in the second wettest year in the UK since records began and resulted in delayed or abandoned harvests, poor crop quality and flooded fields that prevented autumn sowing of many crops.

Far more sinister are the changes caused by man's activities on the planet. These are mostly industrial activities but also involve lifestyle influences. The three most important changes, and those which are exercising the minds of scientists and politicians all over the world, are global warming, ozone depletion and the increase in acid rain. The warming of the globe in particular is blamed for some of the extreme weather we are seeing more frequently in the twenty-first century. The jet stream effect may be due to warming of the upper atmosphere which slows it down and makes it meander more, coming further south at times and then going much further north, bringing drought to certain areas of the world. It also seems to get 'stuck' more often, meaning that any weather patterns last for longer.

In order for the Earth to maintain a fairly even average temperature over its surface, the incoming solar radiation must be balanced by an equal outgoing thermal radiation from its surface and atmosphere. The effect of naturally-occurring carbon dioxide, dust particles and water vapour in the atmosphere is to trap some of the thermal radiation while releasing enough heat to balance the sun's input and to keep the average surface temperature at about 15 °C. This state of relative stability has lasted for millions of years although it has been interrupted by the ice ages. This phenomenon is known as the *natural greenhouse effect*.

Because of human activities such as burning of fossil fuels and deforestation we are now experiencing an *enhanced greenhouse effect* due to the increase in carbon dioxide emissions, and in those of other gases, into the atmosphere. This has the effect of reducing the thermal emissions back into space and, to restore the heat balance, the Earth's surface temperature must rise. It has been speculated that a doubling of the CO_2 content of the atmosphere will mean an average temperature rise of about 5 °C, i.e. the average surface temperature will be 20 °C instead of 15 °C. The level of CO_2 has risen by approximately 30% since the early 1800s but there are now serious global attempts to reduce greenhouse gas emissions. Measurements in May 2013 recorded CO_2 levels of 400 ppm for the first time, levels not thought to have existed for three to five million years.

Agriculture contributes about 8–10% to the emissions and does it in several ways. Fertiliser applications and cultivations release nitrous oxide (N_2O), livestock release methane (CH_4) and machinery, grain dryers, transport of agricultural goods, etc., and burning fossil fuel, release CO_2. All over the world most governments are now taking climate change seriously, following the UN's Kyoto Protocol of 2005 which sets binding obligations on industrialised countries to reduce emissions of greenhouse gases, and encourages developing countries to do likewise. The UK has a Department of Energy and Climate Change which is responsible for managing and monitoring the implementation of the Protocol. Figures produced since 1990 show that total GHG emissions from agriculture have dropped from 68 million tonnes to 54 million tonnes in 2011. Much of the reduction has been driven by a fall in animal numbers and a reduction in the use of artificial fertilisers. Carbon dioxide emissions from agriculture only represent 12% of the total GHG production but farmers are encouraged to use renewable fuel (e.g. biodiesel) wherever possible. The use of renewable fuels is far more important in the transport and energy supply industries of course, which release 64% of the total GHG emissions.

Greenhouse gas and carbon accounting is now big business in all industry sectors. A number of tools are available to farmers but they all do much the same thing. They allow farmers to estimate the levels of GHG emissions from the operations they carry out, as well as estimating the amount of carbon locked up (sequestered) through soil and woodland management. There are also calculators available which work out the amount of GHGs emitted from burning biofuels and organisations such as The Carbon Trust work with businesses, governments and the public sector to cut energy bills and reduce carbon emissions.

Farmers are encouraged to consider their on-farm energy policies using technologies such as photovoltaic cells (solar panels), wind turbines, ground heat pumps, biomass boilers and anaerobic digesters to produce electricity.

The effects of global warming are a mixture of good and bad. Because of the increased CO_2 levels certain crops (mainly the C3 types) will respond with increased yields. Certain parts of the planet will change their cropping patterns to include crops that otherwise could not be cultivated while other parts will become more habitable. To offset this, however, it is likely that existing arid and semi-arid zones will find it increasingly difficult to produce food because of lack of useable water. These areas are often already overpopulated, with insufficient food, and this can only get worse under the anticipated climate change.

It is likely that many crops can be adapted by genetic manipulation (either through traditional plant breeding or by genetic modification using biotechnology) to thrive in the new conditions. Research organisations such as CIMMYT, which carries out development programmes in wheat and maize, are trying to develop improved varieties which can thrive in infertile and droughty conditions and can still give high yields under increased pest and disease pressure caused by rising temperatures. However, there is more concern about long-term crops such as forests and plantation crops, whose life cycles are over a much longer time scale and so may adapt less well. There is also concern about thermal expansion of the oceanic water mass, the increased rate of polar ice cap melting and the consequent rise in sea levels. Many areas of comparatively fertile agricultural land at or below sea level could be flooded unless serious attempts are made to improve sea defences.

Ozone is an important gas of the upper atmosphere. It helps to prevent dangerous ultraviolet radiation from reaching the Earth's surface and is also one of the natural greenhouse gases. Large amounts of ozone have been destroyed by the use of chlorofluorocarbons (CFCs) in refrigerators, insulation and aerosols. This allows more UV radiation to reach the Earth and, because CFCs are much more effective greenhouse gases than ozone, adds to the problems of global warming. Thankfully the problem has been recognised and concerted efforts by industry have meant that alternatives to CFCs have been developed and are now widely used. Agriculture contributes very little to the CFCs in the atmosphere.

Following the industrial revolution and the burning of high-sulphur fossil fuels, high levels of sulphur dioxide were being released into the atmosphere. Reactions between water and sulphur dioxide produced weak acids in solution which fell to the earth as acid rain. The effects of acid rain were often felt downwind of the areas causing the problem and large areas of vegetation were destroyed. The control of sulphur dioxide emissions in the US and Europe has meant a dramatic reduction in acid rain incidence, but some parts of the world continue to burn high-sulphur fossil fuels and it is likely that these countries will not change their practices for some time. Indeed, sulphur emissions have become so low in the UK that some crops like oilseed rape and grassland need to have sulphur added as a top dressing during the spring, otherwise they will exhibit deficiency symptoms or produce lower than optimum yields.

2.6   Sources of further information and advice

Further reading

Barry R G and Chorley R J, *Atmosphere: Weather and Climate*, 9th edn, Routledge, 2009.
Houghton, J, *Global Warming – The Complete Briefing*, 3rd edn, Cambridge University Press, 2004.

Mendelsohn, R O and Dinar, A, *Climate Change and Agriculture: An Economic Analysis of Global Impacts, Adaptation and Distributional Effects*, Edward Elgar Publishing, 2009.
OECD, *Farmer Behaviour, Agricultural Management and Climate Change*, OECD Publishing, 2012.
Pittock, A B, *Climate Change: Turning up the Heat*, Earthscan, 2004.

Websites

https://www.gov.uk/government/organisations/department-of-energy-climate-change
http://www.iiea.com/environmentnexus/home

3

Soils and soil management

DOI: 10.1533/9781782423928.1.37

Abstract: Soils are very complex and most have developed over a very long period of time. They provide a suitable medium for plants to obtain water, nutrients and oxygen for growth and development. This chapter discusses soil formation as well as characteristics and management of farm soils including soil erosion and liming.

Key words: soil texture, soil structure, soil characteristics, soil management, soil erosion.

3.1   Introduction

Soils are very complex and most have developed over a very long period of time. They provide a suitable medium for plants to obtain water, nutrients and oxygen for growth and development. Most soils also have enough depth to give plant roots a firm anchorage. Mineral soils are formed initially by the weathering of parent rock, often accompanied by deposition of material by ice, water and/or wind. Organic matter is added to the soil from the growth and decay of living material.

Good management of soil is essential to reduce the amount that is lost or damaged due to erosion or compaction. It is also important that land is managed in a way that minimises carbon losses released as carbon dioxide after the breakdown of organic matter. Soils can become much more difficult to manage if the amount of organic matter is reduced. Due to these issues of soil degradation, farmers in England currently have to undertake a soil protection review of their farm in order to receive support payments.

The topsoil is a layer up to 30 cm deep which may be taken as the greatest depth which a farmer can plough or cultivate and in which most of the plant roots are found. The subsoil, which lies underneath, is an intermediate stage in the formation of soil from the rock below. Some deep-rooting plants such as cereals and oilseed rape can grow in the subsoil down to depths of 1.5–2 m.

A soil profile is a section taken vertically through the soil. In some cases this may consist only of a shallow surface soil of 10–15 cm on top of rock such as chalk or limestone. In deeper well-developed soils there are usually three or more definite layers (or horizons) which vary in colour, texture and structure (Fig. 3.1). The soil profile can be examined by digging a pit or by taking out cores of soil from various depths using a soil auger. A careful examination of the layers can be useful for assessing soil texture, structure and compaction as well as the soil cropping potential. The colour of the soil in the various horizons will indicate whether the soil is well or poorly drained.

There are a number of ways of classifying soil for crop production. Soils have been grouped into *soil associations*. Each association consists of a number of *soil series* each of which has distinct characteristics, both of parent material and soil profile. The soil series is usually named after the place where the soil was first described. The same soil series can occur in different regions. Soil characteristics, together with relief and climate and cropping potential, have also been used to classify land for farming (Appendix 7).

3.2   Soil formation

There are very many different types of topsoil and subsoil. The differences are partly due to the kind of material from which they are formed. However, other factors such as climate, topography, plant and animal life, the age of the developing soil material and farming operations affect the type of soil which develops.

3.2.1   The more important rock formations

- *Igneous rocks*, e.g. granite (coarse crystals) and basalt (fine crystals), were formed from the very hot molten material. The minerals (chemical compounds) in these rocks are mostly in the form of crystals. Igneous rocks are very hard and usually weather very slowly.
- *Sedimentary* or *transported rocks* have been formed from weathered material (e.g. silt and sand) carried and deposited by water and wind. The sediments later became compressed by more material on top and cemented to form new rocks such as sandstone and shale.
- The *chalk* and *limestone soils* were formed from the shells and skeletons of sea animals of various sizes. These rocks are mainly calcium carbonate but in some cases also contain magnesium carbonate. The calcareous soils are formed from them (see Section 3.4.4, page 53).
- *Metamorphic rocks*, e.g. marble (from limestone) and slate (from shale), are rocks which have been changed in various ways, such as by heat or pressure.

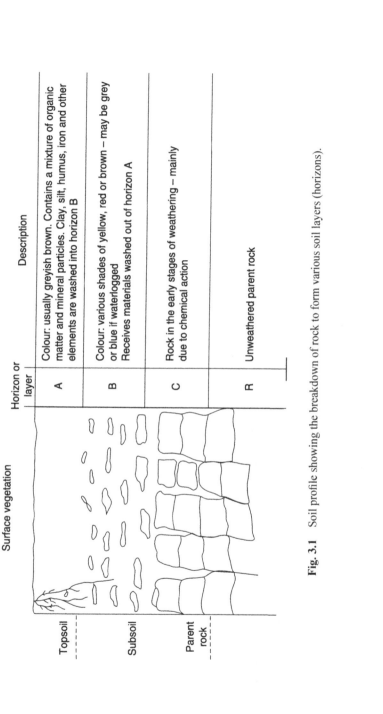

Surface vegetation	Horizon or layer	Description
Topsoil	A	Colour: usually greyish brown. Contains a mixture of organic matter and mineral particles. Clay, silt, humus, iron and other elements are washed into horizon B
Subsoil	B	Colour: various shades of yellow, red or brown – may be grey or blue if waterlogged. Receives materials washed out of horizon A
	C	Rock in the early stages of weathering – mainly due to chemical action
Parent rock	R	Unweathered parent rock

Fig. 3.1 Soil profile showing the breakdown of rock to form various soil layers (horizons).

3.2.2   Some other deposits

- *Organic soils.* Deep deposits of raw organic matter or peat are found in places where waterlogged soil conditions did not allow the breakdown of dead plant material by micro-organisms and oxidation (due to the anaerobic conditions). There are different sorts of peat depending on the type of vegetation and area where it developed.
- *Glacial drift.* Many soils in northern Europe are not derived from the rocks underneath but from material deposited by glaciers, often known as boulder clays. This makes the study of such soils very complicated.
- *Alluvium.* This is material which has been deposited recently, for example, by river flooding. It has a very variable composition. The texture depends on the speed of river flow (e.g. fast rivers – stones and sand, slow rivers – silt and clay).

3.2.3   Weathering of rocks

The breakdown of rocks into smaller particles is called *weathering*. It is mainly caused by physical and chemical processes.

Physical weathering can happen in a number of different ways. One important method is by frost action. Water can cause pieces of rock surfaces to split off when it freezes and expands in cracks and crevices. The pieces of rock broken off are usually sharp-edged, but if they are carried and knocked about by glaciers, rivers or wind, they become more rounded in shape, e.g. sand and stones in a river bed. Expansion and contraction of parent material when exposed to varying temperatures will also cause some rocks to fracture. Plants can also cause weathering by growing through rock cracks and exerting enough pressure to cause some shattering.

Chemical weathering is the breakdown of the mineral matter in a developing soil brought about by the action of water, oxygen, carbon dioxide and nitric acid from the atmosphere, and by carbonic and organic acids from the biological activity in the soil. The soil water, which is a weak acid, dissolves some minerals and allows chemical reactions to take place. Clays are produced by chemical weathering of some primary minerals which then re-crystallise. In the later stages of chemical weathering soil minerals can be broken down to release plant nutrients.

In poorly drained soils, which become waterlogged from time to time, various complex chemical reactions (including a reduction process) occur, referred to as *gleying*. This process, which is very important in the formation of some soils, results in ferrous iron, manganese and some other trace elements moving around more freely and producing colour changes in the soil. Gleyed soils are generally greyish in colour (but may also be greenish or blueish). Rusty-coloured deposits of ferric iron (oxidised iron) also occur in root and other channels and along the boundaries between the waterlogged and aerated soil, so producing a mottled appearance.

3.2.4   Other factors in soil formation

- *Climate.* The rate of weathering partly depends on the climate. For example, the wide variations of temperature and the high rainfall of the tropics make for much faster soil development than would be possible in the colder and drier climatic regions.
- *Topography.* The slope of the ground can considerably affect the depth of soil. Weathered soil tends to erode from steep slopes and build up on the flatter land at the bottom. Level land is more likely to produce uniform weathering.
- *Biological activity.* Plants, animals and micro-organisms, during their life cycles, leave many organic substances in the soil. Some of these substances may dissolve some components of the mineral material. Vegetation such as mosses and lichens can attack and break down the surface of rocks. Holes made in the soil by burrowing animals such as earthworms, moles and rabbits help to break down soft and partly weathered rocks. Biological activity usually increases with higher temperatures and decreases under waterlogged and/or acid conditions.
- *Farming operations.* Deep ploughing and cultivation, artificial drainage and liming can speed up the soil formation processes (Chapter 8).

3.3   The physical make-up of soil and its effect on plant growth

The farmer considers the soil from the point of view of its ability to grow crops. To produce good crops the soil must provide suitable conditions in which plant roots can grow. It should supply nutrients, water and air. The temperature must also be suitable for the growth of the crop.

The soil is composed of:

- *Solids.* Mineral matter (stones, gravel, sand, silt and clay) and organic matter (remains of plants and animals).
- *Liquids.* Soil water (a weak acid).
- *Gases.* Soil air which occupies a variable amount of the pore spaces.
- *Living organisms.* Micro-organisms (bacteria, fungi etc.) and soil fauna including earthworms.

Approximately 50% of the soil volume is made up of the solid matter (mineral and organic) and 50% air and water. The relative amount of air and water and organic matter depends on soil type as well as cropping and cultivation systems.

3.3.1   Mineral matter and soil texture

The description of a soil in relation to the amount of mineral material, clay, silt and sand, is called the *soil texture*. There are large variations in the texture of soils found on farms. Soil texture can be analysed accurately in a laboratory but can also be assessed in the field using hand texturing.

Table 3.1   Particle size and surface area of soils

Material	Diameter of particles (mm)	Surface area
Clay	Less than 0.002	$100\,000 \times a$
Silt	0.002–0.060	$1000 \times a$
Fine sand	0.06–0.02	$100 \times a$
Medium sand	0.2–0.6	$10 \times a$
Coarse sand	0.6–2.0	a
Gravel	More than 2.0	

If mechanically analysing a soil, the organic matter, stones and gravel are first removed. The rest of the mineral fraction is separated using sieving and sedimentation techniques. This gives an accurate measurement of the amount of clay, silt and sand particles present. The particle size for each mineral fraction is shown in Table 3.1. Because these mineral components have very different properties it is important to know the relative amount of each, hence the importance of knowing the soil texture. Soil texture is classified into 11 textural classes in the UK (see Appendix 1).

Hand texturing is a very quick and simple technique; with practice it can be a reasonably accurate method of analysis. It uses the difference in feel and binding properties of moist soil to differentiate between the textural classes. The main characteristics of the three mineral fractions are as follows.

- *Sands.* Feel gritty and are moulded with difficulty.
- *Silts.* Feel smooth and silky. The soil ball is easily deformed.
- *Clays.* Feel sticky and bind to form a strong soil ball.

The texture of the soil has a major influence on many properties including:

- *Soil structure.* Soil structure is the arrangement of the individual particles into larger units or aggregates (sometimes called soil *peds*). It is mainly the clay particles that are charged and attract each other (*flocculate*). Organic matters as well as living organisms such as fungi are also very important in causing soil particles to bind together and form aggregates.
- *Air and water supply.* Clay soils have the largest number of small pores which tend to be filled with water rather than air. Some of this water is unavailable for root uptake. The sandy soils have the greatest number of large pores that tend to be filled with air rather than water. Because of these properties clay soils usually need draining for arable cropping and sandy soils are drought prone. It is the medium-textured silty loams that hold the most water available for crop growth.
- *Cultivation system.* The texture of the soil will affect the cultivation system used to establish crops (Chapter 8). Sandy soils are very easy and require little power to cultivate. Clays, on the other hand, can be very difficult with limited periods when the land can be worked without damaging the structure and with higher power requirements.

- *Lime and fertiliser requirements.* Some soil types have different abilities both to hold and release nutrients. Most clay minerals have a negative charge which attracts positively charged ions such as potassium, magnesium and calcium. This is called the *cation exchange capacity.* The presence of these charged sites affects the availability of nutrients. Sand and silt are relatively inert. Different soil textures with the same level of acidity have different recommendations for the application of lime. This is due to differences in the cation exchange capacity of the soil.
- *Erosion and run-off.* Some of the more poorly structured soils such as the fine sandy loams can suffer from both wind and water erosion and run-off. The peat soils can also be affected by wind erosion.
- *Herbicide use.* Some soil textures are more prone to herbicide leaching, particularly of some soil applied residuals. Recommended application rates of these chemicals are affected by the soil texture in order to avoid any risk of crop damage. On clay soils rates may be higher as some chemicals are adsorbed onto the soil particles and are then unavailable for weed uptake.
- *Cropping.* All these characteristics of the different soil textures determine which crops can be grown successfully and their yield potential.

3.3.2   Soil structure

Soil structure is usually defined in terms of the shape and size of the units or aggregates (Fig. 3.2). Structure, unlike texture, can be altered naturally by weathering (e.g. lumps changed to crumbs by frost action or by alternate wetting and drying), by the penetration of plant roots and very importantly, by cultivations. Soils with a high clay or organic matter content tend to have a more stable soil

Granular structure – only 1–10 mm units
Found in the top soil

Blocky structure – 5–50 mm units
Found in subsoil of well structured
soils

Platy structure – found when soil is
compacted, e.g. if there is a plough pan

Massive prismatic structure, few cracks for water
movement or root growth. Found in subsoils of
heavy clays

Fig. 3.2   Soil structural units.

structure than those containing mostly sand and or silt. Soils containing free calcium carbonate and, to a lesser extent iron oxides, also tend to have a more stable soil structure.

Soils with a good structure through the soil profile have lots of spaces or pores between aggregates with evidence of good root growth and biological activity such as earthworms. Soils with few pores and fissures are said to be compacted. Compacted soils can affect crop root growth and the soil water-holding capacity, reduce nutrient uptake and make the growing crops more prone to pests and diseases. All these effects can lower final crop yields. Compacted soil is much more prone to soil erosion and run-off .

The structure of the soil can be easily damaged by harmful operations, e.g. heavy traffic in wet conditions or ploughing when the soil is too plastic. Overworking poorly structured silty soils can lead to surface capping after a period of intense rain followed by surface drying. Surface capping can lead to crop failure if it occurs after drilling a small seeded crop and also increases the risk of water erosion.

Tilth is a term used to describe the condition of the soil in a seedbed. For example, the soil may be in a finely-divided state or it may be rough and lumpy. Whether a tilth is suitable or not partly depends on the crop to be grown. In general, a small seeded crop requires a finer tilth than large seeds.

3.3.3   Organic matter

Organic matter in the soil is composed of decomposing plant and animal residues and residual organic matter or *humus*. The organic matter content of a soil varies with texture and cropping. It usually increases with clay content, e.g. ordinary heavy soils have about 3–4% organic matter compared with 1–1.5% in very light soils. Mineral soils where arable crops are continuously grown usually have a low organic matter content below 5% compared with 10–20% under permanent grassland. There are varying amounts of organic matter in the 'peat' soils. True peat contains more than 50% organic matter. On draining, liming and cropping these soils have suffered rapid breakdown or *wastage*. For example, the peaty soils only have 20–35% organic matter.

Organic matter may remain for a short time in the undecayed state. However, the organic matter is soon attacked by the soil organisms, such as bacteria, fungi, earthworms and other soil fauna. Sugars, starch and crude protein are broken down first and the complex molecules of cellulose, lignin and waxes last. The remaining residual material (humus) has no fixed chemical composition. It commonly contains 45% lignin, 30% protein and the rest a mixture of fats, waxes and complex carbohydrates. The amount of humus formed is greatest from plants which have a lot of strengthening (lignified) tissue, i.e. straw. Organic matter is broken down most rapidly in warm, moist soils which are well limed and well aerated and have high biological activity. Breakdown is slowest in waterlogged, acid conditions.

The dark-coloured humus produced during the breakdown process is very beneficial in restoring and stabilising soil structure. Like clay, humus is a colloid, i.e. it is a gluey substance which behaves like a sponge; it absorbs water and swells up when wetted and shrinks on drying. The humus colloids are not so gummy and plastic as the clay colloids but they can improve light (sandy) soils by binding groups of particles together. This reduces the size of the pores (spaces between the particles) and increases the water-holding capacity. Humus can also improve clay soils by making them less plastic and by assisting in the formation of a crumb structure (lime must also be present). The humus colloids can also hold bases such as potassium and ammonium ions in an available form.

Plant nutrients, including nitrogen and phosphorus as well as carbon dioxide, are released (mineralised) each year when organic matter breaks down. Soil plays an important role in the sequestration of carbon, especially under permanent grassland or forestry. Currently, many soils in all arable rotations loose more carbon than is returned. Organic matter in the soil may be maintained or increased by growing grass, working-in straw and similar crop residues, farmyard manure, composts and green manures. In areas where erosion by wind and water is common, mineral soils are less likely to suffer damage if they are well supplied with humus. Increasing the organic matter content of a soil will increase its ability to hold water and the amount of water available for crop growth (available water capacity), as well as the cation exchange capacity.

3.3.4   Water in the soil

A productive soil will usually have a very good supply of available water and be well drained. The soil is a mass of irregular-shaped particles forming a network of spaces or channels called the pore space, which may be filled with air or water or both. If the pore space is completely filled with water, the soil is waterlogged. It is then unsuitable for most plant growth because the roots need oxygen for respiration. Ideally, there should be about equal volumes of air and water.

The size of the pores is also very important in determining the amount of water held in a soil. *Macropores* are the largest pores and are important for drainage and aeration. *Mesopores* are medium sized and are the pores containing water available for crop growth. *Micropores* are the smallest pores containing water that is not available for root uptake. A sandy soil has a large proportion of macropores so normally more of the pores contain air rather than water; it is the opposite for the clay soils where there are a larger amount of small pores. The pore space may be altered by a change in the:

- Structure, e.g. by cultivations.
- Amount of organic matter present. The larger the organic matter content the greater the water-holding capacity.
- Compaction of the soil. Compaction reduces the number and size of the pores. It can also affect rooting depth and hence the amount of available water.

Another important factor is the surface area of the particles. Water is held as a thin layer or film around the soil particles. The smaller the particles, the stronger are the attractive forces holding the water. Also, the smaller the particles, the greater is the surface area per unit volume. A comparison of surface area of the various soil mineral particles is set out in Table 3.1. The surface area of the particles in a cubic metre of clay may be over 1000 hectares.

When water falls on a dry soil it does not become evenly distributed through the soil. The topmost layer is saturated first and, as more water is added, the depth of the saturated layer increases. In this layer most of the pore space is filled with water. However, a well-drained soil cannot hold all this water for very long; after a day or two some of it will soak into the lower layers or run away in drains. The soil will eventually reach a stage when the amount of any additional water added by rain or irrigation is matched by the loss of water from the profile through natural or artificial drainage. This is known as *field capacity*. The amount of water which can be held in this way varies according to the texture and structure of the soil. The weight of water held by a clay soil may be half the weight of the soil particles, whereas a sandy soil may hold less than one-tenth of the weight of the particles. The water-holding capacity of a soil is usually expressed in mm per cm, e.g. a clay soil may have a field capacity of 4 mm/cm in depth.

The ways in which water is retained in the soil can be summarised as follows:

- As a film around the soil particles.
- In the organic matter.
- Filling some of the smaller spaces or pores.

Plant roots can easily take up most of this water, but as the soil dries out the remaining water is more firmly held and eventually a stage is reached when the plant can extract no more water. This is the *permanent wilting point* because plants remain wilted and soon die. (At permanent wilting point there is still some water present in the soil.) This permanent wilting should not be confused with temporary wilting which sometimes occurs on very hot days. This is because the rate of transpiration is greater than the rate of water absorption through the roots. In these cases the plants recover as it gets cooler on the same day. The water that can be taken up by the plant roots is called the *available water*. It is the difference between the amounts held at field capacity and at permanent wilting point. In clay soils only about 50–60% of water at field capacity is available; in sandy soils up to 90% or more may be available. It is the silt loams and peaty soils that have the highest total available water capacity.

Water in the soil tends to hold the particles together and thus lumps of soil may stick together. When a clay soil is at, or above, half field capacity, it is possible to form it into a ball which will not fall apart when handled. As a clay soil becomes wetter it becomes more like plasticine. When there is enough soil moisture to roll the soil into a 3 mm diameter 'worm' without breaking it is at the *lower plastic limit*. Cultivating the soil at this moisture content will damage the soil structure. Some of the water in soils with very small pores and channels moves through the soil by capillary forces, i.e. surface tension between the water and the walls of the fine tubes or capillaries. This is a very slow movement and may not be fast enough to supply plant roots in a soil which is drying out.

Water is lost from the soil by evaporation from the surface and by transpiration through plants. It moves very slowly from the body of the soil to the surface; after the top 20–50 mm have dried out, the loss of water by evaporation is very small. Cultivations increase evaporation losses. Plants take up most of the available water in a soil during the growing season and air moves in to take its place. Air moves easily where the soil has large pore spaces, but the movement into the very small pore channels in clay soils is slow until the soil shrinks and cracks – vertically and horizontally – as water is removed by plants. The water which enters the soil soon becomes a dilute solution of the soluble soil chemicals. It dissolves some of the carbon dioxide in the soil and so becomes a weak acid.

3.3.5   Soil aeration

Plant roots and many of the soil animals and micro-organisms require oxygen for respiration. The air found in the soil is really atmospheric air which has been changed by these activities (and also by various chemical reactions); as a result it contains less oxygen and more carbon dioxide. After a time this reduction in oxygen and increase in carbon dioxide can become harmful to the plant and other organisms.

Aeration is the replacement of this stagnant soil air with fresh air. The movement of water into and out of the soil mainly brings about the process, e.g. rainwater soaks into the soil filling many of the pore spaces and driving out the air. Then, as the surplus water soaks down to the drains or is taken up by plants, fresh air is drawn into the soil to refill the pore spaces. Additionally, oxygen moves into the soil and carbon dioxide moves out by a diffusion process similar to that which happens through the stomata in plant leaves (Chapter 1). The aeration process is also assisted by:

- Changes in temperature.
- Changes in barometric pressure.
- Good drainage.
- Cultivations.
- Open soil structure.

Size and number of macropores (pores >0.5 mm) affect the aeration of the soil. Sandy soils are usually well aerated because of their open structure. Clay soils are usually poorly aerated, especially when the very small pores in such soils become filled with water. Good aeration is especially important for germinating seeds and seedling plants.

3.3.6   Soil micro-organisms

There are thousands of millions of organisms in every gramme of fertile soil. They are very important in the breakdown of organic residues and formation of humus. Many different types are found, but the main groups are listed below.

1. Bacteria are the most numerous group and are the smallest type of single celled organisms. Most of them feed on and break down organic matter. They need nitrogen to build cell proteins, but if they cannot get this nitrogen from the organic matter they may use other sources such as the nitrogen applied as fertilisers. When this happens (e.g. where straw is ploughed in) the following crop may suffer from a temporary shortage of nitrogen. Some types of bacteria can convert (*fix*) the nitrogen from the air into nitrogen compounds which can be used by plants (legumes and the nitrogen cycle, see Section 1.7). Numbers of bacteria vary in the soil; they increase near a source of organic matter. In the area next to plant roots (*rhizosphere*) there is an increase in the number of bacteria compared with the rest of the soil. The rhizosphere bacteria may be important in crop nutrient uptake. Bacteria are found in all soils although fewer are present in acid conditions.

2. Fungi are simple types of plants which feed on and break down organic matter. They are the most important organisms responsible for breaking down complex compounds such as lignified (woody) tissue and waxes. They have neither chlorophyll nor proper flowers. The species and number of fungi in the soil are constantly changing, depending on soil conditions and the type of organic material present. Fungi can live in more acid and drier conditions than bacteria; they are the main organisms for decomposing organic material in acid soils. Some fungi form symbiotic associations with plant roots (*mycorrhiza*); these associations can improve nutrient and water uptake and can protect root systems from some pathogens. Mycorrhizas are important in woodland and organic farming systems where levels of available phosphorus and nitrogen are low. As well as the beneficial fungi there are some disease-producing fungi present, e.g. those causing take-all in cereals or club root in brassicas.

3. Actinomycetes are organisms which are intermediate between bacteria and fungi and have a similar effect on the soil. They need oxygen for growth and are more common in the drier, warmer soils. They are not as numerous as bacteria and fungi. Some types can cause plant disease, e.g. common scab in potatoes (worst in light, dry, alkaline soils).

4. Algae in the soil surface are very small simple organisms which contain chlorophyll and so can build up their bodies by using carbon dioxide from the air and nitrogen from the soil. Algae growing in swampy (waterlogged) soils can use dissolved carbon dioxide from the water and release oxygen. This process is an important source of oxygen for crops such as rice. Algae are important in colonising bare soils in the early stages of weathering. Blue-green algae can fix atmospheric nitrogen.

5. Protozoa. These are very small, single-celled organisms. Most of them feed on bacteria, fungi and other microbes. A few types contain chlorophyll and can produce carbohydrates like plants.

3.3.7   Soil fauna including earthworms

In addition to earthworms, there are many species of small animals present in most soils. They feed on living and decaying plant material and other micro-organisms. Some of the common ones are slugs, millipedes, centipedes, ants, spiders, nematodes, mites, springtails, beetles and larvae of various insects such as cutworms, leatherjackets and wireworms. The farmer is mainly concerned with those which damage his crops or livestock and those which are predators of crop pests such as the carabid beetles.

Earthworms have a very beneficial effect on the fertility of soils, particularly those under grass. Several million earthworms per hectare have been recorded under grassland. There are many different kinds found in our soils, but most of their activities are very similar. They live in holes in the soil and feed on organic matter – either living plants or, more often, dead and decaying matter. They carry down into the soil fallen leaves and similar materials. Earthworms do not thrive in acid soils because they need plenty of calcium (lime) to digest the organic matter they eat. Their casts, which are usually left on the surface, consist of a useful mixture of organic matter, mineral matter and lime. The greatest numbers are found in loam soils, under grass, where there is usually a good supply of air, moisture, organic matter and lime. Soil disturbance can have a negative effect on earthworm populations. Numbers are often higher under minimal cultivation systems compared with ploughing. Earthworms are the main food of the mole which does so much damage by burrowing and throwing up heaps of soil.

3.4   Farm soils

The following is a general farming classification based on some of the main soil types – clays, sands, loams, calcareous soils, silts, peats and peaty soils.

3.4.1   Clay soils

These soils have a high proportion of clay and silty material – usually over 60%, at least half of which is pure clay. These so-called heavy soils include the clay, silty clay and sandy clay textures. It is the clay content which is mainly responsible for their specific characteristics. Particles of clay have very important properties:

* They can group together into small clusters (flocculate) or become scattered (deflocculated).
* They usually have a high cation exchange capacity, and can combine with various chemical substances (base exchange) such as calcium, sodium, potassium and ammonia; in this way plant nutrients can be held in the soil.

Grouping or flocculation of the particles is very important in making clay soils easy to work. Clay particles combined with calcium (lime) will flocculate easily, whereas those combined with sodium will not. The adhesive properties of clay are very beneficial to the soil structure when the groups of particles are small (like crumbs). Deflocculation can occur when clay soils are flooded with sea water or worked when wet. If the latter occurs they become puddled and if the weather then becomes dry the clay dries into hard lumps or clods. Frost action, and alternating periods of wetting and drying, will help to restore large lumps or clods to the crumb condition. Some air is drawn into cracks caused by shrinkage. These cracks remain when the clod is wetted again and so lines of weakness are formed which eventually break the clod. In prolonged dry weather, these cracks may become very wide and deep which later may be very beneficial for drainage.

Characteristics

- Clay soils feel very sticky and rolls like plasticine when wet.
- They can hold more total water than most other soil types and, although only about half of this is available to plants, crops seldom suffer from drought.
- They swell when wetted and shrink when dried, so a certain amount of restructuring can take place in these soils depending on weather conditions.
- They lie wet in winter and so stock should be taken off the land to avoid poaching (the compaction of soils by animals' hooves).
- They are very late in warming up in the spring because water heats up more slowly than mineral matter.
- They are normally fairly rich in potash, but are deficient in phosphates.
- Clay soils usually need large infrequent dressings of lime.

Management

Clays are often called heavy soils because for ploughing and subsequent cultivations, compared with light (sandy) soils, two to four times the amount of tractor power may be required. All cultivations must be very carefully timed (usually restricted to a shorter period than on other soil types) so that the soil structure is not damaged. Minimum cultivations are often used to establish crops on these soils and many farms only plough rotationally (Chapter 8). Autumn ploughing, to allow for a frost tilth, is essential if good seedbeds are to be produced in the spring.

Good drainage is very necessary for arable cropping. Originally drainage on clay soils was done using 'ridge and furrows', which were set up by ploughing, making the 'openings' and 'finishes' in the same respective places until a distinct ridge and furrow pattern was formed. The direction of the furrows is the same as the fall on the field so that water could easily run off into ditches. This system of drainage has been replaced by using underground drains (Chapter 8).

In many clay-land areas, especially where rainfall is high, the fields are often small and irregular in shape because the boundaries were originally ditches which followed the fall of the land. The hedges and deciduous trees, which were planted later, grow very well on these fertile, wet soils. Increasing the soil organic matter, such as applying straw-rich farmyard manure, ploughed-in straw or grassland residues, makes these soils easier to work.

Cropping

Because clay soils can be difficult to manage, only a limited range of arable crops are grown, though yield potential is high. In high rainfall areas the ground is usually left in permanent grass and only grazed during the growing season. Autumn sown crops such as winter wheat, winter oilseed rape, winter barley and winter beans are the most popular arable crops for this soil type. Sugar beet and potatoes are occasionally grown in some districts. However, there can be difficulties in seedbed preparation and harvesting these crops, especially in a wet season. The best sequence of cropping with these root crops is following a period under grass; then the soil structure is more stable and the soil easier to work.

3.4.2   Sandy soils

These soils are mainly composed of sand (greater than 70%) with very little clay (less than 15%) or silt. The soil texture is either a sand or loamy sand. Because these soils contain very little clay or organic matter they tend to be very weakly structured. They are very easy to work. No natural restructuring takes place on these soils so any compaction has to be removed using cultivations. As sand is inert the cation exchange capacity of these soils is very low. Because of the large particle size of sand these soils contain a large number of macropores which tend to contain air rather than water.

Characteristics

- They can be worked at any time, even in wet weather, without harmful effects.
- They are normally free-draining, but some drains may be required where there is clay or other impervious layer underneath, or a high water table.
- These soils are droughty as they have a very low available water capacity. For some crops irrigation is required.
- They warm up early in spring.
- These soils are prone to nutrient loss and are very responsive to application of fertilisers.
- They are unstable and can be easily eroded by water (on slopes) and by wind.
- They have little natural structure of their own and often need subsoiling at regular intervals to loosen compacted layers (pans).
- Some crop pests and diseases are affected by the soil type, e.g. slugs are not usually a problem, whereas 'take-all' disease can be very serious in second cereal crops on sandy soils.

Management

In many ways these are the opposite of clays and are often called light soils because, when working them, comparatively little power is required to draw cultivation implements. These soils are not suitable for minimum cultivation systems and topsoil loosening is required, usually by ploughing. After ploughing few cultivations are required to produce a seedbed, often a one-pass system is used. Adequate amounts of fertiliser must be applied to every crop. Liming is necessary but must be used carefully – 'little and often'. Organic matter, especially as humus, is very beneficial because it helps to hold water and plant nutrients in the soil. On properly limed fields, organic matter break down is very rapid. This is because the soil micro-organisms are very active in these open-textured soils which have a good air supply.

Sandy soils are very prone to both wind and water erosion and run-off on slopes, and wind erosion in the spring can destroy a young crop. Various methods are available to reduce the risk of erosion including planting nurse crops, applying organic manures and cultivations.

Cropping

A wide range of crops can be grown, but yields are very dependent on a good supply of water and adequate plant food. Cereal yields are usually low. Market gardening is often carried out where a good sandy area is situated near a large population.

On the lighter sands in low rainfall areas and where irrigation is not possible, less drought-susceptible crops are grown, e.g. rye, carrots, sugar beet, forage maize and lucerne; lupins are grown in a few areas where the soil is very poor and acid. To reduce the risk of drought stress, crops are established early to try and encourage deep rooting; better results are produced from winter- rather than spring-sown cereals. Winter barley is often preferred rather than wheat as it ripens earlier and may be less affected by a summer drought. It is important to ensure there is no soil compaction which would also restrict rooting.

On the better sandy soils, and particularly where the water supply (from rain or irrigation) is reasonably good, the main arable farm crops grown are cereals, peas, sugar beet, potatoes and carrots. Where there are hedges they are often left tall to act as windbreaks. The trees are usually drought-resistant coniferous types. Livestock can be out-wintered on sandy soils with less risk of damage by poaching, even in wet weather.

3.4.3   Loams

Loams are intermediate in texture between the clays and sandy soils and, in general, have most of the advantages and few of the disadvantages of these two extreme types. These soils have developed over a wide range of parent material; they are called medium-textured soils. The amount of clay present varies considerably, and can be up to 35%. Texture classes include the clay loams (resembling clays in many respects), silty clay loams, and sandy loams (resembling the better sandy soils). A large proportion of the sand is coarse and/or medium sized which makes the soil feel gritty. Climate and topography are the main limitations to crop production.

Characteristics

- Average water-holding capacity and so are fairly resistant to drought.
- They warm up reasonably early in the spring.
- They are moderately easy to work.

- Depending how they were formed, some of the loams can contain stones which can affect sowing and harvesting of some crops.
- A potentially fertile soil.

Management
Loams are moderately easy to work but should not be worked when wet, especially clay loams. Minimum cultivation systems can be successful on these soils.

Cropping
Loams are generally regarded as the best all-round soils because they are naturally fertile and can be used for growing any crop provided the depth of soil is sufficient. These soils can be used for most types of arable or grassland farming but, in general, mixed farming is carried on. Cereals, oilseed rape, potatoes and sugar beet are the main arable crops grown, and leys and forage maize provide grazing and fodder for dairy cows, beef cattle or sheep.

3.4.4   Calcareous soils
These are soils derived from chalk and limestone rocks and contain various amounts of calcium carbonate, between 5% and 50%. The depth of soil and subsoil may vary from 8 cm to over a metre. In general, the deep soils are more fertile than the shallow ones. The ease of working and stickiness of these soils depend on the amount of clay and chalk or limestone present; they usually have a loamy texture. Sharp-edged flints of various sizes, found in soils overlying some of the chalk formations, are very wearing on cultivation implements and tyres, as well as being destructive when picked up by harvesting machinery. In some places the flints are found mixed with clay, e.g. clay-with-flint soils.

The soils overlying chalk are generally more productive than are those over limestone because plant roots can penetrate the soft chalk and explore for water. The limestones are harder and mainly impenetrable. Limestone rock pieces, loosened by cultivations, are a more severe problem to the farmer than the pieces of chalk that work their way to the surface on the downland arable farms.

Dry valleys are characteristic of the limestone and chalk downland. The few rivers rise from underground streams and the deliberate flooding of water meadows in the river valleys used to be a common practice. Watercress beds flourish along some chalk streams. The farms and fields on this type of land are usually large, especially on the thinner soils. There are very few hedges and the trees are mainly beech and conifers. Walls of local stone form the field boundaries in some limestone areas.

Characteristics

- The soils are free draining except in a few small areas where there is a deep clay subsoil.
- The soils are often shallow and can be prone to drought (limestone soils).

- The soils are usually found at fairly high altitudes, above 120 m.
- Stones can affect sowing and harvesting of some crops.
- The soils naturally contain low levels of some major nutrients. The alkalinity of the soil can cause some trace elements to be unavailable. Only the deepest and/or sandy outcrops ever need liming.

Management

All, except soils with high clay content, are easy to cultivate. Minimum cultivation systems have been very successful on these well structured soils. As the soils are free draining there is a large window when it is possible to cultivate. Drilling of spring cereals can often take place very early in the spring. Organic matter can be beneficial, but in the alkaline conditions it breaks down fairly rapidly.

Cropping

Cereals are good crops for these soils and malting barley has been very successful. Continuous barley and/or wheat production has been common practice, but this has changed partly due to the difficulty of controlling grass weeds and better returns from other crops. Roots such as sugar beet, fodder beet and potatoes (some for seed) are grown on the deeper soils on some farms. Other crops such as oilseed rape, peas, beans, linseed, and leys for grazing, conservation or seed production provide a break from cereals. On the thin limestone soils it is important to establish the winter cereals early so that they are less affected by any summer drought.

As calcareous soils can be found at fairly high altitudes and the fields are often exposed, great care should be taken when growing crops which shed their seed easily, e.g. oilseed rape. Harvesting must be carried out carefully to minimise yield loss.

Areas of black puffy soil (18–25% organic matter) are found on some chalkland farms and they require special treatment; cultivations can be difficult and often trace element deficiencies, e.g. for copper, need correcting.

3.4.5   Silts

The silty soils can contain up to 80% silt. Silty soils include sandy silt loam and silt loam textures. The sand fraction is mainly of very fine sand particles. They have a very silky, buttery feel. These soils are formed from glacial, river, marine and wind-blown deposits and usually have deep stone-free subsoils.

Characteristics

- Most silt soils need some sort of drainage. In certain coastal areas there has to be a pumped system in order for the water table to be lowered and arable crops grown.
- Potentially very fertile mainly due to their great depth and very high available water capacity.

- Good working properties provided that organic matter is maintained above 3%.
- Water erosion and run-off can be a problem in some areas.

Management
Although very fertile, silts can be difficult to manage. The two main problems are capping and compaction.

Capping occurs when heavy rain falls on a very fine seedbed. Silt and clay particles go into suspension in a surface slurry and, as this dries out, it forms an impenetrable layer on the surface of the soil. It is particularly damaging if seeds have been sown but not yet germinated and emerged. Crops with small seeds, such as brassicas, will sometimes need to be redrilled if capping occurs. Leaving a rougher seedbed and increasing surface organic matter can decrease the risk of capping.

The lighter silts, as with the sands, can be damaged by compaction if worked in unsuitable conditions. Correction by deep cultivations may be necessary.

Cropping
On the heavier silts, mainly grass, cereals or fruit are grown; yield potential is high. However, on the drained light silts there are no limitations to crop growth. A wide range of crops can be grown including wheat, potatoes, sugar beet, vining peas, bulbs and field vegetables. River and marine silts in high rainfall areas, that are liable to flood, are best left in permanent grass.

3.4.6   Peat and peaty soils
Peaty soils contain about 20–35% of organic matter, whereas there is over 50% in true peat. These organic soils are usually very black or dark brown in colour and feel silty.

The blanket bog peats have formed in waterlogged upland areas overlying impermeable parent rock where there is a high rainfall. Plants such as mosses, cotton grass, heather, *Molinia* grass and rushes grow. The dead material from these plants is only partly broken down by the types of bacteria which can survive under these acidic waterlogged conditions. This 'humus' material builds up slowly, about 30 cm every century. These peats are very acidic and are very low in nutrients. A certain amount of improvement can be done by drainage, fertilising and liming but because of their situation they are only suitable for permanent grassland or forestry.

There are two important types of lowland peat, the sedge or fen peats and the moss peats. The fen peats developed in marshy land depressions below sea level, where the water coming into the area was rich in nutrients. Successive layers of decaying plant material built up. The main vegetation consisted of reeds, sedges and woody plants. Old bog oaks are sometimes brought up by cultivations. When drained these peats break down to produce very fine particles. These soils are slightly acidic and are low in phosphorus, potassium and some trace elements. The moss peats developed on boggy land with a high rainfall. The main vegetation was moss which grew and decayed very slowly to produce a raised area (raised moss). When drained these peats have a low pH and are very low in nutrients. Moss peat is more fibrous than fen peat.

Characteristics

- These soils are rich in nitrogen, released by the breakdown of the organic matter, but are very poor in phosphates and potash and also trace elements such as manganese and copper.
- Very fertile soils if drained and fertilised. As well as ditches for drainage those areas below sea level need a pumped drainage system if arable crops are to be grown.
- The soils are naturally acidic and initially need heavy liming. Over liming can lead to trace element deficiencies.
- On drainage the peat initially shrinks. Afterwards the organic matter breaks down very rapidly and the soil level can fall by as much as 2.5 cm per year; this is called 'wastage'. Eventually the mineral subsoil will come close to the topsoil and have a major effect on the characteristics and potential cropping of the soil.
- Wind erosion or 'blowing' in spring is a serious problem due to the light weight of peat when dry. Several plantings of crop seedlings, together with the top 5–8 cm of soil and fertilisers, may be blown into the ditches. Deep ploughing and/or cultivation to mix the underlying clay/mineral subsoil with the organic topsoil can prevent this. Husbandry techniques such as straw planting, nurse crops or strip cultivation are also used to prevent wind erosion.
- Weed control can be very difficult. Weeds grow very vigorously with the high nitrogen levels. Residual or soil-applied herbicides are not normally effective as the chemicals become adsorbed onto the particles of organic matter. Foliar-acting herbicides or mechanical methods are commonly used for weed control on these soils.
- Peat soils have a very high available water capacity for crop growth. Crops rarely suffer from drought on these soils.
- Under grassland these soils are very susceptible to poaching.

Management

Before reclaiming this land for cropping, some of the peat can be cut away for fuel or sold as peat moss for horticultural purposes. Good drainage must then be carried out by cutting deep ditches through the area. Deep ploughing also helps to drain the soil. Most of the lowland peat is below sea level and so the water in the ditches has to be pumped over the sea walls or into the main drainage channels. The depth of the subsequent water table will affect the potential cropping of the area. Once the peat has wasted to a depth of less than 0.9 m, under drainage may be required if the subsoil is clay.

Heavy applications up to 25 t/ha of ground limestone may be required to reduce the soil acidity. Sometimes a very acid reddish brown layer ('drummy' layer) builds up between the organic topsoil and the underlying clay subsoil. Peat soils are easy to cultivate and do not suffer from soil compaction. Under-consolidation of the seedbed is common. The peaty soils with a lower organic matter content suffer less from under-consolidation. Cultivation systems for the organic mineral soils are similar to those for the mineral soils.

Cropping
Drained fen or light peat soils are among the most fertile arable soils. Crops such as potatoes, sugar beet, celery, onions, carrots, lettuce and market garden crops are commonly grown. Cereals produce low yields. On light undrained peats, or where the water table remains high, the main crop is grass. On some of these areas willows are grown. Peaty loams are suitable for growing root crops and cereals. Where the depth of peat is very shallow (due to wastage) cropping is similar to that for the underlying mineral soil type.

3.5   Soil fertility and productivity

Soil fertility is a rather loose term used to indicate the potential capacity of a soil to grow a crop (or sequence of crops). The productivity of a soil is the combined result of fertility and management. The fertility of a soil at any one time is partly due to its natural make-up (inherent or natural fertility) and partly due to its condition (variable fertility) at that time. Natural fertility can have an influence on the rental and sale value of land. It is the result of factors which are normally beyond the control of the farmer, such as:

* The texture and chemical composition of the mineral matter.
* The topography (natural slope of the land) which can affect drainage, temperature and workability of the soil.
* Climate and local weather, particularly the effects of temperature, and rainfall (quantity and distribution).

Soil management can affect soil condition by changing:

* The amount of organic matter and the biological activity in the soil.
* The amount of water in the soil by drainage and irrigation (Chapter 8).
* The loss of soil by erosion (removal by wind and water).
* The pH of the soil.
* The amount of plant nutrients in the soil (Chapter 4).
* The soil structure.

Good management of the above factors should help maintain or increase soil fertility and at the same time be commercially profitable.

3.5.1   Soil erosion and the Soil Protection Review
Soil erosion is the movement of soil particles by natural processes such as wind or water. Soil erosion normally occurs very slowly on undisturbed soils with natural

vegetation. It is only when soils are cultivated for arable cropping or are intensively grazed that soil erosion and degradation can become an issue. On cropped land erosion can cause losses of nutrient rich topsoil as well as (in severe cases of wind erosion), total crop loss and deposition of soil sediment on roads and into water courses. Soil runoff of nutrients, organic matter and pesticides can also affect water quality and habitats. Although soil erosion in the UK is not as severe as found in other parts of the world, it is an increasing problem due to the incidence of more extreme weather events. All farms in England currently receiving the single payment have to undertake a Soil Protection Review as part of their cross-compliance. Farmers have to introduce appropriate soil management practices to try and reduce the impact of erosion and runoff on the wider environment.

Wind erosion and control

Cultivated, dry, sandy and peaty soils are most susceptible in the spring before crops are fully established. Keeping field sizes small and planting tall shelter belts can help reduce wind speeds, but other management measures are also required. Growing nurse crops or planting straw between rows of susceptible crops have been very successful at reducing wind erosion (Chapter 15). Leaving rough seedbeds and minimising cultivations can also help stabilise the soil, as can spreading mulches or even applying some synthetic polymers.

Water erosion and control

Silty and sandy soils are most prone to water erosion, particularly in areas with a high rainfall and where the ground slopes. Fields very susceptible to erosion commonly have shallow channels or rills, or even larger deeper gullies, after heavy rain. Surface water runoff happens when the intensity of rainfall is so great that all the water cannot soak into the soil. Livestock management can also affect the amount of soil erosion and runoff in grassland. Grazing with large numbers of stock in wet conditions can cause serious poaching of the soil and compaction in the topsoil which can lead to an increased risk of runoff.

Some cropping systems are much more prone to water erosion than others. The most susceptible land use is when crops are harvested in the autumn under adverse soil conditions which can happen with forage maize and field vegetables. Long term leys and permanent pasture usually pose little risk. Soils with a good crop cover are much less susceptible than where there are fine seedbeds and or bare ground.

Reducing compaction, increasing soil organic matter, using green manures, avoiding late planted autumn sown or harvested crops, planting buffer strips and establishing grass leys can all help to reduce the risk of soil erosion on susceptible fields.

Soil Protection Review

The Soil Protection Review was introduced to try and reduce soil degradation. Farmers are required to record their current and potential soil problems as well as assessing soil types and erosion risks on their land. Suitable land management measures have to be undertaken to prevent soil erosion and runoff issues. The success of the management options chosen has to be reviewed annually and adapted accordingly.

3.5.2   Liming

Most farm crops will not grow satisfactorily if the soil is very acidic. In acid conditions aluminium, iron and manganese become more readily available. Excessive uptake of aluminium in acid conditions can severally affect crop growth. Some crops are more affected than others. This can be remedied by applying one of the commonly used liming materials.

The chemistry of liming

All substances in the presence of water are either acid, alkaline or neutral. The term 'reaction' describes the degree or condition of acidity, alkalinity or neutrality. Acidity and alkalinity are expressed by a pH scale on which pH 7 is neutral, numbers below 7 indicate acidity and those above 7 alkalinity. (pH of a soil solution is the negative logarithm of the hydrogen ion concentration.) Most cultivated soils have a pH range between 4.5 and 8.5. Soil pH can be assessed in the field by using a soil testing kit, or more accurately in the laboratory.

To give crops the best opportunity to grow well, the minimum soil pH on mineral soils should be as shown in Table 3.2. The optimum pH for mineral soils where arable crops are grown is 6.5 and for continuous grass and clover swards the figure is 6.0.

Indications of soil acidity (i.e. the need for liming) are as follows:

- Crops failing in patches, particularly the acid-sensitive ones such as barley and sugar beet. The plants look yellow and stunted with a stumpy root system.
- On grassland, there are poor types of grasses present such as the bents. Often a mat of undecayed vegetation builds up because the acidity reduces the activities of earthworms and bacteria that break down such material.
- On arable land, weeds such as sheep's sorrel, corn marigold and spurrey are common where a soil is or has a history of being acidic. Also, in acid conditions residual herbicide activity may be reduced.
- Club root disease in brassicas is aggravated in acid soils. Liming is used as a method of control.

Table 3.2   Minimum soil pH requirements

Barley, sugar beet, beans, peas and lucerne	6.5 pH
Maize, oilseed rape, oats, wheat, cabbage and carrots	6.0 pH
Potatoes, rye and apples	5.5 pH
Ryegrass and fescues	5.0 pH

Lime requirement

This is based on the *optimum* pH for the crop to be grown, the soil texture, organic matter content and current pH. Grassland has a lower requirement than arable crops. Heavy (clay) soils and soils rich in organic matter require more lime to raise the pH than other types of soil. For example, to raise the pH from 5.5 to 6.5 on a loamy sand may require about 7 t/ha of ground limestone, but on a clay soil 10 t/ha of ground limestone may be necessary. The sandy soil, however, will need to be limed more frequently than the clay soil. Lime should be applied before growing susceptible crops such as beet or barley but after growing such tolerant crops as potatoes. (Lime requirements are shown in Table 3.3.)

The main benefits of applying lime are:

1. It neutralises the acidity of the soil by removing the hydrogen ions on the soil charged sites and replacing them with calcium or magnesium.
2. It supplies calcium (and sometimes magnesium) for plant nutrition.
3. It improves soil structure and makes the structure more stable. In well limed soils, plants usually produce more roots and grow better. Bacteria are more active in breaking down organic matter and this also usually results in an improved soil structure so the soil can be cultivated more easily.
4. It affects the availability of plant nutrients. Nitrogen, phosphate and potash are freely available on properly limed soils. Too much lime is likely to make some minor nutrients or trace elements unavailable to plants, e.g. manganese, boron, copper and zinc. This is least likely to happen in clay soils and most likely to happen on organic soils.

Lime losses

Lime is removed from the soil in many ways. About 125–2000 kg/ha of calcium carbonate may be lost annually in drainage water. The rate of loss is greatest in areas of high rainfall, well-drained soils and soils rich in lime.

Fertilisers containing ammonium, nitrate, sulphate or chloride ions have the greatest acidifying effect. Every 1 kg of sulphate of ammonia removes about 2 kg of calcium carbonate from the soil. Other nitrogen fertilisers have

Table 3.3   Lime requirement of different soils (t/ha)

Cropping	Soil type	pH 6.0	pH 5.5	pH 5.0
Arable cropping	Light	4	7	10
	Medium	5	8	12
	Heavy	6	10	14
	Organic	4	9	14
	Peat	0	8	16
Grassland	Light	0	3	5
	Medium	0	4	6
	Heavy	0	4	7
	Organic	0	3	7
	Peat	0	0	6

a smaller acidifying affect. (Some organic manures can slightly reduce soil acidity.)

Crops and stock also remove some calcium carbonate, for example a 7 t/ha cereal crop can remove 50 kg and a 500 kg animal sold off the farm removes about 16 kg of calcium carbonate in its bones.

Materials commonly used for liming soils
Ground limestone or chalk (also called carbonate of lime or calcium carbonate $CaCO_3$) is the commonest liming material used at present. These are obtained by quarrying the limestone or chalk rock and grinding it to a fine powder. The finer the particle size the quicker the reduction of acidity.

Some liming materials are obtained from industrial processes where lime is used as a purifying material such as in sugar beet factories, these are called *waste limes*. Sugar beet waste lime is also a valuable source of plant nutrients.

Magnesium or dolomite limestone consists of magnesium carbonate ($MgCO_3$) and calcium carbonate ($CaCO_3$). It is commonly used as a liming material in areas where it is found. Magnesium carbonate has a better neutralising value than calcium carbonate of approximately 20%. Additionally, the magnesium may prevent magnesium deficiency symptoms in crops (e.g. interveinal yellowing of leaves in potatoes, sugar beet and oats).

Calcareous sand or shell sand is a liming material collected from some beaches where there is a high shell content. The neutralising value is lower than for ground limestone. It is a useful material in areas far removed from the limestone quarries.

The supplier of lime must give a statement of the neutralising value (NV) of the liming material which is the same as the calcium oxide equivalent

Cost
The cost of liming is largely dependent on the transport costs from the lime works to the farm. By dividing the cost per tonne of the liming material by the figure for the neutralising value, the unit cost is obtained. In this way it is possible to compare the costs of the various liming materials. Most farmers now use ground limestone or chalk and arrange for it to be spread mechanically by the suppliers. Where large amounts are required (over 7 t/ha) it may be preferable to apply it in two dressings, e.g. half before and half after ploughing, to ensure more even distribution through the soil.

3.6   Sources of further information and advice

Further reading
Ashman M R and Puri G, *Essential Soil Science*, Blackwell Publishing, 2002.
Batey T, *Soil Husbandry*, Soil and Land Use Consultants Ltd, 1988.
Brady N C and Weil R C, *The Nature and Properties of Soils*, 14th edn, Pearson Education, 2007.
Bridges E, *World Soils*, Cambridge University Press, 1997.

Curtis L F, Courtney F M and Trudgill S, *Soils of the British Isles*, Longmans, 1976.
Davis B D, Eagle D J and Finney J B, *Resource Management: Soil*, Farming Press, 2001.
Defra, *Protecting our Water, Soil and Air*, The Stationery Office, 2009.
Defra, *Controlling Soil Erosion*, Defra Publications, 2005.
Dubbin, W, *Soils*, The Natural History Museum, 2001.
Environment Agency, *Think Soils*, Environment Agency, 2008.
Gerard G, *Fundamentals of Soils*, Routledge, 2000.
Kilham K, *Soil Ecology,* Cambridge University Press, 1994.
MAFF Report, *Modern Farming and the Soil*, HMSO, 1970.
Morgan R P C, *Soil Erosion and Conservation*, Blackwell Publishing, 2005.
Rowell D L, *Soil Science: Methods and Applications*, Longmans, 1994.
Simpson K, *Soil*, Longmans, 1983.
Soil Survey for England and Wales, *Soils and their use in East England, Midlands and North Western England, S. E. England, S. W. England and Wales*, SSEW, 1985.
White R E, *Principles and Practice of soil science*, 4th edn, Blackwell Publishing, 2006.
Wild A, *Soils and the Environment: An Introduction*, Cambridge University Press, 1993.

Websites
www.defra.gov.uk
www.landis.org.uk
www.smi.org.uk
www.soils.org.uk
www.soilsworldwide.net

4

Fertilisers and manures

DOI: 10.1533/9781782423928.1.63

Abstract: This chapter explains the importance of feeding crops and ensuring that soil fertility is maintained. It discusses the main elements required by crops, both major and trace. It explains why liming is so important for crop nutrition. It deals with the calculations and sources of information needed to accurately decide on fertiliser application rates, and cost them. It describes the materials used on-farm as fertilisers and explains the differences between straights, compounds and blends, as well as dealing with liquids and solid fertilisers. There is a section on the use of organic manures and slurries and how to make best use of them. Finally, the chapter deals with the effect of fertilisers on the environment, during both their manufacture and application.

Key words: nitrogen, phosphate, potash, lime, fertiliser manual, compounds, straights, farmyard manures, slurry.

4.1   Nutrients required by crops

If good crops are to be grown in a field, there should be at least as many nutrients returned to the soil as have been removed and these nutrients should be readily available to the plant as and when needed. This can be achieved by the sensible use of chemical fertilisers (manufactured inorganic sources of nutrients), where possible in conjunction with organic manures such as farmyard manure and slurry or, as in organic systems, the use of natural inorganic fertiliser materials in conjunction with organic manures and legume crops. It is important that any nutrient supply system minimises the loss of nutrients to the environment as well as avoiding either a deficiency or an excess of individual, required elements. Table 4.1 indicates average figures for nutrients removed by various crops.

Nutrients required by crops can be divided up into major elements and micro (trace) elements (Table 4.2). Micro (trace) elements include boron, copper, iron, manganese, molybdenum, zinc and cobalt, some of which are dealt with later in this chapter.

Table 4.1   Nutrients removed by crops (kg/ha)

Crop (good average yield)	N	P_2O_5	K_2O	Mg	S
Wheat					
Grain 7 t/ha	130	55	40	9	35
Straw 5 t/ha	17	4	31		
Total	147	59	71	9	35
Barley					
Grain 6 t/ha	100	47	34	7	22
Straw 4 t/ha	25	3	25		
Total	125	50	59	7	22
Oats					
Grain 6 t/ha	100	47	34		
Straw 5 t/ha	15	5	59		
Total	115	52	93		
Beans					
Grain 4 t/ha	176	44[a]	48	8	40
Potatoes					
Tubers 50 t/ha	150	50	290	15	20
Sugar beet					
Roots 45 t/ha	80	36	76		
Fresh tops 35 t/ha[b]	120	39	202	27	33
Total	200	75	278	27	33
Kale					
Fresh crop 50 t/ha[b]	224	60	250		

[a] The response to phosphatic fertilisers is greater than these figures suggest.

[b] If sugar beet tops or kale are eaten by stock on the field where grown, some of the nutrients will be returned to the soil.

Table 4.2   Major and micro elements required for crop growth

Major elements	Taken up by plant as:
Nitrogen	Nitrate
Phosphate	Hydrogen phosphate
Potassium	K^+
Sulphur	Sulphate
Magnesium	Mg^{2+}
Calcium	Ca^{2+}
Sodium	Na^+

Nitrogen is supplied by fertilisers, organic manures, nodule *Rhizobium* bacteria on legumes (e.g. clovers, peas, beans, lucerne), and bacteria in the soil which decompose organic matter and produce nitrates in a process called mineralisation. Free-living bacteria (*Azotobacter*) also fix nitrogen in the soil. Some nitrogen is also produced during thunderstorms by lightning strikes and some is contained in rainfall and snow. Nitrogen can be removed from the soil by bacteria which steal the oxygen from nitrates to leave gaseous nitrogen. This is a process called denitrification. Nitrogen can also be lost by leaching (the process of nitrates in solution moving down away from the root zone and eventually into the drains or bedrock).

Phosphate is supplied by fertilisers and organic matter from manures or plant debris. Soluble fertilisers provide hydrogen phosphate for immediate use by the plant while insoluble fertilisers need to be worked on by bacteria and the action of weak acids in the soil before they can be used. Organic matter phosphate goes through a transition phase of microbial phosphate before becoming hydrogen phosphate in soil solution.

Potassium comes from organic manures and plant debris as well as fertilisers. Within the soil there are three types of potassium: water-soluble potassium which is available to plants, exchangeable potassium (an intermediate stage) and a potassium reserve held in the clay lattices within the soil which, by the action of weathering, becomes available over a period of time.

Sulphur is provided by fertilisers, organic matter and as sulphur dioxide and sulphurous acid in rainfall (acid rain). Organic matter sulphur is converted to sulphur and then sulphate by microbial action. Under anaerobic conditions sulphur can be lost from the soil in the form of hydrogen sulphide (a gas smelling of bad eggs). Table 4.3 shows the major plant foods for crop growth.

A decision has to be made as to how much and what type of fertilisers should be used for each crop. For phosphorus, potassium and magnesium the amount applied should be based on an analysis of a soil sample. Soil analysis will show:

- Soil texture.
- pH (usual range 4–8). This is a useful guide to the lime requirement needed to bring acid soils up to an optimum level for the particular crop.
- Available nutrients. This is the level of phosphate, potash and magnesium and it is indicated by index ratings 0–9. A deficiency level is indicated by 0 and an excessively high level by 9 (never reached under field conditions). An index of 2 and 3 is satisfactory for farm crops. As indices drop below 2 more fertiliser is recommended so that the index will go up over a period of time, and as they get above 2 or 3 the recommendations decrease until, at index 4 and over, no applications of phosphorus, potassium or magnesium are required.

The field sampling for analysis must be done in a methodical manner if a reasonably accurate result is to be obtained. Generally, sampling should be carried out every four years, although pH may need to be checked more frequently, especially on light soils where calcium is easily leached. At least 25 soil cores, 15 cm deep (7.5 cm on long-term grassland), should be taken to make up a sample weighing at least 0.5 kg. Unless the area is very uniform, samples should be taken for every four hectares. A 'W' pattern should be walked across the area with sub-samples being taken frequently. Headlands, gateways and minor areas of obvious soil differences should be avoided. The analysis is carried out on the fine part of the soil and this must be borne in mind when interpreting the results for very stony soils. In this case the indices are usually too high.

Table 4.3 The need for and effect of some plant nutrients

Plant nutrient	Crops which are most likely to suffer from deficiency	Field conditions where deficiency is likely to occur	Deficiency symptoms	Effects on crop growth	Effects of excess	Time and method of application
Nitrogen (N)	All farm crops except legumes (e.g. beans, peas, clover); especially important for leafy crops (e.g. grass, cereals, kale and cabbages)	On all soils except peats, and especially where organic matter is low, and after continuous cereals	Thin, weak, spindly growth; lack of tillers and side shoots; small yellow/pale green leaves, sometimes bright colours	Speeds up growth of seedlings and roots; hastens leaf growth and maturity; encourages clover in grassland; improves quality	Lodging in cereals; delayed ripening; soft growth susceptible to frost and disease; may lower sugar and starch content	N fertilisers in seedbed or top-dressed
Phosphorus (P)	Root crops (e.g. sugar beet, swedes, carrots, potatoes), clovers, lucerne and kale	Clay soils; acid soils especially in high rainfall areas, chalk and limestone soils and peats; poor grassland	Similar to nitrogen except that leaves are a dull, bluish-green colour with purple or bronze tints	Increases leaf size, rate of growth and yield; makes leaves dark green	Might cause crops to ripen too early and so reduce yield if not balanced with nitrogen and potash fertilisers	Phosphorus fertilisers applied in seedbed for arable crops; 'placement' in bands near or with the seed is more efficient; broadcast on grassland
Potassium (K)	Potatoes, carrots, beans, barley, clovers, lucerne, sugar beet and fodder beet	Light sandy soils, chalk soils, peat, badly drained soils, grassland repeatedly cut for hay, silage or 'zero' grazing	Growth is squat and growing points 'die-back', e.g. edges and tip of leaves die and appear scorched	Crops are healthy and resist disease and frost better; prolongs growth; improves quality; balances N and P fertilisers	May delay ripening too much; may cause magnesium deficiency in fruit and glasshouse crops and 'grass-staggers' in grazing animals	K fertilisers broadcast or 'placed' in seedbed for arable crops; broadcast on grass-land in autumn or mid-summer

Element	Crops	Soils/conditions	Deficiency symptoms	Function	Likelihood	Treatment
Magnesium (Mg)	Cereals, potatoes, sugar beet, peas, beans, kale	Light sandy soils, chalk soils, often of a temporary nature due to poor soil structure, excessive potash, etc.	Chlorotic patterns on leaves (short of chlorophyll)	Associated with chlorophyll, and potassium metabolism	Unlikely, requirements about same as phosphorus and one-tenth that of nitrogen and potash	Use FYM or slurry; lime with magnesium limestone; Epsom salts; fertiliser containing kieserite
Sulphur (S)	Most crops but especially brassicae, e.g. kale, oilseed rape; grass cut for conservation; bread wheat	Light sandy soils – especially if low in organic matter and intensively cropped; modern fertilisers used; no smoke pollution	Yellowing leaves and less vigorous growth, like nitrogen deficiency	Nitrogen made more efficient	Unlikely to occur	Use FYM or sulphur foliar spray or to soil as gypsum

The recommendations for fertiliser use were reviewed in 2010 and published by the Department for Environment, Food and Rural Affairs (Defra) in their *Fertiliser Manual* 8th edition (RB 209). This is essential reading for anyone involved in crop production. The new recommendations take more account of the economic use of fertilisers, the value of a wider range of organic manures with more emphasis on readily available nutrients, the minimisation of environmental risk associated with the use of all types of fertilisers and manures, and introduce new recommendations for biomass crops.

Particularly important are the changes made to the nitrogen recommendations. Nitrogen is the most important element in terms of crop yield and quality, and deficiencies will lead to a serious reduction in profitability. However, excess nitrogen will be wasteful and could lead to the pollution of the environment. It is important therefore that the nitrogen requirements of individual crops should be calculated as accurately as possible. Nitrogen response curves using value of crop yield plotted against cost of applied nitrogen can be used to find the economic optimum rate of nitrogen to apply. The new recommendations for wheat use a breakeven ratio of 5:1 (i.e. 5 kg of grain are needed to pay for 1 kg of nitrogen). This economic optimum application rate takes into account the amount of nitrogen supplied from the soil and can be provided from inorganic and/or organic sources.

The new Defra nitrogen recommendations continue to use Soil Nitrogen Supply (SNS) as their basis but the system has been fully revised with clearer definitions of soil type and revised index tables. Defra defines SNS as 'the amount of nitrogen (kg N/ha) in the soil (apart from that applied for the crop in manufactured fertilisers and manures) that is available for uptake by the crop throughout its entire life, taking account of nitrogen losses'. It is calculated using the equation:

SNS = Soil mineral nitrogen (SMN) + estimate of nitrogen already in the crop + estimate of mineralisable soil nitrogen

Although the SMN (and thus the SNS) can be measured using laboratory analysis of soil samples, in most situations the SNS index will be determined by a 'Field Assessment Method' based on previous cropping, previous fertiliser and organic manure use, soil type and winter rainfall. The SNS index system is shown in Table 4.4.

The SNS of a field following an arable, forage or vegetable crop will depend on what soil type and in what rainfall area the crop was grown. For example, cereals can leave a field anywhere between SNS Index 0 and 6 depending on where they were grown. A winter wheat crop grown on a light sand soil in a low rainfall area will mean an index 0 field; the same crop grown on a deep silty soil in a moderate rainfall area will mean an index 1 and grown on a peat soil in all rainfall areas will leave the field at between index 4 and 6. The nitrogen requirements for a following crop can then be read from tables using the determined SNS value. For example, oilseed rape at SNS index 0 requires 220 kg/ha of spring N; at SNS index 3 it only requires 120 kg/ha. The SNS system and the recommendation tables can be found in the Defra *Fertiliser Manual* (RB 209).

Table 4.4   Defra soil nitrogen supply index system (SNS)

Variables used to determine SNS	
Rainfall	Soil type
Low rainfall areas: 500–600 mm annual rainfall, up to 150 mm excess winter rainfall	Light sands or shallow soils over sandstone
Moderate rainfall areas: 600–700 mm annual rainfall, 150–250 mm excess winter rainfall	Medium soils or shallow soils, not over sandstone
High rainfall areas: over 700 mm annual rainfall, or over 250 mm excess winter rainfall	Deep clayey soils Deep silty soils Organic soils Peat soils

Source: Defra, 2010.

Table 4.5   Example of SNS indices for a medium or shallow soil (not over sandstone)

Crop	1st crop	2nd crop	3rd crop
All leys with 2 or more cuts annually receiving little or no manure. 1–2 year leys, low N* 1–2 year leys, 1 or more cuts 3–5 year leys, low N, 1 or more cuts	1	1	1
1–2 year leys, high N*, grazed 3–5 year leys, low N, grazed 3–5 year leys, high N, 1 cut then grazed	2	2	1
3–5 year leys, high N, grazed	3	3	2

* Low N = average annual inputs of less than 250 kg N/ha in fertiliser plus available N in manure used in the last two years, or swards with little clover. High N = average annual applications of more than 250 kg N/ha in fertiliser plus available nitrogen in manure used in the last two years, or clover-rich swards or lucerne. Source: Defra, 2010.

SNS indices for sequences of crops following grassland are more complicated. Table 4.5 shows an example of indices for a medium soil or shallow soil (not over sandstone) in any rainfall area.

4.2   Trace elements

The need for trace elements is likely to be greatest:

- on very poor soils;
- where soil conditions, such as a high pH or a very high organic matter, make them unavailable;
- where intensive farming (with high yields) is practised;
- where organic products such as farmyard manure and slurry are not used.

The importance of trace elements in plant nutrition is most appreciated when plants are grown in culture solutions circulated past their root systems (hydroponics). This is a form of horticultural crop production that is rapidly increasing, especially where the solutions can be replenished automatically with nutrients.

Supplying trace elements to plants is not always easy and care is required to prevent overdosing which may damage or kill the crop. To facilitate their application and availability to the plants, trace elements such as copper, iron, manganese and zinc can now be used in a chelated form. These chelates are 'protected' water-soluble complexes of the trace elements with organic substances such as EDDHA (ethylene diamine dihydroxyfenic acetic acid) or EDTA (ethylene diamine tetra-acetic acid). These organic substances form a protective 'crab's claw' around the element, effectively sequestering them and protecting them temporarily. They can be safely applied as foliar sprays for quick and efficient action, or to the soil, sometimes as a supplement with a compound fertiliser, for root uptake without wasteful 'fixation' because they do not ionise. They are compatible when mixed with many spray chemicals. Trace elements may also be applied as *frits*, which are produced by fusing the elements with silica to form glass which is then broken into small particles for distribution on the soil, or as salts in solution.

It is very important, whenever possible, to obtain expert advice before using trace elements. This is because crops that are not growing satisfactorily may be suffering for reasons other than a trace element deficiency, e.g. major nutrient deficiency, poor drainage, drought, frost, mechanical damage, viral or fungus diseases. Trace element deficiencies can be diagnosed by either leaf or soil analysis, depending on the nutrient.

Manganese is the element most frequently found to be deficient in field crops. It is associated with high pH levels and high organic matter, both of which lock up manganese in the soil. It can be remedied using foliar sprays of manganese sulphate or chelated manganese, usually applied just before periods of rapid crop growth.

Boron deficiency is most often seen in root vegetables and sugar beet where it causes a problem of growing-point necrosis. Soils can be tested for boron levels and remedial action should be taken before crops are grown. Sprays of boron or boronated fertilisers can be used where necessary.

Copper deficiency is a problem of peaty, very light sandy and thin organic chalky soils. Symptoms are most often seen in wheat and barley where it produces shrivelled leaf and ear tips. Chelated copper can be applied directly to crops or copper sulphate applied to soils.

4.3   Units of plant food

4.3.1   Plant food requirements

The kilogram is now the unit of plant food in many parts of the world and recommendations are given in terms of kilograms per hectare (kg/ha). For example, the recommendation for a spring cereal crop may be to apply 100 kg/ha N, 50 kg/ha P_2O_5 and 50 kg/ha K_2O. To convert this into numbers of bags of fertiliser per hectare, it is necessary to use the percentage analysis figures for the fertiliser in question. This is clearly stated on each bag, e.g. 20:10:10 means that the particular fertiliser contains 20% N, 10% P_2O_5 and 10% K_2O, always given in that order. The percentage declarations are calculated from the atomic weights of the individual elements. For example, 100 kg of 20:10:10 fertiliser contains 20 kg N, 10 kg P_2O_5 and 10 kg K_2O; therefore a 50 kg bag contains 10 kg N, 5 kg P_2O_5 and 5 kg K_2O, i.e. the number of kilograms of plant food in a 50 kg bag of fertiliser is half the percentage figures.

In the example of the plant food requirements for the spring cereal crop, i.e. 100 kg N, 50 kg P_2O_5, 50 kg K_2O, this would be supplied by 10×50 kg bags of 20:10:10 fertiliser. In some cases, the figures may not work out exactly and a compromise has to be accepted (see for example Table 4.6).

The extra 80 kg of nitrogen shown in Table 4.6 would be top-dressed and may be supplied by approximately 4½ 50 kg bags per hectare of ammonium nitrate (34.5%), i.e. 80/17.25 = 4.6. If bulk fertiliser is used, each tonne contains 10 times the percentage of each plant food:

1 tonne (1000 kg) = 100×10 kg; therefore, 1 tonne of 20:10:10 contains 200 kg N, 100 kg P_2O_5 and 100 kg K_2O.

In the case of liquid fertilisers, the amount applied will depend on the type of declaration of the manufacturer. Some declare their fertilisers on a weight per volume basis and some on a weight per weight basis. The weight per volume declaration is relatively easy to work out, i.e. so many kilograms of plant food per 100 litres of water. A weight per weight declaration means that the user must know the specific gravity (SG), of the product, e.g:

100 litres of water weighs 100 kg
but a 28% N solution has an SG of 1.29
therefore 100 litres of 28% fertiliser weighs 129 kg
28% fertiliser means 28 kg in 100 kg product (wt per wt)
therefore there are 28 kg of N in 100/1.29 = 77.5 litres of product
or, put another way, the weight of N in 100 litres is 36 kg

Table 4.6   Example of recommended and supplied plant food level variance

	N (kg/ha)	P_2O_5 (kg/ha)	K_2O (kg/ha)
Spring cereals recommendation	100	50	50
Four 50 kg bags 10:25:25 combine-drilled, supply	20	50	50
Difference	80		

4.3.2   Kilogram cost

The cost of plant food per kilogram can be calculated from the cost of fertilisers which contain only one plant food such as nitrogen in ammonium nitrate, phosphate in triple superphosphate, and potash in muriate of potash.

If 1 tonne of ammonium nitrate (34.5% N) costs £300 and contains 345 kg N, then

1 kg of N costs 30 000p/345 = 87 pence

If 1 tonne of triple super (45%) costs £345 and contains 450 kg P_2O_5, then

1 kg of P_2O_5 costs 34 500p/450 = 77 pence

If 1 tonne of muriate of potash (60%) costs £310 and contains 600 kg K_2O, then

1 kg of K_2O costs 31 000p/600 = 52 pence

1 tonne of (9:24:24) compound contains:

N	P_2O_5	K_2O
90 kg	240 kg	240 kg

The value of this, based on the costs of a kilogram of nitrogen, phosphate and potash, is:

N	$90 \times 87p = £78.30$	
P	$240 \times 77p = £184.80$	
K	$240 \times 52p = £124.80$	
Total	£387.90	

Normally, the well-mixed granulated compounds cost more than the equivalent in 'straights'. Bulk blends of 'straights' are usually cheaper. Fertiliser prices have varied considerably in recent years; actual costs at any time should be substituted in the calculations shown.

It is also possible to compare the values of 'straight' fertilisers of different composition on the basis of cost per kilogram of plant food, e.g. the cost of a kilogram of nitrogen in ammonium nitrate (34.5% N) is 87 pence from the calculation above. This compares with the cost of a kilogram of nitrogen in urea (46% N) which is:

460 kg N/tonne at £340 per tonne = 34 000p/460 = 74 pence

Therefore, apart from lower handling costs because of the higher concentration of N, the urea is the lower-priced fertiliser although, compared with ammonium nitrate, it does have some limitations.

4.4   Straight fertilisers

Straight fertilisers supply only one of the major plant foods.

4.4.1 Nitrogen fertilisers

The nitrogen in many straight and compound fertilisers is in the ammonium (NH_4^+ cation) form but, depending on the soil temperature, it is quickly changed by bacteria in the soil to the nitrate (NO_3 anion) form. Many crop plants, e.g. cereals, take up and respond to the NO_3 anions quicker than the NH_4^+ cations, but other crops, e.g. grass and potatoes, are equally responsive to NH_4^+ and NO_3^- ions.

The ammonium cation, as a base, is held in the soil complex at the expense of calcium and other loosely-held bases which are lost in the drainage water. This will have an acidifying effect on the soil.

Ammonium nitrate and urea fertilisers are produced by spraying a solution of the fertiliser from a vibrating shower head into a 'prilling tower'. As the droplets fall down the tower against a stream of cold air they become round and solid, producing prills 1–3 mm diameter. Nitrogen fertilisers in common use are:

- *Ammonium nitrate* (33.5–34.5% N). This is a very widely used fertiliser for top-dressing. Half the nitrogen (as nitrate) is very readily available. It is marketed in a special prilled or granular form to resist moisture absorption. It is a fire hazard but is safe if stored in sealed bags and well away from combustible organic matter. Because of the ammonium present, it has an acidifying effect.
- *Ammonium nitrate lime* (21–26% N). This granular fertiliser is a mixture of ammonium nitrate and lime. It is sold under various trade names. Because of the calcium carbonate present it does not cause acidity when added to the soil.
- *Urea* (46% N). This is the most concentrated solid nitrogen fertiliser and it is marketed in the prilled form. It is sometimes used for aerial top-dressing. In the soil, urea changes to ammonium carbonate which may temporarily cause a harmful local high pH. Nitrogen, as ammonia, may be lost from the surface of chalk or limestone soils, or light sandy soils when urea is applied as a top-dressing during a period of warm weather. When it is washed or worked into the soil, it is as effective as any other nitrogen fertiliser and is most efficiently utilised on soils with adequate moisture content, so that the gaseous ammonia can go quickly into solution. In dry conditions in the height of summer it is probably better to use ammonium nitrate. Chemical and bacterial action changes it to the ammonium and nitrate forms. If applied close to seeds, urea may reduce germination.
- *Sulphate of ammonia* (21% N, 60% SO_3). At one time, as a fertiliser, this was the main source of nitrogen. However, sulphate of ammonia is seldom used now. It consists of whitish, needle-like crystals and it is produced synthetically from atmospheric nitrogen. Bacteria change the nitrogen in the compound to nitrate. It has a greater acidifying action on the soil than other nitrogen fertilisers. Some nitrogen may be lost as ammonia when it is top-dressed on chalk soils.
- *Sodium nitrate* (16% N, 26% Na). This fertiliser is obtained from natural deposits in Chile and is usually marketed as moisture-resistant granules. The nitrogen is readily available and the sodium is of value to some market garden crops. It is expensive and is not widely used.

- *Calcium nitrate* (15.5% N). This is a double salt of calcium nitrate and ammonium nitrate in prilled form. It is mainly used on the Continent.
- *Anhydrous ammonia* (82% N). This is ammonia gas liquefied under high pressure, stored in special tanks and injected 12–20 cm into the soil from pressurised tanks through tubes fitted at the back of strong tines. Strict safety precautions must be observed; it is a contractor rather than a farmer operation. The ammonia, as ammonium hydroxide, is rapidly absorbed by the clay and organic matter in the soil and there is very little loss if the soil is in a friable condition and the slit made by the injection tine closes quickly. It is not advisable to use anhydrous ammonia on very wet or very cloddy or stony soils. It can be injected when crops are growing, for example into winter wheat crops in spring, between rows of Brussels sprouts and into grassland. The cost of application is much higher than for other fertilisers, but the material is cheap, so the applied cost per kilogram compares very favourably with other forms of nitrogen. On grassland it is usually applied twice – in spring and again in midsummer – at up to 200 kg/ha each time. In cold countries it can be applied in late autumn for the following season, but the mild periods in winters in this country usually cause heavy losses by nitrification and leaching. At one time it was fairly popular in the United Kingdom. However, because the main marketing source ceased, this is no longer the case, although there is no reason why it should not be used again.
- *Aqueous ammonia* (12% N). This is ammonia dissolved in water under slight pressure. It must be injected into the soil (10–12 cm), but the risk of losses is very much less than with anhydrous ammonia. Compared with the latter, cheaper equipment can be used, but it is still usually a contractor operation.
- *Aqueous nitrogen solutions* (26–32% N). These are usually solutions of mixtures of ammonium nitrate and urea, and are commonly used on farm crops (liquid fertilisers).

Various attempts have been made to produce slow-acting nitrogen fertilisers. Reasonable results have been obtained with such products as resin- or polymer-coated granules of ammonium nitrate (26% N), sulphur-coated urea prills (36% N) (soil bacteria slowly break down the yellow sulphur in the soil), urea condensates and urea formaldehydes (30–40% N). At present these types of fertiliser are considered too expensive for farm cropping but are used in amenity and production horticulture.

Organic fertilisers such as *Hoof and Horn* (13% N), ground-up hooves and horns of cattle, *Shoddy* (up to 15% N), waste from wool mills, and *Dried Blood* (10–13% N), a soluble quick-acting fertiliser, are usually too expensive for conventional farm crops and are mainly used by horticulturists. They can be used in propagating composts only for organic produce.

It should be noted that a significant proportion of the nitrogen now supplied to farm crops comes from compound fertilisers in which it is usually present mainly as monammonium phosphate (MAP) or diammonium phosphate (DAP), as described in the section on phosphate fertilisers.

4.4.2   Phosphate fertilisers

Phosphate fertilisers can be classed as:

* Those containing water-soluble phosphorus.
* Those with no water-soluble phosphorus. The insoluble phosphorus is soluble in the weak soil acids.

By custom and by law, the quality or grade of phosphate fertilisers is expressed as a percentage of phosphorus pentoxide (P_2O_5), a gas equivalent only used in the UK. The rest of Europe uses a declaration based on elemental phosphate. The main phosphate fertilisers used in agriculture are:

* *Single superphosphate.* This contains 18–20% water-soluble P_2O_5 produced by treating ground rock phosphate with sulphuric acid. It also contains 27% SO_3 plus a small amount of unchanged rock phosphate in addition to gypsum ($CaSO_4$), which may remain as a white residue in the soil. It is suitable for all crops and all soil conditions, but is not widely used now.
* *Triple superphosphate.* This contains approximately 46% water-soluble P_2O_5 in a granulated form. It is produced by treating rock phosphate with sulphuric acid, followed by phosphoric acid. It also contains a small amount of elemental sulphur. One bag of triple superphosphate is approximately equivalent to 2½ bags of single superphosphate.
* *Ammonium phosphates (MAPs and DAPs).* These are produced by treating rock phosphate with phosphoric acid and ammonia. They range from 46 to 55% P_2O_5 and also contain useful amounts of nitrogen.
* *Nitrophosphates.* These contain 20% P_2O_5 and the same amount of N. They are made by treating rock phosphate with nitric acid.
* *Ground mineral phosphate* (ground rock phosphate). This contains 25–40% insoluble P_2O_5. It is the natural rock ground to a fine powder, i.e. 90% should pass through a '100-mesh' very fine sieve (16 holes/mm^2). It should only be used on acid soils in high rainfall areas and then preferably for grassland. A softer rock phosphate is obtained from North Africa, sometimes known as 'Gafsa'. It can be ground to a very high degree of fineness, 90% passing through a '300-mesh' sieve (48 holes/mm^2). This means that it will dissolve more quickly in the soil, although it should still only be used under the same conditions and for the same crops as ordinary ground mineral phosphate. To be classified as a soft rock phosphate under EU Regulations, at least 55% of its total P_2O_5 must dissolve in a 2% solution of formic acid. Ground mineral phosphate fertilisers are now granulated.
* *Basic slags.* These contain 5–22% insoluble P_2O_5. They are by-products from the manufacture of steel. However, because of improved methods of steel manufacture, and the shrinkage of the industry itself, very little of this valuable fertiliser is now available to farmers in the United Kingdom. Small quantities of low-grade slag are sometimes imported.

There has been recent speculation about 'peak' fertiliser use and whether we are running out of phosphate in particular. Estimates of our remaining P reserves vary between 100 and 300 years, with 600 years' worth of potash lying in known underground deposits. However, reserves are not the same as resources, and the estimated resource levels of P and K would allow us to keep using them for thousands of years, albeit at a higher cost. There are also advances in recycling phosphate, in particular from sewage.

4.4.3   Potassium fertilisers

These are also known as potash fertilisers, which is short for potassium ash. The quality or grade of potassium fertilisers is expressed as a percentage of potassium oxide (K_2O) equivalent. The main potassium fertilisers used in agriculture are:

- *Muriate of potash* (potassium chloride). As now sold, it usually contains 60% K_2O. It is the most common source of potash for farm use and is also the main potash ingredient for compound fertilisers containing potassium. As a straight fertiliser it is normally granulated, but some is marketed in a powdered form. KCl is found in vast quantities all over the world and is mined from rock deposits left by dried-up oceans. It is nearly always found in conjunction with NaCl and the two are separated by a flotation process.
- *Sulphate of potash* (potassium sulphate). This is made from the muriate and so is more expensive per kilogram than K_2O. It contains 48–50% K_2O and, ideally, should be used for quality production of crops such as potatoes, tomatoes and other market garden crops. However, the cost limits its use. It also contains 27% SO_3.
- *Kainit, sylvanite and potash salts*. These are usually a mixture of potassium and sodium salts and, depending on the source, magnesium salts. They contain 12–30% K_2O and 8–20% sodium (Na). They are most valuable for sugar beet and similar crops, for which the sodium is an essential plant food.

4.4.4   Magnesium fertilisers

- *Kieserite* (26% MgO). This fertiliser is quick acting and is particularly useful on severely magnesium-deficient soils where a magnesium-responsive crop such as sugar beet is to be grown. It also contains 50% SO_3.
- *Calcined magnesite* (80% MgO). This is the most concentrated magnesium fertiliser, but it is only slowly available in the soil. To help to maintain the magnesium status of a light sandy soil which is naturally low in magnesium, 300 kg/ha should be applied, say, every four years in a predominantly arable cropping situation.
- *Epsom salts* (17% MgO). This is a soluble form of magnesium sulphate and is used as a foliar spray where deficiency symptoms may have appeared on a high-value crop. It also contains sulphur.

4.4.5   Sodium fertilisers

Sodium is not an essential plant food for the majority of crops. However, for some, notably sugar beet and similar crops, it is highly beneficial and should replace at least half the potash requirements. The adverse effects it has on weak structured soils such as the Lincolnshire silts should be noted but, on other soils, this should not be a problem. Agricultural salt (sodium chloride, 37% Na) is the main sodium fertiliser used. It is now available in a granular form.

4.4.6   Sulphur fertilisers

Sulphur is an important plant nutrient (involved in the build-up of amino acids and proteins in the plant), a deficiency of which can limit the response of the plant to nitrogen. Sulphur deficiency has become more pronounced in some, but not all, crops in the last two decades. The natural build-up of sulphur in the plant is now much less because purer fertiliser is being used and the plant itself is growing in a less polluted atmosphere. It is, however, generally only necessary in areas away from industry and then perhaps for certain crops such as second and third cut grass for silage and oilseed rape.

The main sulphur fertilisers (which should be applied in the spring) are gypsum (calcium sulphate), potassium sulphate and sulphur contained in compound fertilisers.

4.4.7   Phosphorus and potassium plant nutrients

Increasingly, plant nutrients are now expressed in terms of the elements P (phosphorus) and K (potassium) instead of the commonly used oxide terms P_2O_5 and K_2O respectively. Throughout this book the oxide terms are used, but these can be converted to the element terms by using the following factors:

$$P_2O_5 \times 0.43 = P, \text{ e.g. } 100\,\text{kg }P_2O_5 = 43\,\text{kg P}$$

$$K_2O \times 0.83 = K, \text{ e.g. } 100\,\text{kg }K_2O = 83\,\text{kg K}$$

4.5   Compound fertilisers

4.5.1   Compound fertilisers, complex fertilisers and blending

The main constituents for compound fertilisers used in the United Kingdom are urea, mono and diammonium phosphate and potassium chloride. These compound fertilisers, or compounds, supply two or three of the major plant foods (nitrogen, phosphorus and potassium). Other plant foods, e.g. trace elements, as well as pesticides, can also be added, although this is not commonly done now. The exception to this is sulphur which is increasingly being offered as part of compound fertilisers to overcome the deficiencies in certain parts of the country.

Because of the use of more concentrated basic ingredients, compound fertilisers have become much more concentrated in the last 60 years. For example, in 1948 the total N, P, K content averaged 24% and in 2013 the total N, P, K content averaged 50% with some concentrations at 60%, e.g. 10% N, 25% P_2O_5, 25% K_2O. Approximately 75% of all fertilisers now used are complex compounds or blended compounds. They can exist as either solid or fluid materials.

Complex fertilisers are normally made by drying a wet slurry, containing the appropriate raw materials, on a fluid bed system to produce granules, each containing the declared nutrients in the correct ratio, size 2–5 mm diameter.

Blending of the straight fertilisers to make a mixture is becoming more popular again with a number of UK companies offering bespoke blends tailor made for individual farms. It should not be compared with the farm mixture which was quite common 40 years ago. In modern blends it is important that the individual single ingredients are, as far as possible, matched in physical characteristics (granular size and density). This is to avoid segregation out (separation) of the ingredients and uneven spreading patterns. They are dry mixes.

High quality blended fertilisers can be made to specific plant food ratios. They are cheaper than the complex fertilisers, but generally the handling and, most important, spreading qualities are not as good. However, new blending technology, involving blending towers and computer control systems, means that high quality blends are now being produced without the need for fillers and with accurate matching of ingredients.

Some examples of compound fertilisers and possible uses are shown in Table 4.7.

Table 4.7   Some compound fertilisers

Compound (N : P : K)	Crop	N (kg/ha)	P_2O_5 (kg/ha)	K_2O (kg/ha)	Applied rate of material (kg/ha)
12 : 12 : 18	Potatoes	150	150	225	1250
20 : 10 : 10	Spring cereal	100	50	50	500
0 : 24 : 24	Autumn cereal	0	60	60	208

4.5.2   Plant food ratios

Fertilisers containing different amounts of plant food may have the same plant food ratios (see Table 4.8).

Table 4.8   Example of plant food ratios found in various fertilisers

Fertiliser	Ratio	Equivalent rates of application
(a) 12: 12: 18	1: 1: 1½	5 parts of (a)
(b) 15: 15: 23	1: 1: 1½	=4 parts of (b)
(c) 15: 10: 10	1½: 1: 1	7 parts of (c)
(d) 21: 14: 14	1½: 1: 1	=5 parts of (d)
(e) 12: 18: 12	1: 1½: 1	5 parts of (e)
(f) 10: 15: 10	1: 1½: 1	=6 parts of (f)

4.5.3   Ways in which fertilisers are supplied

Solids

- *50 kg bags*. Although a small amount of fertiliser is supplied as individual bags, most of it is delivered on 30 bag (1.5 tonne) pallets to facilitate handling. Special fork-lift equipment is required, as manhandling into spreaders is slow and can be dangerous.
- *Big bags*, e.g. 500 and 1000 kg bags with a top-lift (hook) facility are easy to load into spreaders, but they can be difficult to stack. It is estimated that 90% of all fertiliser now used on farms is delivered in 500 kg bags. Care must be taken to avoid damage to the bags which can result in the loss of fertiliser, and in lumpy or sticky material reducing the efficiency of the spreader.
- *Loose bulk* (1 tonne occupies about 1 m^3). This system is not widely used now. The fertiliser can be stored in dry, concrete bays and covered with polythene sheets. It can be moved into spreaders by tractor plus hydraulic loaders or augers.

Fluids

In the United Kingdom as a whole the fluid fertiliser share of the market is about 9%. However, in the typical arable areas where the system has, until recently, been concentrated, liquid nitrogen has a share in the range of 10–15%, and fluid compounds account for between 5 and 10% of the total tonnage of compound fertiliser used.

Solutions

These fertilisers are non-pressurised solutions of the same raw materials that are used for solid fertilisers. They should be distinguished from pressurised solutions, such as aqueous ammonia and anhydrous ammonia. At the time of writing, there is very little price differential in the kilogram cost of plant food in the solid or fluid form. Fluid compounds are based on ammonium polyphosphate or ammonium phosphate, urea and potassium chloride, whilst ammonium nitrate and urea are the main constituents for liquid nitrogen fertilisers.

The concentration of liquid nitrogen is, for all practical purposes, the same as solid nitrogen fertiliser as long as both urea and ammonium nitrate are used together in the same solution. Fluid phosphorus and potassium compounds are only about two-thirds the concentration of solids. Potash is the main constraint on the concentration of the compounds as it is the least soluble of the important plant foods. However, fluids have bulk densities in the range 1.2–1.3 kg/l compared with 0.9–1.0 kg/l for solid fertilisers (i.e. 100 l = approximately 125 kg). The higher bulk density plus quicker handling and application of fluids will compensate for the lower concentration. It is an obvious consideration when fertiliser work rates are assessed on the weight of nutrients applied in a given time and not the weight of the product. However, the cost of storage may be higher and special care must be taken during handling, storage and application to avoid environmental pollution. The Fertiliser Manufacturers' Association (now amalgamated into the

Agricultural Industries Association [AIC]) produced a *Code of Practice for Transport, Handling and Storage of Fluids*, as well as one for solids (both updated and published in 1998).

Fluid solutions are stored in steel tanks (up to 60 t capacity) on the farm. A cheaper form of storage – glass-reinforced plastic tanks – is available as an alternative to mild steel. Storage capacity is generally based on up to 33% of the annual farm requirements.

There is a range of equipment available for applying fluid fertiliser. The broadcasters, which range from a 600 litre tractor-mounted to 4000 litre capacity self-propelled applicators, have spray booms 12–40 m wide. Although the broadcaster is specially designed for fertiliser, by changing the jets it can be used for other agricultural chemicals, and so the fluid system fits in well with tramlining because only two pieces of equipment have to be matched. Equipment is also available for the placement of fertiliser for the sugar beet, potato and brassica crops, and a combine drill attachment can be used for sowing the fluid and cereal together. For top-dressing, and to minimise scorch, a special jet can easily be fitted which produces larger droplets for a better foliar run-off. Alternatively, a dribble-bar, which dribbles the fertiliser on to the soil, can be used.

Volume of product to apply

Example, if 100 kg nitrogen/ha is recommended, liquid nitrogen, N 37 kg per 100 litres is used, i.e. $100/37 \times 100$ litres/ha = 270 litres/ha.

Units of sale

Whilst solid fertilisers are applied by weight (kg/ha) and sold by weight as £ per 1000 kg, fluid fertilisers are applied by volume (litres/ha) and sold by volume as £ per 1000 litres ($1 m^3$).

Suspensions

These are fluids containing solids, and they can be produced with a concentration almost up to that of the granular solid. This is achieved either by crushing the solid raw material and adding to it up to 3% of a semi-solid attapulgite clay to help minimise the settling of the soluble salts, or by reacting phosphoric acid and ammonia followed by the addition of potassium chloride and extra nitrogen. For long-term storage, crystallisation may be a problem and occasional agitation is necessary to prevent the crystals from growing whilst in store. Special applicators are also needed and this is why suspensions are, at present, usually applied by a contractor. Suspensions are developing in North America (where, in fact, fluids are more widely used) and to some extent on the Continent.

'Distressed' fertiliser

This is fertiliser that has been damaged, usually in transit from overseas. It is bought by distributors and sold at a heavy discount to farmers who normally make it into a liquid fertiliser, simply by dissolving it in water.

There is no difference in plant growth following the application of fertiliser in solution, suspended or solid form. The benefit of fluid fertilisers seems to be that they can be delivered more quickly and accurately to the plants and urea in fluid form is not as volatile as the solid product, particularly important in drying conditions.

4.6   Application of fertilisers

The main methods used are:

1. *Broadcast distributors* using various mechanisms such as:
   (a) *Pneumatic types.* The principle involved is that the fertiliser is metered into an airflow which conveys it through flexible tubes to individual outlets placed over deflector plates to distribute the fertiliser evenly. Hopper capacity is from 1 to 6 tonnes with a spreading width of 12–36 m (selected to match tramline widths). Application rates vary from 5 kg/ha (for broadcasting seeds) to 2500 kg/ha.
   (b) *Spinning disc types.* These consist of a hopper placed above a rotating disc. The fertiliser is fed on to the disc from where it is distributed. Wide areas can be covered, but the accuracy of distribution can vary quite considerably (see below).
   (c) *Oscillating spout type.* A large nozzle situated at the back of the hopper moves rapidly from side to side throwing the fertiliser to the required width.
   The spread patterns of all broadcast fertiliser spreaders should be checked regularly. This can be done by driving the spreader through a line of collection boxes and then measuring the amount of material in each box. A number of commercial companies now offer this service to farmers. The quality and uniformity of the fertiliser material also has an effect on spread pattern. For nitrogen fertilisers, a standardised Spread Pattern (SP) rating has been adopted by the Agricultural Industries Association. The Silsoe Research Unit in Bedfordshire developed a system to measure the 'throw and flow' characteristics of nitrogen materials. Ratings are on a scale of 1–5 with SP5 being the best.
2. *Combine drills.* Fertiliser and seed (e.g. cereals) from separate hoppers are fed down the same or an adjoining spout. A star-wheel feed mechanism is normally used for the fertiliser and this usually produces a 'dollop' effect along the rows. In soils low in phosphate and potash, this method of placement of the fertiliser is much more efficient than broadcasting and can require less fertiliser to be used per hectare, e.g. P and K in potatoes. It is known as combine drilling and is sometimes referred to as 'contact placement'. Because of possible scorch, combine drilling should only be used for cereal crops.
3. *Placement drills.* These machines can place the fertiliser in bands 5–7 cm to the side and 3–5 cm below the row of seeds. It is more efficient than broadcasting for crops such as peas and sugar beet. Other types of placement drills attached

to the planter are used for applying fertilisers to the potato crop and some brassicas.

4. *Broadcast from aircraft.* Rarely used except in inaccessible hill land and forests. Usually done by helicopter. Highly concentrated fertilisers should be used, e.g. urea.

5. *Liquids injected under pressure* into the soil, e.g. anhydrous and aqueous ammonia.

6. *Liquids (non-pressurised)* broadcast or placed.

4.7   Organic manures

4.7.1   Farmyard manure (FYM)

This consists of dung and urine, and the litter used for bedding stock. It is not a standardised product, and its value depends on:

- *The kind of animal that makes it.* If animals are fed strictly according to maintenance and production requirements, the quality of dung produced by various classes of stock will be similar. But in practice it is generally found that, as cows and young stock utilise much of the nitrogen and phosphate in their food, their dung is poorer than that produced by fattening stock.
- *The kind of food fed to the animal that makes the dung.* The more proteins and minerals in the diet, the richer will be the dung.
- *The amount of straw used.* The less straw used, the more concentrated will be the manure and the more rapidly will it break down to a 'short' friable condition. Straw is the best type of litter available, although bracken, peat moss, sawdust and wood shavings can be used. About 1.5 tonnes of straw per animal are needed in a covered yard for six months, and 2–3 tonnes in a semi-covered or open yard.
- *The manner of storage.* There can be considerable losses from FYM because of bad storage, although it is appreciated that expensive, elaborate storage is no longer viable these days.

Dung from cowsheds, cubicles and milking parlours should, if possible, be put into a heap which is protected from the elements to prevent the washing out and dilution of a large percentage of the plant food which it contains. Dung made in yards should preferably remain there until it is spread on the land, and then, to prevent further loss, it is advisable to plough it in immediately.

Farmyard manure is important chiefly because of the valuable physical effects on the soil of the organic matter it contains. It improves the friability, the structural stability and the water-holding capacity of the soil. It is also a valuable source of plant foods, particularly nitrogen, phosphate and potash, as well as other elements in smaller amounts. An average dressing of 25 t/ha of well-made FYM will provide about 40 kg N, 50 kg P_2O_5 and 100 kg K_2O in the first year. At least one-third of the nitrogen could be lost before it is ploughed in, although this depends on the time of application. All the plant food in FYM is less readily available than that in chemical fertilisers.

Application. The application of FYM will be dealt with under the various crops.

4.7.2 Liquid manure and slurry

The widespread use of cow cubicles and self-feed silage clamps and the need to dispose of dirty water has meant that many dairy farmers produce slurry rather than the traditional FYM from strawed yards. Similarly, intensive pig units with their mechanised waste management systems find it easier to deal with liquids rather than solids.

Slurry must not be allowed to pollute watercourses. At 10 000–20 000 mg/l it has a high Biological Oxygen Demand (BOD). This is the measure (in milligrams per litre) of the amount of oxygen needed by the micro-organisms to break down organic material (pollution of water is caused when the micro-organisms multiply, and so extract oxygen, to deal with the organic material). Problems can also arise from the nuisance of smells and possible health hazards over a wide area when the slurry is applied. If slurry is stored in anaerobic conditions, dangerous obnoxious gases (mainly hydrogen sulphide) are produced and are released when it is being spread. The relevant environmental authorities have a duty to prevent the pollution of streams and rivers with farmyard effluents. Legislation under the Statutory Management Requirements (SMRs), Good Agricultural and Environmental Condition (GAECs) and the Code of Good Agricultural Practice should prevent any problems in this respect, but it does mean that the farmer has to find the best possible way to utilise the slurry produced on the farm.

When applying slurry, relevant recommendations from the Defra Code of Good Agricultural Practice for Farmers, Growers and Land Managers (CoGAP) should be followed:

1. A single slurry application should not exceed 50 000 litres/ha. At least a three-week interval should be allowed between applications. Slurry should not be applied during the 'closed period'. When applying slurry through irrigation lines, the precipitation should not exceed 5 mm/hour. Rain guns should be avoided.
   Too much slurry can:
   (i) restrict aeration leading to partial oxidation products such as methane and ethylene which are toxic to plants;
   (ii) weaken soil structure by the sealing of the soil surface.
2. An untreated strip of at least 10 m width should be left next to all watercourses, and slurry should not be applied within at least 50 m of a spring well or bore-hole that supplies water for human consumption or is to be used for farm dairies.
3. Slurry should not be applied:
   (i) to fields which are likely to flood in the month after application;

(ii) to fields which are frozen hard; surface run-off must be avoided;

(iii) when the soil is at field capacity;

(iv) when the soil is badly cracked down to the field drains, or to fields which have been piped, moled or subsoiled over a drainage system in the preceding 12 months.

Table 4.9 indicates the average amounts and composition of slurry produced by livestock.

- For slurry diluted 1: 1 with water, the figures in Table 4.5 should be divided by 2.
- To estimate the dilutions of slurry, a comparison should be made of the volume of slurry in the store with the expected volume of undiluted slurry.

The figures in Table 4.9 can only be used as a basis and they apply to slurry collected in an undiluted form, e.g. under slatted floors or passageways where washing and rainwater are excluded. The slurry on most farms is obviously diluted and it is not easy to be certain of its composition, even when samples are analysed. Rota-spreaders and similar machines can handle fairly solid slurry, and modern vacuum tankers and pumps are very efficient in dealing with slurry with less than 10% dry matter. The jet from these tankers must not spread slurry more than 4 m high.

The injection of slurry into the soil using strong tines fitted behind the slurry tanker can reduce, but not eliminate, wastage and offensive odours. It can also reduce the volatisation of nitrogen into the air as ammonia gas. However, it is expensive and it can spoil the surface of a grass field where there are stones present.

It is highly desirable that the valuable nutrients in organic manures should be utilised as fully as possible. This is best achieved if the slurry can be applied at a time when growing crops can utilise it, normally in the spring. This means that storage is usually necessary. This is expensive, whether the slurry is stored in a compound (lagoon) or storage tank. However, with the rising costs of fertiliser, the annual cost of storage and handling of slurry is usually less than the plant food value of the slurry, but this calculation should be done before investing in a slurry store.

Nitrogen losses resulting from applying slurry in the autumn/early winter (ground conditions permitting) to the following year's maize field, for example, can be reduced by the use of nitrogen inhibitors such as dicyandiamide (e.g. Didin: Enrich). These inhibit the activity of the nitrifying bacteria which delays the change of ammonia to nitrate. Denitrification and leaching are therefore

Table 4.9   Nature of slurry produced by livestock during housing period

Livestock	Undiluted excreta	Nutrients produced		
		N (kg/m^3)	P$_2$O$_5$ (kg/m^3)	K$_2$O (kg/m^3)
Dairy cow	9.6	48.0	19.0	48.0
Fattening pig	1.5	10.5	7.5	6.0
Poultry (1000 hens)	41.0	660.0	545.0	360.0

reduced. The cost of inhibitors is approximately £45/ha with slurry applied at 50 000 litres/ha. It is claimed that the increased yield with the maize crop following the use of an inhibitor can be valued at £70/ha.

In addition to forage maize, slurry is best applied to:

* *Grass.* At least a six-week, preferably longer, interval should be allowed between a slurry application and taking the crop for silage or hay. This will avoid any disease problems and it should minimise any possible contamination of the conserved crop. Slurry can be applied to grazing swards but an interval of at least four weeks should elapse before animals are allowed access. Animals should then be carefully monitored for symptoms of hypomagnesaemia (staggers), as the high levels of potash in slurry could lead to reduced herbage magnesium content.

* *Kale.* As this crop (like maize) is not normally sown until late April or May, there is usually a good opportunity to apply slurry from the winter accumulation. There should be no need for any phosphate and potash fertiliser following a slurry application, although extra nitrogen will normally be necessary. This fertiliser recommendation also applies to forage maize.

Although it is not so usual, slurry can also be used on cereals and other forage crops. Band spreaders have been developed which make it possible to spread across the whole width of tramlines. These also reduce the ammonia emissions by 30–40% compared with traditional slurry tankers.

Slurry can be separated into solid and liquid fractions by screening or centrifugal action. The solid part (12–30% dry matter (DM)), can be handled like FYM, and the liquid portion is much less likely to cause tainting of pastures; the disease risk is also reduced. Slurry separators are, however, expensive and they are not easy to justify. Table 4.10 shows the constitution of cattle slurry.

4.7.3   Pig slurry

The composition of pig slurry depends on the feeding system, i.e. dry meal feed has a 10% DM; liquid feed has a 6–10% DM and whey 2–1% DM (see Table 4.11). Supplementation to the diet means that pig slurry can contain large amounts of copper and zinc. These elements will build up slowly in the soil, but eventually there could be crop toxicity problems. It should be noted that sheep are particularly susceptible to copper poisoning.

Table 4.10   The constitution of cattle (dairy) slurry (in kg/m^3)

1000 kg slurry (6% DM), unseparated			Mechanically separated slurry (4% DM)			Weeping-wall separated (3% DM)		
N	P_2O_5	K_2O	N	P_2O_5	K_2O	N	P_2O_5	K_2O
2.6	1.2	3.2	3.0	1.2	3.5	2.0	0.5	3.0

See also Defra *Fertiliser Manual.*

Table 4.11 Comparison of pig slurry

	% dry matter	Total N (kg/m^3)	Available nutrients (year of application) (kg/m^3)	
			P$_2$O$_5$	K$_2$O
Pig slurry	2	3.0	0.5	1.8
	4	3.6	0.9	2.2
	6	4.4	1.3	2.5

Note: The availability of nitrogen will depend on the timing and method of application, the winter rainfall and the soil type (see also Defra *Fertiliser Manual*)

4.7.4   Poultry manure

Poultry manure refers to:

- Fresh poultry manure.
- Broiler manure.
- Deep litter manure.
- Dried manure.

Table 4.12 shows the nutrient value of poultry manure.

- *Fresh poultry manure* from battery cages or from slatted or wire floors is free from litter. It is semi-solid, but rather sticky and, to reduce any public nuisance from smell, it should be spread as soon as possible. To reduce nitrogen losses, the preferred time is in spring and summer. Application rates are similar to those for farmyard manure. Copper and zinc toxicity could be a problem if very high rates of fresh poultry manure are made.
- *Broiler manure* is the droppings mixed with litter. This produces a bulky manure, relatively dry and friable and easily handled. It should be treated like farmyard manure.

4.7.5   Sewage sludges (biosolids)

As a means of disposal, most water companies offer treated sewage sludge to farmers and growers. There is a section of the Defra Code of Practice on the use

Table 4.12 Comparison of poultry manures

	% dry matter	N	Available nutrients (year of application)* (kg/tonne)	
			P$_2$O$_5$	K$_2$O
Battery layers	35	9.5	8.4	8.6
Broilers	60	10.5	15	16.2

* Assumes soil incorporation within 24 hours.

of sewage sludge in agriculture and its recommendations should be followed. There is also the 'Safe Sludge Matrix' which has been produced by agreement between Water UK and the British Retail Consortium and which goes beyond the cropping and grazing restrictions contained within the Code of Practice.

Untreated sewage sludge has not been allowed on agricultural land since December 2005. All sewage sludge is now treated and, as such, it is quite valuable, e.g. 20 tonnes of dried digested sludge can contain 75 kg N, 90 kg P_2O_5 and 30 kg K_2O available in the year of application. New Class 'A' biosolids are now being produced by digestion, thermal drying and pelleting. This is known as enhanced treated sludge, is free from Salmonella and is 99.9999% pathogen free. This can be used safely on all crops including fresh salad vegetables and fruit, but there is still a ten month harvest interval for produce eaten fresh.

4.7.6   Seaweed
Seaweed is sometimes used instead of farmyard manure for crops such as early potatoes in coastal areas, e.g. Ayrshire, Cornwall and the Channel Islands. However, it is expensive to handle which is why it is not used to any great extent now. It can be a useful source: 10 tonnes contain about 50 kg N, 10 kg P_2O_5 and 140 kg K_2O; it also contains about 150 kg salt. The organic matter in seaweed breaks down rapidly because it is mainly cellulose. It should therefore be collected, spread and ploughed in immediately. In this way the loss of potassium, particularly, will be reduced.

4.7.7   Cereal straw
Cereal straw is lignified material and can be a useful source of soil organic matter for maintaining or improving soil fertility. It is particularly beneficial for light, sandy and silty soils in which organic matter breaks down rapidly. However, the amount of organic matter in a soil is closely related to the texture of the soil, and repeated incorporations of straw on various soil types over many years have had little or no effect on the percentage of organic matter in the soils.

The average yield of straw (4 t/ha) supplies about 15 kg N, 5 kg P_2O_5 and 35 kg K_2O, and so the soil should benefit when the straw is ploughed in or otherwise worked into the soil. However, with the sudden influx of straw, the soil bacterial population will start to multiply to cope with the problem of breaking down the straw. In so doing, much of the nitrogen in the soil gets used up for the bodily needs (the protein) of the bacteria. The carbon: nitrogen ratio increases to 50 or 60 : 1 from the normal 10 : 1. The protein is released as the straw decomposes and eventually the ratio returns to 10 : 1. But in the meantime, unless extra nitrogen is added (20–25 kg/ha), the following crop can suffer from shortage of nitrogen. This extra nitrogen is only necessary under naturally low nitrogen conditions.

Incorporation of straw is easier to achieve if it is chopped and spread from the combine. Ploughing is the best method, preferably using fairly wide furrows (30 cm) and 15–20 cm deep. There is seldom any great advantage in premixing;

increased costs can result. Tines and/or discs can only do the job well with chopped straw, and the result is not as satisfactory as good ploughing. Trash and clods left near the surface encourage slugs which can cause damage to following crops by eating the seed and young plants. However, with the proliferation of non-inversion tillage methods (sometimes called Min-Till) straw residue is often left close to the surface and problems can occur with both slugs and weeds. The control of these must be factored in to any gross and net margins.

Repeated shallow incorporations can build up high organic matter levels in the surface layers of some soils. These can be beneficial to the development of young crop plants, but can also cause problems with the adsorption of soil-acting herbicides. If the straw is incorporated at a wet time, anaerobic conditions may develop and the decomposing straw will release toxic substances, such as acetic acid, which can check or kill seedling crop plants.

4.7.8   Green manuring

This is the practice of growing and ploughing in green crops to increase the organic matter content of the soil. It is normally only carried out on light sandy soils. White mustard is a very commonly grown crop for this purpose. Sown at 9–17 kg/ha it can produce a crop ready for ploughing within 6–8 weeks. Fodder radish is becoming more popular for green manuring and, like mustard, it can also provide useful cover for game birds. However, it has been shown that a short ley has very little benefit towards building up organic matter in the soil. There must, therefore, be even less effect from quick-growing crops, which break down equally quickly in the soil.

4.7.9   Green cover cropping

This describes the practice of growing a green crop with the primary objective of absorbing soil nitrogen over the winter months to prevent its leaching. This can benefit the environment as well as saving on expensive fertiliser nitrogen.

The cover crop is sown in late summer to hold any nitrogen in the autumn and winter, to be released in the spring for the following spring-sown crop. A number have been trialled and it would appear that those taking up most nitrogen in November (and this month can be significant) do not necessarily hold it until March. For example, because of their rapid growth, white mustard and forage rape are very effective in trapping autumn nitrogen, but natural senescence and frost kill mean that the nitrogen can be lost before any following crop can take it up. Rye and ryegrass hold nitrogen well in the autumn and winter and retain it until the spring. They can also be used for early cropping in the spring.

4.7.10   Waste organic materials

Various waste products are used for market garden crops, partly as a source of organic matter and partly as a means of releasing nitrogen slowly to the crop.

They are usually too expensive for ordinary farm crops.

- *Shoddy* (waste wool and cotton) contains 50–150 kg of nitrogen per tonne. Waste wool is preferable, and the recommended application rate is 2.5–5 t/ha.
- *Dried blood*, ground *hoof and horn* and *meat and bone meal* are also used; the plant food content is variable.

4.8   Residual values of fertilisers and manures

The nutrients in most manures and fertilisers are not used up completely in the year of application. The amount likely to remain for use in the following years is taken into account when compensating outgoing farm tenants.

Up to 70% of nitrogen in soluble nitrogen fertilisers (e.g. ammonium nitrate and some other compounds) is used in the first year. For phosphate in soluble form, e.g. triple superphosphate and compounds containing phosphate, allow two-thirds after one crop, one-third after two and one-sixth after three crops. For phosphate in insoluble form, e.g. bone meal and ground mineral phosphate, allow one-third after one crop, one-sixth after two and one-twelfth after three. For potash, e.g. muriate or sulphate of potash and compounds containing potash, allow a half after one crop and a quarter after two crops. For lime one eighth of the cost is subtracted each year after application.

4.9   Fertilisers and the environment

When applying fertilisers, it is economically and environmentally important to use the optimum amounts and not just those required for a maximum yield. Accuracy of application, timing and handling are also important in order to reduce the losses, either by leaching or run-off. Only small amounts of phosphorus will be lost by run-off, but with potassium, losses can be greater both by leaching and run-off. Nitrogen as nitrate is the nutrient which is most likely to be lost by leaching and this is affected by soil type, cropping, cultivation timing and rainfall.

There are now directives from the European Union (EU) concerning nitrates in drinking water: a maximum of 50 parts per million has been set. In 1990, Nitrate Sensitive Areas (NSAs) were designated to study the effects of husbandry on nitrate leaching. Initially ten areas were chosen – from Somerset across to Lincolnshire – where the water exceeded or was at risk of containing more than 50 parts per million nitrates. Although the scheme was voluntary, most farmers in the designated areas joined, receiving payments for complying with the regulations. There were 212 NSA contracts in England alone, but it has now ended and no more contracts have been issued.

The NSA scheme involved small changes in husbandry; only the economic optimum, or less, of nitrogen for some crops was allowed, but the timing and amount applied at any one time were restricted. Application of organic manures was also affected and a careful record had to be kept. To hold the nitrogen in the soil, winter fallows were avoided by planting cover crops. There were restrictions on timing of cultivations on grassland. Hedgerows and/or woodland could not be removed unless replaced by an equivalent area.

Superseding the NSAs, and coming into force in 1998, was the designation of Nitrate Vulnerable Zones (NVZs) and the NVZ Action Programme Regulations. There were originally 68 designated NVZs covering 600 000 na of land in England and Wales and the scheme was and still is run by the Environment Agency. The regulations required the careful management of fertiliser and manure use and required farmers to keep careful and accurate field records. They also imposed restrictions on nitrogen applications (organic and manufactured) by specifying closed periods (when nitrogen could not be applied) and 12 month limits on total nitrogen use. These applied to both grassland and arable areas. There were also spreading and storage controls. The European Union did not consider that England had fully implemented the 1991 Nitrates Directive and in 2009/10 the NVZ areas were extended to cover most of the agricultural area of England, parts of Scotland and some of Wales. The maps showing where these are can be obtained from the relevant departmental websites e.g. GOV.UK for England and Wales (see link below).

Reviews of the regulations have been recently carried out and new updates have been added to the scheme in May and November of 2013. A NVZ is designated on all land contributing to 'polluted' waters, i.e. surface or ground water that contains, or is likely to contain if no action is taken, more than 50 mg/litre of nitrates, or is, or likely to become, eutrophic.

If farmers do not comply with the rules of NVZs they can be prosecuted, fined and may lose some of their Single Payment. If their farm is in an NVZ they must:

- Plan their use of livestock manure and manufactured nitrogen fertilisers to ensure that they do not apply more nitrogen than their crops require.
- Produce a risk map for any land where they intend to spread organic manure.
- Comply with the field limit, the Nma (crop nitrogen requirement) limit, closed periods and spreading controls for spreading manufactured nitrogen fertilisers and organic manures.
- Comply with the livestock manure N (nitrogen) farm limit.
- Provide adequate storage capacity for livestock manures.
- Keep records of the nitrogen applied to each of their fields, and some records and calculations relating to their farm as a whole.

(GOV.UK, 2013)

As part of compliance needed to claim their Single Payments, farmers are also required to complete soil protection reviews (SMR 1), manure and nutrient management plans (SMR 4) and comply with GAEC 19 (no spread zones). These are designed to help farmers to meet the NVZ regulations in a manageable way.

Derogations are available each year for grassland farmers who wish to apply more organic manure to their grass than the current 170 kg/ha N equivalent limit, if their farm is at least 80% grassland and they do not exceed 250 kg/ha of manure N.

The manufacture of fertilisers has also come into focus as a source of greenhouse gases (GHGs) and new technologies have meant a reduction in GHGs. These types of products are called 'abated' fertilisers and are helping to reduce emissions, particularly ammonia and nitrous oxide.

4.10   Sources of further information and advice

Archer, J, *Crop Nutrition and Fertiliser Use*, 2nd edn, Farming Press, 1988.

Defra. Protecting our Water, Soil and Air. *A Code of Good Agricultural Practice for Farmers, Growers and Land Managers.* 2009.

Simpson, K, *Fertilisers and Manures*, Longmans, 1986.

Defra. *The Fertiliser Manual.* RB 209, 8th edn, 2010.

Managing Livestock Manures Booklet 1 – *Making better use of livestock manures on arable land*, ADAS Gleadthorpe.

Managing Livestock Manures Booklet 2 – *Making better use of livestock manures on grassland*, ADAS Gleadthorpe.

Managing Livestock Manures Booklet 3 – *Spreading systems for slurries and solid manures*, ADAS Gleadthorpe.

GOV.UK website for Nitrate Vulnerable Zone Information – https://www.gov.uk/nitrate-vulnerable-zones#how-to-comply-with-nvz-rules

Guidance on complying with the rules for NVZs in England for 2013–2016 – https://www.gov.uk/government/uploads/system/uploads/attachment_data/file/261371/pb14050-nvz-guidance.pdf

5

Weeds of farm crops

DOI: 10.1533/9781782423928.1.92

Abstract: Weeds are plants that are growing where they are not wanted. Weeds can significantly affect crop yields by shading and smothering the crop as well as by competing for plant nutrients and water. This chapter outlines the identification and biology of some of the more important grass and broad-leaved weed species. Methods of control as well as issues with herbicide resistance are discussed.

Key words: weed biology, weed identification, methods of weed control, herbicide resistance.

5.1 The impact of weeds

Weeds are plants that are growing where they are not wanted. Even crop plants can be serious weeds in other crops, e.g. volunteer potatoes or weed beet. Weeds can significantly affect crop yields by shading and smothering the crop as well as by competing for plant nutrients and water. The competitiveness of weeds varies both between species and at different times of the year. Some crops are also more affected by weed competition than others. Potatoes produce a dense crop canopy which can be very effective at competing with weeds compared with sugar beet which is precision drilled at a low population, and has many weeks when there is little crop cover. As well as affecting crop yield, weeds also affect crops in other ways:

- They can spoil the quality of a crop and so lower its value, e.g. wild oats in seed wheat, black nightshade berries in vining peas.
- They can act as host plants for various pests and diseases of crop plants, e.g. couch grass is a host for take-all and eyespot of cereals.
- Weeds such as bindweed, cleavers, couch and thistles can affect combine harvesting. All the green material other than grain going through the harvester reduces the work rate. These weeds can also cause a crop to lodge, which again can affect speed of harvesting.

- Weed contamination at harvest can increase the cost of drying and cleaning the grain.
- Thistles, buttercups, docks and ragwort in grassland can reduce the grazing area and feeding value of pastures. Some grassland weeds may taint milk when eaten by cows, e.g. buttercups and wild onion.
- Weeds such as ragwort, horsetails, nightshade, foxgloves and hemlock are poisonous and if eaten by stock are likely to cause poor growth or even death. Fortunately, most stock animals normally do not eat poisonous weeds in the field although they will if such plants are conserved in hay or silage.
- Weeds can be very important for farmland biodiversity; they may provide a source of pollen and nectar for bees and butterflies or seed for many bird species.

When planning a weed control management plan it is important to correctly identify the weeds, understand their life cycle, how competitive they are, incidence of herbicide resistance as well as importance for biodiversity.

Some of the most important weed species such as cleavers, common chickweed, wild oats and black-grass are annuals. Only a small number of biennials are important weeds such as spear thistle and ragwort and they are more of a problem in permanent grassland. There are a number of important perennial weeds such as couch grass and field bindweed. Seed production is not necessarily the main method of propagation, for perennials underground stems (rhizomes) or roots are very important for the spread of some of these weeds.

5.1.1   Success of plants as weeds
There are a number of reasons why some plant species have been more successful as weeds than other plants.

Seed production
Some weed species are able to produce thousands of seeds per plant. Examples of prolific seed producers include corn poppies and mayweed species. The seed reservoir (weed seed bank) in some soils can be as high as $40\,000/m^2$. Not all the seed produced in one year will germinate the next year; the percentage emergence may only be around 2–6% of the weed seedbank. Many species have some sort of seed dormancy mechanism that has to be broken before they will germinate. Once dormancy has been broken environmental conditions must also be correct for germination; this accounts for some of the variation in weed populations between years. Losses of seed and seed viability are taking place all the time. Depth of burial in the soil, number and type of cultivations, soil type and weed species affect the rate of decline. The seed viability of some species such as fumitory, charlock, black bindweed, wild oats and corn poppy declines very slowly compared with the rapid decline of some grasses such as barren brome.

Table 5.1 Germination times for some common weed species

Weed species and main period of germination	Calendar month											
	Jan	Feb	Mar	Apr	May	Jun	Jul	Aug	Sep	Oct	Nov	Dec
Early spring												
• Knotgrass		*	***	***	**	*	*	*				
Spring												
• Black bindweed			*	***	***	**	*					
• Orache				***	***	**	*	*				
Early summer												
• Black nightshade				*	***	***	*	*				
Mainly spring												
• Wild oat (spring)	*	*	***	**	**			*	*	*		
Mainly autumn												
• Black-grass		*	*					*	***	***	**	*
• Cleavers								*	***	***	*	
Autumn												
• Barren brome										*	*	
All year												
• Annual meadow-grass	*	*	**	**	**	**	**	**	**	**	*	*
• Common chickweed												
• Mayweed												*

Key: * small amount of germination possible; ** some germination; *** main germination period.

Seed spread

As well as being able to multiply in one field, weeds can spread from field to field. Before the implementation of some of the current very strict seed standards one important method of spread of some weeds such as wild oats, corn cockle and cornflower was in the seed. These weeds set their seed at the same time as the cereal crop is harvested. The weed seed was then resown with the cereal seed. Commercial seed is now virtually weed free but weed seed can still be spread around fields by farm machinery or in straw. Other methods of spread include wind, water and birds, depending on the characteristics of the seed. Some seed, such as of docks, can remain viable in slurry for many weeks and so can be spread around the different fields where the slurry is applied.

Time of germination

Only a few weed species such as annual meadow-grass, common chickweed and mayweeds can germinate throughout the year and so can be a problem in a wide range of crops. Most other species have specific germination periods (Table 5.1); some are mainly autumn germinating such as black-grass whereas others such as black-bindweed, redshank, knotgrass and fat hen germinate in the spring.

Continuous cropping of the same crop, sown at the same time each year, will tend to encourage those weeds that germinate at the same time. Examples include autumn-germinating bromes and black-grass in winter cereals, and late spring/early summer germinating black nightshade in forage maize.

Same family as crop

Some important weed species are in the same family as the crop plant, e.g. charlock in oilseed rape or fat hen in sugar beet. Weed control using herbicides is then often difficult as there is not enough selectivity in the herbicide activity to kill the weed without injuring the crop.

Weed competitiveness

Weed species differ in how they compete with a crop. Some weeds such as speedwell compete early in the growing season and then die back before harvest. Others, such as cleavers, compete later and carry on growing until harvest. Cleavers is a very vigorous plant and much more competitive than speedwell. Examples of weed competitiveness in a cereal crop:

- Very competitive weeds
  - cleavers
  - wild oats
  - bromes.
- Moderately competitive weeds
  - mayweeds
  - common chickweed

- – corn poppy
- – black-grass
- – rough meadow-grass.
- • Least competitive weeds
  - – common field-speedwell
  - – ivy-leaved speedwell
  - – field pansy
  - – annual meadow-grass.

5.1.2   Assessing weed problems in the field

- • Monitoring and keeping records of the weeds present in the field in recent years can be a very good guide of likely weed problems. Cropping, time of drilling, weather, soil texture and condition will affect the weed species and populations that emerge each year.
- • Assessing the numbers of weeds present, e.g. per m^2 or per ha, and correct identification. Some weeds are very important for wild life and at low populations have a negligible affect on crop yields. Young grass seedlings are particularly difficult to identify correctly, but most are not important for in-field biodiversity.
- • Weed populations tend to be very patchy over the field. It is very time-consuming to map weeds manually. Currently, automated systems for weed mapping are being developed. One option for some of the major weeds such as wild oats (because weeds spread by only a few metres per year) is to assess the weeds in early summer before harvest and to produce a map. This map can then be used for patch spraying in future years.
- • Effectiveness of control needs to be assessed each year especially with the increase in cases of weed populations that are resistant to herbicides.

The effects of weeds on yield will depend on the type and number of weeds, how competitive they are as well as on the density and vigour of the crop. Other effects, which must be considered, include ease of control in the different crops being grown in the rotation.

5.1.3   Weeds and crop biodiversity

Weeds vary in their importance for in-field biodiversity; some species are very valuable for invertebrates and others for seed eating birds. Weeds that are considered valuable for wild life include annual meadow-grass, charlock, common chickweed, and the spring germinating fat hen, knotgrass, redshank and black-bindweed. Many of the environmental stewardship options include management options where these species may be encouraged (see page 231).

5.2   Weed types and identification

For optimum control of weeds, both culturally and chemically, it is very important to be able to identify the weeds correctly. There are two main groups of weeds, the monocotyledons or grass weeds and the dicotyledons or broad-leaved weeds.

5.2.1   Grass weeds

Grasses in the vegetative stage are very difficult to identify but it is at this stage that control is most effective. There are several features which can be studied in order to identify grasses correctly. Usually each grass has one or two distinct characteristics that aid identification (Table 5.2):

Table 5.2   Guide to the recognition of some common arable grass weeds by their vegetative characters

1. Shoots flattened	
Leaves with boat-shaped tips and 'tramlines' on upper surfaces	
A. Young leaves with non-glossy under-surface; non-stoloniferous; ligule tall and rounded.	Annual meadow-grass
B. Young leaves glossy on under-surface; stoloniferous; ligule pointed; heading mid-May onwards.	Rough meadow-grass
2. Shoots cylindrical	
A. Leaves and leaf sheaths hairy.	
(i) Auricles and rhizomes present; heading June.	Couch grass
(ii) Auricles and rhizomes absent.	
Basal leaf sheath with red coloration restricted to veins; dense short hairs; heading May.	Yorkshire fog
Basal leaf sheaths often red coloured; leaves hairy; jagged ligule; heading from mid-May.	Barren brome
(iii) Auricles absent, bulbous base present; slightly hairy leaf blades; tufted habit.	Onion couch
(iv) Auricles absent; no stem-based coloration; hairs present on leaf margin; long blunt ligule.	Wild oats
B. Leaves, leaf sheaths and stems non-hairy, no auricles.	
(i) Prostrate habit; stoloniferous; internodes with strong colour; ligule long; heading July.	Creeping bent
(ii) Upright habit; rhizomatous; ligule long and blunt; heading June.	Black bent
(iii) Upper surface of leaves with distinctly marked veins; no rhizomes or stolons; leaf sheaths purplish; ligule blunt; heading from May.	Black-grass
(iv) Annual; ligule long and oblong; mainly found in the east and south east of England.	Loose silky bent
(v) Similar to black-grass; leaf sheath base pinkish; red sap oozes out of stem base when squeezed.	Awned canary grass

- *Stem/root characteristics.* Annuals spread by seed, whereas perennials are spread either by seed or by pieces of vegetative material such as above ground creeping stems or *stolons* or below ground creeping stems or *rhizomes* (Fig. 5.1). Couch grass has rhizomes whereas rough meadow-grass has stolons.
- *Leaves.* Some grasses have folded leaves in the leaf sheath though most are rolled (Fig. 5.2); this characteristic is used as the starting point in many grass identification keys; perennial ryegrass and meadow-grasses are examples of grasses with leaves folded in the stem.
- *Auricles* (Fig. 5.3) are leaf extensions where the leaf sheath meets the stem. Auricles are only present in a few grass species so a very important structure to look for when identifying grasses.
- *Ligules* (Fig. 5.4) or the membrane at the back of a leaf are usually only assessed when identifying grasses in the same family such as the meadow-grasses. Rough meadow-grass has a large ligule and smooth meadow-grass has a short ligule.
- Presence or absence of *leaf/stem hairs.* A few grasses have very hairy leaves, such as barren brome and Yorkshire fog. Wild oats only have hairs on the leaf margin. Most grasses are usually hairless.
- *Leaf shape* (Fig. 5.5). Leaf shape and size can be very different, from short with distinct barge-shaped leaf ends of the meadow-grasses, to long and twisted in the bromes. Ryegrasses have spear shaped leaves and the bent grasses taper to a sharp point.
- *Stem base coloration.* A few important grasses have coloured stem bases; ryegrass is a deep red, black-grass has a purplish blotch and awned canary grass is pink.
- *Ear characteristics* are distinctive for most of the grasses as detailed below. Normally control is carried out before ear production. Ear number is important when assessing effectiveness of control that year; weed mapping for patch spraying and checking for development of resistant plants.

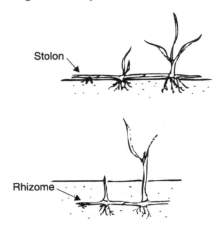

Fig. 5.1 Stolons and rhizomes.

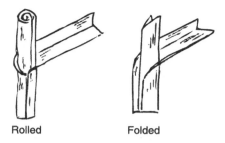

Rolled   Folded

Fig. 5.2   Leaves rolled and folded in stem.

Fig. 5.3   Auricles.

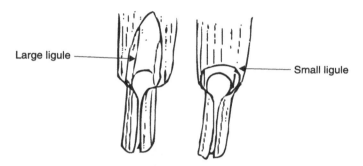

Large ligule ———

Small ligule

Fig. 5.4   Ligules.

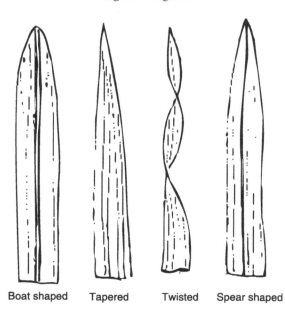

Boat shaped   Tapered   Twisted   Spear shaped

Fig. 5.5   Leaf shape.

The following are some important characteristics of the major grass weeds.

Couch grass (Fig. 5.6a and b) is the most common grass weed and spreads rapidly by rhizomes, especially on light soils. Black bent is also sometimes called couch as it also spreads by rhizomes, but the leaf characteristics are very different. Couch grass leaves are usually slightly hairy, have auricles unlike the bent grasses, and the ear is a spike not an open panicle. Cultivations such as ploughing and rotovation can cause an increase in populations due to cutting up of the rhizomes. Each small bit of rhizome, including a node, is able to produce a new plant. If there is an opportunity for a fallow period then rhizomes can be dragged to the soil surface with harrows and left to dry out. Stubble treatment or pre-harvest use of glyphosate has been very successfully used in many crops so that the weed is not now so important. If there are several clones of couch in a field then viable seed may be produced, although this is not the main method of propagation for this weed. Many of the black-grass herbicides will control the seedlings.

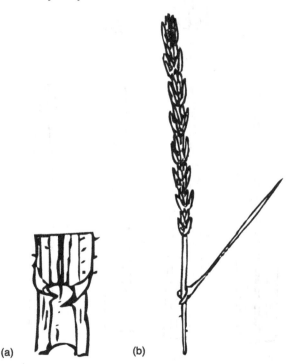

· Auricles and rhizomes present
· Variably hairy

(a)   (b)

Fig. 5.6   Ligule (a) and ear (b) couch grass.

Onion couch (false or tall oat-grass) (Fig. 5.7a and b) is a more serious problem to deal with than couch grass, though occurring less frequently. It is mainly found in cereal-growing areas on the thin alkaline soils. It has swollen bulbous nodes at the stem base, each of which can produce a new plant. Tall or false oat-grass is the same species as onion couch; it is commonly found in hedgerows but does not have bulbils and is not a weed in field crops. The ear of onion couch looks like a small upright wild oat, although the spikelets are much smaller. Unlike couch grass, onion couch can produce much viable seed. Continuous winter cereals and the use of minimum cultivations have encouraged this weed as there are few effective herbicides. Onion couch senesces too early for the pre-harvest glyphosate technique to be effective in some crops, e.g. wheat. Re-growth in stubble can be slow, so control in the stubble can also be poor. Some of the wild oat herbicides can help reduce bulbil numbers and seedling populations.

- Bulbous base
- Leaves slightly hairy
- No auricles

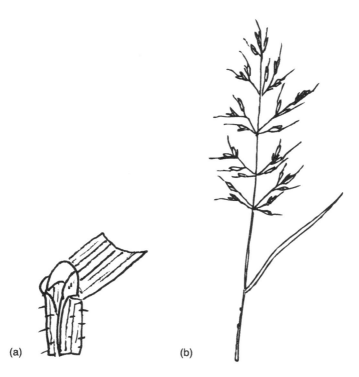

Fig. 5.7   Ligule (a) and ear (b) onion couch.

Creeping bent (Fig. 5.8a and b) is a common grass especially in grassland. It is a perennial that is spread by seed and surface rooting stolons which can be controlled by good ploughing or spraying when it is green in the stubble. Black bent is very similar but has rhizomes like couch grass and control is the same. The bent grasses have characteristic tapered leaves with no leaf hairs or auricles and an open panicle. Sometimes stems are coloured red.

- Stolons present
- No auricles
- Leaves tapered
- Variably coloured red stems

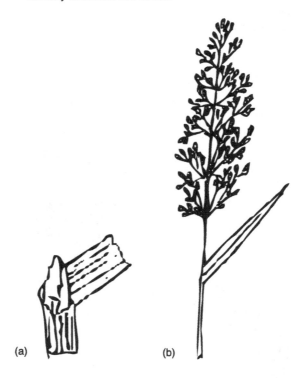

(a)   (b)

Fig. 5.8 Ligule (a) and ear (b) creeping bent.

Meadow-grasses (Fig. 5.9a and b) are some of the most commonly occurring grass weeds and can cause problems when present in large numbers. Rough meadow-grass is much more competitive than annual meadow-grass. Rough meadow-grass is a perennial and spreads by both seed and stolons. It is a common weed on mixed arable and grass farms. All meadow-grasses have their leaves folded in the stem and lack hairs or auricles and the leaf tip is barge shaped. Rough meadow-grass has shiny undersides to the leaf, unlike annual meadow-grass. All meadow-grasses have open panicles. As the seed is small it germinates in the top 5 cm of soil, so ploughing can help reduce populations. These grasses can be effectively controlled by many of the grass weed herbicides.

- Leaves folded in stem
- Boat-shaped leaf tips
- No auricles
- Small grass

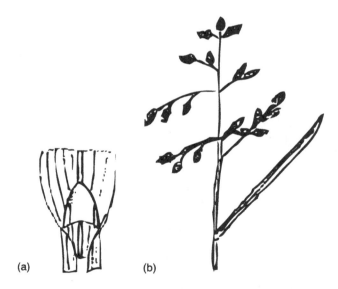

(a)   (b)

Fig. 5.9   Ligule (a) and ear (b) annual meadow-grass.

Wild oats (Fig. 5.10a and b) are found on a large percentage of cereal-growing farms and is one of the most important weeds economically. There are two main types of wild oat in the UK. The spring wild oat is the most common, germinating mainly in the spring with a small amount of seed germinating in the autumn. The winter wild oat is found mainly in southern and eastern areas; it mostly germinates in the autumn. These annual weeds look just like tame oats in the vegetative state with no auricles but they do have a few hairs along the side of the leaf blade. The leaves often have a characteristic anti-clockwise twist. The spikelets on the open panicle have long awns. Once established in a field, no matter what control measures are taken, wild oats are likely to persist for a very long time because of dormant seed in the soil. This dormancy problem is made worse by stubble cultivations after harvest which buries the seed. Most of the shed seeds, if left on the surface, are destroyed or disappear. Deeply buried seed, which may have fallen down cracks on clay soils, for example, can remain viable for over 20 years and germinates from up to 25 cm depth of soil. The seed can arrive in a field in many ways, e.g. in the crop seed, dropped by birds, in farmyard manure made with infested straw, and from the combine.

The first wild oat to appear in a field should be removed by roguing. Hand roguing can be justified when numbers are fewer than 500 plants/ha. Later, if numbers are allowed to increase, it may be necessary to use herbicides. It is difficult to decide when this can be justified. It can be based on yield response or likely grain contamination or to prevent a build-up of numbers in the future. Left uncontrolled, wild oats can increase threefold each year. There are now fields where wild oats have been found to be resistant to many of the commonly applied herbicides.

- No auricle
- Leaves twist anti-clockwise
- Leaves sparsely hairy
- Large plant
- All spikelets awned

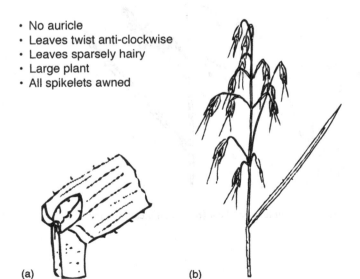

(a)   (b)

Fig. 5.10   Ligule (a) and ear (b) wild oats.

Cereal yield reductions are very high from wild oats as they are one of the most competitive grass weeds. Premiums can be lost for seed, malting and bread-making. The presence of wild oats on a farm may even incur a high dilapidations claim at the end of a tenancy.

Black-grass (Fig. 5.11a and b) economically is the most important annual grass weed on many cereal farms (13 plants/m^2 can reduce yield by 5%) in the UK. It used to be associated with heavy, wet soils where winter cereals were grown, but is now widespread on most types of soil where autumn-sown cereals predominate. As a seedling black-grass is not very distinct, with no auricles or hairs on the leaves. The stem base is often partially coloured purple. The ear is a long narrow spike with awns and is coloured purple/black, hence the name black-grass. Black-grass produces a very great number of viable seeds (typically, individual plants can produce 2 to 20 ears with approximately one hundred seeds/ear). Most seed germinates in the first three years but some can remain dormant for up to nine years. The seeds germinate mainly in early autumn, but in heavily infested fields spring germination can also be important. If uncontrolled, populations can increase 30 fold per year, so a high percentage control must be achieved to contain the problem; this usually means that cultural practices are necessary to supplement herbicides. Ploughing, to bury the seed deeply, is preferable to shallow cultivations. Black-grass only germinates in the top 5 cm of soil. Spreading the seeds to clean fields should be avoided by using weed-free seed and thorough cleaning of the combine. Vigorous crop competition is very important. Delaying drilling, or changing to spring cereals or even including a fallow can be very important in helping reduce weed populations. A number of herbicides are available and the cost can be justified in cereals where there is more than two black-grass plants/m^2. Care has to be taken with choice of herbicide as there are an increasing number of fields where black-grass is resistant to the main products.

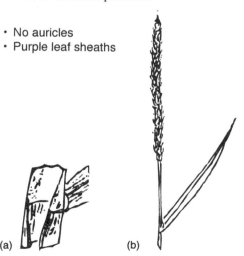

- No auricles
- Purple leaf sheaths

(a)   (b)

Fig. 5.11   Ligule (a) and ear (b) black-grass.

Barren or sterile brome (Fig. 5.12a and b) became a serious problem on many cereal farms after the very dry years in the mid-1970s. It spread from hedgerows and waste corners and was encouraged by increased early sowing of winter cereals, established using minimal cultivations. The seeds are anything but barren and germinate very readily in the autumn. The seed has little dormancy. There are five different bromes that can cause problems in field crops; barren brome and great brome are in the *Anisantha* family. This group has large open drooping panicles with long awns. Great brome has awns that are twice the size of those found in sterile brome. Rye brome, meadow brome and soft brome are in the *Bromus* family; they have more upright heads with broader spikelets. It is very difficult to distinguish between the different bromes at the vegetative stage. All the bromes have hairy leaves with no auricles. The main difference at the young plant stage is in size and shape of the ligules.

Ploughing can be very effective in controlling this weed if the seed is buried more than 13 cm. Stubble cultivation, in a damp autumn, can also aid control.

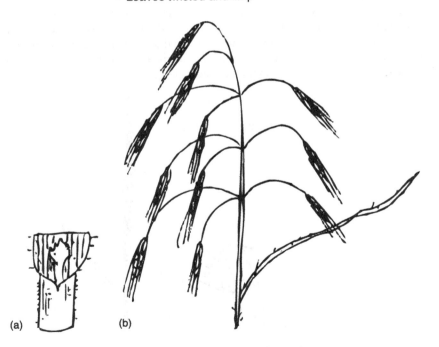

- Very hairy soft leaves
- Leaves twisted and limp

(a)   (b)

Fig. 5.12 Ligule (a) and ear (b) barren brome.

Field margins need to be managed to ensure that viable brome seeds are not produced. It is much better to have a permanently cropped field edge than a bare area which can quickly be colonized by brome. There are only a limited number of herbicides for use in cereals and control can be variable, although a number of herbicides used in other crops are effective.

The incidence of other brome species, such as rye brome and meadow brome, is increasing and such weeds can be more difficult to control than barren brome.

Other grass weeds

Awned canary grass is a relatively new problem in the UK although it is more common in other parts of the world. It is not widespread but it can cause serious yield losses and harvesting problems where it does occur. The heads resemble large Timothy heads. At the seedling stage the plant looks very like black-grass. The seedling has a pink stem base and when broken a red sap is often released. The seed has a limited amount of dormancy. Deep ploughing and spring cropping can help reduce weed incidence. Care must be taken with herbicide choice as only a few of the commonly used grass weed herbicides control awned canary grass.

Loose silky bent is a local problem on sandy or light loam soils in eastern and southern England and many parts of Europe. As with many other annual grass weeds, deep ploughing and spring cropping can be very good methods of control.

Occasionally in maize crops some of the important tropical grass weeds such as common millet and green bristle grass have emerged but never at populations that would affect yields. Any changes in climate could affect incidence of these grasses in the future.

5.2.2   Broad-leaved weeds

There are a large number of different broad-leaved weeds that can be important in field crops in the UK; most are annuals although there are a few perennial broad-leaved weeds that can cause problems, especially in grassland. How important economically a broad-leaved weed species is depends on the crop being grown. None other than cleaver are as competitive in cereals as some of the grass weeds. Numbers of species found tends to depend on the rotation; rotations including spring crops usually have the widest range. Some broad-leaved weed species favour different soil types; redshank is often associated with the heavier wetter soils whereas the mayweeds are often associated with the lighter soils. Corn spurrey is found on the light to medium textured soils, especially those that are naturally acidic. Populations of broad-leaved weeds are often higher where soils are ploughed rather than where only shallow cultivations are used. Control, both chemical and mechanical, of annual broad-leaved weeds is usually most effective at the seedling stage. In order that the correct product is chosen it is important to identify the weed correctly. Many broad-leaved weeds have very characteristic cotyledons (seed leaves) and true leaves which aid identification. Size, shape, colour, if stalked, presence of hairs or prominent leaf wax all help recognition.

Figure 5.13a–d shows the characteristics of some common broad-leaved weeds.

Cleavers (Fig. 5.13a) is the most competitive weed in winter cereals and winter oilseed rape, more competitive than wild oats. It can be confused at the seedling stage with ivy-leaved speedwell (Fig. 5.13a). Note that cleavers has a notched cotyledon and the cotyledons are not stalked. Cleavers germinates at fairly low temperatures mainly in the autumn and winter, with some in the spring.

Charlock (Fig. 5.13a) is a brassica showing the typical large kidney-shaped cotyledons. It is the shape and presence or absence of hairs on the true leaves that identifies the different brassicas. Charlock used to be a major weed problem in cereals but with the introduction of more effective chemical control methods it is now less important. It is an important plant for different invertebrates.

Speedwells (Fig. 5.13b) have increased in recent years. Ivy-leaved speedwell is mainly autumn germinating whilst common field-speedwell germinates all the year round. All the speedwells have spade-shaped cotyledons except ivy-leaved speedwell. It is important to be able to distinguish between these weeds, as chemical control is different for many of them. Speedwells are low growing and are not very competitive.

Fat hen (Fig. 5.13b) is in the same family as sugar beet and so can sometimes be difficult to control in beet. It germinates in the spring and is often associated with the application of dung. It has a very distinct waxy/mealy appearance on the cotyledons and first true leaves. The stalk and underside of the cotyledons are red. It is an important weed for invertebrates and seed eating birds.

Polygonums (Fig. 5.13b) including knotgrass, redshank and black bindweed are a very important group of spring germinating weeds. Knotgrass is the first to germinate; it has narrow cotyledons in the shape of a 'V'. All the polygonums seedlings have a red stem. They are important for in-field biodiversity.

Mayweeds (Fig. 5.13c) germinate in most months, although spring is the peak period. Unlike many broad-leaved weeds, mayweeds remain green all season and so compete with the crop for a long time. They are prolific seeders and the seed can remain dormant for many years. Mayweeds have very small cotyledons and very distinctive shaped first true leaves. They are also important for invertebrates.

Field pansy (Fig. 5.13c) has become more abundant in recent years, partly due to some commonly applied herbicides having little control. It is, however, a very poor competitor. The seedlings have distinct small, notched cotyledons and hairless first true leaves.

Common chickweed (Fig. 5.13d) is the most common broad-leaved weed found in virtually all arable fields in the UK. Chickweed favours fertile soils, germinating all the year round and it readily overwinters. This weed is very successful, partly because it can produce more than one generation of plants a year and thus a lot of seed. At the seedling stage it is a bright green colour and has true leaves with the same shape as the cotyledons.

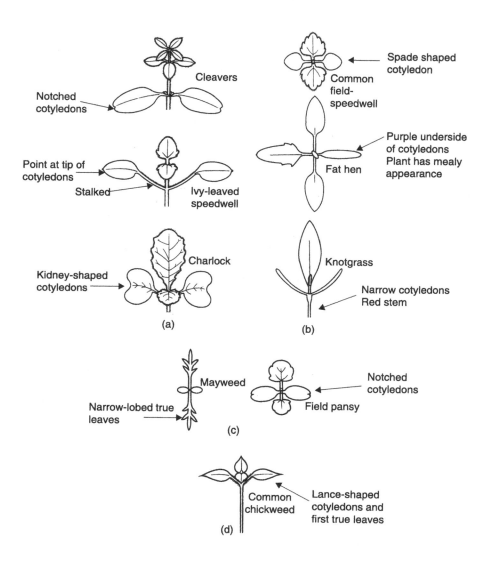

Fig. 5.13 Identification of common broad-leaved weeds: (a) large cotyledons; (b) medium sized cotyledons; (c) small cotyledons; (d) similar-shaped cotyledons as true leaves.

5.3   Control of weeds: general

The control of weeds with herbicides is standard practice on most farms; nevertheless, it is worth remembering that other husbandry methods play an important part, especially with the increase in herbicide resistance problems and the limited introduction of new products. Non-chemical methods of weed control are the only ones available to organic farmers.

5.3.1   Methods of weed control

- Crop hygiene. It is particularly important to ensure that only clean seed is sown when using home-saved seed. Hand roguing can be a very effective method of preventing weeds becoming a problem in the first place. Avoidance of machinery contamination can also help to prevent weed increase, e.g. wild oats.
- Cultivations. Ploughing can be very effective at containing or reducing some annual grass weed problems such as black-grass and barren brome; ploughing can bury freshly shed seed to 15/20 cm and will normally bring to the surface fewer viable weed seeds. Most seeds only germinate from the top 5 cm of soil. Black-grass populations can be reduced by 60–70% after ploughing. With heavy clay soils, where minimum cultivations are used, it is recommended to rotationally plough once every three or four years to help reduce the grass weed burden. Weeds such as wild oats that have large seeds that can germinate from plough depth are unaffected by ploughing; sometimes populations are worse after ploughing compared with the use of minimum cultivations.

  In some years, if there is enough soil moisture, stubble cultivations and the 'stale' seedbed technique can be a useful aid to weed control. The soil is cultivated ready to sow the next crop; weeds are allowed to grow and are then killed before planting the crop. Inter-row cultivations are still used in some crops, particularly if there is a difficult weed to control, such as weed beet in sugar beet, or where there are limited herbicides approved for use, as in field vegetables. There is some interest in the use of in-crop weeders as a method of reducing herbicide inputs.
- Cutting, e.g. bracken, rushes, ragwort and thistles. This weakens the plants and prevents seeding. The results are often disappointing if not repeated, however control of spear thistle can be very effective from cutting as long as the timing is correct, although creeping thistle is more difficult to control.
- Drainage. This is a very important method of controlling those weeds which thrive in waterlogged soils. Lowering the water table by good drainage will help to control weeds such as rushes, sedges and creeping buttercup.

- Rotations. Growing leys and various arable crops that are planted at different times of the year usually leads to a different weed flora. There is also an opportunity to use different types or groups of herbicides. This method is useful if there are difficult weeds such as barren brome, volunteer potatoes, herbicide resistant black-grass and weed beet to control.
- Maintenance of good fertility. Arable crops and good grass require a high level of fertility, i.e. the soil must be adequately supplied with lime, nitrogen, phosphates, potash and organic matter. Under these conditions crops can compete strongly with many weeds.
- Crop seed rates. High seed rates and good crop establishment all help to reduce the impact of weeds.
- Time of drilling. Delaying drilling cereals from September to October can significantly reduce the number of black-grass and brome seedlings that emerge. Later drilling will also allow more stubble treatments to be undertaken.
- Competitive crop/variety. Some crops, such as winter oilseed rape and potatoes, are very competitive and can help smother any weeds present. There can also be a difference in competitiveness of different crop varieties due mainly to differences in crop height and leaf attitude.
- Burning. The use of propane burners can be an effective method of weed control; it is mainly used in organic systems. When farmers were able to burn crop stubbles this was very effective at reducing the weed seed burden.
- Mulches. There are a number of mulching materials that can be used to prevent weed growth in some small-scale plantings, e.g. amenity plantings, fruit and vegetable crops. Black polythene is the most commonly used material. As well as preventing weed growth it also helps to conserve soil moisture.
- Chemical control. There are many different herbicides and products on the market, although several have been withdrawn recently due to EU legislation. New herbicides cost a lot to develop and to get approval for use, so currently few new products are coming on to the market. The following is a summary of modes of action of some of the currently approved herbicides.

5.3.2   Herbicides – modes of action

Most of the chemicals used have a *selective* effect, i.e. they are substances which stunt or kill weeds and have little or no harmful effects on the crop in which the weeds are growing. A severe check of weed growth is usually sufficient to prevent seeding and to allow the crop to grow away strongly. There are a limited number of *non-selective* or *total* herbicides such as glyphosate. They kill or check all vegetation and are usually used in non-cropped areas before the crop emerges, or as crop desiccants.

Herbicides are usually sold under a wide range of proprietary names which can be very confusing, but the common name of the active material must always be stated on the container. In the text of this book the common name of the chemical is used when referring to herbicides. The selectivity of herbicides depends on a number of factors, including the amount of the active ingredient (ai) applied, its formulation with an adjuvant (normally a type of surfactant), as well as the quantity of carrier (water, oil or solid).

The chemicals now commonly used as herbicides can be classified according to their chemical group and biochemical mechanism of action. Most simply they can be classified by their basic mode of action as follows:

- *Contact-acting herbicides*. These will kill most plant tissue by a contact action with little or no movement through the plant; shoots of perennials may be killed but re-growth from the underground parts usually occurs. Effective control relies on good coverage of green plant material by the herbicide. Some examples of contact-acting herbicides are glufosinate-ammonium, carfentrazone-ethyl and phenmedipham.
- *Soil-acting or residual herbicides*. These chemicals act through the roots or other underground parts of the plant after being applied to the soil surface. Some examples are propyzamide, pendimethalin, triallate, flufenacet and prosulfocarb. These herbicides vary in their persistency in the soil from weeks to months. Some residual herbicides also have some contact activity, such as metribuzin.
- *Translocated herbicides* are those which can move through the plant before acting on one or more of the growth processes. Some of these herbicides can be very effective at controlling perennial weeds, e.g. the control of couch by glyphosate. Good spray cover is not so important with this type of herbicide.

Herbicides work by affecting one or several different biochemical pathways or physiological processes within the plant. The speed of action reflects the mechanism of action. Some of the earliest developed herbicides are similar to substances (hormones) produced naturally by plants. Susceptible plants initially show distorted growth and then will take a few days to die back. They are mainly used for controlling broad-leaved weeds in cereals and grassland, for example mecoprop-P.

A large number of herbicides affect photosynthesis. Symptoms are usually seen as a rapid yellowing or *chlorosis* of the leaves followed by leaf death or *necrosis*. Other herbicides limit or stop cell division and elongation of the growing point. These chemicals can be very slow acting as seen with propyzamide (a residual amide) in oilseed rape.

Several of the newer chemicals, such as the sulfonyl ureas, are very specific and only affect the production or synthesis of single compounds such as amino acids. Again these herbicides tend to be slower acting than those that affect photosynthesis. Examples include amidosulfuron, thifensulfuron-methyl and rimsulfuron.

It is vital to check the manufacturer's recommendations before buying and applying pesticides and to follow the Code of Practice for Using Plant Protection Products. Only products approved by the Chemicals Regulation Directorate (CRD) can be sold or applied in the UK. It takes many years for new chemicals to be approved as the CRD has to have adequate data to ensure the products are safe, do what they are supposed to do and have limited environmental impact. The

European directive (91/414/EEC) was introduced to harmonise all the different pesticide regulations within the EU. The aim of the legislation was to further reduce risk and hazard to both human health and the environment. This legislation has led to the loss of several commonly used pesticides in the UK.

5.3.3   Herbicide choice

Herbicide choice will be affected by a number of factors including:

- *Weeds present and their growth stage.* The efficacy of herbicides is often affected by the growth stage of the weed. Control of annual broad-leaved weeds is usually most effective when the weeds are small. Control of annual grass weeds is usually more effective before the 4-leaf stage. The range of species controlled is also greatest when the weeds are small. Herbicide recommended rates are often lower when the weeds are small. A broad-leaved weed growth stage key has been produced which is used on chemical product guides. There is also an annual grass weed key which follows the key for cereals (Table 13.2 in Chapter 13). A description of annual broad-leaved weed growth is shown in Table 5.3.
- *Crop/variety and crop growth stage.* Some herbicides are only recommended for use on a limited range of crops, e.g. phenmedipham on beet and strawberry crops only, whereas triallate is recommended for wild oat control on a wide range of crops. Other herbicides are only recommended for use on some varieties of crops as there can be problems with crop damage, i.e. metribuzin in potatoes.

  The crop growth stage is important partly because of crop damage. The hormone herbicides such as mecoprop-P are not recommended after the early stem extension stage of cereals. Crop growth stage is also important with the contact-acting herbicides as they usually work best when the crop and weed are growing well. But note that if the crop is too advanced, and is shading the weeds, then it is very difficult to get good spray coverage on to the weeds.
- *Soil type and condition.* In order for the residual chemicals to work effectively the soil tilth should not be too cloddy. Activity is affected by the amount of soil

Table 5.3 Description of annual broad-leaved weed growth stages

Pre-emergence	Plants up to 50 mm across/high
Early cotyledons	Plants up to 100 mm across/high
Expanded cotyledons	Plants up to 150 mm across/high
One expanded true leaf	Plants up to 250 mm across/high
Two expanded true leaves	Flower buds visible
Four expanded true leaves	Plant flowering
Six expanded true leaves	Plant senescing
Plants up to 25 mm across/high	

moisture present as this will affect movement of the chemical to the germinating weeds. Persistency of the chemical in the soil will be affected by the rate of chemical applied, and speed of breakdown. The main method of breakdown is by microbial activity, which is affected by soil moisture and temperature.

The rate of application of some residual chemicals is affected by the soil texture. Some are not recommended on sandy soils as there can be too much leaching down to where the crop is growing, which can lead to crop damage.

Herbicide choice is restricted on soils with a high organic matter or those that have a high adsorption coefficient. Residual activity is reduced on soils with a high organic matter (usually greater than 10%). Trash on the soil surface can also have this effect; the chemical becomes attached to the charged sites on the organic matter (*adsorbed*) and is then unavailable for weed uptake. The activity of residual herbicides can also be affected by soil pH.

- *Weather conditions.* Weather conditions, including rainfall and temperature, can affect activity and/or efficacy of a treatment, as well as crop damage. Control of many weeds using foliar applied chemicals is most effective when the weather conditions are optimal for weed growth. Some herbicides are rain-fast fairly soon after application whereas glyphosate requires at least six hours dry weather after application to give the best results.

- *Tank mix compatibility.* Some chemicals only have limited compatibility with other pesticides; this can be due to a problem with the formulations or with the activity of the chemical being affected. To get optimum weed control in some crops mixtures of chemicals are required or sometimes the use of sequences/stacking (products applied in close succession) is more effective.

- *Cost.* Pesticides (including herbicides, fungicides and insecticides) are expensive when they are first marketed, partly to cover the very high development costs. Once they are off patent costs usually fall. Cost can significantly affect product choice and rate used, especially if crop prices are low.

- *Following crops.* There are a few restrictions on following crops and intervals between applying the chemicals and following crops, especially with residual chemicals; always check the label. Some chemicals also have recommendations to plough before planting the next crop.

- *Water buffer zone requirements.* Many pesticides have a buffer zone requirement when spraying next to water courses. There are not as many herbicides as other pesticides that are affected in this way, but it can influence choice of chemicals. If required, a Local Environmental Risk Assessment for Pesticides (LERAP) should be carried out; this takes into account type of water course, rate of chemical used and type of nozzles on the sprayer. Undertaking a LERAP can reduce the buffer zone requirement.

- *Resistance.* Herbicide resistance is an increasing problem, particularly on mainly cereal farms. Herbicide resistance is the inherited ability of a weed to survive rates of chemicals that normally control the weed. Black-grass, wild oats and ryegrass are grass weeds that have developed a problem of resistance to some of the most commonly applied grass weed herbicides. Once herbicide resistance has been diagnosed then chemical choice will be severely affected.

5.4   Herbicide resistance

Cases of herbicide resistance have developed more slowly than insecticide and fungicide resistance mainly because most weeds only have one generation each year, unlike diseases and some pests like aphids. Cases of resistance started being found in the early 1980s in black-grass on farms where populations were very high, continuous winter cereals were grown (established using minimum cultivations) and the same herbicide was often applied both in the autumn and spring every year. Currently in the UK, black-grass resistance is present to some extent on all fields where there has been regular application of black-grass herbicides. Resistant Italian rye-grass is also fairly widespread, wild oat resistance is less common and there are a few confirmed cases of broad-leaved weed resistance. Poor results after applying a herbicide may not necessarily be due to herbicide resistance. Seed samples can be collected in mid-July and tested if resistance is suspected. Other indicators for resistance are good control of other susceptible species and healthy plants next to dead plants of the same species.

5.4.1   Mechanisms of herbicide resistance
There are two main mechanisms of herbicide resistance

1. *Enhanced metabolism resistance (EMR)*. This is the most common mechanism and develops slowly over many years. Herbicides affected by enhanced metabolism are broken down (detoxified) more rapidly in resistant than susceptible plants. Each time the same herbicide is sprayed more of the resistant plants will remain so that gradually the weed population will be dominated by the more resistant plants.
2. *Target site resistance (TSR)*. Some of the grass weed herbicides (the 'fop', 'dim' and 'den' graminicides) mainly work by affecting a single biochemical pathway. Resistant weeds are able to block this site of action so the herbicide has no activity. This type of resistance develops very quickly. Target site resistance is also affecting the sulfonylurea herbicides.

5.4.2   Control of herbicide resistant weeds
Once resistance is confirmed it will always be present on the farm. Recently there has been a reduction in the number of different herbicides available and few new herbicides are coming on the market, so management of resistant weed populations is becoming more difficult.

- Non-chemical control methods must be used in combination with herbicides. Rotational ploughing, delayed drilling and spring cropping can help reduce grass weed populations.
- Use of pre-emergence treatments reduces risk of developing resistance compared with many of the post-emergence products applied in cereals.
- Use mixture or sequences of herbicides with different modes of action. Do not rely on the same chemical or chemical group.

- Growing other crops than cereals will give an opportunity to apply totally different and effective chemicals.
- Continue to monitor effectiveness of control measure to ensure resistance problems do not increase.

5.5   Spraying with herbicides: precautions

This is a skilled operation and should only be carried out by trained operators. A recognised certificate of competence is required by all who store, sell, recommend or apply pesticides. Method of application and timing can significantly affect herbicide activity. The following precautions should be taken when spraying:

1. Make a careful survey of the field to determine the weeds to be controlled. Choose the most suitable and safest chemical and the best time for spraying. The risks associated with the use of any herbicides should be assessed as defined in the HSE regulations (COSHH – *Control of Substances Hazardous to Health*). Appropriate measures should be taken to control any risk.
2. Wear the correct protective clothing. Read carefully the instructions issued for that product when handling the concentrate and when spraying.
3. Check carefully the amount of chemical to be applied and the volume of water to be used. Droplet size is very important when using post-emergence herbicides, whereas with pre-emergence herbicides medium size droplets are just as effective as small droplets. When weeds are small, droplet size needs to be small to ensure the highest deposition on the plants. Angling the nozzles can increase deposition on small weeds.
4. Only use approved tank mixes. Follow the manufacturer's recommendations for mixing.
5. Control spray drift. Wind speed, boom height, speed of application, atmospheric conditions, and water volume and spray quality can all affect the amount of spray drift. Fine sprays are most affected by drift. The optimum wind speed is when there is a light breeze and tree leaves are just rustling. Often, air-induction sprayers produce smaller droplets than conventional sprayers, but these sprayers can give as good results as the medium droplets produced by conventional sprayers and produce less spray drift.
6. Make sure that the boom is level. The lower the boom the less the drift.
7. Keep pesticides out of water courses and other environmentally sensitive areas. Due to EU and UK water quality legislation this is a major issue particularly with some herbicides such as propyzamide. Pesticides can get into a water course from spray drift, spillages of the concentrated chemical, sprayer washings or from surface run-off and, importantly, through drains. Establishing grass buffer strips of at least 6 m next to water courses can significantly reduce the risk. Not spraying when field drains are running or when the soil is very dry and cracked can also help stop herbicides entering water courses.

8. Check buffer zone requirements for all herbicides, not just the sensitive ones, when spraying next to water courses. (If necessary undertake a Local Environmental Risk Assessments for Pesticides (LERAP) assessment.) A LERAP assessment is not required if there is a 5 m no-spray zone or buffer strip next to the watercourse.

9. Wash out the sprayer and pesticide containers thoroughly after use. Follow the guidelines for safe disposal.

10. Record keeping. By law, records must be kept of all sprays applied to crops grown for human or animal feed. Details of operator, field, date of spraying, chemical used, and rate of application and water volumes plus weather conditions when sprayed must be kept. It is also good practice to note effectiveness of the treatment, especially with the increases in pesticide resistance.

5.6   Sources of further information and advice

Further reading

Anderson W P, *Weed Science, Principles and Applications*, Waveland Press, 2007.
Bayer Crop Science, *Weed Guide*, Bayer Crop Science, 2009.
British Crop Protection Council, *Using Pesticides*, BCPC, 2007.
British Crop Protection Council, *IdentiPest cd*, BCPC, 2001.
Caseley J C, Atkin R K and Cussans G W, *Herbicide Resistance in Weeds and Crops*, Butterworth-Heinemann, 1991.
Cobb A H and Reade J P H, *Herbicides and Plant Physiology*, Wiley-Blackwell, 2010.
Cooper M R, Johnson A W and Dauncey E, *Poisonous Plants and Fungi*, TSO, 2003.
Cremlyn R J, *Agrochemicals, Preparation and Modes of Action*, Wiley, 1991.
Defra, HSE, *Pesticides – codes of practice for using plant protection products*, Defra, 2006.
Defra, *Identification of Injurious Weeds*, TSO, 2011.
Garthwaite D G, Barker I, Parrish G, Smith L, Chippindale C and Pietravalle S, *Pesticide Usage Survey Report 235, Arable Crops in the UK 2010*, Defra, 2011.
Gwynne D C and Murray R B, *Weed Biology and Control, Batsford Technical*, 1985.
Naylor R E L, *Weed Management Handbook*, 8th edn, Blackwell Publishing, 2002.
Hanf M, *The Arable Weeds of Europe*, BASF, 1983.
HGCA, ADAS and BASF, *Encyclopedia of Arable Weeds G47*, HGCA, 2009.
HGCA, *Identification and Control of Brome Grasses IS06*, HGCA, 2009.
HGCA, *Managing and Preventing Herbicide Resistance in Weeds G10*, HGCA, 2003.
HGCA, *Managing Weeds in Arable Rotations – a guide G50*, HGCA, 2010.
Hubbard C E, *Grasses*, Penguin, 1984.
Lainsbury M A *The UK Pesticide Guide*, BCPC and www.cabi.org, published annually.
Mathews G, *Pesticides, Health, Safety and the Environment*, Blackwell Publishing, 2006.
Williams J B and Morrison J R, *Weed Seedlings* (colour atlas), Manson Publishing, 2003.
Wilson M F, *Optimising Pesticide Use*, John Wiley and Sons Ltd, 2003.

Websites

www.agricentre.basf.co.uk
www.bayercropscience.co.uk
www.cropprotection.org.uk

www.ewrs.org
www.fera.defra.gov.uk
www.hgca.com
www.hracglobal.com
www.pesticides.gov.uk
www.voluntaryinitiative.org.uk

6

Diseases of farm crops

DOI: 10.1533/9781782423928.1.119

Abstract: The main living organisms causing disease in crop plants are fungi, bacteria and viruses. Mineral deficiencies and physiological disorders are often classified as diseases and are included in this chapter. The chapter outlines the types of disease-causing organisms and factors that affect the development of a disease epidemic. Methods of disease control are included, as is a section on fungicide resistance. Finally, there is a summary of the main diseases found on UK crops, their symptoms and methods of control.

Key words: diseases of crop plants, disease symptoms, disease cycles, disease control, fungicide resistance.

6.1 Introduction to plant disorders

The main living organisms causing disease in crop plants are fungi, bacteria and viruses. Mineral deficiencies and physiological disorders are often classified as diseases (abiotic problems) and are included in this chapter. Diseases, like pests, annually cause millions of pounds worth of damage and loss to the agricultural industry. Diseases can significantly reduce crop yields and quality as well as increasing losses of produce in store.

- *Parasites*
  The organisms that cause plant diseases (pathogens) are called *parasites* as they obtain their food from the infected crop plant. There are several types of parasites:
  (a) *Obligate parasites* (or *biotrophs)* are dependent on the living host; they are responsible for causing many plant diseases. They can only grow and reproduce in living hosts. Examples include the rust and mildew fungi, and viruses.
  (b) *Non-obligate parasites* can live on either living or dead tissue (most fungi and bacteria).

(c) *Facultative parasites* or *semi-parasites* kill the host tissues and live on the dead cells and are sometimes called *necrotrophs*.

- *Saprophytes*
  These live on dead organic matter and are often present in plants attacked by parasites or plants that have reached maturity and died. They are also found on leaves coated in aphid honeydew. Saprophytes play an important part in helping to break down plant remains into organic matter.

Most pathogens only attack one crop species or family, but a limited number have a wider host range. For a pathogen to be successful it must be able to invade and colonise a plant, grow and reproduce, be able to spread effectively and have some method of survival. This is called the disease cycle. Each pathogen has a distinct cycle. It is important to understand the disease cycle so that the most effective methods of control can be used at the optimum time.

Knowledge of a particular disease cycle can also help when predicting likelihood of a disease epidemic. Conditions have to be right for diseases to develop, so not every year has the same disease importance. Disease development is affected by many factors and they are all interlinked:

- *The host plant.* The vigour and growth stage of the crop can affect the amount of infection that is likely to occur.
- *The pathogen.* Amount and type of infection material and ability to colonise the host plant.
- *The environment.* Each pathogen has specific requirements (temperature, moisture/rainfall and humidity) for infection, colonisation, reproduction and spread. Because of the environmental differences across the country, disease incidence is often different. For example there is a much higher risk of septoria tritici in the west in wheat and of yellow rust in the east of England and Wales (Fig. 6.1).
- *Farming activities.* Rotation, variety choice, time of sowing, seed rate, fertiliser use, irrigation and cultivations can all affect the amount of disease that is present.

6.2   Symptoms

Symptoms of many diseases are characteristic for that disease and crop. Often the disease can be identified without having to send them on to a plant pathology laboratory for diagnosis using microscopy, culturing or even serological tests. Some diseases can be identified visibly, i.e. by seeing the fungal growth or fruiting bodies, as with powdery mildew and the rusts. For other diseases it is the type of lesion that helps identification, as follows:

- *Yellowing or chlorosis.* Many of the viruses, such as beet virus yellows, cause foliar chlorosis as do some mineral deficiencies. Where the yellowing is on the leaves, the shape of the infected area may help with identification.

- *Death of tissues/necrosis.* Many diseases can cause the death of the plant tissue, including leaves and stems. Again, size and shape of the lesions are important. Leaf spots are well-defined lesions and sometimes the dead tissue falls away to leave a shot-hole effect. Leaf blotches tend to be of variable size. Blights can cause total death of leaves and stems and finally to all above-ground parts of the plant.
- *Abnormal growth.* Some diseases cause the infected tissue to enlarge either by an increase in cell numbers or size, e.g. potatoes with common scab.
- *Stunted growth.* This is seen in severe cases of barley yellow dwarf virus and with some root diseases such as take-all in cereals.

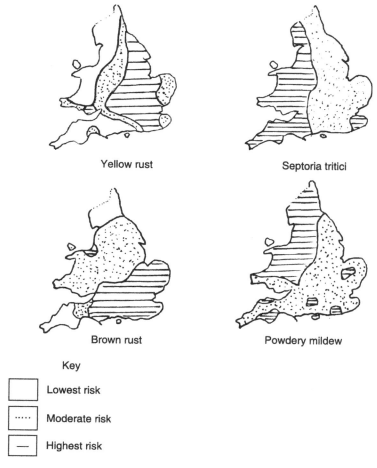

Yellow rust   Septoria tritici

Brown rust   Powdery mildew

Key

Lowest risk

Moderate risk

Highest risk

Fig. 6.1 Generalised maps of England and Wales showing variation in risk of foliar diseases in wheat.

- *Wilting.* There are a number of bacterial and fungal diseases that cause wilting, normally late on in the development of the disease.
- *Tissue disintegration.* This type of damage is associated with many of the root or foot rots and storage rots. The affected cells are broken down and release liquid (wet rot) or become dry and brittle (dry rot).

6.3   Some important types of plant pathogens

6.3.1   Fungi

There are many thousands of different species of fungus, the majority of which are invisible to the naked eye and it is only a small number that are important as plant diseases. Fungal diseases usually have a common name plus an internationally agreed scientific name. As understanding of diseases and their life cycles has developed some names have been changed, see Appendix 4. Naming can get confusing as sometimes the original scientific name has been incorporated into the common name, then the scientific name changes as has happened with septoria tritici in wheat. The fungal disease cycle is outlined in Fig 6.2.

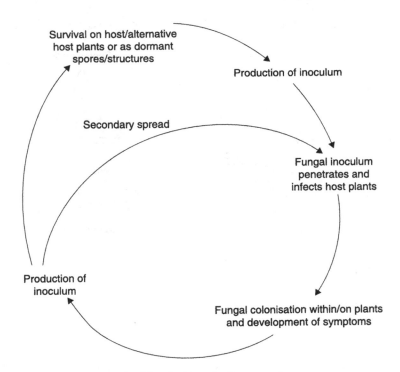

Fig. 6.2   Outline of a disease cycle.

Fungal disease cycle

Fungal infection

A typical fungus is composed of long thin filaments (made up of single cells) termed *hyphae*. Collectively, these are known as *mycelium*. Fungi enter the host plant surface either by direct penetration of surface cells (main form of entry) or through wounds or natural openings. Not all fungi that enter plants develop into a disease, sometimes the fungus is killed within the host plant, leaving a visible necrotic spotting or hypersensitive reaction. .

Once within a susceptible plant the fungus can grow and spread between cells. Some fungi produce extensions to the hyphae which enter the host cells and act as nutrient-absorbing organs (*haustoria*). In the case of most parasitic fungi, the mycelium is enclosed within the host (only the reproductive parts protruding), although some fungi are only attached to the surface of the host, e.g. powdery mildew. Sometimes, even after a fungus has entered the host, plant symptoms may not be visible at first. This lag time is called the *latent infection period* and once the symptoms have appeared it can be almost too late to treat with a fungicide.

Reproduction

Fungi can reproduce simply by fragments of the hyphae dropping off, but usually reproduction is by production of spores. Spores can be compared to the seeds in ordinary plants, but they are microscopic and occur in vast numbers. Asexual spores are usually important for the spread and increase of a population or epidemic. Sexual spores are important for fungal survival.

Fungal survival and dispersal

Not all fungi have the same mechanisms for spread and survival; this knowledge will help in deciding on disease prevention and control methods.

* *The seed.* Some fungi are carried from one generation to the next by surviving on the seed coat or inside the seed itself, e.g. smut diseases of cereals.
* *The soil.* The spores or even resting bodies such as *sclerotia* (which are a dense mat of hyphae) drop off the host plant and remain in the soil until another susceptible host crop is grown in the field. A suitable rotation will go a long way to check these diseases such as those caused by *Sclerotinia* sp.
* *The air.* Spores of many foliar diseases are dispersed on dry air currents. Spores of some of the cereal rusts have been known to travel hundreds of kilometres in air currents.
* *Water* is important for the dispersal of some fungal spores over short distances.
* *Alternate hosts.* Some diseases have the ability to overwinter on alternative hosts as seen with some of the rusts.
* *Infected plant material.* Many diseases survive from one season to another either on plant debris or volunteer crops. Potato blight often starts from infected potato dumps.

6.3.2   Viruses

The virus was discovered in the nineteenth century. It is a very small sub-cellular organism and cannot be seen using a normal light microscope; only by using an electron microscope can it be seen and identified. Serological and molecular techniques are now commonly used to aid identification. Viruses consist of nucleic acids (DNA or RNA) usually surrounded by a coat of protein. Some viruses can cause the host cell to produce various crystalline structures which can also be used to aid identification, as can shape of the outer virus coat.

Once a virus is in a host plant it causes the host cell to produce more nucleic acids and proteins from which more virus particles are formed which can then move around the plant in the phloem. All viruses are obligate parasites. They are not known to exist as saprophytes. The virus is generally present in every part of the infected plant except the seed, pollen and often the apical meristem. Therefore, if part of that plant, other than the seed, is propagated, the new plant is itself infected, e.g. the potato. The tuber is attached to the stem of the infected plant, and infection is carried forward when the tuber is planted as 'seed'.

In most plant virus diseases, the infection is transmitted from a diseased to a healthy plant by vectors; insects are the main vectors although nematodes and mites are known to transmit some virus diseases. Fungi can also be vectors, e.g. the mosaic viruses in cereals are transmitted by the soil-borne fungus *Polymyxa graminis*; rhizomania in sugar beet is transmitted by *Polymyxa betae*. Viruses can overwinter on perennial infected crops that might not show any visible symptoms and on volunteer crops. One of the main methods of survival is in alternative host plants such as weeds. Viruses tend to have a fairly wide host range. The host species must also be suitable for the vector. The only way viruses can survive in the soil is in a living organism such as a nematode or fungus.

Viruses cause a variety of symptoms from plant yellowing to distortion of growth. The name of a virus usually describes the symptoms and a main host plant such as barley yellow dwarf virus. Virus names are abbreviated using the first letter of each name, e.g. barley yellow dwarf virus, BYDV.

6.3.3   Bacteria

Bacteria are very small single cell organisms, only visible under a microscope. They are of a variety of shapes, but those that cause plant diseases are all rod-shaped. About 12% of the identified bacteria are plant pathogens. Like fungi, bacteria feed on both live and dead material. Although they are responsible for many diseases of humans and livestock, in arable crops in the UK they are usually of minor importance compared with fungi and viruses.

Bacteria overwinter in a similar way to fungi except that no resting spores or bodies are produced. They can survive in infected plants, seeds and tubers, in plant debris and, just a few, in the soil. They mainly enter plants through wounds or occasionally through openings like the stomata. Bacteria, causing wet rots, often gain entry into their host root crops after mechanical damage during harvest. Once in the host plant bacteria can reproduce themselves simply by the process of splitting into two. Under favourable conditions this division can take place about every 30 minutes. Thus bacterial diseases can spread very rapidly once established.

6.4   Other disorders

6.4.1   Lack of essential plant foods (mineral deficiency)

When essential plant foods become unavailable to particular crops, deficiency symptoms will appear. These symptoms, such as chlorosis and necrosis, are characteristic for that nutrient and crop but can often look like a foliar disease (see Chapter 4). Distribution of the symptoms in the crop can be useful in deciding if it is likely to be a disease causing the problem or a nutrient deficiency. Fungal diseases often start with a primary foci and then spread out, whereas nutrient deficiency symptoms are more widespread or confined to certain distinct soil types or farm operations. Soil and or leaf tissue analysis can be used for diagnosing a nutrient deficiency. If a crop is suffering from a nutrient deficiency, it can make it more susceptible to attack by some pests and diseases.

6.4.2   Physiological diseases (stress)

Physiological diseases are often triggered by adverse environmental conditions which can upset the normal physiological processes of the plant. Usually this is only temporary, but there may be occasions when the effect is more permanent.

- Temporary conditions, e.g. a high water table in the early spring. This will cause yellowing of the cereal plant as its root activity is restricted, considerably reducing its oxygen and plant food intake. When the water table falls, the plant is able to grow normally once more, assuming a healthy green colour.
- Permanent conditions, e.g. where the soil has become compacted the root activity of the plant can be restricted. This will result in poor stunted growth with the plant far more vulnerable to pest and disease attack. The yields will be reduced. Periods of very dry conditions followed by wet weather can cause root cracking and splitting in crops such as carrots and potatoes; it can also cause secondary tuber production in potatoes. Hail damage can often be confused with a disease. It can cause leaf and flower spotting and even damage the flower parts. In cereals, hail and heavy rain at flowering have been known to increase the number of blind grain sites in the ear.

6.4.3   Herbicide damage

Herbicide drift or application of a spray contaminated with a herbicide not recommended for that crop can occasionally cause crop damage. Symptoms vary depending on the chemical involved. Usually it is very obvious that the problem is not caused by a disease because of distribution in the field.

6.5   The control of plant diseases

Before deciding on control measures it is important to know what is causing the disease. Once identified then knowledge of its life cycle and factors affecting the outbreak and spread (*epidemiology*) will help decide on the most appropriate method of control. Depending on the disease (see Table 6.2 at the end of this chapter) there are many non-chemical methods of control that will help reduce the risk before any agrochemicals are required. With the increase in number of cases of fungicide resistance, it is more important than ever that alternative control methods are included in the programme in order to maintain the effectiveness of currently available chemicals.

6.5.1   Non-chemical methods of control

- *Crop rotation*
  A good crop rotation, avoiding growing too many years of the same crop or related species, can help to avoid an accumulation of disease inoculum, particularly those that are soil borne. In many cases the organism cannot exist except when living on the host so if the host plant is not present in the field then the disease will decline. Some diseases take years to die as they have resting spores in the soil waiting for the next susceptible crop to be planted again, e.g. club root that affects plants in the brassica family.
- *Soil fertility*
  A crop under stress from low nutrient status can be more prone to disease attack. Too lush a growth can also encourage disease. Early application of nitrogen in winter cereals can cause increased infection from foliar diseases. The acidity or alkalinity of the soil can affect both the availability of nutrients as well as some diseases. Maintaining a high soil pH of 7 is very effective at reducing incidence of club root in brassicas. In potatoes, common scab is usually worse if the pH is high so liming a soil should always be done after growing this crop.
- *Crop hygiene*
  Diseases should be discouraged by avoiding sources of infection on the farm, e.g. blight from old potato clamps, and virus yellows where sugar beet was stored prior to being sent to the factory. Good stubble cleaning will minimise certain cereal diseases being carried over from one cereal crop to the next year's cereal crop. Some parasites use weeds as alternative hosts. By controlling the weeds, the parasite can be reduced, e.g. cruciferous weeds such as charlock are hosts to the organism responsible for club root.
- *Seedbed conditions*
  Puffy seedbeds can increase the risk of take-all in cereals. Over-consolidation or compaction can lead to poor growth and other diseases.
- *Clean seed*
  Seed Certification Schemes ensure that seed lots meet strict standards for presence of seed-borne diseases. These schemes have been very successful at reducing the incidence of some diseases such as bunt or stinking smut in

cereals. Home-saved seed should be tested to avoid poor crop establishment and growth due to seed-borne disease.

With potatoes it is essential to obtain clean 'seed', free from virus. In some districts where the aphid is very prevalent, potato seed may have to be bought each year. Under the Seed Certification Scheme it is mandatory to have the field inspected and the seed virus tested.

- *Resistant varieties*

In plant breeding, although the breeding of resistant varieties is better understood, it is not by any means simple. For some years plant breeders concentrated on what is called single or major gene resistance. However, with few exceptions, this resistance is overcome by the development of new races of the fungus to which the gene is not resistant.

Some breeding programmes are now concentrating on multigene or 'field resistance', which means that a variety has the characteristics to tolerate infection from a wide range of races with little lowering of yield. Emphasis is now on tolerance rather than resistance.

Variety resistance is the main method of control for some fungal-transmitted viruses such as barley yellow mosaic virus.

- *Variety diversification*

The risk in any year of a serious disease infection can be reduced on farms growing cereals if a range of varieties are grown that have different disease resistance ratings and resistance to different races of the disease. NIAB produces variety diversification tables for yellow rust in wheat and mildew in barley.

- *Variety mixtures (blends)*

This is a natural extension of diversification in that varieties from different diversification groups are grown together in the same field. In this way a disease carrying spore from a susceptible variety, but within a blend of varieties (two or three varieties) making up the crop, has less chance of successfully infecting a neighbouring plant than in a pure crop. Yields of blended crops can be more reliable than pure crops, and this may be achieved with the use of fewer fungicides. The problem comes when selling a mixed crop, and in practice usually only grass seed is sold as mixtures.

- *Time of sowing*

Time of drilling can affect the amount of disease infection in many crops, partly due to the amount of inoculum present and environmental conditions that favour the disease. Early-drilled winter cereals are more likely to be infected by eyespot, take-all and some foliar diseases and early-drilled spring barley is more susceptible to *Rhynchosporium*, but late drilled barley is more susceptible to mildew.

- *Crop density*

Many rain-splash diseases, and those that require a high humidity, are more common where there is a high plant population compared with a more open crop with a low population.

- *Control of insects/virus vectors*

Some insects are very important vectors of viruses, e.g. control of the green fly (*Myzus persicae*) in sugar beet using insecticide-treated seed is very effective at controlling beet yellows virus. Other virus vectors such as soil fungi are very difficult to control so that other methods such as use of resistant varieties is the main option for arable farms. (Soil sterilisation will kill both nematode and fungal vectors but is only used on high-value horticultural crops.)

- *Legislation*
  A number of countries have introduced legislation to ensure that some diseases do not spread and become a serious threat. In the UK potato wart disease, brown rot and ring rot in potatoes are three examples of notifiable diseases which are subject to quarantine measures. Potato wart disease has been controlled by these statutory measures and some contaminated fields where potatoes were prohibited from being grown are now being found to be free from the disease and are being de-scheduled.
- *Irrigation*
  Irrigation can often encourage disease development by altering the crop environment so that it is more conducive for disease spread and survival; one exception is in potatoes where irrigation applied as the tubers are initiated can be very effective at controlling common scab.
- *Biological control*
  Biological control is the use of other organisms to control the disease organisms. Very few treatments are currently available commercially as often in the open field results can be very variable. One successful treatment is the fungus *Coniothyrium minitans* for control of soil-borne sclerotia of *Sclerotinia sclerotiorum*.

6.5.2   Chemical control – fungicides

Over the last 50 years there has been a large increase in both the number of fungicides available and their subsequent use on crops. They are used in the following ways:

- The dressing of seed with a fungicide; this is carried out to prevent certain soil-borne and seed-borne diseases. Various fungicides can be used, depending upon the disease to be controlled and the crop. In many cases an insecticide is also added to help prevent attacks by soil-borne pests.
- Foliar application to the plant. Different treatment programmes involving the use of fungicides are now considered as an essential part of many crop production programmes.

Fungicides are used when it is considered that a specific disease has developed to a point (the economic threshold) which will actually cause a loss of yield that will pay for the cost of treatment and application. There are now many established thresholds for application of fungicides. Often the rate of the fungicide applied will be adjusted depending on the disease risk and variety resistance. This is called an appropriate dose rate. Some computerised models have been or are being developed to aid decision making.

With some diseases such as potato blight, treatment needs to be applied before symptoms are seen and is based on blight warnings. These warnings rely on weather conditions and forecasts and the likelihood of disease developing.

Protectant fungicides
The first chemicals developed for control of fungal diseases were the inorganic compounds such as sulphur and copper compounds. These chemicals do not move in the crop plant (are non-systemic), they simply protect the crop plant from disease infection. They are called protectant fungicides. Good crop coverage is essential for this type of product. The chemicals affect a number of biochemical processes in the fungi so are called multi-site fungicides. Some of the first fungicides produced after these inorganic compounds, such as the dithiocarbamates, have very similar characteristics, e.g. they are protectant, multi-site fungicides. Although these chemicals are not as effective as some of the newer systemic compounds they still have some uses today, particularly in programmes where there is a high risk of disease resistance, e.g. control of potato blight.

Systemic fungicides
Systemic fungicides have been developed since the 1960s and are now the most commonly used fungicides. On entry into the crop plant, they can move to a certain extent within the crop, usually in the xylem vessels, to the site of infection. These fungicides tend to affect a single biochemical pathway within the pathogen and are called site-specific.

Many systemic fungicides can be applied after the initial infection period, before symptoms appear (the latent period); these treatments are called curative. When disease symptoms are visible then an eradicant fungicide is required; these chemicals have the ability to eradicate a disease that is already present and then protect the plant for a certain time after application. Persistency, curative and eradicative activity varies between chemicals.

Fungicides are grouped together according to their mode of action and chemical structure. Some frequently used fungicides are included in Table 6.1 which also includes commonly used abbreviations. It is important to know the family group of the fungicide when considering fungicide programmes in order to reduce risk of fungicide resistance. (See the individual crop chapters for further details on

Table 6.1 Summary of some commonly used fungicide groups

Family group (based on mode of action (MOA))	Abbreviated family group name	Chemical family	Example active ingredient (ai)
Demethylation inhibitors	DMI	Triazole	Epoxiconazole
Quinone outside inhibitors	QoI	Strobilurin (strob)	Pyraclostrobin
Succinate-dehydrogenase inhibitors	SDHI	Carboxamide	Isopyrazam
Methyl benzimidazole carbamates	MBC	Benzimidazole	Carbendazim

disease control programmes.) Currently the European Commission is reviewing the use of the triazole group of fungicides. If this group of fungicides is banned in the future it could lead to large losses in crop yields as there are few suitable alternative fungicides available in some crops.

6.6   Fungicide resistance

Disease resistance to fungicides is now widespread. It is a problem with the systemic products that act on one site only of the fungus. When a fungicide controls a fungal disease effectively, the fungus is 'sensitive' to the chemical. However, other strains of the fungus can and do occur over a period of time, and some of these may be resistant ('insensitive' or 'tolerant') to the fungicide which means that the disease is then not controlled adequately. In some cases once there is resistance it is total and the fungicide is not effective. An example of this type of resistance (single step) is seen in the control of eyespot with the MBC fungicides. In other cases there is a shift in the sensitivity of the fungus population to the fungicide but there is still some control. This has been found with control of mildew in cereals using many of the triazoles fungicides, it is called multi-step resistance.

Resistance builds up through the survival and spread of the resistant strains and it is speeded up by repeated application of the same fungicide treatment or fungicides with the same mode of action. There is an increased risk of this happening with fungicides which are site-specific in the fungus compared with multi-site fungicides. Likelihood of resistance developing is also affected by the biology of the target disease.

There are several ways of avoiding a build-up of resistance by a fungus or reducing the risk. It is very important that farmers follow these guidelines in order that currently available chemicals remain effective as few new products are coming on the market.

- Combine application of fungicides with non-chemical methods to reduce disease risk.
- Avoid growing large areas of very susceptible varieties in areas where disease incidence is usually high.
- Where possible, use fungicides with different modes of action (i.e. from different family groups) especially when more than one fungicide application is to be used on the same crop.
- Use approved tank mixtures of fungicides with different modes of action, rather than always relying on single fungicides.
- Apply fungicides only when necessary; use disease forecasts and thresholds to avoid unnecessary treatment. Walk and monitor crops regularly.

Unsatisfactory disease control following the use of fungicides is not always due to fungicide resistance; there are several other reasons, the main ones being wrong timing, wrong chemical, use of too low a dose rate and poor application.

Table 6.2 outlines some of the main diseases affecting farm crops and both chemical and non-chemical methods of control.

Table 6.2 Major plant diseases and their control

Crop attacked	Disease	Causal agent	Symptoms of attack	Life-cycle	Methods of control
Cereals	(1) **Bunt, covered or stinking smut** of wheat. **Leaf spot** of oats (2) **Covered smut** of barley. (3) **Covered and loose smut** of oats	Fungus	Brown or black spore bodies with distinct smell replace the grain contents.	Infected grain is planted; seed and fungus germinate together and thus young shoots become infected. The spores are released when the skin breaks, and so combining contaminates healthy grain. With use of seed dressings, diseases are now rarely seen.	(1) Seed dressing, virtually all the currently available fungicide seed dressings control these diseases
	(1) **Leaf stripe** of barley (2) **Leaf stripe** of oats	Fungus	The first leaves have narrow brown streaks. Subsequently, brown spots appear on the upper leaves.	Infected grain is planted; seed and fungi germinate together and thus young shoots are infected. From the secondary infection, spores are carried to developing grain.	(1) Seed dressing, virtually all the currently available fungicide seed dressings control these diseases
	(1) **Loose smut** of wheat (2) **Loose smut** of barley	Fungus	Infected ears a mass of black spores. They do not remain enclosed within the grain as with the covered smuts.	Similar to the covered smuts, but the fungus develops within the grain. The spores are dispersed by the wind to affect healthy ears.	(1) Resistant varieties (2) Sow certified seed (3) The seed can be dressed with a broad-spectrum fungicide
	Yellow rust	Fungus	Yellow-coloured pustules in parallel lines (stripes) on the leaves, spreading in some cases to the stems and ears. In a severe attack the foliage withers and shrivelled grain results.	The fungus mainly attacks wheat though there are distinct forms that attack barley and oats. Initial primary foci may be found in early spring. From the pustules, spores are carried by the wind to infect healthy plants. During winter, spores are normally dormant on autumn-sown crops. Cool, humid conditions favour disease. Temperatures above 25 °C inhibit growth.	(1) Resistant varieties, although new races appear against which these varieties can soon have reduced resistance (2) Fungicides, e.g. a triazole, strobilurin or SDHI (3) Variety diversification (4) Control volunteer plants between harvest and emergence of next crop to reduce carry-over of inoculum

(Continued)

Table 6.2 Continued

Crop attacked	Disease	Causal agent	Symptoms of attack	Life-cycle	Methods of control
	Powdery mildew	Fungus	On winter cereals, grey-white and brown fluffy mycelium on lower leaves in February. Infection spreads to other leaves and plants. Disease common between May and August. Early infected leaves go yellow. Towards the end of season black spore cases formed among brown fungi.	From self-sown cereals in stubble, winter and spring cereals can be infected. Specific races attack wheat, barley, oats and rye. Warm, humid (but not wet) weather favours disease. Dense crops and those where high rates of N have been used tend to encourage disease development.	(1) Clean-up old stubble (2) Resistant varieties (3) Fungicides available, although field resistance to some fungicides is now common. Better control with some of available fungicides when applied early in disease development (4) Variety diversification
	Rhynchosporium — **leaf blotch** of barley	Fungus	Initially symptoms look like a water soaked area on the leaf. When fully formed lentil shaped blotches (light blue/green/grey with dark brown margins) are seen on the leaves. As disease progresses, blotches coalesce.	A disease of barley, rye and triticale. Fungus over-winters on self-sown barley plants and on crop debris. From here spores are carried to planted barley crops. Disease is spread by rain splash. Early sown crops most at risk.	(1) Clean stubble of all self-sown barley plants (2) Many fungicides are available including some triazoles, strobilurins and SDHIs (3) Select resistant varieties in high risk areas

Disease	Cause	Symptoms	Survival and spread	Control
Net blotch of barley	Fungus	Short brown necrotic stripes on older leaves; on younger leaves in addition to the striping, irregular-shaped dark brown blotches 'spots' or 'netting' symptoms. Can spread to ears.	Can be seed-borne but mainly spread from previously infected crop or volunteers. Disease encouraged by cool, wet weather.	(1) Seed dressing (2) Crop hygiene to clear stubble of previously infected crop (3) Fungicides, e.g. SDHI mixtures (4) Resistant varieties
Halo spot of barley	Fungus	Small pale oval lesions with a dark brown border, scattered across the leaf. Black fruiting bodies (pycnidia) found in lesions.	Fungus survives in seed, plant debris and on volunteer cereals. Cool moist conditions favour the disease Only found in south west England.	(1) Many of currently available fungicide seed treatments and foliar sprays can help suppress the disease
Brown rust of barley and wheat	Fungus	Numerous, very small individual orange-brown pustules on the leaves. Normally does not develop significantly until the summer. A severe attack causes shrivelled grain.	Different species in wheat and barley. The resting spores over-winter on volunteers. From here the pustules are airborne to infect healthy plants. Encouraged by warm, humid weather; spores can be spread by wind.	(1) Crop hygiene to clear stubble of volunteer plants (2) Resistant varieties (3) Many systemic fungicides, e.g. a triazole, strobilurin or SDHI mixture
Crown rust of oats	Fungus	Orange-coloured pustules spread mainly on leaf blade. Later in season black pustules are produced. Severe attack prior to, and including, milk-ripe stage, causes shrivelled grain.	Only affects oats, a distinct race affects other grasses. Spores are air borne from over-wintering volunteer plants, and winter oats, to the spring crop. Disease encouraged by warm, humid conditions.	(1) Crop hygiene to clear stubble of volunteer plants (2) Keep winter and spring oat crops as far apart as possible (3) Some varieties show reasonable resistance (4) Some of the triazoles, strobilurins and SDHI mixtures give good control

(Continued)

Table 6.2 Continued

Crop attacked	Disease	Causal agent	Symptoms of attack	Life-cycle	Methods of control
	Ramularia leaf spot of barley	Fungus	Brown rectangular spotting on upper side of leaf that can be confused with net blotch. Symptoms also seen on awns and stems.	A seed borne disease. Symptoms often appearing when crop stressed. Spores spread following wet weather. A disease mainly found in the north of the UK.	(1) Try to ensure crop not stressed (2) In high risk areas grow more resistant varieties (3) Apply a fungicide, e.g. triazole/SDHI mixture at booting stage
	Septoria tritici – leaf blotch of wheat	Fungus	Mainly a foliar disease. Oval, bleached/grey discoloured blotches of varying sizes and shapes (on which appear minute black dots or fruiting bodies (pycnidia)) seen from late autumn to early summer. Infection of the ear is rare.	Leaf spores are liberated in wet weather, and they over-winter on volunteer crops and then transfer to winter cereals. Symptoms can develop 3–4 weeks after the infection period. Optimum temperatures for the disease to develop are between 15 and 20 °C. Early drilled crops most at risk.	(1) Crop hygiene to clear stubble of volunteer plants (2) Clean seed (3) Tolerant varieties (4) Rotation and time of drilling (5) Many systemic fungicides, e.g. triazole SDHI mixtures
	Septoria nodorum – leaf and glume blotch of wheat	Fungus	Disease (other than seedling blight) does not develop until midsummer. Leaf lesions start as oval yellow blotches. Fruiting bodies are difficult to see without magnification. Glume blotch becomes prominent in July and August. Irregular, chocolate spots or blotches on glumes, beginning at the tips; later ears become blackened with secondary infection.	Glume blotch has a similar life cycle as Septoria tritici but the fungus can also be carried on the seed. The disease develops rapidly in warm weather if there is high humidity. Symptoms develop 10–14 days after an infection period.	Control as for septoria tritici

Tan spot of wheat	Fungus	Causes brown lesions very similar to septoria nodorum. Spots tend to be oval with a yellow halo and a dark centre.	Tan spot is a trash borne disease. High temperatures and wet weather encourage development and spread. Mainly a disease of wheat though can affect barley and rye.	(1) Plough stubble (2) Control volunteers (3) Fungicide treatment, some triazoles are effective
Barley yellow dwarf virus (BYDV)	Virus	Stunted plants in patches or scattered as single plants. Poor root development. Affected leaves in oats are red, yellow in barley and bronze in wheat. Late heading and reduced yield. Yield losses greatest from autumn infection.	Cereal aphid vectors carry the virus, e.g. bird cherry aphid and grain aphid. Volunteers and grasses act as source of infection. Early drilled cereals after grass are most susceptible. Grain aphids less affected by frosts than bird cherry aphids.	(1) Spray off previous crop/green stubble and leave at least 5 weeks between primary cultivations and sowing this reduces aphid carry-over (2) In high-risk areas use insecticide treated seed (3) Apply an insecticide end October/early November. High-risk areas may require earlier treatment in warm autumns (4) Do not drill too early in high risk areas (5) Oats most susceptible, wheat the least
Soil-borne mosaic viruses (barley yellow mosaic virus (BaYMV), barley mild mosaic virus (BaMMV), oat mosaic virus (OMV), oat golden stripe virus (OGSV) and soil-borne wheat mosaic virus (SBWMV))	Virus	Appears in patches in field. Pale green streaks later turning brown, particularly at leaf tip; leaves tend to roll inwards, remain erect to give plant a spiky appearance. Symptoms can disappear during warmer weather but the plants are still stunted and late to mature.	Soil-borne fungus. *Polymyxa graminis* is the vector. Disease encouraged by hard winter. Symptoms appear from January. Found only in autumn sown cereals. May survive in soil for many years.	(1) No chemical control (2) On affected fields grow resistant varieties (3) Reduce the spread of contaminated soil around the farm (4) Widen the interval between barley crops to reduce the risk of BaYMV and BaMMV

(Continued)

Table 6.2 Continued

Crop attacked	Disease	Causal agent	Symptoms of attack	Life-cycle	Methods of control
	Eyespot	Fungus	Often starts as a brown lesion at base of stem later developing into eye-like lesions on stem about 75 mm above ground. Grey mycelium inside stem; can cause crop to lodge. White-heads present at harvest in severe cases. Two strains of eyespot present in the UK.	The fungus can remain in the soil, on old stubble and some species of grasses for several years. It usually attacks susceptible crops in the young stages. Cool moist conditions favour this disease. Disease spread by rain splash.	(1) Use risk assessment key based on region, soil type, previous crop, cultivations and sowing date (2) Avoid growing continuous cereals (3) Sow varieties with good resistance in high risk fields (4) Monitor crop for eyespot in spring (5) Apply systemic fungicides, some triazole/SDHI and triazole/strobilurin mixtures are very effective
	Sharp eyespot	Fungus	Lesions more sharply defined than true eyespot. Brown/purple border followed by cream-coloured area and brownish centre. Lesions more numerous and occur further up stem than true eyespot. Pink/brown mycelium may be found inside the stem. Can cause lodging, white-heads and shrivelled grain.	The fungus is soil-borne but is also found on plant debris. Tends to be more severe on light soils and on early drilled winter crops. Mainly a problem in winter wheat and barley; though oats and rye can be affected.	(1) Dispose of crop residues (2) Delay drilling winter cereals (3) Some of the eyespot fungicides will give some control

Disease	Cause	Symptoms	Notes	Control
Fusarium – brown foot rot and ear blight	Fungus	Emerging seedlings can be killed. Undefined brown discoloration at base of tillers and lower leaf sheaths, dark brown nodes. Interior of stem and or grain shows pink fungal growth with some species. Premature ripening and 'whiteheads' or 'blind' ears. Some species can cause the production of mycotoxins in the grain.	There are a number of Fusarium spp. but infection is either seed-borne or from previously infected stubble and crop remains. Wet weather during flowering increases risk of Fusarium development. Growing maize in the rotation also increases the risk.	(1) Stubble hygiene (2) Seed treatment using e.g. fludioxonil or carboxin (3) Foliar fungicides e.g. some of the triazoles and MBCs (4) Plough maize stubbles
Take-all	Fungus	Black discoloration at base of stem and roots. Infected plants ripen prematurely; in severe cases the ears are bleached and contain little or no grain.	Wheat is most affected though barley, rye and triticale can be infected. Oats are immune. The fungus survives in the soil in root and stubble residues and on rhizomatous grasses. Third wheat crops in a rotation are usually most affected but on light acid soils in wet areas the second wheat crop can also be seriously infected. After four years of continuous wheat growing, infection appears to lessen, take-all decline.	(1) Rotation (2) Extra nitrogen helps the growth of new roots. (3) Ensure good drainage (4) Direct drilling appears to lessen the intensity of the disease as does a consolidated seedbed (5) Drill crops at risk in October rather than September (6) Some fungicide seed dressings can help reduce take-all (7) Controls grass weeds

(Continued)

Table 6.2 Continued

Crop attacked	Disease	Causal agent	Symptoms of attack	Life-cycle	Methods of control
	Pink snow mould (mainly of wheat and rye)	Fungus	Patchy crop in autumn. Stunted seedlings occasionally with white/pink mycelium at base. After snow has thawed withered plants in patches temporarily covered by white-pink mould. Thereafter, infected plants stunted, weak root system and shrivelled grain.	A seed-borne fungus, but contamination of crop is also possible from the old stubble and other plant debris. Fungus is particularly favoured by low temperatures. Mainly a disease of winter barley.	(1) Good stubble hygiene (2) Clean seed (3) Seed treatment
	Snow rot (Typhula rot) mainly of winter barley	Fungus	Thin, poorly tillered crop; plants yellowing and withered in patches. Old leaves often covered by white and pink mycelium and brown resting spores. Young leaves standing erect but eventually yellowed. Weak root system.	Soil-borne fungus which can remain dormant as sclerotia for years; when active, infects emerging cereal plants, developing rapidly in dark and humid conditions, i.e. under snow. Continuous winter barley increases the risk of this disease.	(1) Sow early in the autumn (2) Some varietal resistance (3) Seed treatment (4) Foliar spray with azole fungicide in the autumn
	Ergot	Fungus	Hard black curved bodies up to 20mm long replacing the grain, and protruding from the affected spikelet. Although disease has little effect on yield, ergot is poisonous to mammals (but it does contain medicinal properties).	Ergots fall to ground and remain until next summer when they germinate and produce short stems with globular heads containing the spores which are then air-borne to the cereal flowers, and certain grasses, depending on the species. Open-flowering cereals most affected, e.g. rye. Infection in oats is rare.	(1) Crop rotation (2) Control of grass weeds (especially black-grass) (2) Sow certified seed

	Disease	Cause	Symptoms	Biology	Control
	Manganese deficiency of cereals	Manganese deficiency	Yellowing on leaf veins followed by development of brown lesions. With oats the spots enlarge and can extend across leaf. Thus the leaf can bend right over in the middle. Older leaves wither and die. Can lead to shrivelled grains.	Manganese deficiency associated with high pH soils.	(1) Apply manganese sulphate, treatment may need repeating (2) Analyse plant material to confirm deficiency
Oilseed rape	**Phoma – stem canker** and **leaf spot**	Fungus	Beige coloured circular spots with distinct brown margin (0.5–1.0 cm diam.) on leaves; spores spread to produce brownish/black canker at base. Stem splits and rots causing lodging; rapid stem elongation and premature ripening.	Air borne spores produced from the stubble carry infection to young crops in the vicinity. Fungus spreads from the leaves to stem where cankers develop. The fungus can affect all parts of the plant including the seed. Leaf symptoms seen from Oct to April.	(1) Destroy stubble debris soon after harvest (2) Rotation – avoid close cropping with rape (3) Grow resistant varieties (4) Use a fungicide seed treatment (5) Treat with a fungicide when 10 to 20% of plants have leaf spot in the autumn and or spring
	Alternaria – dark leaf and **pod spot**	Fungus	Circular small brown-to-black leaf spots sometimes coalescing on leaves and later on pods. Premature ripening and loss of seed.	Seed-borne disease although spores can be carried through the air from other infected brassica crops. Favoured by warm, humid conditions.	(1) No resistant varieties (2) Use a seed treatment (3) Foliar treatment for other diseases will aid control of Alternaria – otherwise treat mid to late flowering if necessary

(Continued)

Table 6.2 Continued

Crop attacked	Disease	Causal agent	Symptoms of attack	Life-cycle	Methods of control
	Light leaf spot	Fungus	From November onwards, light green/bleached lesions surrounded by small white spore droplets. Plant population and seed yield can be severely affected.	A trash-borne fungus. Disease is spread from plant to plant by rain splash during wet weather. Flower buds can become infected and killed at early extension stage.	(1) Dispose of infected crop residues (2) Spray with a fungicide at first sign of the disease in the autumn and in the spring if 15% of plants are affected at stem extension (3) Some varieties less susceptible than others (4) Apply a suitable systemic insecticide to reduce spread of virus
	Stem rot – sclerotinia	Fungus	Bleached stem lesions which contain black resting bodies (sclerotia).	Sclerotia left in soil after harvest can survive for 8 years. They germinate in spring and produce spores which infect susceptible crops. Encouraged by warm, wet weather.	(1) Wide rotation of susceptible crops (2) Fungicides such as some triazoles and strobilurins – at early flowering
	Club root	Fungus-like – protozoa	See club root in brassicas below.		
	Verticillium wilt	Fungus	Initially yellow/brown stripes on stem then causes pods to turn pale green and to ripen prematurely. Mainly found in eastern and southern counties of England.	Soil-borne sclerotia germinate and fungus invades susceptible host roots and spreads to xylem vessels. Microsclerotia are then produced on outside of vascular system.	(1) Widen rotation (2) Plant clean seed (3) Sow resistant varieties

	Turnip yellows virus TuYV	Virus	Leaf margins turn red. Reduces crop growth and subsequently reductions in pod numbers and oil content. Wide range of plant species affected.	Virus spread by peach potato aphid. Mild winters favour survival of aphid and disease spread. TuYV is a persistent virus.	(1) Insecticide seed treatment (2) Foliar insecticide treatment if large number of aphids found in the autumn (3) Resistant varieties
Linseed	**Alternaria**	Fungus	Damage affects seedlings as they emerge. Brick red lesions are found on stems and roots. Can also affect mature plant.	A seed-borne disease. Can spread up mature plant during periods of wet weather.	(1) Most effective method of control is a fungicide seed treatment (2) Some evidence of varietal resistance (3) Foliar application of a number of fungicides have been effective against late infection
	Botrytis – grey mould	Fungus	Attacks leaves, stems and seed capsules. Reddish browning on stem base. Plants then become covered in grey mould.	A seed-borne disease encouraged by warm, moist conditions.	(1) Sow certified seed (2) Seed dressings (3) Low nitrogen rates and plant populations tend to reduce problem
	Pasmo	Fungus	Grey/black spots with black pycnidia found on leaves and stems.	A disease of winter linseed. Can be seed borne or spread from plant debris.	(1) Sow clean seed (2) Some triazole fungicides are effective when applied at mid-flowering

(Continued)

Table 6.2 Continued

Crop attacked	Disease	Causal agent	Symptoms of attack	Life-cycle	Methods of control
Peas	**Downy mildew**	Fungus-like – oomycete	Infection on young seedlings can cause plant death. A secondary infection shows as isolated yellow/green spots on the upper surfaces of the leaves with greyish-white mycelium on underside. Can considerably reduce yield with fewer or no seeds per pod.	The disease is soil and seed-borne from where spores infect the growing point of seedlings. A secondary infection can follow, with spores spreading by air currents and rain splash to the developing foliage of other plants. Cool moist conditions favour this disease.	(1) Widen rotation (2) Seed treatment if growing susceptible varieties (3) Some varieties are more tolerant (4) Foliar fungicides give limited control
	Ascochyta – leaf and pod spot	Fungus	Brown sunken spotting on stems, leaves and pods. The lesion is usually surrounded by a dark brown margin. Can cause seedling loss. Peas can become stained.	Disease is mainly seed-borne. Infected seed germinates and lesions develop on first leaves from where spores spread to rest of foliage, including the pods, and to other plants by air currents and rain splash.	(1) Sow healthy seed (2) Use a fungicide seed dressing
	Botrytis – grey mould	Fungus	Grey mould develops where petals have fallen on to pods and leaves.	Fungus grows on many host plants and crop debris. Disease spread by damp and humid conditions during flowering.	(1) Treat with a suitable fungicide at early flowering (2) Avoid over fertilising the crop
	Mycosphaerella – foot rot and leaf spot	Fungus	Small brown spots on leaves and pods. Can cause whole plants to die.	Spread on seed and from sclerotia in the soil or on crop residues. Wet weather favours disease.	(1) Widen rotation (2) Use healthy seed (3) Apply a fungicide at mid flowering

Bacterial blight	Bacteria	Dark brown lesions on leaves, stems and pods.	Spread on seed. Wet windy weather favours disease.	(1) Only sow healthy seed
Pea wilt – Fusarium wilt	Fungus	In late May/June, lower leaves tend to turn grey before rolling downwards. All leaves eventually affected. Death of plant either before podding or before pods have swollen. White mycelium appears on stem after death.	The pathogen is soil-borne. It invades the plant and only returns to the soil from the infected dead plant material.	(1) Rotation (2) Use resistant varieties *Note*: There are other Fusarium species, notably root rot
Pre-emergence damping-off	Fungus – complex of different species	Poor germination and seedling establishment. The seed rots, or if not the stem is soft and dark brown in colour.	Soil-borne pathogens invade seeds and/or plant stems at or just below soil level. Worst when peas sown in cold, wet soil.	(1) Seed protected by seed dressing
Marsh spot	Manganese deficiency	Yellowing of leaves between veins which remain green; with severe deficiency growth is restricted and yield reduced. Brown discoloration (marsh spot) in centre of pea.		(1) Manganese sulphate as soon as in full flower. May need repeating 7 days later
Field beans **Chocolate spot**	Fungus	Starts with small circular chocolate coloured discoloration's on leaves and stems; with an aggressive attack symptoms move to flowers and pods. Spots can coalesce and cause total plant death.	Fungus carried over from previous year on debris of old bean haulm and on volunteer plants. Disease favoured by warm, wet weather. Autumn-sown beans, especially early sown, are more liable to attack than those that are spring sown.	(1) Clean up stubbles containing remains of old crop of beans (2) Apply a foliar fungicide, some strobilurins and triazoles recommended. First treatment in winter beans at early flowering stage if disease present on lower leaves. Treatment should be repeated 3 weeks later

(Continued)

Table 6.2 Continued

Crop attacked	Disease	Causal agent	Symptoms of attack	Life-cycle	Methods of control
	Ascochyta – leaf spot	Fungus	Leaves affected by regular brown to black spots, some up to 2 cm in diameter; spots have slightly sunken grey centres (in which can be seen small black spots with brown margins). Pods and seed also affected, the latter covered with brownish-black lesions.	Infected seeds when sown may produce seedlings with characteristic disease symptoms on stem at soil level or on lowest leaves. In cold moist conditions the disease will move up the plant and on to other bean plants.	(1) Healthy seed – seed can be tested (2) Hygiene – kill any volunteer beans in other fields (3) Seed treatment (4) Grow resistant varieties
	Sclerotinia – stem rot	Fungus	Rotting of shoots and roots. Can cause death in severe attacks on winter beans.	Soil-borne. The strain that attacks spring beans also attacks peas, red clover and oilseed rape. The sclerotia or resting bodies can persist in the soil for 8 years.	(1) Limited fungicide recommendations (2) Use wide rotations
	Bean rust	Fungus	Red/brown rust pustules surrounded by a yellow halo appear on leaves.	Can be spread on seed, trash or volunteers. More of a problem on spring beans and on potassium deficient soils. Disease encouraged by warm temperatures and high humidity.	(1) Hygiene – kill any volunteer beans (2) Apply a foliar fungicide – e.g. some triazoles and strobilurons can be very effective
	Downy mildew	Fungus	Pale green water soaked lesions. Grey fungal growth on underside of leaf.	Seed- and soil-borne. Spring beans most affected. Disease encouraged by cool, wet weather.	(1) Rotation (2) Hygiene – destroy debris (3) Fungicides – treat if risk high

Crop	Disease	Type		Control
Sugar beet, fodder beet, mangels	Virus yellows (beet mild yellowing virus – BMYV, beet yellows virus – BYV and beet chlorosis virus – BChV)	Virus	First seen in June/early July on single plants scattered throughout the crop – a yellowing of the tips of the plant leaves. This gradually spreads over all but the youngest leaves. Infected leaves thicken and become brittle. The yield is seriously reduced by an early attack. The crop is infected by aphids which have over-wintered mainly in clamps. Several green aphids carry the virus, particularly peach potato aphid. After a mild winter and warm spring aphid migration is early and the chances of an epidemic are increased. BMYV and BYV have a wide host range including some common weeds.	(1) All beet/mangel clamps should be cleared by the end of March (2) Use an insecticide seed treatment (3) Aphid warnings issued by British Sugar to aid timing of aphicide (4) Foliar aphicide if no seed treatment applied. Care must be taken with insecticide choice as there is resistance to some of the foliar products
	Powdery mildew	Fungus	Powdery greyish-white mycelium on foliage seen in dry weather in late summer, early autumn. The spores are air-borne and move from diseased plants found at loading sites and from roots left in the field and also from weed beet to infect the new crop. Disease favoured by dry, warm weather	(1) Spray with a fungicide. Several triazoles/strobilurins mixtures recommended. Spray as soon as disease is seen but before mid-September (2) Some varieties are partially resistant (3) Use warnings given by Broom's Barn
	Rhizomania – root madness (beet necrotic yellow vein virus – BNYVV)	Virus	Wilted plants (sometimes in patches in field) showing pale yellowing of veins; development of elongated, strap-like leaves often protruding above surrounding plants. The virus is transmitted by the soil fungus – *Polymyxa betae* which is present in UK soils. Widespread in Europe and now in the UK. *Polymyxa* can survive in the soil for a very long time.	(1) Tolerant varieties are the main method of control (2) Extend the rotation to reduce the risk (3) Reduce movement of infected soil

(Continued)

Table 6.2 Continued

Crop attacked	Disease	Causal agent	Symptoms of attack	Life-cycle	Methods of control
	Rhizomania – BNYVV (continued)		Infected roots smaller than healthy roots, constricted below soil level, usually fanging with a proliferation of small lateral roots (bearding). Inside of root shows brown-streaked tissue from tip of tap root upwards.		
	Ramularia – leaf spot	Fungus	Pale brown circular spots with a pale appearance in centre of lesion. Can eventually cause death of all leaves	Mainly soil-borne disease. Wet weather encourages infection and spread.	(1) Choose more resistant varieties (2) Apply an approved fungicide, e.g. a triazole
	Cercospora – leaf spot	Fungus	Similar spots as those caused by Ramularia but smaller and darker.	A seed and soil borne disease. Initial infection can come from seed, alternate hosts and infected crop residues. High temperatures and humidity favour disease development	(1) Use resistant varieties (2) Bury any infected crop residues (3) Some strobilurin/triazole mixtures are effective
	Speckled yellows	Manganese deficiency	Small yellowish areas between leaf veins, later fuming to buff-coloured angular, sunken spots (speckled yellows) which eventually coalesce.	Symptoms often disappear as root system develops. But disease can be a problem on near-alkaline soils or those soils with high organic matter content.	(1) Foliar spray with manganese sulphate in May and June which may have to be repeated.

Heart rot	Boron deficiency	In young plants the youngest leaves turn a black/brown colour and die off. A dry rot attacks the root and spreads from the crown downwards. The growing point is killed, being replaced by a mass of small deformed leaves.	This deficiency is more apparent on dry and light soils and can be made worse by heavy liming.	(1) Deficiency diagnosed by soil analysis (2) Apply boron if required
Potatoes	**Late blight**	Fungus	Brown areas on leaves. Whitish mould on the underside of leaves. Leaves and stems become brown and die off. Affected tubers may rot in-store due to secondary infection with rotting bacteria.	Infected tubers (either planted, ground keepers, or from dumps) are the main source of blight. Fungal spores are carried by the wind or rain to infect the haulms. Recently resting spores have been found-their importance is unknown. From the haulms the spores are washed into the soil to infect the tubers. Infection can also take place at harvest. The fungus cannot live on dead haulm. Risk of blight encouraged by certain weather conditions – Smith periods – e.g. two days with min. temp above 10 °C and relative humidity at 90% for at least 11 hours/day. Disease more likely to develop once crop meets between rows. New more virulent strains have been isolated.	(1) Ensure that potato dumps do not sprout. Spray off or cover with plastic sheeting and control volunteers (2) Blight spreads rapidly warnings of blight periods are given by various organisations. Early preventative spraying, followed by repeated sprays every 7–14 days, is advisable (3) A wide range of fungicides are available, both protectants (e.g. dithiocarbamates or organotins) and systemics, e.g. phenylamide group. Due to resistance problems, care should be taken with chemical choice and mixtures (4) The haulms should be burnt off before harvest to prevent the tubers becoming infected whilst being lifted (5) There is some difference in susceptibility to foliage and tuber blight between varieties

(Continued)

Table 6.2 Continued

Crop attacked	Disease	Causal agent	Symptoms of attack	Life-cycle	Methods of control
	Potato leaf roll virus – PLRV	Virus	Lower leaves are rolled upwards and inwards; they feel brittle and crackle when handled. The other leaves are lighter green and more erect than normal. Yield is lowered.	The virus is transmitted by aphids from plant to plant. Peach potato aphid is the main vector. Infected tubers (which show no signs of the disease) are planted and thus the disease is carried forward from year to year.	(1) Resistant varieties (2) Use high grade 'seed' (3) Ensure no aphids in the chitting house (4) Apply a suitable systemic insecticide to reduce the spread of the virus (5) Reduce sources of infection – dumps etc.
	Mosaics – potato virus Y (PVY)	Virus	May range from a faint yellow mottling on leaves to a severe distortion of the leaves and distinct yellow mottling. Yield can be seriously reduced by the severe forms.	Peach potato aphid is the main vector of PVY. The virus is non-persistent on the aphid. Symptoms may not appear for 4 weeks after infection.	(1) As for leaf roll (2) Some varieties resistant (3) Using aphicides can give variable results
	Common scab	Filamentous bacteria (several species)	Skin-deep irregular-shaped scabs on tuber: these can occur singly or in masses. With a severe attack cracking and pitting takes place with secondary infection by insect larva and millipedes.	The soil-borne organism attempts to invade the growing tuber (the lenticels) which responds by development of corky tissue to restrict the parasite to the surface layers. Organism re-enters soil when infected seed is planted. Disease is particularly prevalent on light sandy, alkaline soils, in dry conditions.	(1) Avoid liming just prior to planting potatoes (2) Irrigation is main method of control. Irrigate when soil moisture deficit reaches 15 mm during the 6 weeks from the start of tuber initiation (3) Some varieties are more resistant than others

Powdery scab	Fungus like protozoa	Appearance can be similar to common scab, but the spots are rounder and formed as raised pimples under the skin which burst; sometimes cankers and tumours develop.	Spore balls can remain in the soil for many years or may be planted on infected tubers and the zoospores attack the new tubers by way of the lenticels, eyes or wounds, resulting in scab development. Usually more troublesome in wet seasons.	(1) Avoid using infected seed or contaminated FYM (2) Some varieties are very susceptible (3) Check zinc levels. Soils with a high level of soil zinc may have a lower disease risk
Dry rot	Fungus (several species)	Infected tubers are usually first noticed in January and February. The tuber shrinks and the skin wrinkles in concentric circles. Blue/pink or white pustules appear on the surface.	The soil-borne fungus enters the tuber from adhering soil. Infection can only enter through wounds and bruises caused by rough handling at harvest. Favoured by high storage temperatures.	(1) If the potatoes are handled carefully, infection is considerably reduced (2) Use of fungicides at lifting or first grading (3) Varieties vary in their susceptibility
Spraing – tobacco rattle (TRV) or potato mop-top virus (PMTV)	Virus	Foliage – very variable; TRV stem mottling; PMTV yellow blotches and bunching of leaves on short stems – like a mop. Tubers primary (after soil infection): wavy or arc-like brown, corky streaks in flesh of cut tuber. Secondary – from infected tubers: PMTV – badly formed and cracked tubers. TRV – brown spots in flesh.	TRV spread by free-living nematodes in soil, especially in light sandy soils. PMTV – spread by powdery scab fungus and can remain in the soil for years in fungal resting bodies.	(1) Plant only resistant varieties on infected soils (2) Control free living nematodes with a nematicide

(Continued)

Table 6.2 Continued

Crop attacked	Disease	Causal agent	Symptoms of attack	Life-cycle	Methods of control
	Blackleg and **tuber soft rots**	Bacteria	Plants stunted and pale green or yellow foliage; easily pulled out of the ground and stem base is black and rotted. Infected and neighbouring tubers develop a wet rot in the field or in store, especially in damp and badly ventilated (warm) conditions.	The bacteria move to tubers via the stolons and in wet soil to healthy tubers and enter via lenticels or damaged areas. Carried on seed tubers.	(1) Rogue or reject seed crops where the disease shows on foliage (2) Do not plant infected tubers (3) No resistant varieties, although some are less susceptible (4) Monitor store regularly. Ensure good store ventilation
	Gangrene	Fungus	A serious tuber rot which develops in storage, usually late; it shows as grey 'thumb-mark' depressions on the tubers and the flesh beneath rots also, pin-head black spore cases.	The fungus remains alive in the soil and on trash, and can infect tubers in the soil and from tuber to tuber when handling. Cold wet growing conditions favour this disease. More commonly found in cool stores.	(1) Do not plant diseased seed (2) Reduce mechanical damage (3) Tuber treatment with a fungicide reduces incidence in store (4) Varieties vary in their susceptibility
	Silver scurf	Fungus	A superficial disease. Causes a silvery skin finish. Can affect quality for pre-pack market.	Mainly spread from air-borne spores during storage. Symptoms develop in store especially in warm conditions.	(1) Plant disease free seed (2) Apply a fungicide (tuber treatment) (3) Cool, dry storage will reduce risk (4) Some varieties are very susceptible

Disease	Cause	Symptoms	Conditions	Control
Skin spot	Fungus	Tuber symptoms develop during late storage and appear as pimple-like, dark brown, shrunken spots with raised centres. The worst damage is the destruction of the buds in the eyes of seed tubers.	Mainly spread by infected tubers. Tuber infection occurs at lifting and is worse in cold, wet seasons.	(1) Apply a fungicide (tuber treatment) (2) Dry storage conditions (3) Variety resistance
Black dot	Fungus	Similar symptoms on tubers as silver scurf. Small black sclerotia present on the lesions. A problem in potatoes for pre-packing.	Soil- and seed-borne disease. Warm and moist conditions in store increase the amount of infection.	(1) Some varietal differences in susceptibility (2) Plant clean seed (3) Apply a recommended fungicide (4) Store potatoes in cool dry conditions
Stem canker and black scurf	Fungus	If black sclerotia (black scurf) present on seed tuber can lead to poor emergence. On young plants may cause sunken brown stem lesions below ground. Stem infection can lead to yellowing and rolling of leaves. In severe cases shoots may die. White mycelium may be found at base of stems.	Seed-and soil-borne disease. Low soil temperatures encourage disease development especially on light soils.	(1) Widen rotation (2) Resistant varieties (3) Plant clean seed (4) Plant chitted seed to encourage rapid emergence (5) Limit time between burning off and harvesting to reduce black scurf developing (6) Fungicide tuber treatments available

(Continued)

Table 6.2 Continued

Crop attacked	Disease	Causal agent	Symptoms of attack	Life-cycle	Methods of control
	Brown rot	Bacteria	Symptoms only seen in warm conditions. Initially leaves wilt leading eventually to plant death. Brown streaking on the stem. Tubers have brown vascular discoloration and oozes a bacterial slime.	Outbreaks linked to contaminated irrigation water. Woody nightshade is secondary host. Potentially a very serious disease.	(1) A notifiable disease in the EU and UK.
Brassicae (Brussels sprouts, cabbage, kale, swedes and turnips)	**Club root** or **finger and toe**	Fungus-like – protozoa	Swelling and distortion of the roots. Stunted growth. Leaves pale green in colour.	A soil-borne fungus. The fungus grows in the plant roots and causes the typical swellings. Resting spores can pass into the soil, especially if diseased roots are not removed. They can remain alive for several years, becoming active when the host crop is again grown in the field. The spores are more active in acid and wet conditions.	(1) Rotation. With a serious attack advisable not to grow the crop for at least 5 years in the same field (2) Liming and drainage (3) Resistant crops. Kale is more resistant than swedes or turnips (4) Some varieties of swedes and turnips are more resistant than others (5) Test soil if a possible risk
	Black rot	Bacteria	V shaped lesions at edge of leaves. Lesions start yellow then eventually become necrotic. Can lead to leaf loss. Symptoms vary depending on brassica species.	A seed-borne disease. Spreads during propagation and from infected weeds and crop residues. Warm humid conditions favour the disease.	(1) Plant healthy seed (2) Rotate crops with non-host crops

Disease	Cause	Symptoms	Source/Conditions	Control
Powdery mildew	Fungus	Upper surface of leaves show blue/black discoloration; sprouts turn black.	Spores over-winter on infected plants and are carried by air currents to infect the following year's crop. Particularly susceptible crops are Brussel sprouts, swedes and turnips.	(1) Varieties differ in their susceptibility to this disease (2) Apply an approved fungicide at first signs of disease (3) Widen rotation (4) Remove infected volunteers
Downy mildew (e.g. cauliflower)	Fungus	Pale green spots on leaves. Fungal growth/spores can be found on underside of leaves. Badly infected plants may die.	A soil-borne problem very important at the seedling stage especially when producing plants in modules.	(1) Ensure good ventilation of the transplants (2) Apply an approved fungicide at the first sign of disease (3) Variety resistance
Ring spot (sprouts, cabbages and cauliflower)	Fungus	Small spots on leaves initially with a well-defined edge. As the disease develops the lesions enlarge to produce several concentric brown rings.	Infected debris is the main source of infection. Encouraged by cool wet weather.	(1) Rotation (2) Destroy crop debris (3) Apply a recommended fungicide (4) Grow resistant varieties
Brown heart of swedes and turnips (Raan)	Boron deficiency	No external symptoms, but when the root is cut open a browning or mottling of the flesh is seen. Affected roots are unpalatable.		See Heart rot of sugar beet
Stem rot in kale	Boron deficiency	Brown rot in the stem pith followed by stem collapse.		See Heart rot of sugar beet

(Continued)

Table 6.2 Continued

Crop attacked	Disease	Causal agent	Symptoms of attack	Life-cycle	Methods of control
Grass	**Barley yellow dwarf virus (BYDV)**	Virus	Leaves turn yellow, red or brown, discoloration starting at tips and going down leaves. Plants generally stunted, but can produce more tillers. Disease more conspicuous in single plants than in whole sward.	Spread by several species of aphids. Ryegrass and the fescues are the most susceptible.	(1) Some ryegrass varieties more resistant than others
	Ryegrass mosaic (RMV)	Virus	Yellowish/green mottling or streaking of leaves. Severe infection can show more general browning of leaf.	Number of different strains of virus. Spread by wind-borne mite vectors which are favoured by hot, dry weather. Spring sown Italian ryegrass is most susceptible.	(1) Some ryegrass varieties are more resistant than others (2) Autumn sowing (3) Early grazing can reduce mite population
	Crown and brown rust of perennial ryegrass	Fungus	Crown rust usually seen in late summer. Pale yellow leaf flecking, followed by bright orange/yellow oval pustules. Brown rust is similar in appearance but is found in spring/early summer.	Spores are air borne and can quickly infect a clean sward especially in warm, dry weather.	(1) Frequent grazing (2) Some grass varieties are more resistant than others (3) Treat silage or seed crops with a recommended fungicide
	Rhynchosporium – leaf spot	Fungus	Irregular scald-like blotches on leaves and is most apparent in spring and early summer.	Spores are air borne and move from a diseased to a clean sward in the spring. Favoured by cool, wet weather.	(1) Choose resistant varieties in susceptible areas (2) Treat silage or seed crops with a recommended fungicide

Crop	Disease	Cause	Description	Control
	Drechslera – leaf spot	Fungus	Causes necrotic spotting or streaking of leaves which looks similar to net blotch in cereals. Can also cause yellowing of leaves above necrotic area. Grass yield and quality can be severely affected.	Can be seed-borne. Encouraged by wet weather and lush crops. Occurs throughout the year though more commonly seen in the spring and autumn. There are several different species of *Drechslera*.	(1) Ensure adequate levels of potassium in soil (2) Treat silage or seed crops with a recommended fungicide (3) Grow resistant varieties
	Powdery mildew	Fungus	Disease found throughout the country especially in Scotland. Present in early spring onwards, particularly in dense swards of short duration ryegrass. Greyish-white mycelium on leaves.	Spores are air borne, and move from an infected to a clean crop. There are many strains of this disease.	(1) Resistant varieties (2) Treat silage or seed crops with a recommended fungicide
Red and white clover, lucerne, sainfoin	Clover rot (sclerotinia)	Fungus	In autumn necrotic spots found on leaves then spreads to stems. Foliage turns olive-green and then in early spring turns black and eventually dies. The root can also die.	Soil-borne disease. Resting bodies of fungus produced on affected plants in winter and spring. They are small (size of clover seed), white at first and then turning black. Bodies remain dormant in summer but in autumn produce spores which affect other plants.	(1) Clean seed (2) Use resistant varieties where possible (3) Rotation at least an interval of 4–5 years between susceptible crops
Lucerne	Verticillium wilt	Fungus	Usually seen in fairly isolated patches in first harvest year; in next 2 years spreads to many parts of the field. Normally after first cut, lower leaves turn pale-yellow colour and then white, and eventually shrivel from base upwards. Whole plant finally dies.	Disease can be introduced by contaminated seed; spores can also be transported by air, as well as being spread by contaminated fragments of the crop moving from plant to plant and then from field to field by machine.	(1) Where suspected use tolerant/resistant varieties (2) Use clean seed (3) Harvest healthy crops first before moving onto older infected crops (4) Extend interval between lucerne crops

(Continued)

Table 6.2 Continued

Crop attacked	Disease	Causal agent	Symptoms of attack	Life-cycle	Methods of control
Maize	**Fusarium spp. – stalk rot**	Fungus	Base of plants attacked in August/September; foliage grey/green colour and wilting. Pith brown/ pink at base of stem; leads to premature senescence. Some Fusarium spp. produce important mycotoxins.	Soil- and seed-borne disease. Can build up in the soil. Grain maize more likely to be affected because disease develops most rapidly in mature crops.	(1) Differences in varietal susceptibility (2) Fungicide treatment
	Kabatiella – eyespot	Fungus	Round cream spots on leaves surrounded by a dark brown ring and a yellow 'halo'. Causes crop to die prematurely.	Overwinters on crop debris. Airborne spores can be important for spread between crops. Disease encouraged by cool wet weather and growing successive crops.	(1) Widen crop rotation (2) Plough previous crop residues in the autumn (3) A limited number of fungicide have Extensions of Authorisation for Minor Uses (EAMU) (4) Some variety differences in susceptibility
	Smut	Fungus	Large black galls on any of the above ground parts of the plant, including the cob.	A seed- and soil-borne disease. Spores from galls can re-infect other maize plants, or they can remain dormant in the soil, surviving for several years.	(1) Crop rotation – more serious when maize is cropped frequently (2) Resistant varieties

6.7   Sources of further information and advice

Further reading

Agrios G N, *Plant Pathology*, Elsevier Academic Press, 2005.

Alford D V, *Pest and Disease Management Handbook*, Blackwell Science, 2000.

Biddle A J and Cattlin N, *Pests, Diseases and Disorders of Peas and Beans*, Academic Press, 2007.

Carlile W R and Coules A, *Control of Crop Diseases*, Cambridge University Press, 2012.

FRAG, *Fungicide Resistance in Cereals*, FRAG-UK, 2012.

HGCA, BASF, *The Encyclopaedia of Cereal Diseases*, HGCA, 2008.

Kavanagh K, *Fungi, Biology and Applications*, John Wiley & Sons Ltd, 2005.

Koike S T, Gladders P and Paulus A O, *Vegetable Diseases*, Manson Publishing, 2007.

Lainsbury M A, *The UK Pesticide Guide*, BCPC and www.cabi.org, published annually.

Lucas J A, *Plant Pathology and Plant Pathogens*, Blackwell Science, 1998.

Mathews G, *Pesticides, Health, Safety and the Environment*, Blackwell Publishing, 2006.

Murray T D, Parry D W and Cattlin N D, *Diseases of Small Grain Cereal Crops*, Manson Publishing, 2008.

NIAB, *Diseases of Peas and Beans*, NIAB, 1990.

Parry D W, *Plant Pathology in Agriculture*, Cambridge University Press, 1990.

Thind T S, *Fungicide Resistance in Crop Protection, Risk and Management*, CABI, 2012.

Wale S, Platt H W and Cattlin N, *Diseases, Pests and Disorders of Potatoes*, Academic Press, 2008.

Walters D, *Disease Control in Crops – Biological and Environmentally Friendly Approaches*, Wiley-Blackwell, 2009.

Websites

www.agricentre.basf.co.uk
www.bayercropscience.co.uk
www.blightwatch.co.uk
www.cropmonitor.co.uk
www.fera.defra.gov.uk
www.hgca.com
www.inra.fr/hyp3/diseases.html
www.pesticides.gov.uk
www.potato.org.uk
www.syngenta-crop.co.uk
www.voluntaryinitiative.org.uk

7

Pests of farm crops

DOI: 10.1533/9781782423928.1.158

Abstract: Animal pests are responsible for millions of pounds of damage to agricultural crops every year, although only a relatively small number of species in the animal kingdom are responsible. This chapter describes the structure of some common pests, identification characteristics, life-cycles and the type of damage that they can cause. There is a discussion on the various methods used to control pests as well as issues with pesticide resistance. Finally, there is a summary of the main pests found in UK crops, their symptoms and methods of control.

Key words: pests of farm crops, symptoms of pest damage, pest life-cycles, pest identification, methods of pest control.

7.1   Introduction

Animal pests are responsible for millions of pounds of damage to agricultural crops every year, although only a relative small number of species in the animal kingdom are responsible. Many animals are in fact very beneficial to crops, acting as pest predators or even as parasites within the pests (parasitoids), and some are important pollinators.

Some pests are a regular problem and occur most years, although not always at population levels that cause economic damage. Other pests only occur occasionally, depending on several factors including rotation, weather conditions and crop growth stage. Several pests are very specific to individual crop species whereas others damage a large number of different crops. Many animals are successful as pests partly because they can reproduce very quickly, with many generations a year, have an effective method of spread and are adapted to changing environmental conditions. The group of animals that contains some of the most important crop pests is the insects. Nematodes are the next important group, followed by the molluscs, birds, mammals and mites.

Before discussing the various methods used to control pests, it is important to understand something of the structure of the pest, identification characteristics, life-cycles and type of damage that they can cause.

7.2   Insect pests

The main groups of pests are the insects, including the plant bugs (e.g. aphids and capsids), butterflies and moths, beetles (including weevils), as well as sawflies and true flies. Insects are invertebrates, i.e. they belong to a group of animals which do not possess an internal skeleton. A hard external covering – the exoskeleton – supports their bodies. It is composed chiefly of chitin, and is segmented so that the insect is able to move. The segments are grouped into three main parts (Fig. 7.1):

1. The head, on which is found:
   (a) the antennae or feelers carrying sense organs for touching and smelling;
   (b) the eyes; a number of simple and a pair of compound eyes are present in most species;
   (c) the mouth parts (Fig. 7.2); two main types are found in insects:
       (i) The biting type used for grazing on foliage;
       (ii) The sucking type – insects in this group suck the sap from the plant and do not eat the foliage.
2. The thorax, which carries:
   (a) the legs; there are always three pairs of jointed legs on adult insects;
   (b) the wings – two pairs; these are found on most, but not on all, species. Aphids often go through a wing-less phase and the true flies only have one pair of wings as the hind wings are modified into club-shaped structures (halteres).
3. The abdomen; this has no structures attached to it except in certain female species where the egg-laying apparatus may protrude from the end.

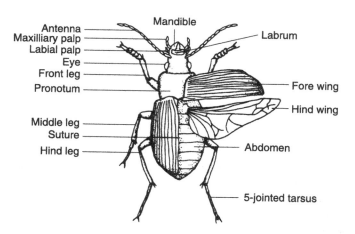

Fig. 7.1 Structure of an insect.

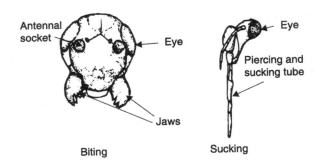

Fig. 7.2   Insect mouth parts.

Life-cycle of insects

Knowledge of the life-cycles of insects is important when identifying the pest and deciding on the best method of control. Most insects begin life as a result of an egg having been laid by the female. What emerges from the egg, according to the species, may or may not look like the adult insect.

There are two main types of life-cycle:

1. The *complete* or four-stage life-cycle (flies, beetles, butterflies, moths and sawflies) (Fig. 7.3).
   (a) The egg.
   (b) The larva or the immature growth stage is entirely different in appearance from the adult. This is the active eating and growing stage. In order for a larva to expand they must shed their skin and replace it with a new, larger one. The stage between each moult is called an instar. Most larvae will moult at least 3 or 4 times (go through 3 or 4 instars). The larva usually possesses biting mouthparts, and it is at this stage, in many insects, that they are most destructive to the crops (Fig. 7.2).
   (c) The pupa – the resting state. When mature the larva pupates and undergoes a complete change, or metamorphosis from which emerges –
   (d) The adult insect – this may feed on the crop, e.g. flea beetles, but in many cases it is the larval stage that causes the major crop damage, not the adults.

Insects vary in the length of time it takes to complete their life-cycle; for example click beetles (larvae – wireworm) may take five years, whereas crane fly (larvae – leatherjackets) take a year. The time of year when larvae are found also varies between species.

There are distinct differences in larval appearance for the different insect groups. Identification characteristics include size, colour, presence or absence of legs on the chest, presence or absence of false or prolegs on the abdomen and development and colour of the head (Fig. 7.4).

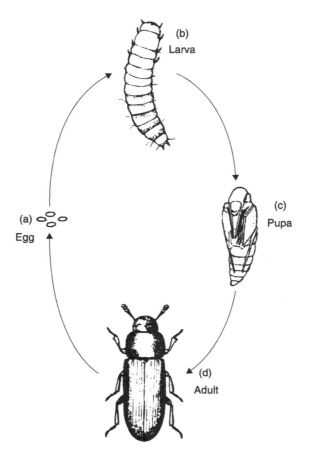

Fig. 7.3 Four-stage life cycle. (a) The egg. (b) The larva or the immature growth stage, entirely different in appearance from the adult. This is the active eating and growing stage. The larva usually possesses biting mouthparts, and it is at this stage, in many insects, that they are most destructive to the crops (Fig. 7.2). (c) The pupa – the resting state. The larva pupates and undergoes a complete change, or metamorphosis, from which emerges (d) the adult insect – this may feed on the crop, e.g. flea beetles.

2. The *incomplete* or three-stage life-cycle as found in plant bugs (aphids and capsids) (Fig. 7.5).
   (a) The egg.
   (b) The nymph – this is very similar in appearance to the adult, although it is smaller and may not possess wings. It is the active eating and growing stage. Again the nymphs will go through several instars before reaching maturity.
   (c) The adult insect – invariably this stage will also feed on and damage the crop, e.g. aphids.

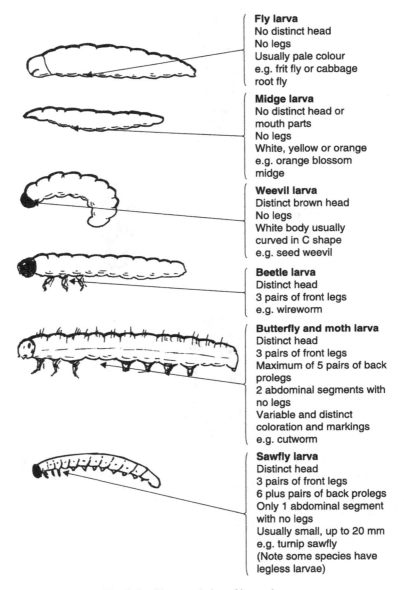

Fly larva
No distinct head
No legs
Usually pale colour
e.g. frit fly or cabbage root fly

Midge larva
No distinct head or mouth parts
No legs
White, yellow or orange
e.g. orange blossom midge

Weevil larva
Distinct brown head
No legs
White body usually curved in C shape
e.g. seed weevil

Beetle larva
Distinct head
3 pairs of front legs
e.g. wireworm

Butterfly and moth larva
Distinct head
3 pairs of front legs
Maximum of 5 pairs of back prolegs
2 abdominal segments with no legs
Variable and distinct coloration and markings
e.g. cutworm

Sawfly larva
Distinct head
3 pairs of front legs
6 plus pairs of back prolegs
Only 1 abdominal segment with no legs
Usually small, up to 20 mm
e.g. turnip sawfly
(Note some species have legless larvae)

Fig. 7.4   Characteristics of insect larvae.

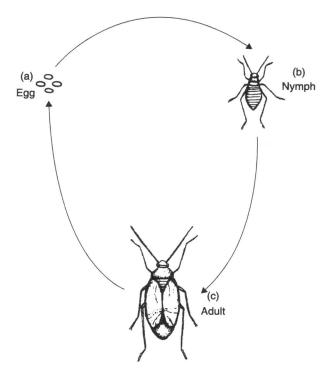

Fig. 7.5 Three-stage life cycle. (a) The egg. (b) The nymph – this is very similar in appearance to the adult, although it is smaller and may not possess wings. It is the active eating and growing stage. (c) The adult insect – invariably this stage will also feed on and damage the crop, e.g. aphids.

Some aphids at certain times of the year have an even shorter life-cycle giving birth to live young; this can lead to a very rapid increase in pest numbers.

Not all insects are harmful to crops; some are beneficial as they prey on or parasitise crop pests. Some of these, especially at the larval stage, can look like some crop pests so it is important to be able to identify accurately. Ladybirds, hoverflies, ground beetles and lacewings are particularly important beneficial insects.

7.2.1   Nematodes (eelworms)

Nematodes are microscopic and have non-segmented elongated worm-shaped bodies surrounded by a tough cuticle. They vary from 0.1 mm to 2.0 mm in length with an average size of 1.0 mm. As far as is known, most species of nematodes are free-living and beneficial, but there are a number of important species which are either parasitic within crop plants (endoparasitic) or feed on the surface tissues of crop roots (ectoparasitic). Most nematode pests only attack or cause damage to specific host species. The nematodes themselves have a limited ability to move very far, the most important means of spread is by movement in water or soil. Farm machinery and contaminated planting material can be very important in spreading the pest further afield.

The mouth parts of these plant-feeding nematodes consist essentially of a cavity – the stoma – in which is positioned the mouth spear or stylet (Fig. 7.6). It is this which pierces the cellular tissue for sucking out the cell contents of the parasitised plant.

The life-cycle is relatively simple, consisting of an egg hatching into a larval form. There are several juvenile stages (shedding a cuticle each time) before reaching the sexually mature male or female stage. In most cases, part of the life cycle takes place in the soil. With many pest species there is a stage in the life cycle (e.g. a cyst containing eggs) which may remain dormant, sometimes for several years, only becoming active again when conditions are suitable. Some species, however, have more than one generation in the year, and are only inactive during the winter months. Both the juvenile and adult stages are collectively the cause of damage to the host plant. The symptoms, supported by plant and soil analysis, allow the identification of the species involved and the formation of future control strategies. No curative control is yet available to check damage to an already infected crop.

The major nematode pests are the root-attacking cyst nematodes. Other agriculturally important nematodes are the stem nematodes. These can have a wide host range but on each host they cause varying forms of necrosis and deformity of the shoot systems. These stem nematodes are often spread in contaminated seed or plant debris.

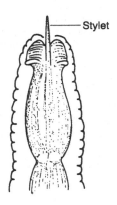

Stylet

Fig. 7.6   Mouth parts of plant-feeding nematode.

7.3   Other pests of crops

7.3.1   Molluscs

Slugs are the most important mollusc pests of field crops and most crops can be affected. There are several species of slugs, including the field, garden and keeled slugs which can feed on crops throughout the year, although activity is reduced during periods of drought or frost. Crops growing on heavy clay soils are most at risk; previous crop and soil cultivations and soil consolidation can affect activity. Each slug is able to lay up to 500 eggs which usually hatch either in early autumn or late spring, depending on weather conditions. On hatching, the young slugs resemble the adults, except in size. Initially the young feed on organic matter in the soil, later on plant material both underground and on the soil surface.

7.3.2   Birds

Generally, birds are more helpful than harmful, although this will depend on the district and type of farming carried out. To the grassland farmer, birds are not such a problem as they are to the arable farmer. Although birds will eat some cereal seed, most of them help the farmer by eating many insect pests and weed seeds; the diet of some also includes mice, young rats and other rodents. Sometimes when some birds have caused a lot of damage to an emerging cereal crop, it is damage caused while the birds are looking for other insects such as leatherjackets.

The wood pigeon does far more harm than good. Not only will it eat cereal seed and grain of lodged crops, it also causes considerable damage to young and mature crops of peas and brassicae. The only effective ways of keeping this pest down are by properly organised pigeon shoots and the use of various bird scarers.

7.3.3   Mammals

There are several mammals which can cause damage to crops. Rabbits and deer can be very serious pests; they eat many growing crops, particularly young cereals. Mice and rats can eat and damage both growing and stored crops.

The harmless mammals, as far as crops are concerned, are: hedgehogs which eat insects and slugs, foxes which kill rats and rabbits; and squirrels which eat pigeons' eggs.

7.3.4   Mites

Mites are microscopic animals. Many mites are important for breaking down organic matter but a few can be major crop pests. Mites differ from the insects in having four pairs of legs on the adult (and nymphs), unlike the three pairs found in insects, and not having wings (Fig. 7.7). Mites have mouth parts (chelicera) that are adapted for piercing or biting. Normally the life-cycle is incomplete. A 6 legged larva emerges from the egg, this passes to the 8 legged nymph stages before becoming an adult. Mites can reproduce very quickly and so by the time damage is noticed the pest population can be very high.

7.4   Types of pest damage

It is often the damage to a crop that is noticed in the field first before a pest is found and identified. The type of damage, distribution in the field, time of year and

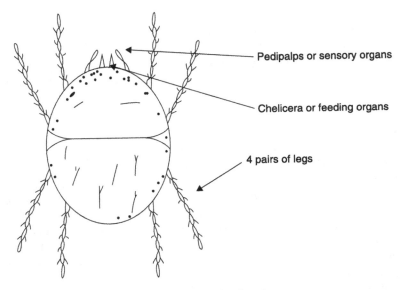

Fig. 7.7   Structure of a mite.

the crop are important aids when identifying a pest and deciding on whether control measures are required.

- *Total plant loss.* Some pests eat out the seed before it even germinates. For example, slugs can hollow out the grain of cereals, or mice can eat all the seeds in a row of peas. At other times the seed germinates but then subsequently dead plants are found eaten off at ground level or just below the soil surface. Slugs as well as leatherjackets, cutworm, swift moth larvae, vine weevil larvae and the cockchafer grubs can cause this type of damage.
- *Loss of the central growing shoot.* A few pests feed on the central growing shoot causing it to die, producing dead hearts. Frit fly in grass and cereals and wheat bulb fly produce these symptoms. The dead stem can be split open to find the larva.
- *Holes and/or leaf damage.* Leaf damage can sometimes be superficial and not worth controlling. Adult weevils cause characteristic notching around the outside of crop leaves whereas slugs cause leaf shredding. Flea beetle adults cause shot holing in the leaves of brassicas. This damage can severely reduce crop vigour and even result in total crop loss.
- *Distorted leaves.* Aphids and capsids feeding on crop leaves can cause the crop leaves to become distorted.
- *Loss of flowers and/or buds.* A few pests enter flowers and/or flower buds and their feeding damage can cause the buds to abort. Pollen beetle causes this type of damage especially in spring oilseed rape.

- *Stunted growth.* Nematodes such as potato cyst nematode and stem nematodes can cause severe crop stunting.
- *Transmission of viruses.* some animals act as vectors for some important virus diseases. Examples include free-living nematodes transmitting spraing in potatoes, aphids transmitting barley yellow dwarf virus in cereals and mites that transmit ryegrass mozaic virus.
- *Damage to harvested crop.* Several pests can damage the crop just before harvest or even after harvest. Grain aphids can affect wheat grain filling so that the crop is thin and shrivelled. Slugs and wireworm can eat into potatoes so that the quality is severely affected and the crop becomes unsaleable. Grain weevils and mites are pests that can cause significant damage to grain in store.

See Table 7.1 at end of this chapter for further details on crop pests, crop damage, life-cycles and methods of control.

7.5   Methods of pest control

7.5.1   Non-chemical control methods

With the increasing number of cases of pesticide resistance, the limited production of new chemicals and worries over the environmental impact of pesticides it is important that farmers adopt a more integrated approach to pest management. There are many methods of pest control that can help reduce the risk of pest damage and requirement to apply a pesticide.

- *Rotations*

  Having a more diverse cropping rather than mono-cropping, e.g. with cereals, can have a variable effect on some crop pests. Introducing grass into an arable rotation can increase the risk of frit fly and leatherjackets, whereas continuous cereals encourage wheat blossom midges. A short rotation with potatoes, e.g. one year in four or less, encourages potato cyst nematode (PCN).
- *Time of sowing*
  (a) A crop may sometimes be sown early enough so that it can develop sufficiently to withstand an insect attack, e.g. wheat bulb fly in the wheat crop.
  (b) A crop can be sown late to avoid the peak emergence of a pest, e.g. drilling winter cereals late to avoid aphid-transmitted barley yellow dwarf virus (BYDV) or not drilling a susceptible crop until at least six weeks after ploughing up a ley to reduce the risk of frit fly.
- *Cultivations*

  Ploughing and other cultivations can expose pests such as wireworms, leatherjackets and caterpillars, which are then eaten by birds. Well-prepared seedbeds encourage rapid germination and growth which will often enable a crop to grow away from the pest attack. Ploughing can have a negative effect on a number of beneficial organisms.

Non-ploughing techniques have been found to reduce aphids as BYDV vectors in cereals as well as reducing the risk from frit fly and yellow cereal fly; slugs are usually more of a problem.

A consolidated seedbed can help reduce the impact of some pests. Rolling is a useful control method for slugs and leatherjackets in crops growing in cloddy seedbeds.

- *Encouragement of growth*
Good quality seed should be used which will germinate quickly and evenly. It is also important that the crop is not checked to any extent, for instance, by lack of a plant food. A poor growing crop is far more vulnerable to pest attack than a quick growing crop. A top-dressing of nitrogen, just as a crop is being attacked, may sometimes help. Aphids are often more frequently found on thick, well-fertilised crops.

- *Clean farming*
It is important to control volunteer crop plants and those weeds that are alternative hosts to pests in other crops so that the pest life-cycle can be broken.

To reduce the spread of pests such as cyst nematodes, contaminated soil should be returned to the same field after grading and cleaning crops such as potatoes or sugar beet. Care should be taken when moving machinery from heavily infested land to clean land to try and reduce movement of contaminated soil.

Grain stores should be thoroughly cleaned before harvest to reduce the incidence of grain pests.

- *Resistant varieties*
Some crop varieties show resistance to certain pests, as found in potatoes. There is a wide variation in susceptibilities to slug damage and to cyst nematodes between varieties.

- *Biological control (biopesticides)*
A parasite or predator is used to control only the target species. This is a very important method of control of many pests in glasshouses and has reduced the reliance on pesticides. There area few examples in field crops:
  (a) predatory mites are used successfully to control the red spider mite in cucumber production under glass;
  (b) *Bacillus thuringiensis* for bacterial control of caterpillars. The crop is sprayed and the bacterium invades the target organism; it is very effective at controlling caterpillars including those of the cabbage white butterfly.

- *Barriers*
Fleece or net crop covers will stop egg laying of some pests such as carrot fly. On a field scale it is expensive although a very effective option. In swedes this is the only method available to growers to control cabbage root fly as there are currently no suitable recommended chemicals.

- *Legislation*
Government legislation can be important for either reducing the risk of entry of potential pests into the country or by controlling spread of pests by seed.

Legislation has been successful to date in the UK at stopping the introduction of Colorado beetle and potato flea beetle. Land growing potatoes for certified seed production must officially have no potato cyst nematodes.

- *Trap cropping*
  Trap cropping involves growing a susceptible plant species before the main crop or growing it in strips around a crop. This sacrifice plant/crop is used to attract mobile predators before they attack the crop; these pests can then be killed so that the pest life-cycle is not completed.
- *Encourage beneficial insects*
  Many insects are predators of crop pests. For example a hoverfly larva can eat 400 aphids in its life time. Some insecticide treatments are not very specific so can affect the beneficial insects as well as the pests. Beneficial insects can be encouraged on the farm by ensuring that there are suitable diverse habitats as encouraged by many of the options available in the current environmental stewardship schemes.
- *Irrigation*
  Intense rainfall or irrigation can be very effective at reducing the impact/controlling some pests such as cutworm.

7.5.2   Chemical control methods

A wide range of pesticides are available for pest control. A large number of them achieve activity by affecting different parts of the pest's nervous system. Currently worldwide over 24 different chemical modes of action groups are available, although in the UK the number is smaller. Due to the types of modes of action, most insecticides have a higher mammalian toxicity than many of the other types of pesticides such as the herbicides.

Pesticides are applied in a number of different ways and formulations. Most are applied as sprays or as seed dressings, though a few are applied in granular form or as baits. Fumigants are used in grain stores for the control of beetles and weevils.

There are two main ways in which pesticides kill pests:

1. *By contact*
   (a) The pest is killed when it comes in contact with the chemical, such as when:
   (b) it is directly hit by the spray;
   (c) it picks up the pesticide as it moves over foliage which has been treated;
   (d) it absorbs vapour;
   (e) it passes through soil which has also been treated.
2. *By ingestion*
   Stomach poisons need the pest to eat the foliage treated with the pesticide, or if the chemical is used in a bait for the bait to be eaten. A *systemic* compound is one that is applied to the foliage or to the soil around the base of the plant and then is taken up by the plant and translocated (usually in the xylem) within the plant. Systemic chemicals are most effective against sucking

pests such as aphids. Some chemicals only show limited movement in a plant such as from one side of a leaf to the other side, these are said to have *translaminar* activity.

Most pesticides kill by more than one method, which makes them very effective. But many of them are extremely toxic to animals and humans and, by law, certain precautions must be observed by the persons using them, both when handling and applying. These chemicals have a statutory requirement for a 5 m buffer zone when spraying next to water courses.

7.6   Classification of pesticides

A wide range of pesticides are available for pest control. The definition of an insecticide is one that only kills insects, in practice this term is used for many types of pesticides that kill invertebrates, not just insects, including the acaricides (mite control), nematicides (control nematodes) and molluscicides (control of slugs and snails). A large number of them achieve activity by affecting different sites of activity at the target species nerve endings.

7.6.1   Insecticides

- Organic compounds – *fatty acids, soaps*. These compounds are used in horticulture against sucking insects. Pests must be sprayed directly to achieve any control.
- Organochlorides. These insecticides were all stomach and contact poisons. Many of the products had high mammalian toxicity and were not very specific, they also had very detrimental environmental effects. There are now no approved products; they have been banned and withdrawn from the market now that safer products are available.
- The organophosphates. As a group, the organophosphates are also dangerous to use. This group of insecticides mainly affects the pests' nervous system. They are generally non-persistent and do not accumulate in the environment. They should be handled strictly in accordance with the manufacturer's instructions. Examples of some of the organophosphate insecticides in use are:
  - *Non-systemic- Pirimiphos-methyl*: for control of stored grain pests;
  - *Systemic – Dimethoate*: this is used for the control of aphids on many agricultural crops. This insecticide is not very selective and is dangerous to the environment and many beneficial insects.
- The carbamate compounds. These compounds also affect the pests' nervous system. They tend to work quickly and have a reasonable rate of breakdown. Examples include:
  - *Pirimicarb*: for control of aphids. This insecticide is very specific and has little effect on beneficial insects.

- Synthetic pyrethoids. These chemicals have a particularly high insecticidal power whilst generally being safe to humans and farm animals. They are efficient contact insecticides with a rapid knockdown action and some are thought to act as antifeedants. These properties are seen as being valuable in the chemical control of plant viruses by controlling the insect vectors, e.g. cypermethrin and λ-cyhalothrin.
- Neonicotinoids. The newest group of insecticides developed from nicotine. Mammalian toxicity is lower than for the organophosphates. The chemicals are systemic and are very effective against sucking insects. These insecticides work by affecting the pests' nervous system. In the UK the majority of treatments are applied as seed dressings. There is growing concern about the use of this group of insecticides and decline in bee populations. Research has not been conclusive. In 2013 the European Commission restricted the use of three neonicotinoides (imidicloprid, clothianidin and thiamethoxam) on crops that can be attractive to bees. This decision will be reviewed in 2015.

7.6.2   Molluscicides

- *Metaldehyde.* This is applied as a mini-pellet, ready to use bait or bait concentrate for the control of slugs and snails. It works by dehydrating the molluscs. Recently traces of metaldehyde have been found in catchments where water is abstracted for human consumption. A scheme has been introduced to promote responsible use (Get pellet wise) including not applying within 6 m of water courses and limiting the total amounts applied per year.
- *Methiocarb.* This is applied as a mini-pellet for the control of slugs and snails. It is a stomach-acting carbamate.

7.6.3   Nematicides (including soil sterilants)

Nematicides, such as *fosthiazate*, have been developed for the control of nematode pests of crops. Fosthiazate is a soil-applied organophosphate, contact-acting nematicide that controls potato cyst nematodes. Certain chemicals in this group are classified as soil sterilants/fumigants, e.g. *Dazomet.* It is important to remember that a time interval must be observed between the last application of the pesticide and the harvesting of edible crops, as well as the access of animals and poultry to treated areas. With some pesticides this interval is longer than others. This is another reason for very careful reading of the manufacturer's instructions.

For up-to-date information on pesticides available, and the regulations and advice on the use of these chemicals, reference should be made to the *UK Pesticide Guide*, published annually by CABI/BCPC or through the government website or by subscription to LIAISON run by FERA.

7.7   Resistance

Pests resistant to insecticides have developed with the repeated application of insecticides with the same mode of action, especially in species that have many generations a year and have high multiplication rates. After introduction of new chemicals it usually takes from 2 to 20 years for cases of resistance to appear. Resistant populations develop as susceptible insects are killed leaving those that are resistant or less sensitive to increase by passing on their resistance to their offspring. There are several mechanisms of resistance.

- Metabolic resistance – this is the most common mechanism. Resistant pests are able to detoxify or destroy the toxin faster than susceptible insects.
- Target-site resistance – the site of activity is modified in resistant pests so that the pesticide does not work. This is the second most common mechanism.
- Behavioral resistance – resistant insects appear able to detect the pesticide and leave the treated area.
- Penetration resistance – resistant strains appear to have developed a cuticle which restricts absoption of the chemical into their bodies.

When insects are resistant to some insecticides the resistance may be due to more than one of the above mechanisms.

7.8   Integrated pest management

With increasing awareness of the side effects of pesticides and the increasing number of cases of pesticide resistance, integrated pest management (IPM) is an important method of pest control. IPM involves using cultural methods of control combined with the use of pesticides. Pesticide use is minimised in an attempt to protect and enhance the activities of beneficial insects (natural enemies and pollinating insects) and extend the life of the pesticides that are available.

Economic thresholds have been calculated for many pests and are an important tool in IPM. An economic threshold is the population of pests which if controlled will give a yield return that will pay for the cost of pesticide and application. Where available these have been included in Table 7.1. Another important tool to help with decision making for application of a pesticide is the use of forecasts of pest populations. A number of organisations run forecasting services for some important crop pests such as cutworm, pea moth and carrot fly. These forecasts combined with crop monitoring and use of traps including pheromone traps can all help to reduce the requirement for pesticide applications.

When pesticides must be applied the most selective should be chosen and applied at the correct time and dose rate. It is important to vary the mode of action of pesticides used in order to slow down the development of resistance.

Table 7.1 Major pests and their control

Crop attacked	Pest	Description	Life-cycle	Symptoms of attack	Control	Notes
Cereals	*Adult:* Species of **Clickbeetle** *Larva:* **Wireworm**	*Adult:* Brown, 6–12 mm long *Larva:* Growing to 25 mm long, yellow orange colour.	Larvae hatch out during summer from eggs laid just below soil surface, mainly in grassland. The larvae take 4–5 years to mature, and after pupation in the soil, the adults appear in late summer and lay their eggs the following spring.	Yellowing of foliage followed by the disappearance of successive plants in a row. This is caused by wireworm moving down the row. Larvae eat into the plants just below soil surface. They are usually found in soil around the plants.	Good growing conditions to help the crop grow away from an attack. Insecticide-treated seed should be used in susceptible fields, e.g. after long-term grass. Damage is reduced when the soil is well consolidated. Control grass weeds.	Wheat and oats are more susceptible than barley. Seed treatment is recommended if soil counts find more than 750,000 larvae/ha. Recently wireworms have been found in some fields in arable rotations.
	Adult: **Cranefly** (mainly *Tipula* sp.) *Larva:* **Leatherjacket**	*Adult:* 'Daddy longlegs' *Larva:* Leaden in colour up to 40 mm long.	Eggs laid on grassland or weedy stubble in the autumn from which the larvae soon emerge. They feed on the crop the following spring, pupating in the soil during the summer.	Crop dies away in patches, root and stem below ground having been eaten. Larvae found in soil.	If possible, plough the field before September to prevent the eggs being laid. Rolling and nitrogen top-dressing helps crop to recover.	Spring cereals after grass are most affected. Chemical treatment may be worthwhile in spring cereals if 5 or more leatherjackets are found 1 m row.
	Wheat bulb fly	*Larva:* Whitish-grey, up to 12 mm long. Larva has a blunt tail end.	Eggs laid on bare soil or on soil between row crops in July/August. Eggs start hatching from early January. Larvae feed on the crop until May. Pupation then follows either in the soil or plant.	Central shoot of plant turns yellow and dies in early spring, 'deadheart' symptoms. Larva found in base of tiller.	Late sown wheat after vegetables, potatoes, sugar beet or peas are most at risk. Increase seed rates in susceptible fields. Use insecticide treatments if thresholds reached. (i) Seed dressing (ii) Spray at egg hatch (iii) Deadheart spray.	Wheat bulb fly is a problem mainly of central and eastern counties. Use a seed treatment if late drilled and > 2.5 million eggs/ha. Time of egg hatch is monitored by several companies to assess areas most at risk and to enable correct spray timing.

(Continued)

Table 7.1 Continued

Crop attacked	Pest	Description	Life-cycle	Symptoms of attack	Control	Notes
	Frit fly	*Larva*: Whitish, 4–6 mm long	Three generations in the year; the first in spring when eggs are laid on host plant and larvae feed on crop in May and June. The second generation damages the oat grain, whilst the third generation over-winters on grass, but when the latter is ploughed for autumn cereals the larvae move on to the emerging cereals.	In early summer, the central shoot of the oat plant turns yellow and dies – 'deadhearts', but the outer leaves remain green; blind spikelets and shrivelled grains are caused by second generation; autumn cereals show 'deadheart' symptoms.	Sow spring oats before mid-April, and try and get them past the 4-leaf stage as quickly as possible. Allow a 6 week interval between ploughing grass and sowing the winter cereal. In high risk situations use an insecticide seed dressing. Only use a foliar spray at the 1 to 2 leaf stage if damage threshold reached.	Spring oats and maize are particularly susceptible. Mainly a problem in rotations with grass. Chemical control likely to be cost effective if 10% of plants are damaged at the 1 to 2 leaf stage. Some companies monitor frit fly emergence from grassland as this can give an indication of risk of damage.
	Yellow cereal fly	*Larva*: Yellowish; slender, up to 7 mm long, pointed at both ends.	Eggs laid near wheat plant in October and November, hatching early in new year; larvae move to between outer leaves to feed on main tiller. Pupation in early summer, adults appear in June and a month later they migrate to hedgerows before returning to wheat field in late autumn.	Circular or short spiral band at base of tiller producing brownish scar and then death of shoots – 'deadhearts'. Larvae only inhabit one tiller and do not move to others (unlike frit fly).	Early sown wheat in the eastern counties at greatest risk. Crop loss is usually low. Establish a good population of at least 200 plants/m². Chemical treatment in November can give good control.	Winter barley rarely attacked; spring sown cereals virtually immune. There is no established threshold for treatment.

Gout fly	*Larva:* Legless, yellowish-white, 6 mm long	There are two generations a year. Eggs are laid for the first generation in May and September for the second generation. Larvae hatch and tunnel into the plant tillers. The larvae pupate within the plant.	Infected tillers are swollen and stunted. Crops affected usually compensate by producing new tillers. The first generation in spring cereals can cause the most economic damage.	Time of drilling is important, September drilled winter crops are most at risk. Insecticide seed dressings will help protect plants. Encourage tillering in the spring by application of an early nitrogen treatment.	Barley and wheat are mainly affected in England and Wales. Oats are not attacked. Crops can compensate for up to 25% of plants damage. If 50% of winter cereal plants have eggs present then consider a foliar insecticide treatment.
Cereal aphids	Various species; mainly bird cherry, grain and rose grain aphids. Green, yellowish or reddish brown in colour.	Winged females found feeding on cereal crops in May and June. Wingless generations produced which continue feeding during summer. Most species overwinter on woody hosts and some grasses and cereals, depending on aphid species.	Damage is either direct feeding or transmission of virus – BYDV. The grain aphid causes empty and/or small grain; by puncturing the grain at the milk-ripe stage, the grain contents seep out. This also reduces the weight of the grain. Rose grain aphids affect the leaves not the ear and damage is usually less important. Virus transmission is only a problem before stem extension.	The grain aphid causes most concern in the summer. Spray only when threshold levels are reached. Apply predator safe products. In high risk BYDV areas see Table 6.1 for control.	Threshold for grain aphids at flowering is when two thirds of the ears are infected. Do not treat if more than 5% of aphids are mummified. Threshold for leaf aphids is when 50% stems are infected.

(Continued)

Table 7.1 Continued

Crop attacked	Pest	Description	Life-cycle	Symptoms of attack	Control	Notes
	Orange wheat blossom midge and yellow wheat blossom midge	*Adult:* – 3 mm orange midge. *Larvae* – up to 2.5 mm bright orange (yellow wheat blossom midge – yellow adult and larvae).	Adults appear in May and June. Adult if present can be seen at dusk. Eggs are laid in wheat ears. Eggs hatch within a week and the larvae start feeding on the developing grain.	Larvae can cause yield loss due to shrivelled grain. Hagberg Falling Number can also be reduced so of particular importance in quality wheat crops. Yellow wheat blossom midge affects wheat yield more than quality as they feed on the anthers and prevent grain fertilisation.	Treat within a week of finding threshold populations of adults at ear emergence. Use an approved insecticide. Some varieties of wheat are resistant. Once crops have reached the flowering stage they are past being at risk of damage.	Orange wheat blossom midge are mainly a pest of wheat in the main arable areas of the UK and are more important than yellow wheat blossom midge. Thresholds for feed wheat is one midge found per three ears. Thresholds for milling wheat are one midge per six ears. In the field it can be useful to put pheromone traps or yellow sticky traps to assess numbers of midges.
	Cereal cyst nematode	Dark-brown lemon-shaped cysts about 1 mm long.	Live and breed in the roots. White-looking cysts (female containing large numbers of eggs) are found on roots. Later these cysts (now dark brown) become free in the soil to infect the host plant again.	Crop shows patches of stunted yellowish/green plants. Root system very bushy. Cysts visible on roots from June onwards.	Avoid growing oats too often in the same field. Grow resistant varieties when necessary.	Oats mainly affected. Populations of this pest are declining, damage is rarely seen.

Pest	Description	Life cycle	Damage	Control	Notes
Saddle gall midge	*Adult* – red up to 5 mm. *Larva* – white to red 5 mm.	Adults emerge from soil in May. Eggs are laid on leaves. Larvae emerge and burrow down to leaf internodes. Larvae migrate to soil in July to over winter.	Saddle shaped galls appear on stems where larvae feeding. Can cause stems to break or lodge. Reduces nutrients movement to ears.	Crop rotation is very important; including crops other than cereals can be successful at controlling this insect. Avoiding late drilling of winter cereals.	Affects wheat, barley and some other grasses. This pest is very occasionally a problem in some of the more intensive cereal growing areas on heavy soils in the UK.
Slugs and snails	Several species cause damage. The grey field slug is the most common. Lightish-brown in colour, about 40 mm long.	Most active in moist and humid conditions. An attack can be more serious when the seed is direct-drilled if the slit has not been properly covered or if the seedbed is loose, cloddy and trashy. Populations and damage are highest on heavy textured soils.	Wheat grain damaged by being eaten in the ground before it germinates. Young cereals can be completely grazed off by a severe autumn attack.	Consolidating the seedbed by rolling after drilling can help reduce the problem. Use a molluscicide if risk of damage is high. Rolling and early nitrogen top-dressing can help a damaged crop to recover in the spring.	Winter wheat chiefly attacked. Often worse after oilseed rape. Use slug traps to check likely risk with a bait such as chicken layers mash (not slug pellets). A catch of 4 slugs/trap suggests a high risk of damage.
Stored grain **Saw-toothed grain beetle**	*Adult:* Dark brown, 3 mm long *Larva:* White and flattened.	Eggs are laid on the stored grain; larvae feed on the damaged grain. Pupation takes place in the grain or store.	The grain heats up rapidly; it becomes caked and mouldy. This is seen with the appearance of the beetles.	Clean stores thoroughly before use and apply an insecticide or use physical pest control using diatomaceous earth (DE). Seal any gaps/cracks in building. A number of insecticide treatments are approved for use on grain in store.	Use traps to identify pest species and numbers present. Check store temperature regularly, maintain grain at <15% moisture content and below 15°C.

(Continued)

Table 7.1 Continued

Crop attacked	Pest	Description	Life-cycle	Symptoms of attack	Control	Notes
	Grain weevil	*Adult:* Reddish-brown, about 3 mm long with an elongated snout.	During autumn the weevils bore into the stored grain to lay their eggs. The larvae feed inside the grain where they also pupate.	Hollow grains. Sudden heating of the grain. Weevils found a few feet below the surface of stored grain.	As for saw-toothed grain beetle.	
	Rust-red grain beetle	*Adult:* Shiny red, up to 2.5 mm in length with long antennae. *Larva* 4 mm.	Eggs are laid on grain. Larvae emerge and burrow into grain. Pupation takes place on grain or other surfaces.	Mainly feed on embryo.	As for saw-toothed grain beetle.	Favoured by high humidity.
	Mites	Adults and larvae are too small to be seen with the naked eye.	Eggs are laid on the stored grain. The larvae hatch out and go through several nymph stages. The time taken to complete the life-cycle is affected by environmental conditions.	Causes grain to heat up. Can taint flour.	Grain should be dried and cooled as quickly as possible. Grain should be stored at 14% moisture content or below. Chemical control as for other storage pests.	Storage mites are highly allergenic.
Maize	**Frit fly**	*Larva:* Whitish, up to 4–6 mm long.	As for frit fly on cereals.	Twisting of leaves surrounding the growing point. In severe attacks this is killed, leading to secondary tillers.	Leave at least 6 weeks from ploughing up grass to planting maize to reduce the risk. Use an insecticide seed dressing if necessary.	
	Wireworm	See wireworm on cereals.			Use a seed treatment.	Maize after grass at greatest risk of damage.

| Beans | **Black bean aphid** (black fly) | Small oval body, black to green colour. Up to 3 mm long. | There are many generations in a year. In summer winged females feed on the crop; wingless generations are then produced which continue to feed. Eventually a winged generation flies to the spindle bush on which eggs are laid for overwintering. | On all summer host plants, colonies of black aphids are seen on the stem leaves, the flowers and developing pods. Damage caused by direct feeding. Infected plants can become stunted and with a heavy infestation may be killed. Aphids can be important for transmission of viruses. | Apply an insecticide if threshold reached. | Mainly a pest of spring beans. It also attacks sugar beet and mangels. Populations of aphids and amount of colonisation and predation by beneficial insects vary each year. Warning systems are available. The threshold for treatment is 5% of plants infected in headland on windward side of field pre-flowering to 2.5% infected across field and colonies spreading onto pods. |
| | **Bean seed beetle** (bruchid beetle) | *Adult:* 3–5 mm mainly black oval body. *Larva:* 6 mm creamy white with a brown head. | Adult lays eggs on the surface of the developing pod. The larvae hatch and burrow a hole into the pod and enter developing seed. Larvae pupate *in situ*. | Circular holes in seed caused by emerging adults. Value of seed for export, seed and human consumption are affected. | Apply an insecticide if threshold reached on susceptible crops where quality an issue. | Inspect crop for adults as first pods develop. If adults found and there have been 2 consecutive days of at least 20 °C may be worth treating. Thresholds still being developed. |

(*Continued*)

Table 7.1 Continued

Crop attacked	Pest	Description	Life-cycle	Symptoms of attack	Control	Notes
	Stem nematode	*Adult:* clear and slender only 1.2 mm in length.	Young adults can over-winter in the soil before moving in moist soil to invade suitable hosts in late spring. Nematodes then reproduce rapidly within the plant; they can move up plant and infect the seed.	Infected plants show red/black patches on stem base, causes plants to be stunted or lodge.	No chemical control. Ensure use only clean seed. Have any home saved seed tested. Include non-susceptible crops in rotation such as wheat and barley. Ensure wild oats are controlled as also host to the nematodes.	There are several different races or strains which infect and breed in a number of legumes and weed species. Once soil infected will remain so for many years.
Peas and beans	**Pea and bean weevil**	*Adult:* light brown with lighter stripes, 6 mm long. *Larva:* 6 mm white with brown head.	During early spring eggs laid in the soil near plants. Larvae feed on roots, whilst adults feed on leaves. Pupation takes place in the soil in midsummer.	Seedling crops checked. U-shaped notches at the leaf margins caused by adults. Larvae eat root nodules.	Some pyrethroid are approved for use when damage severe particularly in backward crops when conditions are dry. Damage is worse if there is a cloddy seedbed.	Spring sown crops are more affected than winter sown. A monitoring system is now available to give recommendations if spray treatments are likely to be warranted.
Peas	**Pea moth**	*Adult:* Dull greyish-brown, about 6 mm long with a wing span of 15 mm. *Larva:* Yellowish-white with darker head; up to 14 mm long.	Eggs laid June to mid-August, hatch in a week. Larvae enter pods and feed on peas until fully grown. Larvae leave pod and make way to soil; pupate in spring and adult emerges in early summer.	Holes in peas caused by larvae. Heavy infestations can significantly affect pea quality.	One or more sprays of a synthetic pyrethroid insecticide. The timing is important. PGRO run a pea moth monitoring service. Growers can use pheromone traps to aid spray decisions.	Early and very late sown crops suffer less damage. Dry harvested peas for human consumption or seed are most likely to require treating. Threshold is >10 moths caught in one of a pair of pheromone traps on 2 consecutive monitoring days up to flowering. (Monitor 3 times per week.)

Crop	Pest	Description	Damage	Control	Notes
	Pea aphid	Large green aphid up to 4 mm long	Direct feeding causes distortion of plant and pods and direct yield loss. Can also transmit a number of viruses.	Treatment with an approved pyrethroid is worthwhile only if thresholds reached.	Adults and eggs overwinter on leguminous crops. Adults migrate to peas from mid-May. Numbers peak in June/July.	A predictive model has been developed. Treat when 15% of plants affected. Note level of pest predators and parasitoids.
Oilseed rape	Cabbage stem flea beetle	*Adult:* 4 mm, metallic green/black. *Larva:* Up to 7 mm, body white with black head.	Adult causes characteristic shot-holing in leaves. Larvae burrow into leaf petioles and cause stunting and leaf loss. In severe cases can cause plant loss.	Seed treated with an insecticide can control the adults and reduce egg laying. Use an approved pyrethroid if treatment threshold reached.	Adults move into rape crop in September and lays eggs. Larvae invade plant from October to March then larvae migrate to soil to pupate.	Either use water traps in September/October or count larvae/plant in November to decide if a treatment required. Treat when 35 beetles are caught/trap or when 2 or more larvae are found per plant.
	Pollen beetle (Blossom beetle)	*Adult:* Metallic-blue/black in colour, up to 3 mm long	Damaged buds wither and die, and the number of pods set is reduced.	Treat with approved pyrethroid if threshold numbers reached. Best results obtained by treatment at green/yellow bud stage. Extreme caution should be taken to ensure that no serious damage occurs to pollinating insects. Note resistance to some pyrethroids has been found in the UK.	Adults emerge from hibernation during spring to feed on buds and flower parts. Eggs laid and then larvae hatch to also feed on buds and flowers. When larvae fully grown they drop to soil to pupate.	Spray thresholds for winter and spring crops are based on beetle numbers/plant and crop population. If more than 70 plants/m^2 then the threshold is 7 beetles/plant compared with 25 beetles/plant with thin crops of less than 30 plants/m^2.

(Continued)

Table 7.1 Continued

Crop attacked	Pest	Description	Life-cycle	Symptoms of attack	Control	Notes
	Bladder pod midge	*Adult*: 2 mm greyish-brown. *Larva*: Whitish-cream up to 2 mm.	Females can lay up to 60 eggs/pod. Eggs are inserted into pods through holes made by seed weevil, other insects or mechanical damage. Larvae feed on developing seed and walls of pod; the larvae pupate in soil after 4 weeks.	The feeding larvae cause pods to swell and ripen prematurely. Seed is shed early.	As for seed weevil. A wide rotation will help reduce build-up of this pest.	Winter oilseed rape most affected. No thresholds. Most damage is usually found in the crop headland.
	Cabbage seed weevils	*Adults*: Lead-grey in colour, about 3 mm long. *Larva*: Up to 5 mm, creamy white with a brown head.	From hibernation near previous year's crops adults lay a single egg in the young pods. Larvae feed on seeds in developing pods; they leave the pods and fall to the ground where they pupate in the soil.	Seeds destroyed in pods by larva. Damage also caused by adult which makes holes for the pod midge to enter.	Spray with an approved insecticide, when thresholds reached during flowering.	Treatment thresholds are 1 weevil found per plant. This is reduced to 1 weevil for every 2 plants on average in northern Britain. Note the headland usually has higher numbers compared with the rest of the field.
Brassicas (cabbage, kale, swedes, turnips)	**Flea beetles** (Various species)	*Adults*: dark coloured 1.5–3 mm. Jump like a flea.	Adults emerge from hibernation during late spring to feed on crops. Eggs are laid, but larvae do little damage. Pupation takes place in the soil during the summer.	Very small round holes are eaten in the cotyledons, first true leaves and stems of the plants. In severe cases plants are killed.	Sow the crop either early or late, i.e. avoid April and May. Good growing conditions to get the crop quickly past the seed leaf stage. Only a very limited number of insecticides are approved for use. Crop covers will also help reduce damage.	An increasing problem especially in dry springs.

Pest	Description	Life cycle	Damage	Control	Notes
Cabbage root fly	*Adult:* 6–7 mm grey/black. *Larva:* Up to 8 mm creamy white.	There are usually 2 or 3 generations a year. First generation eggs are laid next to the brassicas plants during May and second and third generations during July to September; these 2 generations may overlap.	Affected plants are often stunted and may even die especially if invaded when young.	A very limited number of chemicals are approved for use. Crop covers can be very effective. Control brassica weeds which can be alternative hosts.	A very serious pest of brassicas. Winter oilseed rape can be attacked by the third generation of flies. A weather based forecasting system is available.
Turnip sawfly	*Adult:* 6–8 mm, black head and orange body. *Larva:* Grey wrinkled caterpillar up to 18 mm.	There are 3 generations a year. Adults emerge when warm enough from late spring. Eggs are laid on leaves of host. Larvae hatch and when fully grown drop to soil to pupate.	Larvae eat leaves and can totally strip plants in a short time.	Insecticide seed treatments can give some protection in oilseed rape. Crop covers can be effective.	Recently numbers of this pest have increased in the UK. Mainly a problem in turnips, oilseed rape and swedes.
Mealy cabbage aphids	Adult: blue/grey covered mealy wax.	Aphids over winter as adults or as eggs on brassica host. Eggs hatch in spring and aphid move to underside of leaves and shoots. Winged forms migrate from end of May to infest other crops.	Damage caused by direct feeding weakening plants. Damage most serious on young plants which can lead to plant death. Marketability affected in some vegetable brassicas. Important for transmission of viruses, e.g. cauliflower mosaic virus.	Destroy remains of harvested over wintered brassica crops by beginning of May. Several insecticides approved for use in oilseed rape. Some EAMU recommendations in other brassica crops.	Not all brassicas affected. Several weeds are hosts to this aphid. Populations vary widely each year due to weather conditions. Threshold for treating winter oilseed rape is 13% plants infected before petal fall.

(Continued)

Table 7.1 Continued

Crop attacked	Pest	Description	Life-cycle	Symptoms of attack	Control	Notes
Sugar beet, fodder beet and mangels	**Mangold fly**	*Adult*: Greyish brown 6mm. *Larva*: Yellow-white, up to 7mm long.	White oval-shaped eggs are laid in the underside of leaves in May. Larvae bore into the leaf tissue and after about 14 days they drop into the soil where they pupate. There are 2 or 3 generations a year.	Blistering of leaf which can become withered. Retarded growth and in extreme cases death of the plant.	Good growing conditions to help the crop pass an attack. Insecticide seed treatments are very effective against this pest.	This pest is now not so common. Once crops are established damage is less important economically. The use of insecticide seed treatments have reduced the incidence of this and many other soil and foliar pests.
	Aphids (peach-potato aphid – but also potato aphids and black bean aphids)	The green-fly (peach potato aphid) has an oval-shaped body of various shades of green to yellow.	Adults of peach potato aphids over winter on a number of host plants. During spring winged aphids migrate to the summer host crops. Multiplication can be very rapid reaching a peak in July.	Infestations do not normally cause much damage by direct feeding but these aphids, especially the peach potato aphid, transmit virus yellows.	The current insecticidal seed dressing are very effective at controlling peach potato aphids. Insecticide resistance is now a problem with some post emergence insecticides.	See Table 6.1 for alternative methods of control and thresholds. Seed dressings can remain active for up to 10 weeks.
	Beet cyst nematode (BCN)	Cyst forming nematode cysts lemon shaped.	Cysts remain dormant in soil until a suitable host plant is present. Larvae hatch and enter host roots. Once fertilised the females swell with the developing eggs to form cyst on outside of roots. There can be 2 generations a year.	Crop failing in patches. Plants which do survive are very stunted with a proliferation of lateral roots (bearding). Yield losses can be very high	Wide rotations are effective at reducing the risk as well as having a rotation with few if any brassica crops or host weed species. Grow tolerant varieties. No nematicides are currently approved for use in the UK.	Most commonly found on organic and lighter soil types. If necessary the soil can be tested for viable eggs and larvae. Other crops such as red beet, fodder beet and oilseed rape can be hosts for BCN.

Wireworm	See wireworm on cereals.		The roots of seedling plants are bitten off.	Seed treatments will give some protection.	Wireworm is traditionally a problem in grass rotations. There are an increasing number of cases of wireworm in arable rotations on the chalk soils. About 15 % of national crop at risk. Insecticide treatments will also give some control of other soil and foliar pests.
Docking disorder – migratory nematodes	Caused by needle and stubby-root nematodes.	The free living nematodes move to germinating seedlings and feed on the host plant roots. There can be several generations per year. Nematodes are most active in moist soils.	Causes irregularly stunted plants with fangy root growth or even root galls. Damage usually in patches or along rows. A problem of sandy soils.	Improve soil structure. Apply a granular carbamate insecticide in the seed drill furrow.	
Carrots **Carrot fly**	*Adult:* About 7 mm long; shiny black body with reddish/brown head and yellowish wings. *Larva:* When fully grown creamy-white, 8–10 mm long.	Usually two generations a year. Eggs laid on soil surface near carrot in April/June or July/September. After hatching the larvae eventually burrow into root, forming 'mines'; after third moult pupate in soil close to tap root. Some of the second, and possibly a third, generation over winter in the roots, emerging the following spring	Brown and rusty tunnels (mines often with larvae protruding) becoming progressively worse as season proceeds. In severe cases plants may be killed. Makes the mature carrot unmarketable.	Rotation and avoid sowing susceptible crops close together. Hygiene round edge of field to cut down shelter for adult. Some varieties are partially resistant. Crop covers are an option on small areas. Sowing late to avoid the first generation or harvesting early help reduce the damage. The number of recommended insecticides has been reduced; there are a limited number of EAMU recommendations.	Celery, celeriac, parsley and parsnips also attacked. A weather based forecasting system is available to optimise the effectiveness of the insecticide programme.

(Continued)

Table 7.1 Continued

Crop attacked	Pest	Description	Life-cycle	Symptoms of attack	Control	Notes
Potatoes	**Potato cyst nematode (PCN)**	Two species present in UK, white and golden.	Eggs within the cysts in the soil hatch when susceptible host plants are grown. The juvenile nematodes invade the roots. New cysts are formed from fertilised females. These cysts can remain viable in the soil for up to 10 years.	The most important pest problem in potatoes. Causes plants to be stunted, yellow and to have a 'hairy' root system that restricts nutrient and water uptake.	An integrated approach needed. Where soil analysis indicates, grow resistant varieties. There are many varieties that are resistant to the golden nematode but only a few varieties that show partial resistance to the white nematode. Wide rotations help keep populations low. Plant classified seed. The effectiveness of nematicides depends on nematode populations and type of product. The use of trap cropping is being investigated.	Regular soil sampling for cysts should be undertaken to aid the management and control of this major pest. Nematodes are not very mobile, aim to restrict movement of contaminated soil. Check NIAB/TAG guide to varieties of potatoes for susceptibility ratings.
	Peach potato aphid (green fly)	See aphids on sugar beet, fodder beet and mangels.		A bad infestation (5 aphids per compound leaf) will check the growth of the plant. The main problem is virus spread (PLRV and PVY) in seed crops by aphids.	A number of insecticides are approved for use. Many of these insecticides are not controlling these aphids due to resistance. Currently there is no resistance to the neonicotinoides.	There are three types of resistance mechanisms in populations of this aphid to insecticides. Care must be taken in chemical choice and programme of treatment.

Pest	Description	Biology/life cycle	Damage	Control	Notes
Wireworm	See wireworm on cereals.		Causes tunnelling in tubers and tubers to be unmarketable.	Sow early maturing varieties. Lift the crop in early September if possible if field at risk.	Pre-crop sampling or bait trapping can help identify a potential problem.
Slugs	See slugs on cereals.		Maincrop potatoes damaged by slugs eating holes/cavities in the tubers. Crops grown on heavier textured soils are at greater risk.	Some varieties are more susceptible than others. Use bait traps to assess the need for treatment with a molluscicide (see slugs in cereals). Early lifting will help reduce amount of damage.	Use 9 slug bait traps per field; in fields greater than 20ha increase to 13. Leave overnight and count slugs following morning. There is a risk of damage if at least 4 slugs are caught/trap.
Cutworm	*Adult:* Several moth species but mainly the turnip moth. *Larva:* Caterpillars up to 35 mm long; vary in colour dull/greyish brown to green.	Eggs are laid on weeds and hatch in approx. three weeks. After early feeding on leaves, caterpillars go into soil and feed (mostly at night) on stem above and below ground level. Most damage is caused in July. Most, but not all, are fully fed in the autumn, and over winter to pupate in the spring in the soil.	Plants cut off at base of stem at soil level, plants wilt and can die. Larvae eat out large hollows in roots. Damage worst on light soils in very dry conditions.	Keep land free of weeds to reduce egg laying. Irrigation can reduce the problem in dry years. A number of insecticides are approved for use. Treat when the caterpillars are small and still feeding above ground.	Also attacks carrots, leeks, turnips, onions, lettuce, parsnips and red beet. A cutworm warning system is available.
Leatherjackets	See leatherjackets in cereals.		Grass dying off in patches, the roots having been eaten away. Larvae found in soil.	Sowing new seeds before middle of Aug and having a consolidated seed bed reduce the problem. Spray with an insecticide if risk of damage high.	Field populations can be assessed. The threshold numbers for economic damage is one million/ha.
Grass					

(Continued)

Table 7.1 Continued

Crop attacked	Pest	Description	Life-cycle	Symptoms of attack	Control	Notes
	Frit fly	See frit fly on cereals.	The third generation larva can reduce the chances of a successful establishment of some autumn-sown grass seed mixtures.	Larvae bore into stems of establishing grass and cause 'deadheart' symptoms.	When reseeding, allow at least 4 weeks from the destruction of the old sward to the sowing of the new ley. Chemical treatments are available.	Seedling Italian ryegrass is more susceptible than perennial ryegrass and where grass follows grass. Mid to late August drilling most affected.
	Slugs	See slugs on cereals.		Seed destruction below ground is particularly serious.	Test bait to assess need for molluscicide treatment.	Direct drilled grass is most at risk.
	Wireworm	See wireworm on cereals.		Base of plant chewed just below soil surface; plant wilts and turns yellow.	Damage least when time interval between ploughing and drilling is small. No chemical treatments are currently available.	
Red clover	**Stem nematode**	*Adult:* Slender and colourless, 1.2 mm long.	Lives and breeds continuously in plant; passes into soil to infect other plants. Can remain dormant in hay made from an infected crop, becoming active again when conditions are suitable.	Thickening at base of stem; some distortion of leaves, petioles and stems. Plants are stunted; infested patches increase in size each year.	Some varieties of red clover show resistance. Rotation – several years' break from red clover.	Most forage legumes are affected including white clover and lucerne. There are a number of distinct races. Some legumes are host to more than one race.

7.9   Sources of further information and advice

Further reading

Alford D V, *Pest and Disease Management Handbook*, Blackwell Science, 2000.

Alford D V, *A Textbook of Agricultural Entomology*, Blackwell Science, 1999.

Bayer CropScience, *Pest Spotter, A Guide to Common Pests of Farm Crops*, Bayer CropScience, 2011.

Dent D, *Integrated Pest Management*, Chapman and Hall, 1995.

Gratwick M, *Crop Pests in the UK*, Chapman and Hall, 1992.

Gurr G M, Wratten S D and Snyder W E, *Biodiversity and Insect Pests, Key Issues for Sustainable Management*, Wiley-Blackwell, 2012.

HGCA, *Beneficials on Farmland: Identification and Management Guidelines*, HGCA, 2008.

HGCA, *Pest Management in Cereals and Oilseed Rape – a guide*, HGCA, 2003.

Jones F G W and Jones M G, *Pests of Field Crops*, Edward Arnold, 1984.

Lainsbury M A, *The UK Pesticide Guide*, BCPC and www.cabi.org, published annually.

Mathews G, *Pesticides, Health, Safety and the Environment*, Blackwell Publishing, 2006.

Van Emden H F and Service M W, *Pest and Vector Control*, Cambridge University Press, 2004.

Websites

www.agricentre.basf.co.uk
www.bayercropscience.co.uk
www.cropmonitor.co.uk
www.cropprotection.org.uk
www.fera.defra.gov.uk
www.hgca.com
www.inra.fr/hyppz/pests/htm
www.irac-online.org/
www.pesticides.gov.uk
www.syngenta-crop.co.uk
www.voluntaryinitiative.org.uk

Part II

Crop husbandry techniques

8

Cropping techniques

DOI: 10.1533/9781782423928.2.193

Abstract: This chapter discusses some of the fundamental techniques underlying successful crop production – including drainage, irrigation and soil alteration or enhancement. It explains what a seedbed should provide for crop establishment and goes on to look at various methods of seedbed preparation. There is a brief overview of some of the machinery used, including deep cultivators for soil structure management. Cultural weed control is discussed and there is a section on crop management by the use of targets, rotations or sequences, and break crops.

Key words: drainage, irrigation, seedbeds, cultivations, sub-soiling, targets, rotations.

8.1   Introduction

The practical on-farm management of soil, the setting of targets and the monitoring of target achievement and the decisions taken about what crops to grow and when they should appear in the cropping sequence are fundamental to successful crop production.

The management of individual soil types has been dealt with in Chapter 3. Farmers have a responsibility to care for the soil for future generations. This means that, while they are trying to grow profitable crops, they must also take into consideration the long-term effects of their practices on soil health, fertility, structure and stability. This concept now comes under the umbrella of 'sustainability'. This chapter attempts to examine some general management practices which could apply to several different soil types depending on individual situations.

8.2   Drainage

8.2.1   The necessity for good drainage

Normally, the soil can only hold some of the rainwater which falls onto it, particularly during the winter months. The remainder either runs off, is evaporated from the surface, is used by the plants if still growing actively, or soaks through

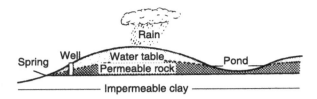

Fig. 8.1   Position of water table and the effect on water levels in wells and ponds.

the soil to the subsoil. If surplus water is prevented from moving through the soil and subsoil, it soon fills up the pore space; this will kill or stunt the growing crops, and any living organisms in the soil, as air is driven out of the profile.

The water table is the level in the soil or subsoil, below which the pore space is filled with water. This is not easy to see or measure in clay soils but can be seen in open textured soils (Fig. 8.1).

The water table level fluctuates throughout the year and in the UK is usually highest in February and lowest in September. This is because of a higher amount of evaporation, more transpiration from growing crops and usually lower rainfall in the summer.

In chalk and limestone areas, and in most sandy and gravelly soils, water can drain away easily into the porous subsoil. These are free-draining soils. On most other types of farmland some sort of artificial field drainage is necessary to carry away the surplus water and so keep the water table at a reasonable level. For most arable farm crops the water table should be about 0.6 m or more below the surface but for grassland 0.3–0.45 m is sufficient. The withdrawal of grants for drainage work, the greater protection of some wetland environments and the rising costs of specialist contractors have meant a large reduction in the area of land being drained compared with the immediate post-war period and the 1970s and 1980s.

Indications of poor drainage are:

- Machinery is easily 'bogged down' in wet weather.
- Stock grazing pastures in wet weather easily poach the sward.
- Water remains on the surface for many days following heavy rain.
- Weeds such as rushes, sedges, horsetail, tussock grass and meadowsweet are common in grassland. These weeds usually disappear after drainage. Peat forms in places which have been very wet for a long time.
- Young plants are pale green or yellow in colour and have generally poor growth.
- The subsoil is often coloured in various shades of blue or grey (mottled), compared with shades of reddish-brown, yellow and orange in well-drained soil.

The benefits of good drainage are:

- Well-drained land is better aerated and the crops grow better and are less likely to be damaged by root-decaying fungi.

- The soil dries out better in spring and so warms up more quickly and can be worked earlier.
- Plants are encouraged to form a deeper and more extensive root system. In this way they can often obtain more plant food and are better able to survive periods of drought.
- Grassland is firmer, especially after wet periods. Good drainage is essential for high density stocking and where cattle are out-wintered if serious poaching of the pasture is to be avoided.
- Disease risk from parasites is reduced. A good example is the liver fluke. Part of its life cycle is in a water snail found on badly drained land.
- Inter-row cultivations and harvesting of root crops can be carried out more efficiently.
- The soil has more potential to support a wider range of crops in the rotation.
- The soil is able to carry heavier machinery without suffering compaction damage.

The main methods used to remove surplus water and control the water table are:

- Ditches or open channels.
- Underground plastic pipe drains.
- Mole drains.
- Ridge and furrow (see page 50).

8.2.2   Ditches and open drains

Ditches may be adequate to drain an area by themselves, but they usually serve as outlets for underground drains. They are capable of dealing with large volumes of water in very wet periods. The size and shape of a ditch varies according to the area it serves and the surrounding soil type (Fig. 8.2). Ditches ought to be cleaned out to their original depth once every three or four years. The spoil removed

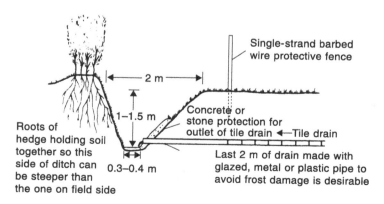

Fig. 8.2 Section through a typical field ditch and tile drain.

should be spread well clear of the edge of the ditch. Many different types of machines are now available for making new ditches and cleaning neglected ones.

Small open channels 10–60 m apart are used for draining hill grazing areas. These are either dug by hand using a spade designed for the purpose, or made with a special type of plough drawn by a crawler tractor. Similar open channels are used on low-lying meadow land where underground drainage is not possible.

8.2.3   Underground drains

The distance between drains which is necessary for good drainage depends on the soil texture. In clay soils the small pore spaces restrict the movement of water. Therefore the drains must be spaced much closer together than on the lighter types of soil where water can flow freely through the large pore spaces (Fig. 8.3 and 8.4).

In the past, many different materials have been used in trenches to provide an underground passage for water (e.g. bushes, straw, turf, stones and flat tiles)

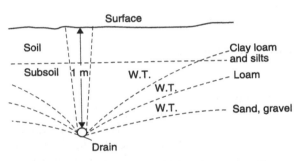

Fig. 8.3   The steepness of a water table (W.T.) varies with different soil types.

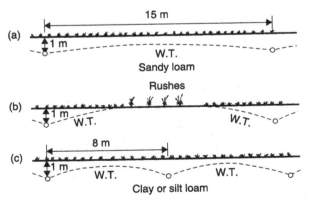

Fig. 8.4   Diagram showing the effect of spacing of drains on the water table (W.T.); (a) sandy loam, (b) and (c) silt or clay loam.

before cylindrical clay tiles became popular. (Concrete pipes have also been used.) However, the flexible plastic pipe has replaced the clay tile for all new drainage work. Under-drainage is very expensive and cost must be related to likely benefits.

Consequently, when designing a scheme, the spacing of the drains must be carefully considered. Digging some holes and examining the subsoil can be very helpful in deciding how the field should be drained. If cost is a limiting factor, it would, for example, be advisable to lay the drains about 40 m (two chains) apart in an area where 20 m would be an ideal spacing. This should be quite satisfactory over most of the field and only a few extra drains may be required later in the wetter areas. This should not be difficult if a detailed map of the drainage layout has been made.

The slopes on a field determine the way the drains should run to be efficient in removing water (Fig. 8.5). Water will not run uphill, unless it is pumped, and so the drains must have a gradient. If the slope is more than 2% (1 : 50) the laterals (side drains) should run across the slope. Ideally, the minimum fall on laterals should be 1 : 250 and, in mains, 1 : 400. It is usually preferable that the laterals go into a main drain before entering a ditch; this means fewer outfalls to maintain. In flat areas, and where the soil is silty (causing silting up of the drains over a long period), it is preferable to let each lateral run straight into the ditch.

A high-pressure jetting device can be used to clean them periodically. Iron ochre deposits can also be cleared in this way.

Where the land is low lying and there is no natural fall to a river, the area can be drained in the usual way and the outlet water from the drainage system can lead to a ditch from which it is pumped into a river or over a sea wall. Underground drainage is best carried out in reasonably dry weather. It is fairly common practice now to lay drains through a growing crop of winter cereals in a dry period in the spring. With modern machinery the little damage done to the crop is more than offset by the beneficial effects of the drains.

Underground piped drains

Plastic pipes come in various configurations. Most of them are corrugated with slits cut at intervals to allow water to enter easily. They are usually supplied in 200 metre rolls which are easily transported. Various diameter sizes are available, e.g. 60 or 80 mm for laterals, 100, 125, 150 or 250 mm for mains.

The size of pipe required will depend on the rainfall, the area to be drained by each pipe, gradient and soil structure. In the case of main drains, it will depend on the number of laterals entering.

Various types of machine are available for laying plastic pipes, and the porous backfill (e.g. washed gravel, clinker) required on the heavier soils. The depth of porous fill is more important than the width in the trench. However, it is very expensive and may constitute 50% of the total cost of the drainage scheme.

Modern machines can lay pipes very rapidly, especially when a laser beam system is used automatically to adjust the gradient on the drain, and a GPS system allows very accurate measurement of spacing within the field.

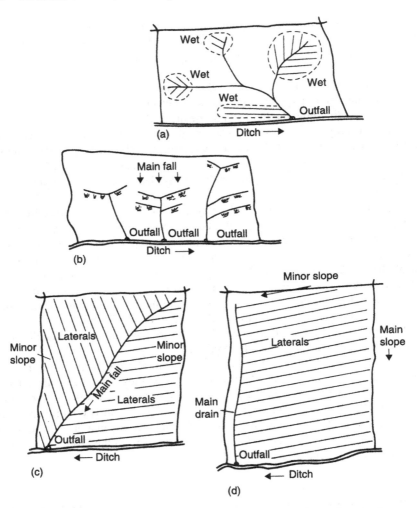

Fig. 8.5 Drain layouts: (a) drainage of wet areas; (b) intercepting springs (rushy areas); (c) herringbone system in valley (note separation of junctions); (d) grid system.

Underground drains – mole drainage

Mole drainage is a cheap method which can be used in some fields. Although mole drains are sometimes used on peat soils, the system is normally applied to fields which have:

- A clay-rich subsoil with no stones, sand or gravel patches.
- A suitable gradient (a fall of 5–10 cm over 20 m).
- A reasonably level surface.

A mole plough, which has a torpedo or bullet-shaped 'mole' plus an expander attached to a steel coulter or blade, forms a cylindrical channel in the subsoil along which water can travel (Fig. 8.6).

The best conditions for mole draining occur when the subsoil is damp enough to be plastic and forms a good surface inside the mole channel. It should also be sufficiently dry above to form cracks as the mole plough passes (Fig. 8.7). Furthermore, if the surface is dry the tractor hauling the plough can get a better grip. The plough should be drawn slowly (about 3 km/h) otherwise the vacuum created is likely to spoil the mole. Reasonably dry weather after moling will allow the surface of the mole to harden and so it should last longer.

Mole drains are drawn 3–4 m apart usually at right angles to the direction of the plastic drain pipes. For best results the moles should be drawn through the porous backfill of a plastic main or lateral drain (Fig. 8.8 and 8.9). A new set of mole drains can be drawn every three to ten years as required. They will last much longer in a soil with a high clay content than in one with more sand or silt where the sides of the moles collapse quite quickly.

Installing a new drainage system is the first step in removing surplus water from a field. However, the structure of the soil and subsoil must, if necessary, be improved so that water can move easily to the drains (see Section 3.4, pages 49–57).

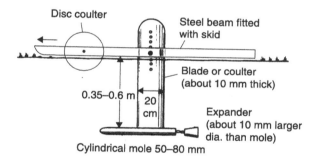

Fig. 8.6   Mole plough.

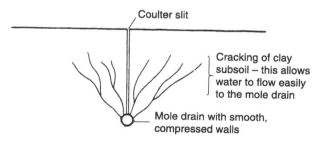

Fig. 8.7   Section through mole drain and surrounding soil.

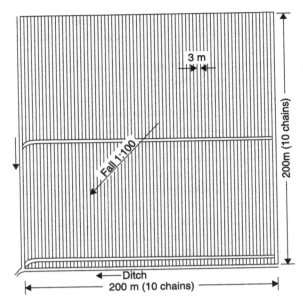

Fig. 8.8   Layout of mole drainage (with field main) in a 4-ha area.

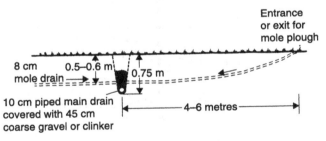

Fig. 8.9   Section through the tile drain shown in Fig. 8.8 to show how water from the mole drain can enter the pipe drain through the porous backing.

8.3   Irrigation

8.3.1   Soil water availability

Irrigation is the term used when water is applied to crops which are suffering, or about to suffer, from drought. It is usual to measure irrigation water, like rainfall, in millimetres:

- 1 mm on 1 hectare = 10 m³ = 10 tonnes.
- 25 mm on 1 hectare = 250 m³ (a common application at one time).

- 100 mm on 1 hectare = 1000 m³ = 1 dam³ or 1000 tonnes (about a season's requirement).
- 100 mm on 50 hectare = 50 dam³ (a guide to the amount of water required). [dam³ = cubic decameter]

Green plants take up water from the soil and transpire it through their leaves (see Section 1.2.2 Transpiration, page 5), although a small amount is retained to build up the plant structure. Green crops which cover the ground have a water requirement. This varies from day to day and place to place according to climate and local weather conditions. It is called potential transpiration (Table 8.1).

When the rainfall is well distributed, crops are less likely to suffer than in wet periods followed by long dry periods. During such dry periods the crops have to survive on the available water held in the soil (see Section 3.3.4, Water in the soil, page 45). The difference between the actual amount of water in the soil and the amount held at field capacity is known as the soil moisture deficit (SMD). It is one of the main factors in determining the need for irrigation.

An approximate soil moisture deficit can, at any time, be calculated from the rainfall and the figures given in Table 8.1. For example, if the soil were at field capacity at the end of April, with a green crop covering the ground, and with no rain in May, then the SMD would be 80 mm at the end of May allowing for losses due to evaporation and transpiration. However, if 50 mm of rain fell during May, then the deficit would drop to 30 mm. The amount of water available to a crop at any time depends on the capacity of the soil per unit depth, and the depth of soil from which the roots take up water. The figures in Table 8.2 are average rooting depths. They would be greater on deep, well-drained soils and less on

Table 8.1 Water requirement for green crops, southern UK

Period	Per day		Per month	
	mm	m³/ha	mm	m³/ha
April and September	1.5	15	50	500
May and August	2.5	25	80	800
June and July	3.5	35	100	1000

Table 8.2 Rooting depths of common crops

Crop	Rooting depth (cm)
Cereals, pasture	35–100
Potatoes (varietal differences)	30–70
Beans, peas, conserved grass	45–75
Sugar beet	55–100
Lucerne (alfalfa)	Over 120

shallow, badly drained or panned soils. This can be checked in the field by digging a soil profile. Normally, most of the plant roots are concentrated in the top layers and only a small proportion penetrates deeply into the soil.

More than three-quarters of all the farm soils in this country have available water capacities between 60 and 100 mm (12–20% by volume) within the root range of most crops (about 500 mm depth). Sandy, gravelly and shallow stony soils have less than 60 mm (12%), whereas deep silty soils, very fine sands, warpland, organic and peaty soils have well over 100 mm (20%) of available water within root range. The available water capacity of the topsoil is often different from that of the subsoil. This should be taken into consideration when calculating the amount present within the root range of a crop.

It is estimated that the roots of a plant cannot take up much more than half of the available water within the depth of its deepest roots. The top half of the rooting zone contains 70% of roots and they do not penetrate all the pore spaces and so, where available, irrigation should be started when about half the available water has been taken up. This is known as the critical deficit.

8.3.2   Timing and sources of irrigation

Decisions on when to start irrigating must be influenced by the time it will take to apply 25 mm of water over the area involved, usually 5 to 10 days. Some companies and organisations provide helpful computerised predictions on when irrigation should be started. These take into consideration all the relevant factors and are tailored for each field and its crop. Farmers can work out their own daily water balance sheet fairly easily, using rainfall records and standard transpiration data to get a SMD figure. The greatest need for irrigation is in the eastern and south-eastern counties where the lower rainfall and higher potential evapo-transpiration means that it would be beneficial about nine years out of ten. The need is much less in the wetter western and northern areas.

Water for irrigation is strictly controlled by abstraction licences issued by the Environment Agency. You will need a licence if you intend to abstract more than 20 m³ per day in the UK. The charge for the water varies between areas and it usually costs up to ten times more for taking water directly from rivers in summer than for taking it to fill reservoirs in winter. Licences are increasingly difficult to obtain and even existing licences are often subject to abstraction restrictions.

Water is generally very scarce in summer in the areas which need it most, but *annual* rainfall is adequate for most irrigation needs provided the winter surplus is stored in reservoirs or underground aquifers. The main sources of irrigation water are rivers, canals and streams, ponds and lakes, reservoirs and boreholes. Some is taken from mains water but is very expensive. Boreholes are also expensive; they may have to be lined, e.g. those dug through sand or gravel, and require special deep well pumps and sometimes filters. However, they may be the most reliable source in a very dry summer. Above-ground reservoir construction is specialised work and must be designed and supervised by a qualified engineer if they are over 25 000 m³ in volume and lie above adjacent land. (A hole in the ground is simpler and may be partly replenished by groundwater.) Lining of a reservoir may be necessary and this could double the cost.

Reservoirs with a capacity of over 25 000 m³ above the natural level of adjoining land must now be registered with the Planning Authorities. Inspection and repairs can be very expensive to ensure that they are safe.

All water used for irrigation should be tested for impurities which may be harmful to crops, particularly salt, heavy metals and bacterial contamination.

8.3.3   The effect of irrigation on particular crops

Irrigation can be expensive, so it is important to use it as efficiently as possible.

Early potatoes will respond with earlier and more profitable production if irrigation is used when the SMD reaches 25 mm. Second earlies and maincrops are not normally irrigated until the tubers are fingernail size, otherwise some varieties may produce too many small tubers and a low saleable yield. On the better soils (with a high available water capacity) irrigation can be delayed until a 40 mm deficit occurs. On light soils where common scab causes problems, irrigation is recommended each time a 15 mm deficit occurs when the tubers are forming. If the soil is very dry and hard at harvest time, some irrigation can make harvesting easier and therefore reduce tuber damage. It can also be used to improve the efficacy of residual herbicides and protect from late frosts.

With **sugar beet** it may be necessary to irrigate a very dry seedbed to obtain a fine tilth, an even germination and more effective use of soil-acting herbicides. After this it is not usually necessary to irrigate until the leaves start meeting across the rows (sugar beet is deep-rooted). It can start with applications of about 25 mm, and increase to 55 mm (in one application) on the high available water capacity soils when the SMD is over 100 mm. The idea is to prevent the leaves from wilting, which reduces their photosynthetic ability, compromising sugar production.

For **peas**, unless the soil is very dry during early crop development, it is preferable to wait and apply the water at the very responsive stages. At the start of flowering 25 mm should be applied (increases numbers of peas) and 25 mm when the pods are swelling (increases pea size), always provided that the SMD is about 35 mm at these times.

With **cereals**, irrigation can only be justified on very light soils in low rainfall areas and only if higher value crops have had their water need satisfied. The most critical time is about GS 37 when the flag leaf is showing and the ear is developing rapidly. An application of 25 mm should be adequate.

Grass and other **leafy crops** respond throughout the season and about 25 mm of water can be applied every time the deficit reaches 30 mm. Although intensive dairying is the most profitable way to utilise the extra production, it is unusual to irrigate the grass crop in the UK.

In the above suggestions, for times and amounts of water to apply, it is assumed that it is a dry period with little or no sign of rain. To irrigate up to field capacity shortly before rain is to risk nitrogen losses by leaching.

8.3.4   Application of irrigation

Different textured soils absorb water at different rates (Table 8.3). When irrigation water or rain falls on a dry soil it saturates the top layers of soil to full field capacity before moving further down. However, if the soil is deeply cracked (e.g. clay) then some of the water will run down the cracks. Very dry soils can absorb water at a faster rate than shown in Table 8.3, but only for a short period.

Irrigation water is applied by:

- *Rotary sprinklers.* Each rotating sprinkler covers an area about 21–36 m in diameter. The sprinklers are usually spaced 10.5 m apart along the supply line which is moved about 18 m for each setting. Special nozzles can be used to apply a fine spray at 2.5 mm per hour for frost protection of fruit crops and potatoes. Icicles form on the plants, but the latent heat of freezing protects the plant tissue from damage.
- *Rain guns.* These are now the most widely used machines. They are used on all types of crops. The droplets are large and the diameter of area covered may be between 60 and 120 m. Some types are used for irrigation of slurry or dirty water.
- *Travelling rain gun with hose reel.* This is similar in many respects to a mobile machine, but the hose reel remains stationary – usually on a headland – while the rain gun and its hose (nylon/PVC) are towed out to the start of the run (up to 500 m). It irrigates as it is wound back, using water pressure to turn the hose reel. At the end of the run it switches itself off automatically, and it can then be moved to the next position. About 80% of all irrigation water is now put on using this system. It can be inaccurate and the large droplets can cause soil erosion problems on uncovered ground, but it has relatively low labour requirements.
- *Centre pivot and linear irrigators.* These are large machines which cover up to 160 hectares. They are mainly centre pivots which move round a pivot (the water source) in a complete circle or part of a circle.
- *Spraylines.* These apply the water gently and are used mainly for protected horticultural crops.
- *Surface trickle.* Water is applied through small bore perforated pipes on or near to the soil surface. They give a much more efficient use of water than sprinklers or rain guns and are often used in perennial horticultural crops or vineyards.
- *Sub-surface trickle.* Often used in orchards or vineyards in hot, dry climates like California where water is becoming a scarce resource. The pipes are buried below ground to reduce evaporation waste and make water use even more efficient.
- *Surface channels.* This method requires almost level or contoured land. It is wasteful of land, water and labour, but capital costs are very low.
- *Underground drainage pipes.* On level land, water can be dammed in the ditches and allowed to flow up the drainage pipes into the subsoil and lower soil layers. This is drainage in reverse!

Table 8.3   Time of absorption of water by various soils

Soil type	Time (hr) required to absorb 25 mm of water
Sandy	1–2
Loams	3
Clay	4–5

On sloping fields, problems can arise with run-off if water is applied too quickly. It may run down between potato ridges, for example, without really wetting the ridges where the crop roots are growing. There is less run-off if a bed system is used (see '*The bed system*' subsection in Section 15.3.6).

The need for irrigation can be reduced by improving soil structure, breaking pans, correcting acidity (soil and subsoil), increasing organic matter in the soil and selecting crops and varieties with deep-rooting habits.

8.4   Warping

This is a process of soil formation where land, lying between high and low water levels, alongside a tidal river, is deliberately flooded with muddy water. The area to be treated is surrounded by earth banks fitted with sluice gates. At high tide, water is allowed to flood quickly onto the enclosed area and is run off slowly through sluice gates at low tide. The fineness of the material deposited will depend on the length of time allowed for settling. The coarse particles will settle very quickly, but the finer particles may take one or more days to settle. The depth of the deposit may be 0.4 m in one winter. When enough alluvium is deposited, it is then drained and prepared for cropping. These soils are very fertile and are usually intensively cropped with arable crops such as potatoes, sugar beet, peas, wheat and barley.

Part of the land around the Humber estuary is warpland and the best is probably that in the lower Trent valley. Most of this work was done in the nineteenth century. It is too expensive to consider now.

8.5   Claying

The texture of 'blowaway' sandy soils and Black Fen soils can be improved by applying 400–750 t/hectare of clay or marl (a lime-rich clay). If the subsoil of the area is clay, it can be dug out of trenches and roughly scattered by a dragline excavator. In other cases, the clay is dug in pits and transported in special lorry spreaders. Rotary cultivators help to spread the clay. If the work is done in late summer or autumn, the winter frosts help to break down the lumps of clay. As a practice, it is no longer considered economical.

8.6   Tillage and cultivations

8.6.1   The necessity for cultivations

Cultivations are field operations which attempt to alter the soil structure. The main object is to provide a suitable seedbed in which a crop can be planted and will germinate (where applicable – transplanted vegetable or salad crops for instance need to be able to grow their roots into the seedbed quickly) and grow satisfactorily. Cultivations are also used to kill weeds and/or bury the remains of the previous crop. The cost of the work can be reduced considerably by good timing and the use of the right implements. Ideally, a good seedbed should be prepared with the minimum amount of working and the least loss of moisture. On heavy soil and in a wet season, some loss of moisture can be desirable. On the medium and heavy soils full advantage should be taken of weathering effects. For example, ploughing in the autumn will allow frost to break the soil into a crumb structure (a frost tilth); wetting and drying alternately will have a similar effect. The workability of a soil is dependent on its consistency, which is a reflection of its texture and moisture content.

8.6.2   Seedbed requirements

Listed below are the seedbed requirements for some important crops.

- *Cereals.*
  - (a) *Autumn planted.* The object here is to provide a tilth (seedbed condition) which consists of fine material and egg-sized lumps. It should allow for the seed to be drilled and easily covered with the surface remaining rough after planting. The lumps on the surface prevent the siltier soils from 'capping' in a mild, wet winter and they also protect the base of the cereals from the harmful effects of very cold winds. Larger clods of soil will cause problems with residual herbicides if used and could make it easier for slugs to move through the soil and cause damage to seeds and seedlings. Harrowing and/or rolling may be carried out in the spring to break up a soil cap which may have formed. It will also firm the soil around plants which have been heaved by frost action.
  - (b) *Spring planted.* A fairly fine seedbed is required in the spring (very fine if grasses and clovers are to be undersown). If the seedbed is dry or very loose after drilling it should be consolidated by rolling. This is especially important if the crop is undersown and where the soil is stony.
- *Root crops, e.g. sugar beet, swedes and carrots: also kale.* These crops have small seeds and so the seedbed must be as fine as possible, level, moist and firm. This is very important when precision drills and very low seed rates are used. Good, early, ploughing with uniform, well-packed and broken furrow slices will considerably reduce the amount of work required in the spring when, if possible, deep cultivations should be avoided to keep frost tilth on top, leaving unweathered soil well below the surface.
- Direct seeding and/or reseeding of *grasses* and *clovers*. The requirements will be the same as for the root crops.

- *Spring beans and peas.* The requirements will generally be similar to the cereal crop, although the tilth need not be so fine. Peas grown on light soils may be drilled into the ploughed surface if the ploughing has been well done.
- *Winter beans* These are often broadcast onto stubble and ploughed in. The surface is then levelled to make travel across the field more comfortable and to kill any weeds which have emerged.
- *Potatoes.* This crop is usually planted in ridges 90 cm wide or in beds, so deep cultivations are necessary. The fineness of tilth required depends on how the crop will be managed after planting. A fairly rough, damp seedbed is usually preferable to a fine, dry tilth which has possibly been worked too much. Early potato crops are often worked a few times after planting – such as harrowing down the ridges, ridging-up again, deep cultivations between the ridges and final earthing-up. The main object of these cultivations is to control weeds, but the implements often damage the roots of the potato plants. Valuable soil moisture can also be lost. Most annual weeds can be controlled by spraying the ridges when the potato sprouts start to appear. This can replace most of the inter-row cultivations and reduce the number of clods produced by the rubber-tyred tractor wheels.
- *Oil seeds.* Winter oilseed rape is sown in August and moisture conservation at this dry, hot time is of paramount importance. It is usual to carry out minimum cultivations to achieve a fine seedbed for the very small seed. Spring oilseed rape and linseed also need fine seedbed conditions to give intimate contact between soil and seed. It is usual to plough in late autumn and then work the resultant frost tilth carefully with straight-tined shallow implements to avoid bringing up unweathered soil.

8.6.3   Non-ploughing techniques

Direct drilling (slit-seeding, zero tillage) and minimal cultivation techniques have now become popular and widely used alternatives to conventional mouldboard ploughing on many farms and on a wide range of soil types, including difficult clays. Sets of heavy discs or tined implements are now pulled by high horsepower tractors directly across stubble to produce autumn seedbeds. Drills which also cultivate are used, sometimes directly into the stubble. These operations are collectively known as 'min-till' systems, although a better name for them might be 'non-inversion' as there is often nothing minimal about them! The original meaning of the word 'minimal' was to reflect that they were cultivation systems where only the minimum depth of soil is moved to allow drilling to take place. They are quicker than traditional plough-based systems, but not necessarily cheaper.

Several types of special drills are now available for either non-inversion drilling or direct drilling in the right conditions, using such developments as heavily weighted discs for cutting slits, strong cultivator tines or modified rotary cultivators. With cereals, true direct drilling should only be used on very clean stubbles where the previous crop has been cut low and the straw removed from the field. Even in these circumstances there will be a small risk of disease spread from the stubble to the new crop. It can still be seriously considered for seeding grass and for crops such as kale.

8.6.4   Tillage implements
The main implements used for tillage are described below.

Ploughs
Ploughing is still the first operation in seedbed preparation on many farms and is likely to remain so for some time yet. However, some farmers are now experimenting with mechanically driven digging or pulverising machines, as alternatives to the plough. Good ploughing is still the best method of burying weeds and the remains of previous crops and it moves 100% of the topsoil profile down to plough depth. It can also set up the soil so that good frost penetration is possible. Fast ploughing produces a more broken furrow slice than slow steady work. Smaller ploughs are fully mounted, while the larger ones are semi-mounted. Large tractors are now used to pull 8 or 9 furrows, so work rate is increased. General-purpose mouldboards are commonly used. The shorter digger types (concave mouldboards) break the furrow slices better and are often used on the lighter soils and slatted mouldboards are recommended for light and fluffy soils such as peat. Deep digger ploughs are used where deep ploughing is required, e.g. for roots or potatoes. Most commercially used ploughs are reversible. They have right-hand and left-hand mouldboards and so no openings or finishes have to be made when ploughing; the seedbed should therefore be at least slightly more level. Round-and-round ploughing with the ordinary, non-reversible plough has almost the same effect, although this is not often used these days.

The proper use of skim and disc coulters and careful setting of the plough for depth, width and pitch can greatly improve the quality of the ploughing. The furrow slice can only be turned over satisfactorily if the depth is less than about two-thirds the width. The usual widths of ordinary plough bodies vary from 20 to 35 cm. If possible, it is desirable to vary the depth of ploughing from year to year to avoid the formation of a plough pan. Very deep ploughing, which brings up several centimetres of poorly weathered subsoil, should only be undertaken with care: the long-term effects will probably be worthwhile but, for a few years afterwards, the soil may be rather sticky and difficult to work. Buried weed seeds, such as wild oats which have fallen down cracks, may be brought to the surface and may spoil the following crops. 'Chisel ploughing' is a term used to describe the work done by a heavy duty cultivator with special spring or fixed tines; unlike the ordinary plough, it does not move or invert all the soil. Disc ploughs have large saucer-shaped discs instead of shares and mouldboards. Compared with the ordinary mouldboard plough, they do not cut all the ground or invert the soil so well, but they can work in harder and stickier soil conditions. They are more popular in dry countries. Double mouldboard ridging ploughs are sometimes used for potatoes and some root crops in the wetter areas.

Cultivators

These are tined implements which are used to break up the soil clods (to ploughing depth). Some have tines which are rigid or are held by very strong springs which only give when an obstruction, such as a strong tree root, is struck. Others have spring tines which are constantly moving according to the resistance of the soil. They have a very good pulverising effect and can often be pulled at a high speed. The shares on the tines are of various widths. The pitch of the tines draws the implement into the soil. Depth can be controlled by tractor linkage or wheels. The timing of cultivations is very important if the operation is to be effective. Many tined cultivators also have rows of discs on them as well, so that a one pass operation can produce a seedbed in good condition e.g. the Vaderstad Top-Down.

Harrows

There are many types of harrow. Harrow tines are usually straight, but they may vary in length and strength, depending on the type of harrow. These implements are often used to complete the work of the cultivator. Besides breaking the soil down to a fine tilth, they can have a useful consolidating effect due to shaking the soil about and rearranging particle distribution. 'Dutch' harrows have spikes fitted in a heavy frame and are useful for levelling a seedbed as well as breaking clods. Some harrows, e.g. the chain type, consist of flexible links joined together to form a rectangle. These follow an uneven surface better and are useful on grassland to remove dead, matted grass at the end of the season. Most chain harrows have spikes fitted on one side.

   Rotary power harrows can result in much better movement of the soil in one pass. They are most valuable on the heavier soils when preparing fine seedbeds for potatoes, sugar beet and other crops. Power harrow/drill combinations are used by some farmers as one-pass operations to prepare seedbeds for cereals. They can be used in conditions where ordinary drills may not be appropriate.

Hoes

These are implements used for controlling weeds between the rows in root crops. Various shaped blades and discs may be fitted to them. Most types are either front-, mid- or rear-mounted on a tractor. The front- and mid-mounted types are controlled by the steering of the tractor drive. The rear-mounted types usually require a second person for steering the hoe.

Disc harrows

These consist of 'gangs' of saucer-shaped discs between 30 and 60 cm in diameter. They have a cutting and consolidating effect on the soil. This is particularly useful when working a seedbed on ploughed-out grassland; some discs have scalloped edges to improve the cutting effect. The more the discs are angled, the greater will be the depth of penetration, the cutting and breaking effect on the clods and the draught.

Disc harrows are widely used for preparing all kinds of seedbeds, but it should be remembered that they are expensive implements to use. They have a heavy draught and many wearing parts, such as discs, bearings and linkages, and so should only be used when harrows would not be suitable. They tend to cut the rhizomes of weeds such as couch and creeping thistle into short pieces which are easily spread. Discing of old grassland before ploughing will usually allow the plough to get a better bury of the turf. Heavy discs, and especially those with scalloped edges, are very useful for working in chopped straw after combining and preparing a seedbed under a non-inversion system.

Rotary cultivators (e.g. rotavator)

This type of implement consists of curved blades which rotate around a horizontal shaft set at right angles to the direction of travel. The shaft is driven from the power take-off of the tractor; depth is controlled by a land wheel or skid. This implement can produce a good tilth in difficult conditions and, in many cases, it can replace all other implements in seedbed preparation. A light fluffy tilth is sometimes produced which may 'cap' easily if wet weather follows. The fineness of tilth can be controlled by the forward speed of the tractor; for example, a fast speed can produce a coarse tilth. It is a very useful implement for mixing crop remains into the soil. The rotating action of the blades helps to drive the implement forward and so extra care must be taken when going down steep slopes.

In wet, heavy soils the rotating action of the blades may have a smearing effect on the soil. This can usually be avoided by having the blades properly angled. Rotavating of ploughed or cultivated land when the surface is frozen in winter can produce a good seedbed for cereals in the spring without any further working or loss of moisture. Narrow rotary cultivator units are available for working between rows of root crops.

Rolls

These are used to consolidate the top few centimetres of the soil so that plant roots can keep in contact with the soil particles, and the soil can hold more moisture. They are also used for crushing clods and breaking surface crusts. Rolls should not be used when the soil is wet; this is especially important on the heavier soils. The two main types of rolls are the flat roll, which has a smooth surface, and the Cambridge or ring roll, which has a ribbed surface and consists of a number of heavy iron wheels or rings (about 7 cm wide) each of which has a ridge about 4 cm high. The rings are free to move independently and this helps to keep the surface clean. The ribbed or corrugated surface left by the Cambridge roll provides an excellent seedbed on which to sow grass and clover seeds or roots. Also, it is less likely to 'cap' than a flat rolled surface. Flat rolls are more often used on grassland before cutting for silage to bury stones or molehills. They are sometimes filled with water to increase their weight.

A *furrow press* is a special type of very heavy ring roller (usually with three or four wheels) used for compressing the furrow slices after ploughing. It is usually attached to and pulled alongside the plough. On light soil it can be used to prepare a surface suitable for drilling.

8.6.5   Pans and soil loosening

A pan is a hard, cement-like layer in the soil or subsoil which can be very harmful because it prevents surplus water draining away freely and restricts root growth. Such a layer may be caused by ploughing at the same depth every year. Such ploughing produces a 'plough pan' and is partly caused by the base of the plough sliding along the furrow. It is more likely to occur if rubber tyred tractors are used when the soil is wet, and there is some wheelslip which has a smearing effect on the bottom of the furrow. Plough pans are more likely to form on the heavier types of soil. They can be broken up by using a subsoiler or by deeper ploughing.

Pans may also be formed by the deposition of iron compounds, and sometimes humus, in layers in the soil or subsoil. These are often called chemical or iron pans and may be destroyed in the same way as plough pans. Clay pans form in some soil formation processes.

Soil loosening (which is justified only when necessary) is a term now used to cover many different types of operation, including subsoiling. This aims to improve the structure of the soil and subsoil by breaking pans and having a general loosening effect. An ordinary subsoiler can be used. It is very effective when worked at the correct depth. A subsoiler is a very strong tine (usually two or more are fitted on a toolbar) which can be drawn through the soil and subsoil (about 0.5 m deep and 1 m apart) to break pans and produce a heaving and cracking effect. This will only produce satisfactory results when the subsoil is reasonably dry and drainage is good. The modern types with 'wings' fitted near the base of the tine produce a very good shattering effect.

Other implements which may be used include the Shakaerator which has five strong tines which are made to vibrate in the ground by a mechanism driven from the tractor. Modified rotary cultivators with fixed tines attached can also be described as soil looseners.

The soil loosening process is very necessary in situations where shallow cultivations – up to 150 mm – have been used for many years and the soil below has become consolidated and impermeable to water and plant roots. Rain water collects on this layer and causes plant death by waterlogging and/or soil erosion. When attempting to loosen a compacted soil/subsoil, it is advisable to start with shallow cultivations, then to work deeper with stronger tines and then, if necessary, with a winged subsoiler. This can be achieved by several passes over the field. However, the one-pass method can also be used. The unit has a toolbar fitted with shallow tines in front, strong spring tines in the middle and winged subsoilers attached to the rear. It obviously requires greater tractor power, but it is an energy-efficient method. This work should only be carried out when the soil crumbles and does not smear.

8.6.6   Control of soil erosion

Soil erosion by water is an increasing problem on many soils, especially on sandy, silty and chalk soils which are in continuous arable cultivation, and where the organic matter is below 2%. Rain splash causes capping of such soils and heavy rain readily runs off instead of soaking into the soil. There may also be panning problems. Erosion can be serious on sloping fields (especially large fields) where the cultivation lines, crop rows, tramlines, etc., run in the direction of the slope, and the wheelings are compacted. In these situations, in a wet time or a thunderstorm, deep rills and gullies, up to a metre deep, can be cut in the fields and up to 150 tonnes/ha of soil washed away. Sometimes crops can be covered at the lower end of a field with soil washed from the upper slopes.

Erosion by wind can cause very serious damage on Black Fen soils and on some sandy soils. Various controls can be used, e.g. straw planting by special machines, sowing nursecrops such as mustard or cereals which are selectively killed by herbicides when the crop (e.g. sugar beet) is established, claying or marling and applying sewage sludge. With all potential erosion problems, the secret is to try and prevent it from happening by ensuring that there is always sufficient ground cover, especially during the winter months. Farmers are expected to produce management plans for their soil under the Defra Code of Practice (Protecting our Water, Soil and Air) and Good Agricultural and Environmental Conditions (GAECs) regulations under the Single Payment Scheme (SPS), in order to protect the structure of the soil and to minimize run-off and erosion from wind and water.

8.6.7   Soil capping and its prevention

A soil cap is a hard crust, often only about 2–3 cm thick, which sometimes forms on the surface of a soil. It is most likely to form on soils which are low in organic matter and have a high silt content. Heavy rain or large droplets of water from rain guns, following secondary cultivations, may cause soil capping. Tractor wheels (especially if slipping), trailers and heavy machinery can also cause capping in wet soils.

Although a soil cap is easily destroyed by weathering (e.g. frost, or wetting and drying) or by cultivations, it may do harm while it lasts because it prevents water moving into the soil, as well as preventing air moving into and out of the soil in wet weather. It also hinders the development of seedlings from small seeds such as grasses and clovers, roots and vegetables.

Chemicals such as cellulose xanthate can be sprayed on seedbeds to prevent soil capping. They enhance seedling establishment and do not affect herbicidal activity in the soil, but are very expensive.

8.7   Control of weeds by cultivation

Historically, the introduction of herbicides reduced the importance of cultivation as a means of controlling weeds. Cereal crops were regarded by many farmers as the cleaning crops instead of the roots and potato crops, mainly because chemical spraying of weeds in cereals was very effective, if expensive. Now that many weed species are showing resistance to a wide range of herbicidal active ingredients, and there is pressure on farmers to reduce their use of pesticides, they are looking to return to cultural methods of controlling weeds in order to achieve satisfactory control. Cultivations are now an important part of an integrated crop management system (sometimes known as ICM), with weeds controlled between crops by the production of stale or false seedbeds, and inter-row cultivations important in row crops such as potatoes and sugar beet. In organic farming the use of in-crop comb weeders is an important control measure (see below). The use of the occasional fallow, and growing spring-sown crops, is also an important management tool for the control of noxious weeds such as creeping thistle and couch grass by cultivations.

Annual weeds can be tackled by:

• Working the stubble after harvest (e.g. discing, cultivating or rotavating) to encourage seeds to germinate. These young weeds can later be destroyed by harrowing or ploughing. Unfortunately, this often allows wild oats to increase.
• Preparing a 'false' seedbed in spring to allow the weed seeds to germinate.
• Killing them by cultivation before sowing a spring root crop. Inter-row hoeing of root crops can destroy many annual weeds and some perennials.

It should be noted that, in organic farming, mechanical weed control is the only option and various fine-tined implements have been developed for use in the growing crop.

Perennial weeds such as couch grass, creeping thistle, docks, field bindweed and coltsfoot can usually be controlled satisfactorily by the fallow (i.e. cultivating the soil periodically through the growing season instead of cropping). However, this is very expensive. A fair amount of control can be obtained by short-term working in dry weather.

The traditional method of killing couch has been to drag the rhizomes to the surface in late summer. This is then followed by rotary cultivation – three or four times at two to three week intervals – to chop up the rhizomes. To get a good kill of the plant, adequate growing conditions are necessary for the couch rhizome to respond to being chopped up by sending out more green shoots and thus to hasten its exhaustion. A much more reliable method is to use glyphosate but, of course, this cannot be used in organic systems.

The deeper-rooted bindweed, docks, thistles and coltsfoot cannot be controlled satisfactorily just by cultivations, although periodic hoeing and cultivating between the rows of root crops can generally reduce the problem.

Thorough cultivations which provide the most suitable conditions for rapid healthy growth of the crop can often result in the crop outgrowing and smothering the weeds.

8.8   Crop management: key issues

Any good management technique should follow the management cycle:

1. Set targets.
2. Assess progress.
3. Adjust inputs.
4. Monitor success.

This applies equally both to car production and to crop production. The main difference will be the effect of the natural environment on *crop* production, with factors such as the weather being largely outside the farmer's control. However, the cycle can still be useful.

8.8.1   Setting targets

Most farmers set targets of one kind or another. Arable farmers set targets of profitability, of cash flow and of capital investment. From an agronomy point of view it is possible to set targets for crop structure so that maximum profit is obtained. For instance they might set a target of 600 fertile ears/m^2 for a wheat crop at harvest. This can be achieved by sowing the correct number of seeds allowing for targeted losses in the field and manipulating tiller numbers by the judicious use of nitrogen and plant growth regulators. They would set targets for the amount of fertiliser and agrochemicals to be used to *maintain* this desired crop structure. They would set drilling date and harvest date targets as well as cultivation targets for the next year's crops. They might also set financial input limits to try and maintain margins. These targets should be realistic and achievable. They should involve an element of risk management as well as trying to satisfy the *personal* goals of the farmer or farm manager.

8.8.2   Assessing progress

This is often the hardest thing to do. It requires a quantitative as well as a qualitative analysis of the crop's progress, and this in itself requires good monitoring techniques. It is similar to the actual figures on a forward cash flow being compared to the budgeted figures. It needs frequent crop walking and a good recording system for factors such as leaf emergence, pod formation, ear size, tiller numbers, disease, weed and pest incidence, rainfall and soil moisture deficit. It also needs a knowledge of the science which underlies the decision making process. Very often, farmers are too busy to do this successfully and will employ people such as crop consultants and agronomists to do it for them.

8.8.3   Adjusting inputs

This is reactive management to problems that have occurred during the growing season. Inputs can be raised or lowered or changed according to crop progress.

This is where the external factors such as weather can have a big part to play. There is no such thing as an average season in agriculture so no management blueprints can ever be 100% successful without some kind of modification. Heavy summer rainfall will mean an increase in disease on most crops; a drought may cause problems of tiller survival; late frosts may mean loss of grain sites in early-sown cereals. The farmer must react to these influences by adjustment of inputs in order to maintain the targeted margin.

8.8.4   Monitoring success

Ultimately the bottom line measurement of success is the profit of the farm as a whole, but it is possible to evaluate the contribution given by each enterprise by the use of gross margins. A comparative analysis can look at the margins between different crops on the farm, between different varieties of those crops, between different fields on the farm, between those crops and standard data and between those crops and potentially new crop introductions to the farm system. A farmer can then see which crops are contributing positively or negatively to the business, how much they are contributing and which elements within the margin could be improved. Decisions for the following seasons can then be made. Management programmes for the farm computer can now make this analysis speedy and accurate, although it must be remembered that the output is only as good as the input, and a system of data collection and management should be in place on most farms these days.

8.9   Break crops and crop rotations

8.9.1   Break crops

Since the mid-1980s, when the price of cereals began to fall, farmers have been looking for ways of diversifying their cropping systems. During the 1970s and early 1980s, good prices for wheat and barley meant that high inputs such as sprays and fertilisers could be afforded because the extra yield obtained would easily cover the material and application costs. Cereals were grown year after year on the same land. As margins were squeezed alternative crops were introduced onto farms. Collectively, these were known as break crops and they fall into two categories: (a) combinable break crops such as oilseed rape, peas and beans; and (b) non-combinable break crops such as sugar beet, potatoes, field scale vegetables and grass.

Alternative crops in a break crop system should do several important things:

- They should break the cycle of weeds, diseases and pests found in the cereal crops.
- They should improve the soil condition in terms of structure, organic matter levels and nutrition, especially nitrogen.
- They should make money; certainly, they should make more money than the alternative continuous cereal system of production.

- They should spread risk. This could be a marketing risk or the risk of crop failure, or another type of risk.
- They should spread the labour and machinery workload throughout the year.
- They should take the pressure off the storage, cleaning, drying and handling system used on the farm.
- They should allow the farmer to carry out more effective cultural control of weeds, diseases and pests, thus saving on pesticide costs.
- They should allow the use of different active ingredients in pesticides to try and overcome the problems of resistance build-up, and to provide cheaper control of the serious grass weeds which can build up in continuous cereal production.

All these factors should give advantages to arable farmers. They should be more profitable, they should have a more interesting system and it should bring environmental advantages to the farm as a whole. However, there are some disadvantages. The cropping system can become too complicated to the point where the crops are not managed well. Small amounts of different crop commodities do not put farmers in a strong marketing position and there can be a problem with broad-leaved weed and volunteer build-up over the course of the rotation.

Some break crops are better at fulfilling the above requirements than others. Oilseed rape is a very good break crop. It breaks the cycles of pests, diseases and weeds, it leaves the soil in good condition, it is drilled and harvested at different times than cereals and there are approved herbicides for use on rape which are effective graminicides. Linseed, on the other hand, is a less effective break crop in some areas. It is not very competitive against weeds, there are few approved herbicides for the crop, it leaves little nitrogen in the soil and is often harvested late and in difficult conditions so that drilling of the following crop is delayed. Legumes such as peas and beans fix atmospheric nitrogen and so provide nutrition for the following crop. However, both species can leave weed problems in the field and can be difficult crops to predict in terms of yield and profit. However, they currently attract a protein supplementary payment as part of the SPS, unlike oilseeds, so their popularity as break crops continues.

During the previous incarnation of long runs of cereals, one of the most difficult problems to overcome was that of the root disease take-all. This disease builds up in second, third and fourth wheats and then declines to manageable levels thereafter. Because it was soil-borne it was impossible to control other than by rotation and other cultural methods. Seed dressings are now available which can help to control take-all and this may lead to a revival of longer runs of wheat growing in the main arable areas, subject to CAP rules on diversification of cropping.

8.9.2   Crop rotations

The rotational use of crops in a 'fixed' sequence around the farm is now the norm on most arable farms. Rotations can be flexible in terms of crops used and length of the sequence. There is nothing new about the use of break crops in rotations with white straw crops. Farmers have been using fodder crops, roots, grass, legumes and cereals since organised agriculture began. Fine tuning of the rotations has taken place over the centuries as we have developed better cultivation and harvest equipment, better varieties, better fertilisers and better crop-protection materials. The principles remain the same: you need a mixture of exhaustive and restorative crops in an attempt to introduce as much sustainability into the cropping system as possible.

There are nearly as many rotations as there are farmers. A typical combinable crop rotation might be:

Year 1 – oilseed rape
Year 2 – first wheat
Year 3 – winter beans
Year 4 – first wheat
Year 5 – peas
Year 6 – first wheat
Year 7 – winter barley

and then back to oilseed rape and the cycle begins again.

This rotation can be lengthened by the inclusion of some second wheats or three-year leys, or can be shortened by the removal of a break crop and one of the first wheats.

A typical root crop rotation might be:

Year 1 – potatoes
Year 2 – first wheat
Year 3 – second wheat
Year 4 – sugar beet
Year 5 – first wheat

and back to potatoes.

The important factors are that the rotation should be agronomically sound and should not cause management problems with such issues as timeliness of operations, clashes of harvest or lack of separate storage facilities.

8.10   Sources of further information and advice

Bailey R, *Irrigated Crops and their Management*, Farming Press, 1990.
British Agrochemical Association, *Integrated Crop Management*, BAA, 1996.
Castle L, McCunnall R and Tring B, *Field Drainage – Principles and Practices*, Batsford, 1984.
Clarke J H, *Rotations and Cropping Systems*, AAB, 1996.
Davies B, Eagle D and Finney J, *Resource Management: Soil*, Farming Press, 2001.
Davies B, and Finney J, *Reduced Cultivations for Cereals: research, development and advisory needs under changing economic circumstances.* HGCA, 2002.
http://www.cropprotection.org.uk/media/2236/integrated_crop_management.pdf

9

Sustainable crop management

DOI: 10.1533/9781782423928.2.218

Abstract: Agricultural systems have intensified in order to produce crops and livestock at the lowest unit cost in a competitive global market. This can only be achieved in the long term through sustainable production systems that take account of the impact of farming on the natural environment and resources – water, soil and air. The Common Agricultural Policy is introducing new rules on having at least three crops in the rotation and a 'greening' requirement of 7%. Targets to improve water quality by reducing diffuse pollution have focused attention on soil management and are driven by the Water Framework Directive.

Key words: integrated farm management, crop rotations, soil management, wildlife and conservation, environmental stewardship.

9.1   Introduction

The agricultural industry has experienced a period of great change in the last decade following a relatively stable period through the 1980s and 1990s. The catalyst was the review of the Common Agricultural Policy (CAP) brought about by the Accession Countries joining the European Union in May 2004. The inclusion of a further ten countries required a change in the arrangements for agricultural support in order for the CAP to be sustainable. The reforms moved from subsidy payments targeted at production finally to a regional flat rate payment per hectare in 2012. These payments are linked to environmental and biodiversity targets. These changes mean that farming systems are now fully exposed to global food markets, and as a consequence prices for agricultural products are more volatile. The impact of these changes has encouraged farmers to produce food at the lowest unit price, which generally results in specialisation of enterprises and an increase in the area farmed to spread overhead costs. However, agricultural systems are still closely aligned to the concept of integrated farm management and sustainability, and the adoption of this philosophy will allow farmers to meet both cross-compliance and statutory management requirements as well as maintaining a viable farming system.

9.2   The Common Agricultural Policy (CAP)

The reform of the Common Agricultural Policy in 2004, called the Mid Term Review, has had a major impact on rural businesses. The aim of the reforms was to decouple support payments from a production basis. It was hoped that this would remove the financial incentive to intensify both arable and livestock enterprises by producing more grain per hectare or increasing stocking rates respectively. The Single Payment Scheme (SPS), (sometimes referred to as the Single Farm Payment (SFP)), started initially as a historic payment in England based partially on support payments received over the period 2000 to 2002. By 2012 these payments levelled out to a flat rate regional payment per hectare. The system was managed slightly differently in Wales and Scotland.

The concept of cross-compliance was introduced as part of the new measures. Farmers only receive the SPS if they demonstrate their land is in Good Agricultural and Environmental Condition (GAEC) and they comply with a number of specified legal requirements relating to the environment, public and plant health, animal health and welfare, and livestock identification and tracing. These are termed Statutory Management Requirements (SMRs). Farmers who are found not to comply can have their single payment reduced. For example, farmers are required to have a field margin of at least two metres which is uncropped from the boundary midpoint, and must produce a soil management plan for the farm which considers the risk of soil erosion and manure application to arable or grassland.

Modulation was introduced to provide the funds necessary to support agri-environment schemes in the UK, including Environmental Stewardship. Each year a certain percentage of the SPS is deducted at source by the Government to fund the England Rural Development Plan (ERDP). The funds deducted from the SPS are matched by an equivalent amount from the UK Treasury to provide the funds for Environmental Stewardship. From 2009 to 2012, 19% was deducted from the SPS (made up of a 5% statutory deduction set by the EU plus 14% set by the UK Government). Further information on SPS can be accessed on the Defra website and the Rural Payments Agency (RPA) website where booklets providing up-to-date information for each region on the SPS and Cross Compliance can be downloaded.

The EU has published proposals for further reforms to the CAP to take effect from January 2014 to 2021. Key changes in the proposals include replacing the current Single Payment Scheme (SPS) with the Basic Payments Scheme (BPS). This will be made up of a basic payment per hectare equating to 70% of the total payment and a greening payment comprising the remaining 30%. The greening of the CAP will require farmers to maintain an ecological focus area of 7% of the farm area, cultivate at least three crops and maintain permanent pasture at the level declared in 2014. An 'active farmer' definition has also been proposed, whereby landowners will only qualify for their payment if they are rearing or growing agricultural products, maintaining land in a condition that allows cultivation or grazing and carrying out a minimum amount of farming activity that will be set by each Member State. Whilst these reforms are proposed for the period 2014–2021, it now appears unlikely that they will be introduced before 2015 or even 2016.

9.3   Sustainable agriculture

Although there is no single definition for sustainable agriculture, most are similar to those for Integrated Farm Management (IFM) with the addition of social issues or resources, such as soil, air, water, biodiversity and energy. A definition for sustainable agriculture or crop production is 'ensuring the continuing availability to the consumer of adequate supplies of wholesome, varied and reasonably priced food, produced in accordance with accepted environmental and social standards'.

In order to be sustainable, the farming system adopted must be profitable, and there is no virtue in introducing all the beneficial features demanded by pressure groups if the business is not sustainable in the long term. The concept of sustainability draws attention to the importance of food safety in agriculture and that producing food on the farm is an integral part of the food chain, together with the processors, distributors and retailers. Currently, there is much more emphasis on the provenance of the food that we eat, and there has been a corresponding growth in farmers markets providing locally sourced food, which has prompted supermarkets to promote both regional and British produce. This has also stimulated a debate concerning the carbon footprint of the food we eat, with retailers starting to provide details on the products supplied. This can be confusing to the consumer depending on the stages of production, processing and transport that are considered. A true comparison can only be gained from a life cycle analysis, which is detailed and time consuming. The discussion has highlighted that as a nation we have become accustomed to a large range of fresh produce all year round, compared with the seasonal supply of fruit and vegetables accepted previously. There is awareness that food production impinges on the environment, and that water quality, soil erosion, wildlife and conservation are included as key aspects of farm management plans. The concern regarding greenhouse gas emissions has highlighted the contribution of methane from ruminants and the release of nitrous oxide from the soil in microbial processes converting nitrogen to different forms in crop production systems. These issues demonstrate the current tensions between implementing sustainable crop production systems and the requirement to grow sufficient food for an expanding global population.

Whilst Integrated Farm Management is put forward as a farming system that will fit the bill for sustainable agriculture, the role of Organic Farming Systems cannot be overlooked in this context. Many would argue that organic farming is the sustainable system that should be adopted to deliver safe, wholesome food and environmental benefits. Organic produce fulfils a significant market requirement that is currently estimated to be around 8% of food production in the UK in terms of monetary value and around 4% in terms of land area.

Other issues could be highlighted which are important components of the overall crop production system. For example, IFM involves the use of traditional farming practices alongside new techniques. Some commentators argue that there is nothing new in IFM, but when the use of precision farming techniques, involving the use of Geographic Information Systems (GIS) for yield and nutrient mapping or disease and insect forecasting are considered, it is clear that modern technology is being embraced in current farming systems. The use of cultural control methods and the reduction in agrochemical use is a key part of IFM philosophy. Alternative strategies for pest control are considered before pesticides are used, which introduces the idea that IFM requires informed decision making. Similarly, in animal production systems this technology allows monitoring and recording of individual animals that facilitates health and welfare as well as optimising production. There is no doubt that the approach requires a high management capability and a commitment in both time and effort.

9.4   The development of Integrated Crop Management (ICM), Integrated Farm Management (IFM) and sustainable crop production

Integrated Crop Management, or ICM, was first introduced in 1991 in an attempt to improve the public's perception of farming. Linking Environment and Farming (LEAF) promoted ICM in an attempt to reassure consumers that produce from British agriculture was safe to eat by informing them of the underlying philosophy and the methods of food production. This was in direct response to the criticisms that agriculture was facing following adverse media coverage of issues such as nitrates in drinking water, organophosphate residues in food, salmonella and *E. coli* contamination of food, the BSE crisis and the Foot and Mouth outbreak in 2001. Integrated Crop Management aimed to introduce a standard for crop production which would publicise the high standards of farming in the United Kingdom. However, it was apparent that although these aims were correct and laudable, it was not sufficient just to demonstrate and promote the principles of ICM. The fresh produce sector realised that further action was required to protect their industry and meet the demands of their customers – the supermarkets. The introduction of Assured Fresh Produce in 1996 gave the field vegetables and salad crops sector an externally verified assurance scheme that covered the health and safety aspects of growing these crops. The Assured Combinable Crops Scheme followed in 1997 to cover the small grain crops such as cereals, oilseed rape, peas and beans. The LEAF Marque has since been introduced to provide greater recognition in the market place of the high standards achieved by farmers in producing food.

There was no universally accepted definition of Integrated Crop Management. However, a definition would generally refer to the financial performance of a business that is producing food that is wholesome and safe to eat and includes the implications of the farming system on the environment in terms of wildlife and pollution. ICM was the first attempt to propose the concept of a holistic or whole farm approach to crop production. Each farm is unique, not only in respect of the physical landscape, features and climate, but also in its management and the farming system. Thus, the implementation of ICM principles will be different on each farm and there cannot be a blueprint.

It became apparent in the late 1990s that the term Integrated Crop Management (ICM) was not entirely appropriate when considering a holistic or whole farm approach. On livestock farms, crops are grown for feed and bedding, with animal manures returned to arable and grassland fields. In addition, issues relating to animal health and welfare have been given a much higher profile on farms to reflect their importance in current production systems. For these reasons LEAF introduced the term Integrated Farm Management as a more appropriate term for modern farming systems. This has been defined by LEAF as: 'A whole farm policy aiming to provide efficient and profitable production which is economical, viable and environmentally responsible.' Whilst integrated farm management is wholly relevant today, the terms sustainable crop production and sustainable agriculture are synonymous with IFM.

9.4.1   Crop rotations

Sound crop rotations are a cornerstone of sustainable crop production because they provide the opportunity to control pests, diseases and weeds by cultural means rather than relying solely on chemical control methods. They provide a mixture of exhaustive crops such as cereals, which remove valuable nitrogen and potassium from the soil, and restorative crops such as legumes, which fix nitrogen from the air into organic compounds. These are released to the following crop after the residues are incorporated into the soil. This reduces the requirement for inorganic nitrogen fertiliser that is not only expensive to produce but also a major energy input to growing crops because the manufacturing process involves using high temperature and pressure to produce nitric acid and then ammonium nitrate.

Set-aside was introduced to control the over-production of agricultural products and became an integral part of arable cropping systems. There was a great deal of flexibility in how set-aside was incorporated into arable cropping systems; it could be rotated or placed strategically adjacent to water courses or conservation areas. Leaving stubbles through the autumn was very valuable in providing a food source of weed and crop seeds, as well as invertebrates on volunteers and weeds for foraging birds. The removal of set-aside was welcomed by the farming community because the concept of leaving land uncropped was never accepted. The loss of habitats, such as overwintered stubbles and buffer strips next to water courses, has raised concerns from environmental organisations. This prompted the Campaign for the Farmed Environment where farmers were encouraged to voluntarily maintain or even exceed the benefits provided by set-aside, by establishing wild bird seed mixes or field margins with nectar and pollen species. A recent survey by Defra indicated that 37% of farmers in 2011/12 had left 82 000 ha of land out of cultivation for environmental benefit, with overwintered stubbles being the preferred option. This falls a long way below the original target of 600 000 ha, and hence the current CAP proposals are for 7% of the farm area to have an ecological focus to ensure that farmers allocate land to wildlife and conservation. Those farmers implementing IFM do manage the arable area of the farm to ensure profitable crops are produced, and consider carefully the options for the non-cropped areas on their farms taking advantage of Environmental Stewardship to achieve their objectives.

9.4.2   Soil management

Good soil management has always been recognised as an essential element of growing high yielding, profitable crops. Since the introduction of IFM much more emphasis has been placed on the importance of soil management in alleviating some of the environmental problems that are a consequence of agricultural systems. The Environment Agency aims to reduce diffuse pollution into both surface and groundwater from agricultural sources. Soil management practices to reduce nutrient leaching and soil erosion are key aspects that need addressing to achieve these aims. The current focus on soil management is strengthened with it being a cross-compliance requirement in the Single Payment Scheme. A Soil Protection Review, which is a documented soil management plan, had to be completed by 31 December 2010 in England. Welsh farmers had to provide this information by March 2005. The soil management plan examines the risk of soil erosion, diffuse and point source pollution on the farm, and then details management practices that can be implemented to reduce or alleviate the problems. The Soil Protection Review must be updated annually and implemented as part of the Cross Compliance requirements – GAEC1 (Good Agricultural and Environmental Condition). Whilst the emphasis is currently on individual farms, this approach will only be successful if implemented on a wider scale considering the farming systems in a geographic region or river catchment. In February 2005 the concept of Catchment Sensitive and Environmentally Sensitive Farming was introduced by Natural England. This aims to deliver practical solutions and targeted support to enable farmers and land managers to take voluntary action to reduce diffuse water pollution from agriculture to protect water bodies and the environment. Sixty five priority catchments and nine catchment partnerships have been identified, and catchment sensitive farming officers or river basin coordinators have been employed to provide advice and guidance to farmers in these areas. This is a key part of the strategy to achieve the aims of the Water Framework Directive. This requires that all inland and coastal waters within defined river basin districts reach at least 'good' status by 2015. The water quality of surface waters (lakes, rivers, estuaries and coastal waters) and groundwaters in the UK is assessed in relation to biological, chemical and physical parameters.

9.4.3   Cultivations

The aim of soil cultivations is to produce a seedbed which is firm enough to provide good soil to seed contact but has spaces between aggregates that allow free movement of water and air with no impedance to root growth. Establishing the desired plant populations and conditions that facilitate uptake of nutrients and water will provide the potential for a high yielding crop provided disease is carefully managed. The importance of soil structure and the role of soil organic matter in the aggregation of soil particles into stable entities is fundamental to these processes. Therefore, cultivations are a key component of IFM because soil management must produce the desired conditions without damaging the soil structure. This will result in surface run-off, erosion and environmental pollution from leaching of nutrients and pesticides.

Traditionally, ploughing has been the primary method of cultivation, inverting the soil to bury weeds and alleviate surface compaction. This is a costly operation, both in terms of the time taken and also the amount of energy required, with contractors currently charging around £60 per hectare. In the 1990s combination drills, incorporating a power harrow to produce a seedbed and a pneumatic seed drill, were used successfully as a one pass system. However, with the reduction in the grain price to £60 per tonne in 2004 severely affecting the profitability of winter wheat, interest in minimal cultivation and direct drilling increased again.

These crop establishment techniques fit comfortably with the philosophy of sustainable crop production. The use of minimum cultivation or conservation agriculture, as it has been referred to recently, has the benefit of reducing the amount of energy used to establish a crop and therefore meets the aim of reducing off-farm energy inputs. There are also additional benefits to invertebrates and wildlife. Trials carried out in the 1970s and 1980s comparing direct drilling, minimal cultivation techniques and ploughing showed that earthworm populations were inversely related to the amount of cultivation carried out. The slicing and inversion associated with ploughing kills earthworms resulting in fewer earthworms per metre squared than either minimum cultivation or direct drilling.

9.4.4   Crop nutrition

While good crop establishment is essential for successful and profitable crop management, the provision of nutrients at the correct rate and timing allow a crop to grow to its full potential. However, the inappropriate use of fertilisers, by applying at too high a rate or at times of the year when they are not utilised efficiently, can lead to losses to water and air causing pollution problems. Nutrient planning can be carried out on a field-by-field basis, tailoring the crop's requirement based on recommendations given in the *Fertiliser Manual*: RB209. Estimates of the soil nitrogen supply (SNS) are used to adjust nitrogen recommendations taking into account rainfall, soil type and previous cropping. The soil mineral nitrogen (SMN) content of the soil, which is a measure of the amount of ammonium and nitrate nitrogen available to the crop, can also be used on an individual field basis to predict the need for nitrogen fertiliser. Where there is a fixed cropping rotation on a farm, the nutrient management plan can be improved by carrying out nutrient balance calculations. From records of crop yields, the nutrient removal of phosphorus in the form of phosphate and potassium as potash can be calculated from standard values. These are then totalled over the rotation and compared with the recommendations calculated from RB209. In many situations there will be imbalances between off-takes and inputs that can be corrected. This should be carried out by trained agronomists or FQAs (FACTS Qualified Advisers), who hold the Fertiliser Advisers Certification and Training Scheme (FACTS) qualification, and can make adjustments according to individual crop needs.

Technological developments have resulted in precision farming becoming an integral part of sustainable crop production systems. The concept of canopy management for winter wheat and oilseed rape crops together with developments of weigh cells on fertiliser spreaders, yield monitors on combines and crop nitrogen sensors has led to the use of variable rate fertiliser application on many farms. This has been extended with soil conductivity measurements being used to prepare soil maps that can then be used for variable seed rate applications depending on seedbed conditions and time of drilling. These are offered as commercial services to farmers.

IFM places great emphasis on the utilisation of organic manures in cropping systems and the assessment of their nutrient contribution when calculating inorganic fertiliser requirements. For an individual crop this means analysing the manure to assess the nutrient content or using standard values from RB209 and subtracting the quantity of available nitrogen, phosphate and potash from the fertiliser recommendation. 'Planet' is available as an on-line nutrient management software tool that is freely available for use by farmers and advisers to calculate the fertiliser requirement of different crops taking into account the availability of nutrients in manures.

Records should be kept of all operations, including primary cultivations and applications with operator name, date, field, crop, soil and weather conditions. These should also include the type of fertiliser or organic manure applied and storage details. In Nitrate Vulnerable Zones (NVZs) records of all organic and inorganic nitrogen applications must be kept. All staff carrying out field operations should be fully trained and have access to records including the Nutrient Management Plan and the Codes of Good Agricultural Practice for Air, Soil and Water. Careful planning of crop nutrition is essential for profitable crop production, but this can only be achieved through the correct timing of application, machinery maintenance and calibration to avoid waste and pollution.

9.4.5   Crop protection

The EU Sustainable Use Directive (Directive 2009/128/EC) is intended to set high and uniform requirements for good practice in the use of pesticides. This Directive was transposed into UK law by the Plant Protection Products (Sustainable Use) Regulations 2012. National Action Plans have been produced by Member States designed to reduce the risks relating to the use of pesticides and to reduce their use wherever possible. There is a requirement that all professional users, distributors and advisers had access to appropriate training by November 2013. A further requirement is the adoption of appropriate measures to protect the aquatic environment and drinking water supplies which provides a link with the Water Framework Directive (2000/60/EC). The Sustainable Use Directive also requires that measures are taken to promote low pesticide input pest management, giving priority to non-chemical methods. The concept of Sustainable Crop Production (or IPM, ICM, IFM), which aims to utilise cultural control methods where the opportunities arise and reduce the reliance on chemical control, is very much in line with this policy and is fully supported by the agrochemical industry. Many publications, leaflets and the Defra website promote IFM as a farming system that will deliver many of their objectives.

The history of agricultural production has been inextricably linked to the control of weeds, pests and diseases in crop production practices. There is no doubt that the production of consistently high yielding, high quality crops requires the use of pesticides. Although organic production methods provide a viable alternative the yields achieved are lower and the quality of the produce can be more variable. The aim, therefore, of an IFM system is to utilise the cultural control methods that are known to reduce the incidence of pests on crops, and then use crop protection chemicals as a final resort.

Whilst it has been stated that rotations are a cornerstone of IFM, there can be little doubt that the choice of variety selected to be grown is an integral part of the philosophy. If the aim is to reduce the application of crop protection chemicals the starting point must be to choose varieties from the HGCA Recommended Lists that have resistance to disease. The next important step in the implementation of a crop protection policy involves the correct identification and evaluation of a pest problem. This has been recognised by the requirement of all advisers recommending crop protection products to hold the BASIS Certificate of Competence.

Once the need for a plant protection product has been justified the choice of product, rate and timing are all very important decisions to make. The crop protection chemical selected will not only depend on its ability to control the pest, but also on its environmental profile. An environmental information sheet (EIS) is produced for each pesticide approved, and these are held on the Voluntary Initiative website. Properties such as the ease with which the chemical leaches to water, persistence in the environment, volatility and effect on non-target organisms can be taken into account. A major challenge for the industry involves the development of strategies to overcome pesticide resistance and avoidance of plant protection products polluting groundwater.

Integrated farm management also places great emphasis on operator training for all aspects of work carried out on the farm. The area of pesticide application provides an excellent illustration, with the spray operator being required to hold a PA2A certificate. These certificates are issued by the National Proficiency Tests Council to show professional competence in calibrating and operating a boom spray applicator. Further training would be required to understand and follow 'The Code of Practice for Use of Pesticides on Farms', which would involve carrying out a COSHH (Control of Substances Hazardous to Health) assessment of the risks to health from using a pesticide before work starts. This includes assessing the level of personal protective equipment which the spray operator is required to wear.

The spray operator would also need to be familiar with the Health and Safety Executive (HSE) guidance on storing pesticides. There are also issues involved in the disposal of pesticides, such as the EU Groundwater Directive, which requires farmers to have designated areas for the disposal of pesticides. Knowledge of the sprayer, and how it operates, can reduce risk to the operator and environment through the use of close transfer filling systems or low drift nozzles to reduce spray drift. The field of crop protection demonstrates the requirement for a well-trained workforce, conversant with the relevant legislation and codes of practice, as well as the practical skills necessary to carry out certain tasks, in order to conform to IFM principles.

9.5   Food quality and safety in the food chain: farm assurance schemes

The farmer is the first link in the food chain. He determines the way crops are planted, the inputs applied to the growing crop and the method of harvesting and storage. Therefore, the farmer has a significant impact on the safety and quality of a food product, especially where fruit and vegetables reach the consumer with very little further processing. The farm assurance schemes set standards for good agricultural practice, and require farmers to provide evidence of the methods used in producing their crops. A further quality control imposed is the maximum residue level (MRL) of a pesticide in food on the supermarket shelf. This is not a safety limit, but indicates if a pesticide has been applied in accordance with the manufacturers' recommendations, and is monitored and tested for by The Food Standards Agency.

The management of food safety in the food chain is being addressed by assurance schemes, with the use of HACCP (Hazard Analysis of Critical Control Points) systems being seen as the preferred approach to management at all stages of food production, including agricultural production. The HACCP system identifies hazards which can cause the consumer temporary or permanent injury, and assesses risk as the probability of a hazard occurring. Current EU hygiene regulations include reference to HACCP proposing that it should be extended to the primary production of food on farms to provide effective control of pathogens and their hazards, fulfilling the key aspects of the Farm Assurance Fresh Produce Scheme (previously Assured Produce) or EUREP GAP Schemes.

The first crop assurance scheme to be introduced in 1996 was Assured Produce for growers supplying vegetable and salad crops. This was an initiative by the producers themselves, putting in place a system, which not only recorded farming operations and inputs, but also allowed external scrutiny or verification of the production systems. The voluntary introduction of the scheme aimed to demonstrate to supermarkets that field vegetables and salad crops were grown to professionally agreed protocols, and provide traceability. This has given rise to several quotes, such as from 'Plough to Plate' and from 'Field to Fork' indicating that the whole food supply chain is included. Assured Produce has provided an industry standard for the main UK supermarkets. The Assured Combinable Crops Scheme was launched in 1998 to address the same issues for arable crops. Whilst not placed directly onto supermarket shelves like many vegetables and salad crops, wheat grown for bread making is processed and baked at the miller's prior to being sold in retail outlets. It is therefore just as important for these to be subjected to the same quality assurance standards as other food.

Both the Assured Produce and Assured Combinable Crops Schemes concentrated on ensuring that good farming practice is followed when growing crops. These schemes are now integrated into Assured Food Standards. Although it may address some environmental issues, there is no requirement for farmers to undertake positive habitat management that will benefit wildlife. One UK supermarket chain did address this issue by offering preferred growers the opportunity to provide a plan for the development and improvement of wildlife habitats on the farm. This has been taken further by developing a Biodiversity Action Plan (BAP) for species that are present on the farm and are highlighted as important species for conservation in the locality. Assurance schemes have a cost to the farmer, which depends on the area of land being used for crop production. The fee pays for the administration of the scheme and includes a visit by the external verifier.

The EUREP GAP Protocol was introduced in November 1999 to standardise the requirements of retailers in the EU and to bring benefits to the food supply industry. The Euro-Retailer Produce Working Group (EUREP) is a technical working party aimed at promoting and encouraging good agricultural practice (GAP) in the fruit and vegetable production industry. The scheme aims to fulfil the same role as the Assured Produce and Assured Combinable Crops Schemes established in the UK, but actually goes further in requiring the producer to have an environmental management plan that addresses wildlife and conservation as well as health and safety issues. EUREP GAP has become established for fresh produce and is currently being introduced for other crops such as cereals.

9.6   Wildlife and conservation

Farmers have been seen as the custodians of the countryside because of the way they have managed and developed landscapes and habitats over the centuries. However, recent surveys have shown that the numbers of breeding birds in farmland habitats have declined substantially. This has focused attention on the farming practices in an attempt to find the reasons and halt or reverse these declines. Loss of habitats, such as hedgerows, woodland and rough grazing, drainage of wetlands, change of cropping practice from spring sown crops to predominantly winter planting, intensive cropping systems dependent on the use of pesticides, fertilisers and increased mechanisation are considered to be the main factors causing farmland birds to decline.

Integrated Farm Management acknowledges that the countryside is our heritage and that it is incumbent on everyone, including farmers and the farming industry, to work towards its conservation. The first step in any on-farm improvement is to carry out a survey of the site to determine what landscape features are present, record the habitats and make an assessment of the species present, possibly quantifying the numbers of selected indicator species. This would then form the basis of a management plan for both the cropped and uncropped areas on the farm.

On many farms in the UK this would focus on field margins, and particularly hedgerows, as a habitat that is a prominent component of the rural landscape. Field margins have been viewed by some farmers as a source of pernicious weeds, such as cleavers, couch and sterile brome. The previous answer was to prevent spread into the cropped area by spraying into the hedge bottom with a broad spectrum herbicide. This was not only expensive and time consuming as it was carried out each year, but also destroyed the field margin vegetation and frequently reduced the growth of the hedge. IFM advocates leaving a minimum one metre grass strip of non-invasive perennial grass species (eg. mixtures of cocksfoot, Yorkshire fog, Timothy and red fescue) adjacent to the hedge or field margin (Fig. 9.1). On many farms this grass margin has been extended to widths varying from 2 to 12 metres using options from Environmental Stewardship. The biodiversity in these margins can be improved further by the addition of native wild flowers to encourage beneficial insects, such as bees, hoverflies and lacewings (Fig. 9.2).

Conservation headlands have been promoted to further increase the biodiversity adjacent to the field margins. A grassy strip adjacent to the hedgerow could

Fig. 9.1   Grass field margin.

Fig. 9.2   Buffer strip with wild flowers.

provide suitable nesting habitat for grey partridge. However, in cereal crops treated with residual herbicides, very few weeds survive to provide habitat for insects and particularly saw fly larvae, which are a preferred food source for partridge chicks. A conservation headland is typically the outer six metres of a field that is selectively sprayed to control serious grass weeds (wild oats and black-grass) and cleavers, but allows other broad-leaved weeds that are hosts for invertebrates to survive. This concept has been shown to be successful in promoting biodiversity and increased partridge chick survival, but does increase the weed seed burden in the harvested crop from the field headland.

Beetle banks are grassy strips created in the centre of large arable fields to provide suitable habitat for predatory insects as well as nesting habitat for partridge or skylark. They are designed to provide a habitat of tussocky grasses (e.g. cocksfoot) where ground beetles can overwinter in relatively dry conditions. The predation of aphids is more effective because the beetles move into the crop from the beetle bank in the centre of the field and the field margin round the outside.

These measures, together with the sympathetic management of existing field margins such as hedgerows, ditches and associated ponds, can have a major impact on wildlife. For example, the introduction of a management plan for cutting hedges would advocate rotational cutting on a two to three year cycle to allow berry development. The actual cutting would be carried out in late winter, where possible, to allow birds to forage through the autumn and winter when food is scarce. Similarly, cereal stubbles could be left undisturbed until spring before sugar beet or maize is sown to provide valuable feed for a wide variety of farmland birds. These are some of the options in the current stewardship schemes that all farmers can consider undertaking to enhance the biodiversity of arable farming landscapes.

Environmental Stewardship was introduced with the CAP Reforms in 2004 as the new agri-environment scheme, which consists of Entry Level Stewardship, Organic Entry Level Stewardship and Higher Level Stewardship, that will eventually replace the old schemes, such as Environmentally Sensitive Areas and Countryside Stewardship Schemes. Entry Level Stewardship (ELS) is a whole farm scheme open to all farmers and land managers who can meet the scheme's requirements. Organic Entry Level Stewardship (OELS) is also a whole-farm scheme open to all farmers who manage all or part of their land organically. Payments are double those for ELS and there is more for the first two years if in-conversion. Higher Level Stewardship (HLS) is combined with ELS options and aims to deliver significant environmental benefits by improving the environment for wildlife, improving water quality and reducing soil erosion, maintaining and enhancing landscape character and protecting archaeological features. Payment in the HLS scheme is based on options delivered. Payments are available for a range of capital works, unlike in ELS and OELS.

There are over 65 management options to choose from that provide a range of environmental benefits. Several have been developed to provide nesting and foraging for farmland birds, such as fallow plots or uncropped areas for ground nesting birds or overwintered stubbles to provide winter foraging for birds and mammals (e.g. finches and the brown hare). Wild bird seed mixtures can be tailored to benefit particular bird species; for example teasels (Fig. 9.3) provide

Fig. 9.3   Field corner with teasels.

Fig. 9.4 Grass buffer strip next to a water course.

seed for goldfinch, sunflowers provide seed for greenfinch, linseed and cereal mixtures provide suitable habitat and seed for the grey partridge and kale provides seed in the second year for linnets. Buffer strips and grass margins prevent surface water run-off, soil erosion and reduce spray drift into water courses (Fig. 9.4). The benefit to wildlife of any grass option is improved greatly if wild flowers and legumes are included to provide pollen and nectar for insects, such as bees and butterflies.

The national target is for 70% of farms to enter ELS agreements. In many cases farmers will not have to change their farming practice significantly to accumulate the points required. However, some of the measures being introduced could have a major impact. For example, the introduction of the Soil Protection Review, requiring a soil management plan, means that a soil erosion risk assessment must be carried out on the farm. On many farms this may change agricultural practices where soil water erosion is a problem, for example in wide-spaced row crops such as forage maize, grown on light sandy or silty soils with moderate slopes.

Applications for HLS generally require the farmer/landowner to have made an application for, or already be in, Entry Level Stewardship – ELS or OELS. The first step in making an HLS application is to complete a Farm Environment Plan (FEP), comprising a whole-farm assessment of biodiversity, historic, woodland and landscape features. The FEP will identify the environmental features that need management and their current status in order to help select the HLS options appropriate for the specific farm. Completion of an FEP will require specialist knowledge and in most cases needs to be carried out by an external consultant/agent, with the cost claimed from the scheme. HLS is discretionary and applications will only be successful if they deliver significant environmental benefits in high priority areas.

9.7   Key points

- Sustainable crop production is delivered by adopting systems based on Integrated Farm Management and Organic Farming.
- Reform of the Common Agricultural Policy together with exposure to global markets has had a major impact on rural businesses, leading to further specialisation and a drive towards reducing the unit cost of production to remain competitive.
- Targets to improve water quality by reducing diffuse pollution have focused attention on soil management and are driven by the Water Framework Directive.
- Crop Assurance Schemes are an integral part of the food supply chain required to ensure food quality and safety of farm produce.
- Environmental Stewardship aims to deliver widespread environmental benefits through ELS and OELS, with HLS aimed at high priority situations and areas, providing an income source to farmers and landowners for these agreements.

9.8   Sources of further information and advice

Further reading

Benckiser G and Schnell S, *Biodiversity in Agricultural Production Systems*, Boca Raton: Taylor Francis Group, 2007.
British Agrochemical Association, *Integrated Crop Management*, BAA, 1996.
Defra, *Fertiliser Manual* (RB209), TSO, 2010.
Defra, *Pesticides – Code of Practice for Using Plant Protection Products*, Defra, 2006.
Defra, *Code of Good Agricultural Practice: Protecting Our Water, Soil and Air – A Code of Good Agricultural Practice for Farmers and Land Managers*, Defra, 2011.

HSE *Guidance on Storing Pesticides for Farmers and other Professional Users*, HSE, 2012.

HGCA, *Arable Cropping and the Environment*, HGCA, 2002.

HGCA, *Managing Uncropped Land to Encourage Biodiversity*, HGCA, 2010.

HGCA, *Field margins – Guidelines for Entry Level Stewardship in England*, HGCA, 2005.

HGCA, *Beneficials on Farmland: Identification and Management Guidelines*, HGCA, 2008.

HGCA, *Enhancing Biodiversity – Six Practical Solutions for Farms*, HGCA, 2007.

LEAF, *Simply Sustainable Soils*, LEAF, 2011.

Natural England, *Entry Level Stewardship: Environmental Stewardship Handbook*, Natural England, 2013.

Natural England, *Higher Level Stewardship: Environment Stewardship Handbook*, Natural England, 2013.

RPA, *The Guide to Cross Compliance in England*, RPA, 2012.

Websites

www.defra.gov.uk

www.rpa.gov.uk

www.naturalengland.org.uk

www.planet4farmers.co.uk

www.voluntaryinitiative.org.uk

www.rspb.org.uk

www.gwct.org.uk

www.cfeonline.org.uk

www.leafuk.org

www.saffie.info

www.cropprotection.org.uk

www.assuredfood.co.uk

10

Precision farming

DOI: 10.1533/9781782423928.2.235

Abstract: This chapter looks briefly at the history of precision farming techniques and how they have evolved up to the present day. It covers the costs and benefits of the technology and techniques, and then describes the various mapping methods that can take place on-farm these days. It discusses the use of auto-steering and variable rate application in particular. It also discusses the difficulties sometimes encountered when interpreting data. Finally it looks at the possible future of precision farming, with its potential use of robotics and data management.

Key words: cost-benefit, auto-steering, controlled traffic, real-time kinetics, variable rate application, robotics.

10.1   Introduction

10.1.1   History

The availability and development of technology in agriculture has always been a driver for change. From horses to tractors, from churns to bulk tanks, from paint marks to electronic tags, technology has been adopted by the innovators and, once it is seen to work, becomes widespread and the 'norm'. In the case of arable farming, this technology has been used for a number of reasons – quicker operations, reduction of labour costs, more effective husbandry, etc. Now arable farming is looking to technology to make its operations more efficient and precise. As long ago as the 1960s, crop researchers had access to such things as GPS signals, but they were a new technology, not particularly accurate and expensive, so there was no commercial uptake. Crop researchers were also doing the things that mainstream farmers do now – they recorded and gathered data in order to analyse it and use it to make informed decisions about inputs.

The technology developed rapidly during the 1980s and 1990s and it became cheaper, for instance, to receive satellite signals for GPS. This, combined with yield monitors on combines and on-the-move nitrogen testing, followed by developments at institutions such as Silsoe (now Cranfield) into variable rate application of fertilisers and sprays, saw the beginnings of precision arable farming as we now understand it.

As more and more data was collected, the innovators and early adopters demanded more and more analysis so companies began to provide commercial systems which were integrated, analytical and reliable. Software was developed to help analyse data and present the results in a user-friendly way and, during the late 1990s and up to the present day miniaturisation of hardware meant that the components were much easier to fit onto machines and, as we have now, many could be carried around in the hand by agronomists, just like a mobile phone.

Nowadays the term often used for the integrated use of precision farming technology, better models and algorithms for decision making, a committed consideration of the environment, food security and food safety, combined with shrewd business decisions is 'smart farming'.

10.1.2   Definitions

There are a number of definitions of precision farming but it might be useful to narrow it down to precision agronomy, which could be described as 'the matching of agronomic inputs and practices to localised conditions within a field and the improvement of the accuracy of their application'.

The Home Grown Cereals Authority (HGCA) define precision farming as being 'management of farm practices that uses computers, satellite positioning systems and remote sensing devices to provide information on which enhanced decisions can be made.'

The United States Department of Agriculture (USDA) call this kind of agriculture 'as needed' farming and define it as 'a management system that is information and technology based, is site specific and uses one or more of the following sources of data: soils, crops, nutrients, pests, moisture or yield, for optimum profitability, sustainability and protection of the environment'.

The latter is quite wordy and is not specific about weeds and diseases, but it encapsulates quite well the meaning and objectives of the system, as does the Defra sponsored Shuttleworth Precision Farming Alliance, which describes precision farming as follows:

> The precision farming procedure can be summarised as follows. Data pertaining to yield and potential yield-affecting factors are initially collected, and then analysed to determine which factors are actually affecting the yield. If yield is being affected, a farm manager decides the type, distribution and amount of treatment to apply. Remedial measures can then be carried out to ensure that the correct treatment is applied at the required rate and to the appropriate area within a field. In effect, the spatial variability in field is managed through the manipulation of inputs such as fertilisers and pesticides.
>
> Variable application of inputs may not always increase yields, but simply hold them constant whilst reducing input costs. Precision farming enables the farmer to reap increased profit through better management, and the

application of more appropriate/reduced chemical treatments also helps to preserve the environment.

Like any system that is designed to provide improvement, there is a cycle of events which should take place to monitor its effectiveness and helps people to understand some of the definitions above (Fig. 10.1).

Like any new technology there is an associated new language and jargon. The HGCA provide a useful glossary of terms on their precision farming website. See 'Sources of further information and advice' below.

10.1.3   Benefits of precision agronomy

Benefits can be grouped in two main areas – financial and environmental.

Financial

The most obvious financial benefit is the potential savings in costs of production, although there is an argument to suggest that yields and quality of crops can be improved by more effective use of inputs. There can be costs savings on all three major inputs to crops – seeds, fertilisers and agrochemicals.

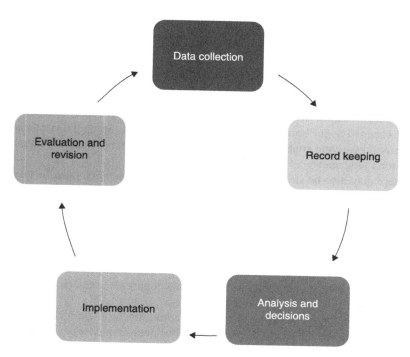

Fig. 10.1   The decision-making cycle.

Seeds

GPS systems, with in-field triangulation, are accurate to within a few centimetres these days so, with the right machinery, seed overlaps can be made a thing of the past. This is especially important when sowing high value crops with expensive seed such as field vegetables or sugar beet. Even with 'cheaper' seed such as cereals or oilseed rape, as long as the individual coulters on the drill can be controlled by the on-board computer, drill overlaps can be substantially reduced, even with the drill travelling at much higher speeds than precision drills.

As well as the physical distribution of the seed, the technology can also allow differential seed rates to be changed on the move to allow for the differences across a field in soil type, condition and nutrition status. This allows optimisation of seed rates to allow, we hope, higher yields and less seed wastage.

Fertilisers

As more and more restrictions are placed on how much, and where, fertilisers can be applied there are good financial reasons for being precise in the distribution of these products. Not only are farmers trying to avoid wasteful use of these ever increasingly expensive materials, but they also risk losing payments and incurring fines if they do not meet their obligations under, for instance, the Single Payment Scheme's cross-compliance regulations in the UK. The most obvious savings can be made when spreading nitrogen, and the technology exists to adjust rates of both liquid and solid N on the move, either based on previously gathered data such as yield mapping or on-tractor monitoring of crop colour or near infra-red reflection.

Sprays

Just as with fertiliser, sprays are expensive and costs of purchase, application and disposal are rising all the time. Data collected by agronomists using hand-held GPS systems can be used to target specific problems, with rates and mixtures being changed as the sprayer drives down the tramlines. Spray overlaps can be minimised by giving control to the on-board computer, rather than relying on the operator's judgement.

There may also be savings of labour costs as field operations can be quicker with, for instance, steering guidance. Costly mistakes may also be reduced.

Environmental

A reduced risk of pollution from nitrates, phosphates and agrochemicals is fairly obvious but less inputs used more effectively means fewer road miles for delivery, less packaging production and waste and less pressure on disposal of plastics, etc. It could mean fewer passes through the field if the inputs are used more effectively, saving on fuel and reducing GHG emissions.

However, before committing oneself to the investment in the technology it is important to assess the likely benefits to the business by a cost/benefit analysis. HGCA provide a cost : benefit tool which will do this for you, but, before you start using such a tool, there are certain factors which should be taken into account.

Yield limiting factors (YLFs) that cannot be controlled in the field will not be affected by precision farming methods. Natural rainfall, sunlight hours, humidity, CO_2 levels and air temperature are variables that cannot be altered. Soil temperature and soil texture can be affected by husbandry, but only over a long period, sometimes decades.

Soil structure, drainage, acidity, inherent soil fertility, available water content and pests, diseases and weeds are all YLFs that can be managed using precision farming, although most are medium term alterations within the field.

Those YLFs that can be quickly altered are factors such as temporary soil fertility, operator error, timeliness of inputs and the effect of sowing and husbandry on crop architecture. Of course there are interactions between all, or at least most, of them.

10.2   Data collection

The HGCA glossary describes this as 'the gathering of information on fields and crops in digital form by sensors, in addition to data collected manually or visually'. Even the data collected manually or visually need to be entered digitally eventually, so that they can be used by the software.

10.2.1   Yield mapping

Probably the oldest of the data collection technology is yield mapping using data collected on the combine harvester, but the technology today can be retro-fitted to nearly all types of machine, including sugar beet and potato harvesters. Data can be sent wirelessly in real time direct to computers, or stored in Clouds for later download and analysis. Yield maps are then produced on screen and can be printed out or saved and used in the application machinery at a later date. It is important to realise that the more years of data you can collect, the more useful the maps are. This allows for both spatial and temporal variation within a field and masks the 'odd' years when yields were either very high or very low because of uncontrollable YLFs, e.g. the weather.

The accuracy of the maps is only as good as the data generated. The hardware and software are improving with every new generation of machine, but accuracy can still be affected by factors such as sudden speed changes of the harvester, blockages, temporary loss of GPS signal and then, post-harvest, factors such as data aggregation and simplification, and point 'smoothing' during the generation of the contour maps.

10.2.2   Soil mapping

Soil mapping in its crudest sense has been carried out for many years. Lime testing across a field is a form of soil mapping which generates data which allows differential application of calcium carbonate to correct pH. Taking standard soil samples and testing for P and K levels is another form of soil mapping, although, in the past, samples from across a field were usually combined to give a single field result on which applications were based.

Now companies offer a mapping service where samples are taken using an ATV fitted with a GPS sensor and a nutrient or texture map can be produced similar to a yield map. Others offer satellite or aerial photography digital images which show soil brightness indicating texture, soil organic matter and soil moisture content. RGB images can also be produced which indicate crop health and could point to underlying problems such as soil compaction or salinity.

In all cases it is important to understand that the technology must be used in conjunction with other data, such as field history, weather records and geology maps. Ground truthing is also important – RGB images may hint at a compaction problem but the farmer will still have to dig a hole to confirm this!

There are also other measurements that are becoming linked to precision agronomy and soil management. Electrical conductivity meters can be hand held or attached to ATVs and, using probes, can give an indication of moisture and clay content of a soil. Hand held compaction meters are available which measure physical resistance to being pushed through the soil profile and sensors can now be fitted to individual legs of a cultivator to measure soil resistance on the move.

10.2.3 Crop mapping

This can be done by remote sensing or by crop walking. Agronomists can use hand-held GPS units to record patches of weeds and incidences of diseases or pests. They can also record areas of poor growth or of damage from larger pests such as rabbits. Remote sensing can be used to record variable biomass or colour. The cameras can be satellite based, on a plane or drone, or on the tractor/sprayer which records colour changes as it passes along the tramlines. All can be affected by weather conditions, those from the air by cloud cover in particular, and those on the ground by the quality of the sunlight and the angle at which it strikes the crop. The latest technology allows the adjustment of the sensors to take into account light quality, or enables them to generate their own light. Developments in technology mean that ultrasound and RADAR have been tested to determine crop structure, such as tiller numbers or height of crop.

10.3   Data interpretation

This part is probably the most difficult to manage when it comes to precision agronomy. Farmers can be overwhelmed by the amount of data being produced, and the numerous interactions that might be taking place in the field. There are some fundamental questions to be asked when looking at data from a field or a crop:

1. Is there variability?
2. Is it significant and consistent?
3. Can I find out why it is happening?

4. Can I do anything about it?
5. Will it improve the situation if I do something?
6. Can I measure and evaluate that improvement?

If the answer to any of the questions above, except perhaps the last one, is 'no' then the mapping and variable application techniques are probably not worth considering on that field (there may be other precision technology that IS appropriate) until things change, or more data is acquired. If the answer to the first 5 questions is 'yes', it is unlikely that it will be 'no' to question 6. All improvements can be measured using the right technology.

When the precision farming systems were in their infancy, it was often difficult to interpret yield maps, for instance when they were all the farmer had to go on. One common dilemma was: If I have an area of low yield in a field, should I put more fertiliser on it to try and bring the yield up, or accept that it will always be low yielding because of other factors, and I need to put less on, thus saving money on wasted material?

The key phrase in the question is 'other factors' and it is these that we are becoming more sophisticated at measuring and modelling. Companies and agronomists are now able to accurately map differences in soil properties, in incidences of pests, diseases or weeds and have a better understanding of the importance of field history, including old boundaries, manure storage, pipelines, earthworks, spillages, sites of buildings, etc.

Now, the more sophisticated interpretation of data allows what the industry calls 'targeted actions' or 'targeted agronomy'. This targeted response can include variable rate applications of fertilisers or sprays, clearly delineated areas for subsoiling or mole ploughing, alteration of sowing rate or depth, specific rabbit fencing or even more detailed and careful crop walking.

10.4   Auto-steering and controlled traffic farming

One of the other main uses for precision farming technology is to make the operator's life easier by allowing the machines to steer themselves. This can also have the effect of more precise and accurate applications and operations as well as making sure that wheelings only occur in a tightly controlled area within the field. This is especially important in crops that are very sensitive to compaction such as sugar beet or carrots. Ensuring that the crop is not grown above a wheeling cannot only increase yield, but also the quality of the final product, e.g. straight or unfanged carrots, which would be accepted for supermarket pre-packs. It also allows subsequent remedial action to correct compaction to be targeted very precisely, thus reducing costs.

Systems are now available that will perform headland turns as well as in-work steering. The operator sets the initial guidance line (sometimes known as the A-B line) and then tells the on-board computer to steer the machinery parallel to it at a pre-determined distance away from it. The operator can then concentrate on making adjustments to any implements, watching out for problems or observing the crop if it is an operation during crop growth.

10.5   The technology

10.5.1   Data collection

Yield monitors
On combine harvesters these measure grain flow and moisture content as well as recording field position, area covered and combine speed. They can then display the yield map on an on-board screen and make it available for download. Data can be recorded as often as every second if required.

Soil mapping
The technology is not yet available to carry out instant soil analysis of N, P and K in the field so samples are taken from a recorded position and then analysed in a laboratory. pH can be measured on-the-move, as can compaction using probes which measure resistance. Electro conductivity can be instantly measured as well.

Crop mapping
This is one area where developments in miniaturising the technology have made it available to more and more agronomists and farmers, similar to mobile phone technology within the general population. The use of satellite imagery, or aerial digital photography (ADP) carried out by unmanned drones, gives the big picture, but more detailed information can be obtained with tractor- or implement-mounted sensors, or hand-held devices that measure, for instance, chlorophyll colour in leaves, either by light reflection or by physical contact with the leaf sap.

Weather
Farmers can buy small automatic weather stations which can be moved around the farm and automatically record rainfall, temperature, humidity, sunlight and wind speed and then wirelessly transmit it to a base computer. They are run off solar power.

10.5.2   Variable applications

Using variable rate technology (VRT) inputs can be applied to a field using an application map or real-time sensors. Auto section control (ASC) is used to turn on or off individual nozzles or boom sections of a sprayer during the application of pesticides or liquid fertiliser. Automatic machine control uses the computer to adjust automatically the rates of application of seeds, fertilisers or sprays. Direct injection sprayer technology, where the sprayer carries clean water and the chemicals are injected into the pipeline as required, means that targeted tank mixes depending on spatial need can be achieved. Solid fertiliser can be adjusted on the move with a spreader-controller. These are available on newer machines as options and can be retro-fitted to older machines.

10.5.3   Controlled traffic and auto-steering

There are numerous guidance systems available. Reliance on satellites alone can be inaccurate because of atmospherics or satellite drift, so the most accurate systems use real-time kinematics (RTK) where the satellite signals are triangulated with a fixed ground system, either on a building or on a tripod in the corner of the field. These can give accuracy to within ±2 cm.

Guidance systems come in many shapes and sizes, as well as complexity and ability. There are two factors involved – positioning of the machine in the field and steering of the machine once in motion. Systems can be used to position and steer fully automatically or can be set up to provide on-board guidance for a tractor driver. Usually a display of some sort will tell the operator if the tractor or harvester is in the right place and heading in the right direction. This can be as simple as a light bar indicator or as sophisticated as a high resolution computer screen with 3D graphics. The simpler systems can be moved from vehicle to vehicle but the more complex ones are inbuilt and stay with the machine.

These systems are at their most useful when there are no tramlines, i.e. before the crop has been drilled and during the drilling process. They can be used in conjunction with the drill to provide very accurate tramlines which can then be used with or without auto-steering options. A full system which will control not only steering and positioning, but also ground speed, engine speed, rates of application, etc., is called automatic machine control.

10.5.4   Data analysis

Once data is collected or logged over time it has to be analysed by software, and then used to variably apply inputs or make other management decisions. It is often combined with other data, e.g. weather records, to help the analysis. A mass of individual pieces of data are produced from, say, a combine harvester travelling across a wheat field. Some errors will be amongst this data. To produce a yield map the data must be smoothed or 'kriged' and it is the ability of the programmes used to do this that determines how useful and accurate the map will be.

Much of the work today is based on modelling or simulation to try and predict what might be the crop response to a range of inputs and other influencing factors such as weed populations, disease pressure, weather factors, soil conditions, etc. These are all being developed to provide an as accurate as possible decision support system for farmers.

10.5.5   Other applications

As the technology improves, becomes more reliable and gets smaller, so more and more uses are likely to be found for it. Of particular interest (especially where there are restrictions on inputs such as low input, organic or biodynamic farming), are those techniques that allow crop weeding to occur accurately within a crop by mechanical means. Machines have sensors (cameras) mounted to them which detect when there is a plant present and then trigger a rotating arm to harrow the soil around it. A number of these mounted on a toolbar can mean rapid mechanical weeding of high value row crops such as salad crops or field vegetables.

Between-the-row weeding can be carried out using GPS knife-based systems where each knife or tine is independently controlled by the GPS/RTK system. This could be developed for precision drilled crops if the seed position was recorded during planting and this information was used by the weeder to carry out between-plant weeding as well, using only positioning information rather than visually identifying the plant.

Developments in robotics are also in the pipeline. Robotic weeders, crop scouts, roguers, selective harvesters are all under consideration by the scientists and engineers. These are likely to be autonomous, lightweight machines that do minimal soil damage, use solar power for energy and can also be used for micro-application of inputs. Quite a long way into the future maybe, but technology moves quickly and the imperative to make farming more sustainable and self-sufficient is a strong driver for change.

Finally, traceability and recording. The future may see all data produced automatically and sent direct to the customer. It may see information on crop inputs, acreages, varieties sent directly to Defra for its returns to be collated, or for compliance to be confirmed. The use of electronic labels on fertiliser bags or seed bags could allow automatic recording of inputs used within a field. Customers buying salad crops might be able to talk directly to robotic harvesters, which might then be able to harvest in the dark ready for delivery the next morning.

These are just a few of the many possibilities for precision agronomy and production. There is no doubt that, with a population expected to reach over 9 billion by 2075 from the 7 billion today, with shrinking land resources and climate change reducing the production capabilities of many highly populated countries, with dwindling energy reserves and a desire to reduce environmental damage, the pressure is on to make better use of our agricultural capacity and the industry believes that precision farming will play an important role in doing this.

10.6   Sources of further information and advice

Home Grown Cereals Authority Be Precise Resources. Available from: http://www.hgca.com/content.output/6259/6259/Crop%20Management/Crop%20Management/Precision%20farming.mspx

National Centre for Precision Farming, Harper Adams University. Available from: http://www.harper-adams.ac.uk/initiatives/national-centre-precision-farming/

Defra Shuttleworth Precision Farming Alliance. Available from: http://adlib.everysite.co.uk/adlib/defra/content.aspx?id=000IL3890W.16NTBYKEP4S1WN

11

Organic crop husbandry

DOI: 10.1533/9781782423928.2.245

Abstract: Organic farming is a system that avoids the routine use of soluble fertilisers or agrochemicals, whether naturally occurring or not. It is a system that aims to use renewable resources where possible and is considered to be a more sustainable method of farming than standard conventional non-organic farming systems. This chapter will discuss the principles of organic farming including rotations, nutrient supply, and pest, disease and weed control. An outline of husbandry for organic winter wheat, potatoes and winter cabbage is also included.

Key words: principles of organic farming, weed, pest and disease control on organic farms, organic farming and the environment, organic winter wheat, organic potato production.

11.1   Introduction

Organic farming is a system that avoids the routine use of readily soluble fertilisers and/or agrochemicals, whether naturally occurring or not. It is a system that aims to use renewable resources where possible and is therefore considered to be a more sustainable method of farming than standard conventional non-organic farming systems. Most organic farms have a lower energy use than conventional farms, mainly due to no inorganic ammonium nitrate being applied. Organic farming is not a return to pre-1940s farming. It uses current technology and relies heavily on good husbandry. Organic farming aims to maintain soil fertility as well as encouraging biological cycles. The majority of organic farms include grassland in their rotations which can help maintain soil organic matter and increase carbon sequestration. There is a large reliance on legumes for supplying the crop nitrogen requirements. Weeds, pests and diseases are controlled mainly by non-chemical methods. Animal welfare, pollution and the wider social and environmental issues are also very important.

The ideas on organic farming started early in the twentieth century. In 1946 a group of like-minded farmers and scientists, including Lady Eve Balfour and Sir

Albert Howard, founded the first UK organisation, The Soil Association. It was not until the 1990s that there was a large increase in interest in organic farming both in the United Kingdom and in many parts of Europe and the world. About 10 million ha or 5% of land in the European Union is now farmed organically. Just over a quarter of the world's organically farmed land is in Europe, with Spain, Italy and Germany being the largest producers. In the UK, organic farming did not increase until the mid-1990s, mainly due to the introduction of organic support and low conventional commodity prices but also due to several food scares and growing public demand for organic produce. Since the 2000s the area has slightly reduced, partly due to difficulties in marketing some organic produce and partly due to the recession reducing demand for the higher priced organic food. There are now over 700 000 ha of organically managed land in the UK (see Fig. 11.1).

Currently in the EU, all states provide some form of support for organic farming through their rural development programmes. The amount of support varies widely between Members States. In England, payments are received if farmers enter the Organic Entry Level Stewardship scheme (OELS). Payments are highest for the first two years in-conversion. Wales, Scotland and Northern Ireland have broadly similar schemes.

Some organic crops such as cereals are cheaper to grow and grain prices are higher than for conventional crops; but to be viable they cannot be grown too frequently in the rotation. Many field vegetables are much more expensive to

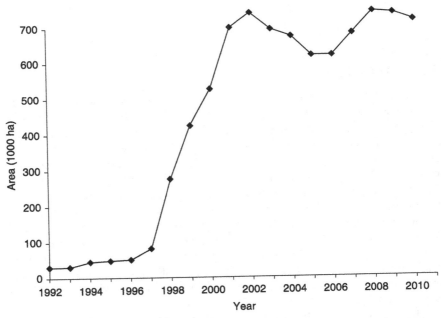

Fig. 11.1   Area of UK organically managed land (*Source*: Defra, organic statistics).

grow than conventionally grown crops especially if hand weeding is required, as in carrots. Yields from organic crops can be very variable and are often only 50–70% of average conventional yields.

For organic farming to work it must be viable financially and the farmer or grower has to be committed to the system. There has to be a change in the approach to the husbandry of growing the crops as agrochemicals and fertilisers cannot be applied if a serious problem arises. The risk of crop loss is much higher than in conventionally grown crops.

11.2   Achieving organic status

All products sold as organic have to be certified. All countries have their own certifying bodies. In the EU all Member States must comply with the EU statutory regulations for organic production. These regulations are underpinned by the guidelines produced by the International Federation of Organic Agriculture Movements (IFOAM).

11.2.1   UK organic standards

In the UK, Defra is responsible for ensuring that the standards set by the certifying bodies comply with EU regulations. There are a number of approved certifying bodies including the Soil Association Certification Ltd, Organic Farmers and Growers Ltd, Organic Food Federation, Scottish Organic Producers Association, Quality Welsh Food Certification Ltd and Irish Organic Farmers and Growers. Each farm is annually inspected by its registered body to confirm that the Approved Guidelines are being followed; Defra carries out random check inspections to ensure standards are maintained.

Standards are constantly being amended; it is important that farmers keep good farm records. Current recommendations include: farmyard manure can only come from stock that has not been fed GMOs (genetically modified organisms) and written proof is required; bought-in farmyard manure has to be composted. As organic seed stocks are limited, farmers can currently obtain derogations from their certifying body to plant non-organic seed. In the future the target is for all seed sown to be organic. A limited number of fertilisers and crop protection products are permitted to be applied to organic crops. Some products have restricted use such as meadow salt (a potassium source) and copper for control of blight. Farmers must have a very good reason to apply these restricted products, and they have to obtain derogation from their certification body before application. The permitted and restricted lists are constantly being amended according to UK and EU regulations.

During the annual inspection, input and output records as well as accounts have to be available. Crop records required include: treatments applied for the last three years for land in-conversion; crop rotation plans and cropping areas; cropping history including yields; application of farmyard manures and other fertiliser, composting treatments, rates and source; source and types of products used for pest and disease control and source and types of seeds and transplants.

After inspection a compliance form is sent to the farmer with details of the certification decision for each enterprise on the farm, and details of any action to be taken. Each farm must provide up-to-date certification details when selling organic produce.

11.2.2   The conversion period

When a farmer decides to convert to organic production methods a conversion plan is produced in order to get certified by an approved body. Depending on the enterprises on the farm the plan includes details of land areas, dates of conversion, cropping and rotations, stocking rates for livestock, forage management and details of stock housing and produce storage. Conversion can start from the date of application of the last non-permitted product. The ground is not fully organic until two years after that date. Some farms convert all their land at the same time; others find it easier for rotational purposes to convert over a few years. At present, in the UK, a farmer does not have to convert the whole farm, unlike most other EU countries.

Support payments for both the two-year conversion period and subsequent organic maintenance payments are run differently in England, Wales, Scotland and Northern Ireland. Currently in England, funding is included in the environmental stewardship schemes organized by Natural England. For farmers to receive funding they must apply for the Organic Entry Level Stewardship scheme (OELS) and comply with their chosen environmental management options.

During the first year of conversion, crops can only be sold as conventionally grown non-organic. Second year in-conversion produce can sometimes be sold at a premium which is usually lower than for full organic products. It is common practice to grow fertility-building crops rather than cash crops during the two-year conversion period. There is usually a reduction in farm returns during the in-conversion period, hence why organic support grants are highest in the first two years.

11.3   Rotations

Rotation is the basis of a good organic system. Initially, when changing to this farming method, during the in-conversion period fertility-building crops such as clover are usually grown. A good rotation will aim to balance nutrient requirements, minimise nutrient losses, maintain soil organic matter and structure and help eliminate or reduce weed, pest and/or disease problems. Organic rotations use more legumes than conventional cropping; red and white clover, lucerne and even vetches are grown to boost soil nitrogen supply.

Green manures such as mustard, vetches or rye are commonly grown after a summer-harvested crop and before a spring-planted crop. Green manures reduce the losses of soil mineral nitrogen during the winter; they also provide plant cover so that the soil is less prone or liable to suffer from soil water erosion. When ploughed-in in the spring the green manures release the readily available nitrogenous compounds for the following crop.

Normally no more than two years of cereals are grown in succession. Cereal prices are highest for milling wheat; often more than £100/t compared with the conventional price. With the increase in organic livestock enterprises there has been a growing demand for organic feed grain and currently the UK is not self-sufficient and has to import. Other crops grown include grain legumes, potatoes and field vegetables. There is no one perfect rotation; each farm will have a rotation to suit the enterprises, soil and climate of that specific farm. Continuous cereals are prohibited and there must be at least a four year break before cropping again with brassicas or potatoes on the same land. An example of a rotation used on mixed organic units is given in Fig. 11.2.

A number of farms practise a stockless system. In stockless rotation, usually two years of cereals will follow a legume. The leguminous crop could either be field beans or even a forage legume for seed. Straw is incorporated and any spring crops grown could follow a green manure. This system has been successfully run for many years on some farms. Other farms have found that livestock needs to be introduced. This is only feasible on farms where there are adequate livestock facilities or where there are local organic livestock units looking for grass keep.

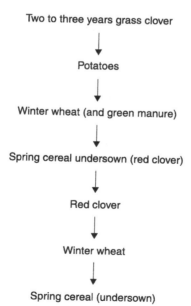

Two to three years grass clover

↓

Potatoes

↓

Winter wheat (and green manure)

↓

Spring cereal undersown (red clover)

↓

Red clover

↓

Winter wheat

↓

Spring cereal (undersown)

Fig. 11.2   An example of a rotation used on mixed organic units.

11.4   Soil and plant nutrition

An important aim in organic farming is to maintain and, if possible, improve the soil – the structure, flora and fauna. The ideal is to balance nutrients removed from the farm, in harvested crops or livestock, with inputs from bought-in feed, straw and seed. There is limited data to suggest that organic farming increases soil bacteria and fungal activity and that these are very important in mobilising some of the nutrient reserves in the soil.

Soil structure can be affected by crops grown, cultivations and application of organic manures. Some crops which have deep tap roots, such as the vetches and field beans, have been shown to improve soil structure. Shallow cultivations tend to be favoured to try and maintain the level of available nutrients in the topsoil. It is not always possible to cultivate shallowly when ploughing out a ley or after a vegetable crop with large crop residues. When ploughing out grass leys there can be an initial flush of available nitrogen released, and it can be difficult to time cultivations and crop drilling to reduce nitrate leaching. Quite often the use of the stale seedbed technique for weed control and avoidance of some pests and diseases leads to late drilling for some crops, and this means that the soil can be liable to nutrient loss before the crop is able to use the available nutrients.

Grazed leys and incorporation of crop remains such as straw and green manure crops help to maintain the soil organic matter.

11.4.1   Inorganic fertilisers

Nutrient off-takes should be balanced by nutrient inputs. There is only a limited range of fertilisers that are permitted for use on an organically grown crop. Most of these products are not very soluble and tend to become available to the crop over a long period. Production should be planned to minimise the need for bought-in nutrients. Lime, e.g. ground limestone or chalk; phosphorus compounds, e.g. natural rock phosphate; magnesium, e.g. magnesium rock or kieserite; and trace elements, e.g. liquid seaweed, are examples of permitted products.

There are limited potassium sources other than farmyard manure (FYM), bought-in straw and animal feeds. Under certain circumstances some restricted products can be applied. Potassium, e.g. meadow salt or Silvinite, is an example of a restricted fertiliser. In the case of Silvinite, the grower has to submit a recent soil analysis result for the field in question. The soil index must be 0 or 1 for derogation to be given. This product is prohibited for use in some other European countries.

Because of the restrictions on fertiliser and manure sources several stockless organic farms have now introduced a livestock system. There are, however, a number of successful stockless farms, which often have started with high levels of soil potassium or have soils with a reasonable clay content. These soils should have sufficient amounts of potassium mineralised each year for crop production.

11.4.2   Organic manures

As the use of inorganic fertilisers is limited, farmers rely on the application of organic manures including approved sources of amenity waste for both soil improvement and provision of crop nutrients. Some organic manures can be brought onto a holding as long as they come from an approved livestock system. The use of plant wastes and animal manures from non-organic sources are restricted and must be approved by the certifying body before use. FYM from non-organic sources has to be composted for three months or stacked for six months. As noted earlier, no GM feed should have been fed to the stock producing the manure.

There are several ways of managing compost. The aim is to limit nutrient loss. FYM should be composted to reduce soil-borne diseases and weed seeds. Turning is encouraged to ensure an adequate breakdown process. Turning usually increases aeration and speeds up the microbial activity, but the more the manure is turned the greater the gaseous loss of nitrogenous compounds. The nitrogen in well-composted FYM is usually fairly stable and not prone to soil leaching. During composting the manure will reduce in volume by 50% so reducing spreading costs. There is a lot of debate about whether to cover the manure pile. Covering can help lower the moisture content and leachate losses, but more commonly the heap is outside, uncovered. If so, the aim is to build the heap so that excess water runs off and, if possible, the leachate is collected.

Manure management is important in order to minimise nutrient losses during the composting period. Care must also be taken to ensure that there is no risk of run-off and pollution of water courses during application. In an organic system, when the manure has been composted, there can be a loss of nitrogen and potassium. As organic feed can be lower in potassium than in conventional sources, normally organic manure contains less available potassium than standard data would suggest. Use of sewage sludge, effluents and sludge-based composts is prohibited. Manure from some livestock systems such as intensive poultry battery systems is also prohibited.

11.4.3   Nitrogen sources

Most of the nitrogen in well-composted FYM is in an organic form and will be slowly released over the following years. Only a relatively small amount will be available in the first season after application. Organic farmers in nitrate vulnerable zones are restricted to applying no more than the equivalent of 250 kg/ha of total nitrogen in FYM on any one field per year. As FYM can vary considerably in N content between farms due to differences in stock, feed and housing it is worth undertaking an analysis. Timing of application of manure can also affect the amount of soluble nitrogen compounds that are present. Losses are usually lowest from a late winter/early spring application.

The major nitrogen source for organic crop production is from nitrogen-fixing legumes. Residues of nitrogen are usually highest after lucerne or red clover. Ploughing-out a legume and planting the following crop can lead to high losses of soluble nitrogen. It is very difficult for an organic farmer to minimise these nitrogen losses as planting date is also governed by weed, pest or disease control needs.

There are a number of pelleted-approved organic fertilisers. These usually contain a very low percentage of nutrients (N, P and K) and are very expensive. Their main use is on specialist horticultural holdings.

11.5   Weed control

In an organic rotation a much broader range of weed species is commonly found. Some of the most difficult weeds to control in organic crops are perennial weeds such as creeping thistle and docks. The importance of different weeds in organic compared with conventionally grown crops is partly due to the different rotations and the soil nutrient status. Many of the most serious conventional arable weeds, such as cleavers and black-grass, tend to be very responsive to applied nitrogen and continuous winter cropping and so are less important in an organic system.

Many growers converting to organic production are often seriously worried about potential weed problems. In practice, there can be problems but there are some very effective methods of control available.

11.5.1   Rotations and cropping

A mixed rotation, rather than mono-cropping, can have a major effect on weed species. A rotation of winter- and spring-sown crops will tend not to favour any weed species in particular, and the inclusion of good cleaning crops such as potatoes or grass leys will also help to reduce weed populations. Some crops themselves have an effect on weeds (allelopathy). Commonly, different weeds grow in different crops planted in the same field at the same time. Oats and red clover are two crops that appear to be very good at suppressing weeds. Growing a mixture of crops such as beans and wheat has also been shown to reduce weed biomass compared with the amount found in the individual crops. Weed problems tend to be less in new grass leys if they are established in the spring as an undersown crop rather than after a cereal crop in the autumn.

Green manures can reduce weeds by ensuring good ground cover during the autumn and winter. Ploughed-in green manure crops such as rye and mustard can also suppress weed (and occasionally crop) germination.

11.5.2   Variety selection

Variety should not be underestimated as a method of weed control. Many varieties of the same crop have very different growth and/or leaf habits. Varieties should be chosen that produce a vigorous leaf canopy, such as *Cara* maincrop potato, which will help suppress weeds. Varieties with horizontal (lax) rather than vertical leaves will also have an effect on weeds. It has been found that the very old variety of wheat, *Maris Widgeon* which has a tall straw and lax leaves, can reduce weed biomass by around half that found in commonly grown short strawed varieties.

11.5.3   Time of sowing

Delayed sowing is often practised on organic farms usually due to the avoidance of pest and disease problems; it can also aid weed control. Delayed drilling can give time for some weeds to germinate and be killed by cultivations before drilling takes place (the stale seedbed technique). When growing some vegetable crops such as leeks and brassicas transplants are usually used. This gives scope to kill weeds before planting the crop. Once the crop is planted it is also able to compete very quickly with most weeds. The combination of transplants, late drilling and in-crop cultivations means that these crops can have a lower number of weeds, so making them excellent cleaning crops.

11.5.4   Seed rates

High seed rates can help reduce weed competition. In winter cereals drilling 500 seeds/m^2 compared with 200 seeds/m^2 can halve the weed biomass. In horticultural crops it is not always possible to plant high populations as this technique can affect crop quality and lead to poor marketable yields.

11.5.5   Cultivations

This is one of the major tools that an organic farmer can use. Often on organic farms more cultivations are required than on conventional farms due to the requirements for weed control. It starts as soon as the first cultivations take place. Most organic farms use the plough more than many conventional farmers. When cropping after grass/clover leys the ground is often rotovated before ploughing as this can help kill the grass. Ploughing is a very good method for controlling some of the grass weeds and some perennial weeds. The 'stale' seedbed technique is commonly used before sowing the next crop. Once sown or planted there is now quite a range of in-crop weeders that can be very effective if the timing and weather conditions are right, the most commonly used being the tine or harrow comb weeders (Fig. 11.3). These can be used on a wide range of crops from cereals to many field vegetables such as the *Brassica*. In cereals these harrows can be very effective especially if the weeds are small; results have ranged from 0 to 80% weed control. Mat-forming weeds such as ivy-leaved speedwell and chickweed can be easily pulled up, even late season, compared with weeds with tap roots such as charlock which are not very well controlled. In winter cereals, autumn weeding (when the crop is not competitive) can lead to more weeds emerging. To get the best results with these weeders there will be some crop damage, but usually there is little effect on crop yield. An effective harrowing may cover the crop plants with up to 30% of soil without causing crop damage.

Fig. 11.3   Harrow comb (Copyright OPICO).

Depending on the season, crops may be treated several times with the harrow comb.

As well as whole-field treatments with the harrow comb some crops are also cultivated with an inter-row hoe (Fig. 11.4). Inter-row cultivators are very important for weed control in organic field-vegetables and root crops. The use of some of these implements is very effective at minimising the amount of hand weeding required. Some of these weeders are self-propelled, some front tractor mounted and others rear mounted with an operator on the back (steerage hoe). A wide variety of hoes is used, from goose-foot shares to brushes or even finger tines. Brush hoes (Fig. 11.5) do appear to work better than conventional hoes in wet conditions. The hoes can have an effect on weeds within the crop row by throwing soil into the rows and so smothering the weeds, especially when the weeds are small. Precision guided weeding systems have been developed which can be much more accurate and work closer to the crop rows than conventional hoes. These weeders use video image analysis techniques to distinguish between crop plants and weeds. To work effectively, the crop plants need to be more dominant than the weeds. A precision weeder is being developed for transplanted crops that can weed both inter-row and inter-plants using the same technology.

As in conventional farming it is important to start the weed control programme in time before the weeds start competing. In many crops, such as beans, carrots, swedes and onions, the first in-crop weeding is the important one. It should be timed when the weeds are small, usually only three to four weeks after 50% crop emergence. This is when the rows are just visible and the crop plants have their first true leaves. In transplanted crops, the first weeding can take place within a

Fig. 11.4   Inter-row cultivator – front mounted on tractor.

Fig. 11.5   Steerage hoeing using a brush weeder.

week or two of planting as long as the transplants are not easily uprooted. Weeding with a tine weeder is usually done as soon as the next flush of weeds is at the cotyledon stage. Dry conditions after weeding will ensure better control.

11.5.6   Flame or thermal weeding

This method of weed control is very important in some horticultural crops. The use of propane gas burners can be effective at controlling a range of weeds. The heat from the burner causes the plant cell walls to burst and the weed normally dries up and dies within two to three days. These burners do not control all species. Weeds such as chickweed, redshank and cleavers are very susceptible even past the two true leaf stage. Charlock is only susceptible at the cotyledon stage and many grass and perennial broad-leaved species are only checked. In crops such as carrots, onions and parsnips, flame weeding is carried out just before the crop emerges. This gives these small-seeded crops a chance to get established before the weeds start to compete. Flame weeders can also be useful to burn off potatoes before lifting.

There are several tractor-mounted models available (Fig. 11.6). The problem with flame weeding is the high energy consumption and slow work rates, but it can reduce hand-weeding costs by half in some horticultural crops.

Fig. 11.6   LPG flame weeder (Copyright Thermoweed).

11.5.7   Other methods of weed control

Hand weeding or hand hoeing cannot be ruled out on organic farms and can be very important when trying to avoid the development of potential serious weed problems such as docks. In some horticultural crops such as carrots and onions, even after flame weeding and inter-row cultivations, hand weeding is still needed. At least 200 man hours/ha are commonly required. To make weeding easier some farms use bed weeders (Fig. 11.7). These are tractor drawn or self-propelled frames on which the weeding gang lie so that both their hands are free for pulling weeds; this ensures that the task is not back-breaking and can make the job much quicker and easier than when hand weeding on hands and knees.

Some perennial weeds such as couch and docks can be difficult to control. One option is to introduce a fallow period and to keep cultivating the ground. If weather conditions are dry then repeated cultivations can be very effective at dragging the weed roots or rhizomes to the soil surface where they will dry out. A mulch such as black polythene is another method of weed control that can be very effective; the problem is cost. Mulches are only economically viable for use on high value horticultural crops. Black polythene is usually laid using tractor-mounted equipment. Polythene mulches have to be removed after harvest; occasionally they can be reused. Currently a number of biodegradable mulches, such as those made out of recycled paper, are being developed.

Fig. 11.7   Bed weeding.

11.6   Disease control

Some diseases common in conventional crops, such as the cereal foliar diseases, are not so serious in organic crops. This is partly because many organic crops are thinner than the conventionally grown crops and this reduces the risk of the spread of rain splash diseases. The thin crops also lead to more air movement and less of a microclimate within the crop canopy. Lush crops tend to be those most prone to damage by diseases. It is suggested that a balanced soil and crop nutrient status will help to reduce the incidence of some diseases. Application of manure has been shown to reduce disease possibly due to affecting the soil micro-flora.

Choice of crop and species is very important. In an organic rotation some crop species are grown less frequently than on many conventional farms; this can significantly reduce disease pressure. A limited number of organic variety trials have been undertaken in a range of crops; however, recommended lists for conventionally grown crops can still be useful when deciding on the most appropriate variety. Growing mixtures of cereal varieties can help to lower disease incidence and lead to small yield increases. Mixtures are only feasible if growing cereals for the feed market, rather than for a specific quality market.

Good crop hygiene is very important for organic farmers to help reduce disease incidence. Seed and storage diseases can be serious; organic growers should only sow, or plant, disease-free seed or transplants and only disease free crops should be stored.

Even with using all these disease-control methods there are still some diseases that can be very serious in organic crops. One example is potato blight, with only a limited difference in susceptibility between varieties, although there are a number of eastern European varieties that are showing promise as regards resistance. Currently in the UK growers can apply for derogation to apply products such as copper oxychloride once blight appears. These copper products have restricted use and are likely to become prohibited in the near future.

11.7   Pest control

As with diseases, there are a number of pest problems that can be reduced by good husbandry practices such as using wide rotations. Increasing the natural predator population can reduce some pests, for example aphids. Organic farmers are encouraged to manage field margins and boundaries sympathetically. On some organic farms higher predatory insect numbers have been found compared with conventional farms. There has been some success from reduced pest damage using companion planting and mixed cropping.

Time of sowing can be very important for reducing pest problems. Cereal growers in high-risk areas normally drill winter cereals after the middle of October to reduce aphid attack and transfer of barley yellow dwarf virus (BYDV). Maincrop carrots are sown at the end of May or in early June to avoid the first generation of carrot fly.

On a limited scale, crop covers are used over susceptible crops when there is likely to be damage. These crop covers form very effective barriers against pests such as carrot and cabbage root fly and caterpillars. Other than cost the problem with crop covers is that they have to be removed before weeding can be carried out.

Pheromone traps are sometimes used as an aid to warn of possible pest problems. There are also a limited number of permitted treatments including soap sprays against aphids and the bacteria treatment, *Bacillus thuringiensis* for control of caterpillars such as the cabbage white. There are several other permitted, naturally occurring, biological control agents mainly for use in glasshouses.

11.8   Husbandry examples

As an organic farmer cannot always control the problems once they appear it is important to get the husbandry correct at the beginning and to realise any potential problems. The following is an outline of standard husbandry for a range of organic crops.

11.8.1   Winter wheat

Organic winter wheat crops usually follow a grass/clover ley or a crop that has been manured, for example potatoes. After grass, at least four weeks are normally left with bare ground to reduce the risk of frit fly. Ploughing should be as shallow as possible, though adequate grass burial is required.

A disease-resistant variety should be chosen for the required market. Feed varieties are higher yielding than the quality milling varieties but the end price is lower as in conventional farming. If the farm is growing any conventional cereal crops it is important that a different variety is grown on the organic area. The crop should not be drilled too early, after mid-October is best to reduce the BYDV risk. A fairly high seed rate should be used, 400–500 seeds/m^2; this seed rate will make up for bird and other pest and disease losses as the seed is untreated. Bird problems can be much more serious when using untreated, compared with standard dressed, seed. In practice, crop establishment can be as low as 40 to 50%.

Weed control will depend on species present and soil conditions. Up to three treatments may be required overall with a harrow comb between November and April. Wild oats and docks will need hand roguing. If the crop is drilled on wide rows then the weeds can be controlled using an inter-row cultivator.

The crop is harvested at a similar time as conventional crops in August. Yields can vary between 3 and 7 t/ha, depending on the fertility of the site. Recently, the major market has been for grain for human consumption, although the feed market is under supplied. Grain quality can be a problem, particularly low protein levels. Prices can be as much as £100/t more than the conventional price.

11.8.2   Potatoes

Potatoes are usually grown after a fertility-building crop such as a grass/clover ley. If the potatoes follow a long-term ley then wireworm could be a potential problem. Potatoes are a very responsive crop to composted manure which can be ploughed into the seedbed. Cultivation systems are similar to those for conventionally grown crops, the only difference being time of ploughing. Autumn ploughing of grass/clover leys can lead to high nitrate losses. In an organic system the aim is to try to limit leaching losses, so cultivations are sometimes delayed.

Varieties should be selected that are suited to the market requirements and with good resistance to foliage and tuber blight. Choose slug-resistant varieties if slugs are a problem. Good quality seed should be used; and organic seed is now readily available; it is important that planted seed is as free of disease as possible. Seed-rate calculations are as for conventional crops, but adjustments should be made for more expensive seed and lower yields. To reduce the effect of blight there is an advantage to chitting the seed so that the crop is earlier to mature.

Potatoes are very competitive against weeds. Weed control is usually completely mechanised as specialised potato ridge cultivators are available. The other option is to chain harrow the ridges, then re-ridge afterwards. These cultivations are usually done just before crop emergence.

Blight is the biggest problem when growing organic potatoes, especially in high-risk areas. Farms should ensure that there are no sources of infection on the farm such as potato dumps and volunteers. Currently, growers can obtain derogation to apply copper compounds once the disease is found. These treatments can help to slow the progress of the disease, but if infection starts early in the summer yields can still be significantly reduced. The potato haulm is usually mown off once blight becomes very serious. Removing the haulm when there is less than 25% infection can reduce yields. The haulm is best burnt off with a flame weeder straight after flailing to reduce the risk of blight spores being washed down to the tubers. (Rounded rather than steep edged ridges can also help reduce the number of blight spores washed through the ridges.) Flame burning will also clean the ridges of many weeds and ease harvesting a few weeks later. Only disease free tubers should be stored.

Potato prices have been as much as double the conventional prices, but yields are very variable. Average marketable yields of organic maincrop potatoes are about 26 t/ha, but can vary between 10 and 40 t/ha.

11.8.3   Field vegetables, e.g. winter cabbage

Organic vegetables are grown on a range of farm types. There are a number of specialist organic vegetable holdings, although more generally they are included in the rotation on mixed farms where there is a supply of compost and nitrogen supplying legumes. Some small-scale producers concentrate on local markets and vegetable box schemes. Other growers market their produce through cooperatives or pack houses, which supply the supermarkets. Marketing needs to be organised before growing the crops.

Meeting the quality requirements for organic vegetables can be much more difficult than for conventionally grown crops, both in relation to size and damage from pest and disease. Storage losses can also be higher than for conventional crops. The procedures used for growing and harvesting are very similar to those for conventionally grown crops except for rotation, crop protection and fertilisers.

With winter cabbages for example, rotations are no more than one year in four as this helps to reduce the incidence of some diseases. Varieties are often chosen not just for yield, disease resistance and competitiveness against weeds, but also for taste. Organically raised transplants are used so that there is plenty of time for seedbed cultivations and weed control. Care has to be taken with plant population; this should be adjusted depending on the fertility of the site so that marketable yield can be maximised. Weed control is best carried out using a harrow comb weeder and inter-row cultivator. Harrow combing (up to three times) should take place at each flush of new weeds followed by an inter-row cultivation; little hand weeding is normally required. Cabbage caterpillars are effectively controlled using *Bacillus thuringiensis*. Crop harvest is sometimes later, depending on soil fertility, than for the same conventional crop. Marketable yields for winter cabbages can be slightly lower than those for the standard crops. Recently prices for organic winter cabbages have not been much higher than for non-organic crops.

11.9   Other systems

Another ecological farming system which is practised by a few growers in the UK, but is more popular in other parts of Europe, is Biodynamic Farming. This is very similar in many ways to the standard organic systems. It was started in the 1920s following the lectures of Rudolf Steiner. It is a whole system involving organic ideals as well as the spiritual approach. Special preparations are used when composting manure. Lunar study can be important when planting crops. Biodynamic produce is sold under the Demeter label.

11.10   Organic farming and the environment

Organic farming principles tend to encourage a variety of habitats on farms. More organic farms have mixed rotations which tend to encourage a greater diversity of flora and increased numbers of fauna compared with some non-organic farms.

In the UK, the Soil Association Certification Ltd (SA) gives recommendations for improving biodiversity on organic farms; such as maintaining at least 5% of land in permanent pasture or un-cropped areas, or ensuring 10% of cropped land is planted after 1 February. Farmers can choose the appropriate stewardship options to ensure that the SA guidelines are achieved.

As organic farming does not use manufactured fertilisers, energy requirements of this farming system and greenhouse gas emissions are subsequently reducing compared with non-organic systems. But as organic yields in t/ha are usually lower than those obtained conventionally, the total energy requirements for every tonne of crop produced is sometimes not very different. The use of manures and mixed rotations including grass and clover leys are very important for increasing the soil organic matter and carbon sequestration on organic farms.

11.11 Sources of further information and advice

Further reading

Applied Plant Research, *Practical Weed Control*, Wageningen, 2006.
Balfour E, *The Living Soil*, Faber and Faber, 1943.
Blake F, *Organic Farming and Growing*, The Crowood Press, 1995.
Briggs S, *Organic Cereal and Pulse Production: A Complete Guide*, The Crowood Press Ltd, 2008.
Davies G and Lennartsson M, *Organic Vegetable Production: A Complete Guide*, The Crowood Press Ltd, 2012.
Davies G, Sumption P and Rosenfeld A, *Pest and Disease Management for Organic Farmers, Growers and Smallholders: A Complete Guide*, The Crowood Press Ltd, 2010.
Davies G, Turner B and Bond B, *Weed Management for Organic Farmers, Growers and Smallholders: A Complete Guide*, The Crowood Press Ltd, 2008.
HGCA, *Organic Arable Farming, Conversion Options*, HGCA, 2008.
Howard A, *An Agricultural Testament* (reprint), The Other India Press, 1998.
Lampkin N, *Organic Farming*, Old Pond Publishing Ltd, 2002.
Lampkin N, Measures M and Padel S, *Organic Farm Management Handbook*, University of Wales, Aberystwyth and Elm Farm Research Centre, 2011.
Newton J, *Organic Grassland*, Chalcombe Publishing, 1993.
Newton J, *Profitable Organic Farming*, Blackwell Science, 1995.
Organic Centre Wales, *Organic Farming*, Aberystwyth University, 2009.
Sattler F and Wistinghausen E, *Bio-dynamic Farming Practice*, BDAA, 1992.
Soil Association publications including *Organic Farming, Living Earth* and *Growers* leaflets.
Younie D, Taylor B R, Welch J P and Wilkinson J M, *Organic Cereals and Pulses*, Chalcombe Publishing, 2002.
Younie D, *Grassland Management for Organic Farmers*, The Crowood Press Ltd, 2012.

Websites

www.biodynamic.org.uk
www.fibl.org
www.gardenorganic.org.uk
www.ifoam.org
www.naturalengland.gov.uk
www.organic.aber.ac.uk
www.organiccentrewales.org.uk
www.organic-europe.net
www.organicfarmers.org.uk
www.orgfoodfed.com
www.orgprints.org
www.organicresearchcentre.com
www.organicresearch.net
www.organicweeds.org.uk
www.organic-world.net
www.soilassociation.org.uk
www.organic-market.info

12

Plant breeding and seed production

DOI: 10.1533/9781782423928.2.263

Abstract: This chapter deals with the principles and processes of agricultural plant breeding as established by Mendel in the nineteenth century and further developed as hybridisation and other enhanced plant breeding techniques, including genetic modification. The role of state-funded and commercial plant breeding and seed multiplication is discussed and the development of variety information systems for farmers by organisations such as NIAB (National Institute of Agricultural Botany). Factors affecting seed quality including generation control and current seed production, inspection and certification processes are covered, together with specific information about seed production techniques for all the major agricultural crops.

Key words: breeding, seed, variety, generation, certification.

12.1 Introduction

The objective of agricultural plant breeding is to develop crop plant varieties which are well adapted to human needs. Most of the crops grown in the United Kingdom today were originally introductions from other parts of the world: potatoes from South America, wheat and barley from the Middle East, oilseed rape from China and maize from Central America are some examples. Early farmers achieved improvements in yields by propagating the most desirable looking plants. These are now sometimes referred to as 'landraces' and the selections made by these farmers were largely based on appearance. The processes of modern plant breeding have enabled further improvements to be made, such as enhanced yield, crop quality and other desirable traits which give some present-day crop varieties only a passing resemblance to their predecessors.

12.1.1 Gregor Mendel

The basis of modern plant breeding lies in the work of a Moravian monk named Gregor Mendel who lived between 1822 and 1884. His work with a variety of organisms established the principles of the heritability of specific characteristics. Mendel demonstrated that plants and animals are inherently variable and from that variability selection can take place. Using peas he showed that by cross-pollinating selected parents it was possible to combine desired characteristics in the offspring in a predictable way. This was the first step in refining the natural process of plant evolution.

12.1.2   Plant breeding in the twentieth century

It was not until the twentieth century and the isolation of the specific parts of the deoxyribonucleic acid (DNA) in the nuclei of plant cells (the 'chromosomes' and the 'genes' which they contain), which predictably convey characteristics from one generation to another, that the significance of Mendel's work was appreciated. It now forms the basis of all conventional modern plant breeding.

Much of the UK plant breeding efforts of the twentieth century were undertaken by publicly funded organisations such as the Plant Breeding Institute (PBI) at Cambridge, the Welsh Plant Breeding Station (WPBS) at Aberystwyth and the Scottish Plant Breeding Station. The Plant Varieties and Seeds Act (1964) introduced Plant Breeders' Rights and a system whereby the breeders of commercial crop plant varieties became eligible for royalty payments. This development helped to fund an increase in the commercial breeding of crops by private sector companies.

12.1.3   The National Institute of Agricultural Botany (now NIAB Ltd) and the Official Seed Testing Station (OSTS)

During the First World War it became apparent that increases in home food production were essential and legislation was introduced in 1917 to establish the Official Seed Testing Station and, significantly, the testing of seed before sale. The NIAB was founded at Cambridge two years later in 1919 and has performed an important function right through to the present day in encouraging farmers to use improved varieties and high quality seeds. The NIAB Recommended Lists and Descriptive Lists of crop varieties became important benchmarks of information for farmers and advisers. Responsibility for publication and dissemination of some of these lists has now been assumed by HGCA (the cereals and oilseeds division of the Agriculture and Horticulture Development Board – for cereals and oilseeds) and by PGRO (the Processors and Growers Research Organisation – for legume crops). With the increasing move towards the use of farm-saved seed in recent years the Defra (previously MAFF) funded Official Seed Testing Station (OSTS) at Cambridge can still perform an important function for individual farmers, testing seed samples for viability, the presence of weed seeds and for important seed-borne diseases.

12.2   Plant breeding methods

Techniques of plant breeding vary according to the particular crop. The production of a new variety is a slow, painstaking and costly operation involving a time scale of up to 12 years or even longer. There is, even then, no guarantee that the variety will become a commercial success.

12.2.1   Conventional plant breeding

Conventional plant breeding usually involves the cross-breeding of specifically chosen parent plants with desirable characteristics to form first (F1) and second (F2) generation populations. Selections are then made, usually from the second generation of plants, which exhibit very large amounts of genetic variation, according to the specific combinations of genes which are present. The breeder selects promising plants from this large pool and progressively, generation by generation, selects lines which conform to more specific requirements. These lines are then 'purified' by growing on for several more generations until the new variety is ready for entry into the official National List trials.

12.2.2   Breeding hybrids

A particular phenomenon sometimes observed in the first (F1) generation cross is known as hybrid vigour or heterosis. This is usually exhibited for one year as improved growth and ultimately yield but may also affect the uniformity or other quality characteristics of the crop. Vegetable (especially members of the *Brassica* family), oilseed rape, forage maize and sugar beet hybrids are commonly grown by farmers. The introduction of hybrid cereal varieties is progressing but so far has not met with great success in the UK.

The technique for the production of hybrids usually involves the production of self fertilised 'inbred' lines which are subsequently crossed to form the F1 hybrid, which exhibits improved characteristics. A major disadvantage of the use of hybrids is that farmers are precluded from saving their own seed and have to purchase a fresh supply each year.

Oilseed rape has received most attention regarding hybridisation in recent years. In 2012 about half of the NIAB recommended and provisionally recommended varieties were hybrids and many showed yield advantages or other desirable traits compared with conventional varieties. These 'restored hybrids' are F1 hybrids produced in a variety of ways, usually from crossing inbred lines but sometimes from more complex breeding operations involving open pollinated varieties.

12.2.3   Enhanced plant breeding techniques

Plant breeders are always seeking ways of reducing the interval between the initial cross-breeding and the final introduction of a commercially viable variety. Accelerated development in environmentally controlled growth rooms, micropropagation and even cross-breeding by the fusion of plant cells (so called protoplast fusion) are ways in which the process can be speeded up. Another way is by the maintenance of common shuttle breeding and establishing multiplication programmes in northern and southern hemisphere countries to reduce production time.

Increasing the chromosome number or 'ploidy' of plants is a phenomenon which occurs naturally at a low level, often as a response to physical conditions such as extreme heat or physical damage. Applied to normal diploid (two sets of chromosomes) seeds or small seedlings, the chemical colchicine (a naturally occurring substance extracted from autumn crocus plants) will readily cause an increase in the chromosome number and the creation of 'polyploid' plants. In comparison with diploids, these are normally larger and lusher in growth but often suffer from reduced fertility levels. Some breeding programmes have been devoted to the production of tetraploids (four sets of chromosomes), particularly in the cases of grasses and forage plants, but they have by no means replaced diploids. Many modern sugar beet varieties are 'triploid' hybrids formed as a result of crossing diploid with tetraploid varieties. Triticale is the result of a breeding programme involving crosses between durum wheat and rye which created a new polyploid crop. Further hybridisation programmes are in train with wheat, barley and rye, although only barley and rye hybrids are commercially available at the time of writing.

Recent advances in plant biotechnology have enabled plant breeders better to identify the genes that determine the characteristics of an individual plant. This is known as the science of 'genomics', DNA profiling or, more colloquially, 'genetic fingerprinting'. It enables breeding programmes to be undertaken with a much greater degree of precision using marker genes for faster selection. It can sometimes be exploited to facilitate the processes of genetic modification (GM) which are covered in the following sections.

12.2.4   Genetic modification

Applications and principles
There is no doubting the greater progress which can be made in plant breeding programmes by genetic modification. These techniques have an almost infinite range of potential applications in the development of new varieties. Enhanced yields, quality, pest and disease resistance and the better tolerance of drought and salinity are just a few of the many examples of what may be achieved in future commercial varieties. New varieties can and have been engineered to be resistant to specific non-selective herbicides (e.g. glyphosate). This enables farmers to control all weeds very easily and substantially to reduce the volume and range of their herbicide applications. Most importantly, of course, it also enables them to reduce their costs. Enhanced nutritional qualities and flavour, extended shelf life and improved appearance are all characteristics which may make foods prepared from GM crops more attractive to the consumer, and some (e.g. genetically modified tomatoes, and especially soyabeans and maize grown for grain) are already being used in large quantities. An important advance was the creation of a new strain of tomato capable of growing in saline conditions. More recently, the International Wheat Genome Sequencing Consortium has made possible important

developments in the breeding of new wheat varieties. Such developments have assumed international importance and agreements by the agricultural ministers of the G20 in 2011 have led to an international research initiative for wheat improvement.

The principles involved in the genetic modification of crop plants have already been alluded to. Having identified from the genome (gene map) of the donor organism the gene which expresses a particular trait, the plant breeder is able to remove it from the DNA using enzymes and then to transfer it into the host plant, often using another organism (e.g. a virus or a bacterium) as a vector. The expectation is then that the same characteristic will continue to be expressed in the new host plant. A specific example concerns genetic material from a bacterium (*Bacillus thuringiensis*) which causes it to produce a toxin called Bt. This has been successfully inserted into the cells of a number of crop species (e.g. maize) as a way of reducing attacks from important insect pests. This can preclude the need for routine applications of broad spectrum insecticides. Such applications are closely regulated and rigorously tested for environmental impact and the need for reduced treatment is an environmental bonus.

There are also mechanical techniques for introducing DNA containing specific genes into plant cells. One of them concerns the use of DNA 'bullets' whereby metal particles (tungsten or gold) combined with DNA fragments are mechanically propelled into host cells where the donor DNA combines with that of the host plant.

Risks and consumer attitudes

In spite of the potential benefits conferred by accelerated breeding programmes incorporating genetic modification there are widespread concerns about the risks associated with these so called 'transgenic' plants. Such concerns have mainly emanated from news media reporting rather than from an understanding of the scientific methods employed but have, nevertheless, prompted the UK government to restrict the growth of GM crop varieties to a small number of farm scale trials designed to assess the likely environmental impact of their more widespread introduction. Furthermore, products containing GM maize or soya have to be appropriately labelled.

There is no doubt that genetically modified crop plants can create difficulties. Organic growers in particular are concerned about the potential of their crops becoming contaminated by pollen from GM crops or by pollinating insects in the neighbourhood. Another example is the almost certainty of 'volunteer' GM plants occurring as weeds in subsequent crops (non-GM oilseed rape is already an important arable weed in the UK and herbicide resistant GM rape would almost certainly follow suit). The suggestion is that cross-pollination with wild plants of the same family may give rise to unique and herbicide resistant 'super weeds'. Further concerns have been expressed about the potential of insect resistant GM crops to reduce populations of other attractive or beneficial species (e.g. Monarch butterflies in the USA) although these experiments have been shown to be flawed. The use of antibiotic resistance as a marker gene in transgenic selection has also been queried and this practice is now ceasing.

However, there is widespread public ignorance and mistrust about GM technology and extremist environmental organisations have not been slow to capitalise upon this. There have been several very well publicised incidents where GM crops have been destroyed by activists and many more where there have been confrontations resulting in farmers being forced to destroy the offending crops as a result of public pressure. However, it must also be said that farmers in other parts of the world are starting to benefit from the use of GM varieties. About 44 million hectares were grown in 2000 (mainly soybeans and maize in North and South America) and there is little, if any, hard evidence of either environmental damage or of health problems among consumers. In 2011 the area had increased to about 160 million hectares planted in 29 countries.

A survey by the British Science Association in 2012 suggested that UK public concerns had reduced to some extent compared with a decade earlier although there was still a substantial majority opposed to GM. In the event of continuing price increases and apparent commodity shortages, it could be argued that farmers may well need to use this technology to maintain adequate and secure food supplies in the future. However a further, largely unspoken, objection may well arise; that of the increasing corporate control of our food production industry.

12.3   Target traits in breeding

12.3.1   Improvements in crop yields

The huge improvements in crop yields observed during the latter half of the twentieth century were due to many factors, including the increased use of fertilisers and pesticides and improved mechanisation techniques to allow planting at optimum times. In addition to these factors, plant breeding has played a major part, not only influencing the harvestable yields (the 'harvest index') of crops, but also, and very importantly, their physical characteristics, pest and disease resistance, maturity times and quality.

The harvest index of wheat has been substantially increased by the breeding of semi-dwarf varieties, and a closely related trait, improved resistance to lodging, has resulted in improvements both to yield and crop quality. Improved yields have been an important trait in all of the other major crops. In some cases improvements have been obtained by plant breeders selecting for winter hardy types which can be autumn sown. Oilseed rape is an important example, others, less successful, have included linseed, peas and lupins.

Crop quality has become increasingly important as farmers attempt to maximise revenue at times of depressed prices. Milling and malting quality of cereals, oil percentage and fatty acid constituents of oilseed rape and juice quality in sugar beet have all been important targets for plant breeders. In the potato crop the suitability of specific varieties for the processing industry (taste, texture, cooking quality, and colorations) are extremely important and related to the levels of enzymes and sugars, or to the storage characteristics of the variety.

Genetic resistance to pests and diseases has also provided important targets for the plant breeding industry; farmers are able to take advantage of the results of this effort to an increasing extent in their quest to minimise the use of all pesticides. Some diseases of great economic importance, such as *Rhizomania* in sugar beet, are the subject of specific programmes and resistant varieties suitable for UK conditions have been introduced. However, some diseases remain perennial problems and can only be controlled by strict adherence to plant health precautions.

12.3.2   Grasses and clovers

One of the main outcomes of the bovine spongiform encephalopathy (BSE) and Foot and Mouth Disease epidemics, in terms of livestock production methods, was a re-evaluation of the systems of production in favour of grass fed animals and some expansion of organic production.

One of the immediate results of these important developments has been an increase in the use of clover seed. The recent products of the Institute of Grassland and Environmental Research (IGER) white clover breeding programme such as *AberHerald* and *AberDai* are by now very well known. Future introductions will concentrate on persistent varieties for harsh upland conditions, and varieties for lowlands which are able to yield well in spite of moderately heavy (<250 kg N/ha) applications of fertiliser. In a further breeding programme involving the use of Caucasian clover it is hoped to introduce the valuable characteristic of drought resistance to the range of white clover varieties.

There has recently been a large increase in interest in the use of red clover and a much needed and very welcome reinstatement of the breeding programme for this legume at IGER (now IBERS – the Institute of Biological, Environmental and Rural Sciences) Aberystwyth. Important objectives for the plant breeders will be increased longevity, coupled with improved resistance to clover rot and stem eelworm.

Bloat remains one of the major inhibitions among stock farmers to the widespread uptake of the use of clovers. Whereas the introduction of a 'non-bloating' clover still seems some way off, there is increased interest in the non-bloating characteristics of legumes such as birdsfoot trefoil and sainfoin. Although of limited interest to farmers in the UK, there is increased interest in the USA in varieties of lucerne (alfalfa) which are being bred for freedom from bloat and suitability for grazing.

Developments in the breeding of grasses have concentrated mainly on the ryegrasses, which constitute more than 80% of the grass seeds sown in the UK. The main thrust, in recent years, has been breeding for improved quality. The introduction of 'high sugar' varieties such as *AberGreen* should lead to much enhanced palatability and intake at grazing, as well as improved silage fermentation. Following on from this should be better performance from stock at grass and a reduced reliance on concentrates.

With increasing concern about diffuse pollution of nitrogen and phosphorus, improving the way grasses and clovers utilise nutrients, both from fertilisers and clover, will also be important in the future. Ryegrass/fescue hybrids ('festuloliums') are being investigated for this purpose and also white clover varieties with reduced phosphorus requirements. Specific grass/legume variety combinations are also starting to be evaluated. These can make substantial differences both to dry matter yield and to the performance of livestock.

12.4   Choosing the right variety

12.4.1   National Lists and Recommended Lists

When a farmer purchases seed of a particular variety within the UK (or any other member country of the EU) it is possible to guarantee its performance potential as a result of the extremely rigorous testing that all varieties are subjected to under the EU Seeds (National Lists of Varieties) Regulations. Varieties which do not appear on the National Lists cannot legally be marketed. All varieties are assessed for their 'Distinctness, Uniformity and Stability' (DUS). These trials also assess the variety's 'Value for Cultivation and Use' (VCU). Many of the varieties submitted by plant breeders do not pass. National List varieties also appear in the EU Common Catalogue of varieties and, subject to plant health requirements, can be traded in other member countries.

The second tier of variety testing in the UK involves the selection of the best varieties, independently recommended, for commercial production. Data from these selections are used to form the HGCA (for all cereals and oilseeds) and PGRO (for all grain legumes), recommended lists for England and Wales and the corresponding lists prepared by the Department of Agriculture and Rural Development (DARD) and Science and Advice for Scottish Agriculture (SASA) for Northern Ireland and Scotland. Further recommended and descriptive lists are produced by the British Potato Council, The British Beet Research Organisation (BBRO) and by NIAB TAG with BSPB (the British Society of Plant Breeders) for forage maize, grasses and forage legumes. These lists are updated annually and varieties will only gain entry to them if they have consistently outperformed existing varieties in one or a number of ways over a three-year period. In recent years these national recommended lists have been further enhanced by the addition of English regional recommendations to take account of differential performance of varieties in specific regions.

12.5   Seed quality

Seed quality is a term which encompasses a number of important criteria, in particular genetic purity, seed physical quality, viability, the presence of weed seeds, pests or diseases, and moisture content. The following section describes the various factors which can impact upon seed quality and their nomenclature.

12.5.1   Classes of seed and generation control

The production of marketable quantities of seed on a commercial scale requires a series of multiplication steps which take place over a number of years. Each time a crop is grown for multiplication there is a danger of some deterioration in the genetic quality of the variety. This may take place through contamination or through unplanned cross-pollination. Cross-fertilised species such as beans and sugar beet are more likely to exhibit variation than self-fertilised crops such as wheat. It is necessary therefore to limit the number of generations over which seeds of a particular species can be multiplied, in order to minimise the level of potential genetic deterioration.

The categories of certified seed which are available for most crops are as follows (each normally represents a single year's step in the multiplication process): 'Breeders Seed', 'Pre-Basic Seed', 'Basic Seed', 'Certified Seed 1st Generation (C1)' and 'Certified Seed 2nd Generation (C2)'. Breeders and Pre-Basic seed are normally produced by the plant breeder or agent whereas Basic seed is produced by the maintainer of the variety. Farmers specialising in seed production would normally only be involved in the growing of Basic seed to produce C1 seed or of C1 seed to produce a crop of C2 seed. Basic seed is normally too expensive to justify its use for growing commercial crops, and C1 and C2 seed are the categories of seed normally purchased and grown commercially by farmers. After the C2 generation, no further multiplication for commercial seed production is permitted. The exceptions are linseed and flax, where a C3 generation is permitted.

The labels of seed bags are important documents for seed growers. They contain evidence of the variety sown and the reference number of the seed lot. It is important therefore that they are retained after sowing and that one example label is displayed in the crop (although these are often vandalised) and that at least one is retained in the farm office with evidence of the name and Ordnance Survey (OS) number of the field where the seed is sown. There is a colour coding of labels which identifies the generation of a particular seed lot: Pre-Basic seed – white with a purple stripe; Basic seed – white; C1 seed – blue; C2 seed – red.

Hybrid varieties are produced from the first generation (F1) of a cross between two lines of Basic seed. The harvest from a hybrid crop is unlikely to exhibit the same characteristics as the original cross and so should never be used for farm-saved seed.

Higher Voluntary Standard (HVS) seed is an enhanced standard which applies to the United Kingdom only and means that the seed has been tested to a higher standard than the prescribed minimum standard in the EU directives. Seed meeting HVS standards is labelled accordingly.

12.5.2   Seed certification

The seed certification authority for England and Wales is the Food and Environment Research Agency (FERA) whose functions are delegated to NIAB at Cambridge. Science and Advice for Scottish Agriculture (SASA) and the Potatoes, Plant

Health and Seeds Division of the Department for Agriculture and Rural Development (DARD) are the certifying authorities in Scotland and Northern Ireland respectively.

12.5.3   Varietal identification and crop inspection

It is very important for farmers and those engaged in the plant breeding and seed production industries to be able accurately to identify specific varieties. This is particularly important when stocks are being multiplied for bulk seed production. All crops must be carefully inspected to ensure a very high degree of varietal purity. If this is not the case the crop may be rejected for seed.

Each individual variety of any crop has specific distinguishing characteristics which, to the trained eye, enables significant deviations to be spotted easily. These characteristics may include such factors as the size, shape and colour of leaves, flowers or even parts of flowers. Hairiness or waxiness of leaves and factors such as the number of seeds per pod or the numbers of tillers or stem branches per plant may also be important. Evidence of rate of maturity (e.g. date of ear emergence or flowering) and of pest or disease resistance may also be taken into account.

All of these identifying features, relevant to the individual species, will be well known to the seed crop inspector who will have received training and updating on the characteristics of new varieties. It is important that crop inspections are carried out at a time when meaningful observations can be made which, for many species, is likely to occur around the time of flowering. The seed crop inspector makes a walk through the crop which should enable him or her to see as much of it as possible. Sample areas of the crop are then selected at random and the numbers of plants that do not conform to the description of the variety being grown are counted and recorded. Other factors, such as the presence of important weeds (e.g. wild oats), diseases (e.g. virus diseases in potato crops) or the proportion of the crop which has lodged, are also recorded. On the basis of one or more inspections the crop will then be accepted or rejected for seed. The penalties for having a crop rejected can be substantial since the grower will have undoubtedly incurred the very high cost of the seed (usually Basic or C1) for multiplication, and a crop rejection will deprive him or her of the extra income which the seed crop can command. It is common therefore to observe the very highest levels of attention to the detail of crop husbandry on farms growing crops for seed.

12.5.4   Identifying seed of particular varieties

In cases involving confusion, dispute, contamination or, increasingly, for routine confirmation, it may become necessary to identify the variety of a particular seed lot. Although seeds of individual varieties have some visual distinguishing characteristics it is very difficult to identify them by eye with any degree of certainty. In such cases the normal procedures which can be carried out are known as electrophoresis tests. These tests involve the extraction of the proteins from the seed under the influence of an electric field and production of a characteristic banding pattern when the proteins are separated out on a gel. These patterns provide the equivalent of a fingerprint for a specific variety.

DNA profiling can also be used to identify varieties but is not used at present to the same extent as electrophoresis.

12.5.5   Maintaining the physical purity of seeds

Farmers, when purchasing seed lots, will not expect them to contain weed seeds, seeds of other crop species, seeds of other varieties of the same species or any extraneous material (e.g. stones, dirt or plant vegetation).

Choice of the field in which seed crops are grown is of the utmost importance since the probable presence of important weeds or diseases is likely to be well known by the seed grower. Furthermore, the likely presence of volunteer plants or groundkeepers from previous crops of the same species is highly predictable from a knowledge of the previous cropping on the farm. Fields where livestock have been fed on hay, straw or coarse rations should be avoided as they often contain cereal or pulse seed, wild oats or grass weeds. There are in fact, for each species and generation of seed, specific requirements about previous cropping laid down in Defra regulations for the production of seed.

Isolation is also an important requirement and the regulations always require seed crops to be separated physically from non-seed crops to prevent physical contamination. Another reason why isolation is important concerns the potential for pollen contamination in the case of cross-pollinated species. An added complication is that such contamination may potentially occur from crops grown on other farms. For each species where cross pollination to any degree is likely to be of concern, specific recommendations are given concerning the minimum isolation gap which is acceptable for the various generations of seed. A simple way in which contamination of seed crops can be minimised and isolation improved is to discard the produce of the headlands. Checking that appropriate isolation distances have been observed is an important function of seed crop inspection.

Farm equipment, e.g. seed drills, combine harvesters, trailers and barn machinery, are all important potential sources of contamination and should be rigorously cleaned before drilling, harvesting or transporting seed crops. Furthermore, mistakes in handling seed in store can occur and seed lots should always be carefully identified and staff made aware of the need to avoid accidental contamination. Inevitably some contamination of seed with weeds or other materials will occur. A variety of techniques involving screening and/or gravity separation can be used to remove most of the extraneous material.

There are specific tolerances allowed in seed samples for the incidence of such items as seeds of other species, weeds, ergot pieces and other contaminants. Certified seed may not legally be sold if it contains more than the specified amounts of contaminants.

12.5.6   Seed viability

Viability refers to the proportion of the seed that is alive. However, this fundamental requirement also describes the ability of seeds to germinate normally, and to produce a vigorous and healthy seedling plant population which in turn is capable of producing an acceptable crop yield.

All seeds marketed must be tested for germination by law and the percentage declared. There are minimum statutory germination percentage levels for each species which vary between 75% and 90%. Germination tests are normally undertaken in satellite laboratories, certified by the OSTS, on seed merchants' premises. Seed samples (usually 100 at a time) are germinated under standard conditions involving specific growth mediums and in environmentally controlled growth rooms. In some cases alternative and more rapid viability testing can be carried out satisfactorily using the chemical 2,3,5,tri-phenyl tetrazolium chloride which, when administered to a seed cut into two, will produce a red stain in a viable embryo.

The fact that seed is alive does not necessarily mean that it will germinate normally and seed dormancy is sometimes an important consideration. In some cases it is necessary to submit seeds to specific conditions (e.g. cooling) in order to break dormancy and to encourage germination. Another important factor which may affect seed germination percentage concerns the drying operation. If excessive temperature has been used during the drying process this can damage the seed and may reduce the germination percentage. Also, in seasons when high temperatures occur during harvest, it may be necessary to cool seeds by ventilation in order to avoid damage.

The use of desiccants, especially diquat, to aid the harvest of seed crops is common. Particular note should be taken of the adverse effect that desiccants, such as glyphosate and glufosinate-ammonium, can have on subsequent seed germination in some crops and the label recommendations and manufacturer's advice should be strictly adhered to.

Seed vigour refers to the total performance of a seed lot throughout the process of germination and seedling emergence. High-vigour seed is beneficial in practice for field establishment in poor conditions. The use of low-vigour seed is likely to result in reduced levels of field emergence and the need to increase seed rates. Electroconductivity is a test that is often carried out on vining pea seed lots. This establishes the degree of physical damage which has occurred to the seed during harvest and storage and which may affect seedling vigour and disease incidence in cold or wet conditions.

12.5.7   Seed size

Seed size has little effect on viability but it is an important factor to take into account when deciding upon seed rates. An important and frequently used aid is the 'thousand seed weight' which can be used accurately to assess the weight of seed required in order to achieve a particular plant population. The following formula can be used to calculate an accurate seed rate:

$$\text{Seed rate (kg/ha)} = \frac{\text{Target plants per m}^2 \times 1000 \text{ seed weight in grams}}{\text{Predicted percentage establishment}}$$

12.5.8   Seed health

Specific diseases such as leaf and pod spot (*Ascochyta fabae*) in field beans and pea bacterial blight (*Pseudomonas syringae*) and a whole range of diseases of cereals and oilseeds, can be carried by seeds. It is important therefore that all seed crops are carefully assessed for them in order to avoid a high level of contamination of the crop. Chemical seed treatments are capable of controlling many seed borne diseases, but not all, and it is sometimes necessary to undertake seed testing in order to establish the levels of infection. Seed potato tubers too can be tested to establish the levels of infection with specific viruses. Standards are set out in the seeds regulations for all crops regarding the proportions of infected seed which are acceptable in commercial certified seed lots.

12.6   Seed production

12.6.1   Cereals

Breeders seed, Pre-Basic, Basic, C1 and C2 generations are produced for wheat, barley and oats. There is no C2 generation for rye or triticale. Basic, C1 and C2 seed can at present be certified at two levels, EU minimum standard and HVS (higher voluntary standard).

The minimum requirements for previous cropping are outlined in Table 12.1. Certain cultural practices, such as early autumn drilling and minimal cultivation, create greater risks of volunteers being carried over to a seed crop and in such circumstances a longer break will be advisable.

Important weeds are wild oats, cleavers, bromes and black-grass. Seed crops should not be grown in fields where these are likely to be a major problem. Light populations of wild oats and other cereal species (e.g. barley in wheat) may be rogued out of a seed crop and should then (in the case of wild oats) be removed from the field in a plastic bag and burned.

Table 12.1   Previous cropping requirements – best practice for growing cereal seeds

Crop	Grade of seed	Previous cropping requirement*
All except rye	Pre-Basic and Basic	No cereals for the previous 2 years
All except rye	C1 and C2	2 years clear of the same species 1 year clear of other cereal species
Rye	Certified seed (termed CS)	1 year clear of rye and triticale

* The minimum requirement for all certified cereal seed production is that no other variety of the same species shall be grown in the field in the previous year.

Isolation is not normally a problem for the self-pollinating wheat, barley and oats and a 2 m gap between crops is all that is required. However, winter barley (especially 6 row varieties) is prone to out crossing and so a greater voluntary isolation gap is often observed. Rye and triticale are both cross-fertilising species and isolation gaps of 250 m for C1 and 300 m for Basic seed are required. Hybrid varieties of rye may require significantly greater isolation dependent on the parent lines, and hybrid barleys, too, have specific requirements for isolation. Since new hybrids are likely to be introduced in future, the reader should check for specific requirements with the relevant certifying authority.

All cereal seed crops are inspected twice for varietal authenticity and for the presence of other cereal species, weeds and diseases. Wild oats are not permitted at more than 7 plants/ha in Pre-Basic, Basic and C1 and C2 HVS crops. The EU minimum standard allows up to 50/ha in C1 and C2 wheat and rye crops and 20/ha in barley. An important seed-borne disease is loose smut of wheat and barley, which, in the case of heavy infestations, can be the cause of the rejection of a crop for seed. Lodging in wheat may also cause problems with seed crops, in particular as it is likely that the grain on lodged plants may start to sprout before harvest. Crops which are badly lodged will almost certainly be rejected for seed.

Harvesting, transport and storage should all be carried out in such a way as to minimise the chances of contamination of the seed crop with other species or varieties. Seed grain should not be dried at a temperature of more than 49°C since high temperatures will adversely affect its germination.

The minimum standard at which certified cereal seed can be marketed is 99% analytical purity (99.7% for C1 and 99.9% for C1 HVS), 85% germination (80% for triticale and 75% for naked oats) and a maximum of 17% moisture content. There are also specific standards for the content of wild oats and ergot pieces and also seeds of corn cockle, couch, sterile brome and wild radish.

12.6.2 Peas and field beans

Certified seed of the Breeders, Pre-Basic, Basic, C1 and C2 generations are produced for varieties grown for harvesting dry for incorporation into livestock rations. There is no HVS standard for these crops. Vegetable (vining pea) varieties are marketed as 'standard seed', although there is a voluntary certification scheme for Pre-Basic, Basic and C1 generations. The regulation governing previous cropping in fields intended for seed production, however, is very straightforward; no pea or faba bean crops (including legume/cereal mixtures for ensiling or dredge corn) in the previous two years, are permitted.

So far as isolation is concerned, there should be at least a 2 m gap (or a substantial physical barrier) between pea or bean seed crops and any other crops. In the case of field beans an isolation gap is required between crops of different varieties. For crops of over 2 ha this has to be 50 m for C2 and 100 m for the C1 and Basic generations. For small crops of less than 2 ha these distances rise to 200 m.

Crop inspections take place at least twice for both pea and bean crops, to assess them for varietal purity standards and freedom from other crop species and weeds.

Bean varieties especially, because of the high degree of cross-pollination (between 30% and 70%) which takes place, are often quite variable in terms of plant height, time of flowering and maturity. Only 2% of off-types are permitted in C2 seed crops, 1% for C1 and 0.3% for Basic seed, and the standards for vegetable varieties are even higher. Wild bees undertake much of the pollination in bean crops but hives of honey bees are also sometimes introduced and are generally thought to be beneficial.

Pea bacterial blight (*Pseudomonas syringae*) is an important notifiable disease of peas. In order to minimise the likelihood of contamination, all pea seed crops must be isolated from other pea crops by at least 50 m.

Leaf and pod spot (*Ascochyta fabae*) is probably the most important seed-borne disease of beans although it affects peas as well. Most disease stems from infected seed, but volunteer plants and the movement of tractors and other machinery from crop to crop can also spread it. The testing of seed stocks for *Ascochyta* is routine. However, modern seed treatments have greatly improved control of this disease.

Stem eelworm (*Ditylenchus dipsaci*) is an important seed-borne pest of field beans. Home-saved seed should always be tested for it and the status of certified seed should be ascertained as well.

An increasingly important pest of bean seed crops is the bean beetle or bruchid beetle (*Bruchus rufimanus*), the larva of which burrows neat circular holes into the near ripe seed which, after combining, can provide a refuge for weed seeds such as wild oats, and may, if infestation is greater than 10% of seeds, adversely affect the germination percentage. Approved insecticide applications to the crop when adult beetles can be found have become common in order to reduce the levels of this pest.

Combining pea and bean seed in excessively dry conditions should be avoided, if possible, because of the increase in damage to the seed which occurs. Drying, if necessary, should be at relatively low temperatures with peas not exceeding 38–43 °C; for beans the most satisfactory system is to ventilate in a bin or on the floor with ambient air or with a very small amount of extra heat. Minimising the damage to both pea and bean seed should ensure adequate germination. Certified C1 and C2 pea and field bean seed has an analytical purity of at least 98% and a minimum germination of 85% (in the case of broad bean seed the minimum germination is 80%). There is a maximum content of seeds of other species of just 0.5% and there are specific standards set for wild oats, docks and dodder.

12.6.3   Oilseed crops

Oilseed rape (sometimes also known as swede rape) and turnip oilseed rape are two very similar species. Breeders, Pre-Basic, Basic and certified (C1) seed generations are available. There is no second generation (C2) of certified seed permitted.

The minimum requirement in terms of previous cropping is that the field should not have grown a cruciferous seed crop in the previous five years. In practice, seed growers are encouraged to choose fields (if available) which have never grown a cruciferous seed crop, and where the incidence of weeds such as charlock and wild radish is at a minimum.

Oilseed rape and turnip rape are 60–70% self-fertile but they will also out-cross readily with other species, if in flower, such as swedes, fodder turnips, fodder rape, black and brown mustards and Chinese cabbage. Volunteer oilseed rape plants in neighbouring fields can be an important source of pollen contamination (especially the high erucic acid varieties). Isolation requirements for seed crops are therefore quite stringent. There should be a physical barrier or at least 2 m fallow between the crop and any other crop likely to cause contamination. In addition, there should be an isolation gap of at least 200 m between a certified seed crop and any source of pollen contamination such as other varieties of oilseed rape or any of the crops mentioned above. For the production of Basic seed this minimum isolation gap should be doubled to 400 m and for hybrid varieties these distances must be increased by a further 100 m.

Field certification standards are also stringent and for the production of certified seed only 0.3% of off-types are allowed and for Pre-Basic and Basic seed this is decreased to 0.1%.

The disease *Sclerotinia sclerotiorum* is important in oilseed rape and a number of other crops as well. It infects the stems of crops during flowering and has a substantial effect on yields. During combining, fragments of the sclerotia of this fungus often break away from infected stems and become mixed in with the harvested seed. There are no field standards for the incidence of this disease but the permissible number of sclerotia in a 100 g certified seed sample is 10 (5 in a 70 g sample for turnip rape varieties).

Crop inspections may take place up to three times, firstly in the vegetative stage when it should be possible to identify the variety through its leaf characteristics. A further inspection at the stem elongation stage should allow for the identification of any off-types in the crop. The final inspection is made around the time of early flowering when the characteristics of the flowers can be confirmed, and the surrounding fields checked for potential sources of pollen contamination.

Certified, Pre-Basic and Basic seeds of oilseed and turnip rape have a minimum analytical purity of 98% and a minimum germination of 85%. The maximum content by weight of other seed species is 0.3%. There are also specific standards applied for wild oats, dodder, dock, wild radish, black-grass and flaxfield ryegrass contamination.

12.6.4 Linseed and flax

Linseed and flax have been heavily subsidised by the Common Agricultural Policy (CAP), and until recently additional funding in the form of Seed Production Aid was available for certified seed production. Breeders, Pre-Basic, Basic, C1, C2 and C3 generations are available.

Fields which are selected for linseed seed production should have had no linseed or flax crops for the previous two seasons. However, best practice suggests that a period of five years should have elapsed. Because linseed and flax are self pollinating only a 2 m fallow strip or a physical barrier is required between a seed crop and any other crops likely to cause contamination. Crop inspections usually take place around the time of flowering when varietal off-types may be readily identified. Field standards for varietal purity allow just 2.5 off-types/ha for C2 and C3 seed, 2/ha for C1 and 0.3/ha for Basic seed.

Linseed and flax are susceptible to some important seed borne diseases. No more than 5% of seeds should be infected with the following: *Botrytis* spp., *Alternaria* spp., *Fusarium* spp., *Colletotrichum lini* and *Ascochyta linicola*. (In the case of flax seed even more stringent conditions apply and no more than 1% of seed should be contaminated with *A. linicola*.) Seed crops are normally sprayed with an approved fungicide before the end of flowering in order to try to achieve these standards.

At harvesting, particular attention should be paid to the cleaning of the combine trailers and any barn machinery. Oilseed rape and linseed can be dried easily to about 8% moisture but if drying on floor it should not be in layers more than 1.25 m deep. Drying air temperature should not exceed 50 °C. Ventilation should be applied to cool the seed down to less than 15 °C. At higher temperatures or moisture contents infestation with insect pests or grain mites can become a problem.

12.6.5 Sugar beet

Basic and certified generations of sugar beet seed are grown very little in the UK. British Sugar at present controls the distribution of all seed in the country and growers place an order for their preferred varieties when they sign the annual contract.

High-quality seed is one of the most important factors in the production of sugar beet. Since the introduction of monogerm varieties in the 1960s the majority of commercial crops have been drilled to a stand without the need for thinning. Many modern varieties are hybrids and more than half of the current varieties offered by British Sugar are triploids, the result of crossing diploid and tetraploid parent lines. Special climatic conditions are required for high-quality seed production, especially during the periods of flowering, maturing and harvesting. This has now largely precluded the growing of the crop in the UK or indeed anywhere else in northern Europe.

12.6.6 Potatoes

Crop production from true potato seed, taken from the fruits of the plant is rare, but it can be done. There are two main advantages: one is that in parts of the world where food is scarce an extra supply is secured by saving the tubers which would normally have to be replanted; the other is that some of the important potato viruses are not transmitted by true seed.

The majority of potato crops, however, are raised from replanted tubers known as seed potatoes. Seed production has traditionally been carried out in Scotland,

Northern Ireland and the hill areas of England and Wales where substantial seed potato enterprises still continue. The main advantages of these areas are that the low temperatures and strong winds keep aphid populations in check. This means that the severe virus diseases (leaf roll and the mosaics) which are spread from diseased to healthy plants by aphids, are less likely to occur. However, recent advances in aphid control and concerns over the quality of seed from some traditional areas have seen successful seed production extended to some of the English arable areas as a profitable break in predominantly cereal and break crop rotations. As with other forms of seed production the certifying authority in England and Wales is FERA, for Scotland SASA (Science and Advice for Scottish Agriculture) and DARD in Northern Ireland. In all cases the same basic Seed Potato Classification Scheme (SPCS) obtains.

The choice of field for seed potato production mainly reflects the need to minimise the possible carry over of groundkeepers from previous crops. Seed potatoes should not, therefore, be grown in the same field for more than one year in five (one in seven is preferable). Fields for seed production must be certified free from potato cyst nematodes (PCN) and seed crops must never be grown on land where potato wart disease has occurred. The main thrust of the practice and the regulations surrounding potato seed production concern the minimisation of the incidence of virus diseases and blackleg. All seed crops must be certified during the growing season. This ensures that they are true to variety but also that they are as free as possible from virus and other diseases. Most growers inspect crops carefully prior to official inspection and any obviously diseased plants are rogued out of the crop. In England and Wales all seed potato crops are inspected by the Plant Health and Seeds Inspectorate (PHSI).

The certification grades of seed potato are set out in Table 12.2. Potatoes produced from tissue culture (TC) are the highest grade and are produced by rooting virus-free plants obtained by tissue culture in sterilised compost in isolated and aphid free glasshouses. Subsequent generations 'super-elite' (SE), 'elite'(E) and 'A' are grown-on in isolation from other stocks to ensure, as far as possible, freedom from viruses and the other important tuber-borne diseases such as gangrene, skin spot and blackleg. Pre-basic and Basic seed is mainly intended for the production of seed crops, while Certified seed (CC) is mainly intended for ware production, although A grade seed is sometimes also used.

Ideally, seed potato crops should be planted with sprouted tubers and at a high seed rate to produce a good yield of seed size tubers (25–50 mm). The haulm should be destroyed chemically, to reduce the risk of blight or virus transmission,

Table 12.2   Grades of seed potato tubers

Pre-Basic	Basic grades	Certified grade
TC (Tissue culture)	Super Elite (SE)	CC
FG (Field grown)	Elite (E)	A

when most of the tubers are of seed size. This reduces not only the risk of spread of virus diseases by aphids but also the spread of blight to the tubers. It is an undesirable practice to let the crop grow to maturity so that additional income can be obtained from sale of large ware-sized tubers. Seed crops should be kept free from weeds and blight infection throughout the growing season and similar attention to detail should continue through to storage. Wounds should be allowed to heal and wet loads kept out of the store. Sprout suppressant treatment should never be used on seed potatoes in store.

Crops grown from Basic seed grades should contain less than 4% of plants with virus infections and those grown from certified (CC) seed, less than 10%. However, growers should not always expect virus infection levels in crops grown from purchased seed to be limited to the tolerances set for seed crop inspection. This is because extra infection may have occurred during the growing season when the inspection took place, or even from re-infection by aphids during chitting.

12.6.7   Herbage seed

All herbage seed production in the UK is overseen by FERA and NIAB. Only Basic seed is used for C1 generation production (termed CS). In some cases it is permitted for seed crops to be taken from the same crop for two seasons, but no C2 generation multiplication is permitted. All seed crops are inspected at around the time of ear emergence to establish variety type and to check for off-types, weeds and isolation. Standards for varietal purity are quite high and in the case of Basic seed only one off-type is allowed per 30 square metres. In the case of CS seed the standard is one per 10 square metres.

In total about 6450 hectares of grasses and clovers were grown for herbage seed in the UK in 2012. Of this over 5500 hectares were ryegrasses of various types. Whereas it is possible to achieve good gross margins from seed production, the vagaries of the weather, the difficulties of combining seed crops at high moisture levels and low prices have brought about substantial reductions in the area grown in recent years.

Seed production from grass requires considerable planning as well as skill and perseverance. Good yields are dependent on good growing conditions and dry weather during the critical harvesting period. Grass seed is grown most successfully in lower rainfall areas such as in the south and east of England, and usually on chalk or limestone based soils. The Italian and hybrid ryegrasses are harvested for one season only, but most of the perennials will produce seed crops for two seasons. Grass for seed production will, in addition, provide threshed hay and some grazing.

It is very important that there are no grass weeds in the field – especially black-grass, couch, rough stalked meadow-grass, sterile brome and other grasses with similar sized seed. Docks too are important contaminants of grass seed. Every effort should be made to rid the field of these weeds prior to growing a seed crop. Herbage seeds should not be grown in fields where dodder (a parasitic plant) is present. High weed infestations can lead to a failure of the crop to pass field inspection and there are high standards for the numbers of weed seeds which are permitted in herbage seed samples. It is also important that the field is not contaminated with volunteer plants of the same species; FERA stipulate up to four years of arable cropping prior to sowing a grass seed crop. Broad leaved weeds can normally be controlled by a range of post-emergence treatments.

Where there is a danger of cross pollination (i.e. when neighbouring grass crops contain the same species), their flowering period overlaps and they are of the same ploidy (e.g. both diploids), then a gap of at least 50 m between the seed crop and its neighbour should be established before flowering begins. If this is not possible then, when flowering is over, a discard strip must be cut out of the seed crop adjacent to the neighbouring field. This material must be discarded and no seed may be used from it. It is important to remember that grass seed crops can be laid down for several years and careful planning is required to avoid cross contamination. Even areas such as waste ground and motorway verges can be sources of pollen.

The ryegrasses and fescues are usually established by undersowing in spring cereals, either in narrow rows or broadcast, whereas cocksfoot and Timothy are usually sown direct in wide rows (about 50 cm). Seed rates are often substantially below (as little as 12.5 kg/ha for Italian ryegrass) those which would be considered normal for forage production. Italian and hybrid ryegrasses are usually cut early for a high quality silage crop or grazed, prior to being laid up for the seed crop. This must be accomplished before the end of April. Moderate levels (about 70 kg/ha) of nitrogen fertilisers are then applied and the seed crop harvested usually in July or August. Perennial ryegrasses receive nitrogen at similar times to the Italians and hybrids, but no cutting or grazing can be carried out. A good ryegrass crop will usually 'lodge' about a fortnight before harvest and this will reduce losses by shedding. The crop may be combined direct or from windrows. The use of a stripper-header, if available, can be very advantageous. The seed must be carefully dried (usually by prolonged blowing with ambient air) and cleaned. With good harvesting and drying procedures most seeds of perennial ryegrass varieties will exceed the minimum 80% germination standard (75% for Italian and hybrid ryegrasses). Yields of cleaned seed range from 1 to 1.75 t/ha but can be extremely variable and considerable skill is needed to achieve high yields.

Red clover. The USA and Canada are the main world centres of production of red clover seed. Only very small areas have been sown for seed in the UK in recent years, but the substantial increase in interest in this legume as well as the improved varieties being introduced, has created new opportunities for herbage seed producers and about 150 hectares were grown in 2012.

White clover. New Zealand, the USA and some South American countries are the primary producers of white clover for seed. In Europe, the largest producer is Denmark. In 2012, including the small area of *Kent Wild White*, just 100 hectares were grown in the UK.

12.6.8   Organic seed

It is important that seed used by organic producers is, wherever possible, of organic origin and this is stipulated by EU regulation 2092/91. Organic seed crops, if entered for UK certification, are currently subject to exactly the same standards as non-organic.

12.6.9   Farm-saved seed

As a way of saving costs many farmers now save seed from a number of species for replanting in the following season. There is relatively little risk associated with this provided that some basic rules are obeyed. The first is to grow the crop as a seed crop with exactly the same precautions as have already been indicated. Many farmers intending to use farm-saved seed will purchase a relatively small quantity of C1 generation seed to grow on. It is important to have the seed cleaned from weed seeds as far as possible and treated professionally with the required approved seed treatment by a reputable mobile seed treatment company. In some cases, if seed-borne disease levels have been confirmed by testing to be absent, it may even be possible to omit the fungicidal seed treatment. The other important rule is to have the seed tested professionally for germination percentage.

When seed is to be tested it is vital that really thorough and rigorous sampling is carried out throughout the seed lot. In the case of oilseed rape, where farmers are intending to use farm-saved seed, it must be tested for glucosinolate levels to conform with EU regulations. Potato growers are allowed to grow on their own 'once grown' seed from a crop grown from CC grade, but a sample of tubers should be tested to ensure that they are as free as possible from viruses and fungal diseases. Farm-saved non-certified seed of any species may not legally be sold. Farmers using farm-saved seed are now required to pay royalties to the breeders of the specific varieties that they are growing. These royalties are collected and distributed on behalf of breeding companies by the British Society of Plant Breeders.

12.7   Sources of further information and advice

Brown J and Caligari P D S, *An Introduction to Plant Breeding*, Blackwell Publishing, 2008.
George R A T, *Agricultural Seed Production*, CABI Publishing, 2011.
Gooding M J and Davies W P, *Wheat Production and Utilization*, CAB International, 1997.
Lammerts van Bueren E T and Myers J R (eds), *Organic Crop Breeding*, Wiley-Blackwell, 2012.
NIAB TAG Seed handbook for agronomists 2012/2013, NIAB TAG, 2012.
Vroom W, *Reflexive Biotechnology Development: Studying plant breeding technologies and genomics for agriculture in the developing world*, Wageningen Academic Publishers, 2009.
Wellington P S and Silvey V, *Crop and Seed Improvement – a History of the National Institute for Agricultural Botany*, Henry Ling Ltd, 1997.

Part III

The management of individual crops

13

Cereals

DOI: 10.1533/9781782423928.3.287

Abstract: Cereals are the most widely grown arable crops in the EU. Wheat accounts for just under 50% of the cereal area followed by barley and grain maize (southern Europe). Other cereals grown to a limited extent include oats and rye. This chapter outlines the identification features, market requirements and husbandry of the main cereals crops grown in the UK.

Key words: cereal identification, grain quality in cereals, cereal growth, cereal yields, cereal inputs, cereal husbandry.

13.1   Introduction

Cereals are the most widely grown arable crops in the EU. Wheat accounts for just under 50% of the cereal area (Fig. 13.1) followed by barley and grain maize (southern Europe). Other cereals grown to a limited extent include oats and rye.

Up to the end of the twentieth century, yields increased most rapidly in wheat compared with the other cereals. In the UK, for example, the average wheat yield is now just below 8 t/ha, compared with 6.3 t/ha for winter barley and 5.1 t/ha for spring barley. The initial increases in yield were largely due to improvements in varieties and husbandry techniques. Fertiliser use, particularly nitrogen and application of crop protection chemicals for the control of pests, diseases and weeds were also important in improving yields. Now, with a number of years of extreme weather conditions from droughts to flooding and increases in pesticide resistance, yields have not improved despite breeding new varieties.

As well as changes in production of cereals, there have been changes in their usage. Over 80% of the wheat used by millers in the UK is now home grown due to improvements in wheat quality and milling technology. Very little Canadian hard wheat is now imported for bread making. Wheat, compared with barley, is also used in greater quantities by the feed compounders. Since the early 1980s,

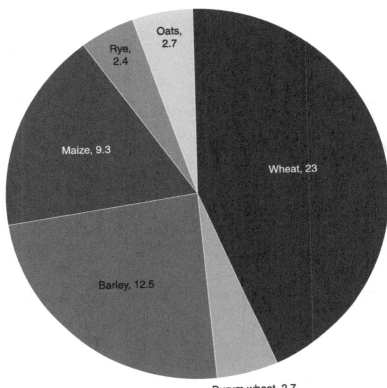

Fig. 13.1 Area of cereals grown in the EU (27) (in millions of ha), 2012.

with the improvements in cereal yields, the EU has become a major exporter of cereals on the world market.

13.2 Cereal identification

The different cereal crops are easily recognised by their grains or flowering heads. Wheat, rye and maize grains consist of the seed enclosed in a fruit coat (the pericarp) and are referred to as 'naked' caryopses (kernels). In barley and oats, the kernels are enclosed in husks formed by the fusing of the glumes (palea and lemma) and are referred to as 'covered' caryopses (Fig. 13.2). Note, naked oat and barley varieties are being developed.

The flowering head of cereals is usually referred to as the ear or spike. In wheat, barley and rye ears the spikelets that contain the grain are attached to the main stem without a stalk (Fig. 13.3–13.6). In oats the spikelets are stalked, giving an open ear or panicle (Fig. 13.7).

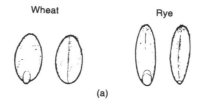

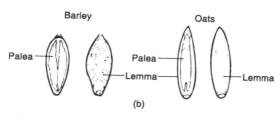

Fig. 13.2 Kernels: (a) 'naked' kernels, (b) 'covered' kernels.

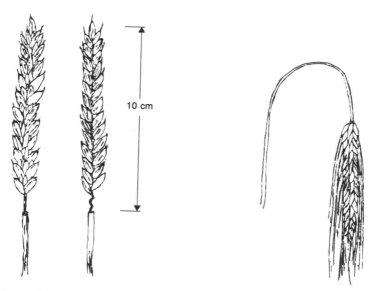

Fig. 13.3 Wheat spikelets alternate on opposite sides of the rachis. 1–5 grains develop in each spikelet. A few varieties have long awns.

Fig. 13.4 6-row barley, all three flowers on each spikelet are fertile. Awns are attached to the grains.

In the vegetative stage, different cereal plants can be identified by their leaves and the presence or absence of auricles, as well as from the size of their auricles (Fig. 13.8). At the seedling stage all cereal leaves twist clockwise except oats and wild oats which tend to twist anti-clockwise.

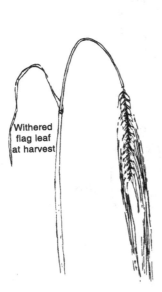

Fig. 13.5 2-row barley heads hang down when ripe: each grain has a long awn; the small infertile flowers are found on each side of the grains.

Fig. 13.6 Grain easily seen in the spikelets of rye.

Fig. 13.7 Oat panicle.

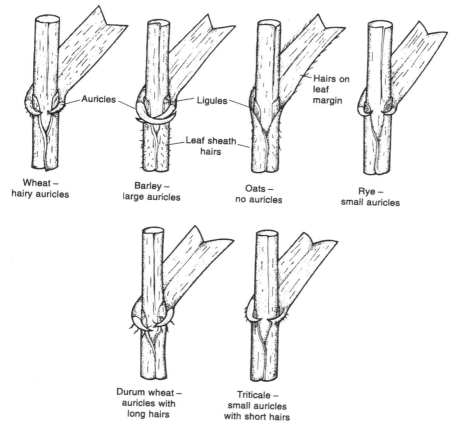

Fig. 13.8   A method of recognising cereals in the leafy (vegetative) stage.

13.3   Grain quality in cereals

There are several markets for cereals, including milling for human consumption, animal feed, malting, seed, export and industrial uses including production of biofuel and bioethanol (Table 13.1). Grain quality requirements will be affected by the proposed market.

Standard tests when selling grain include the following:

- *Moisture content.* This is very important when storing grain. For long term storage it should be about 14%. Too high a moisture content will be penalised or not accepted, but no compensation is given for very dry grain. Grain which is overheated when being dried, or in storage, can be spoiled for seed, malting or milling. Overheating can also damage the germination ability as well as the protein quality.

Table 13.1   EU (27) cereals – main uses 2011/2012 (million tonnes)

	Wheat	Barley	Total cereals
Total crop size	128.7	51.4	285.7
Imports	5.4	0.4	14.4
Exports	14.3	5.7	26.2
Seed	4.7	2.3	9.7
Animal feed	55.2	36.1	167.0
Industrial use	6.1	8.5	20.1
Human consumption	47.9	0.4	65.4
Biofuel/bioethanol	4.6	0.7	9.1

Source: HGCA.

- *Sample appearance and purity.* Good quality grain is clean and attractive in appearance, free from mould growth, pests and bad odours. Sample purity, freedom from contamination with other cereals or weeds will affect the potential value of grain, as will the amount of shrivelled or broken grain. Careful setting of the combine will help to minimise broken grains and produce a cleaner sample. Depending on quality and quantity of grain being sold, some farmers have a drying system with grain-cleaning facilities.
- *Specific weight (bushel weight).* Specific weight is a measure of grain density. A high specific weight is preferred for all markets, particularly export. Wheat has the highest specific weight and oats the lowest. Specific weight can be affected by husbandry (e.g. time of sowing, disease and pest control) as well as by weather conditions during grain fill and harvest.
- *Mycotoxins.* Before harvest some fungi present on grain can produce mycotoxins, e.g. *Fusarium spp.* Toxicity of the mycotoxins varies with fungal species. There are now legal limits for the *Fusarium* mycotoxins deoxynivalenol (DON) and zearalenone (ZON) in wheat destined for human consumptions. Currently there are guidance limits for feed grain. Wet conditions during flowering and wet delayed harvests tend to increase the amount of mycotoxins present.

Other standard tests mainly for milling and/or malting include:

- *Hagberg Falling Number.* This test in wheat measures the amount of breakdown of the grain's starchy endosperm by the enzyme alpha-amylase. A high value indicates low enzyme activity which is required for bread making. If grain has started germinating or sprouting, the Hagberg Falling Number will be low.
- *Protein content.* A high protein content is required for bread making compared with a low percentage of grain nitrogen in malting samples. Nitrogen content

is analysed using the Dumas test or a near infra-red analyser. Results are given at 100% dry matter.

- *Protein quality.* The quality of the protein in the grain affects the baking characteristics (when the protein in wheat flour is mixed with water it produces gluten). Protein quality can be assessed by several tests, including baking, gluten washing, SDS sedimentation, Zeleny index or the Chopin alveograph test. Tests undertaken will depend on the target market.
- *Germination.* Grain for seed or malting is tested for speed and percentage of germination. Germination should be over 97% for malting. Husbandry of the crop, variety and time of harvesting can affect several of the above quality specifications. Once harvested and stored, little can be done to change or improve grain quality. Gravity separators are available which can help to raise specific weights and the Hagberg Falling Number.

Most merchants now only buy grain that comes from farms that have been assured by a registered assurance scheme. These schemes include strict codes on crop production and grain storage.

13.4   Cereal growth, yield and inputs

13.4.1   Growth stages

A decimal growth stage key is commonly used to describe the growth and development of the cereal plant (Table 13.2). There are ten main areas of growth (subdivided into secondary stages) from sowing through to the vegetative leaf and tillering stages, stem elongation, ear emergence, flowering to grain filling and ripening. By dissecting out the growing point or apical meristem, using a microscope, the change from vegetative to ear formation can be seen (Fig. 13.9 and 13.10). These changes in internal development do not always coincide with the same growth stage of the cereal plant. Speed of development will depend on type of cereal, variety, temperature, day length and husbandry factors such as time of sowing.

It is very important to be able to identify growth stages correctly as there are only certain stages when pesticides should be applied if they are needed. The main stem should always be looked at when checking the growth stage. A major difficulty is deciding when the plants have changed from the tillering to the stem extension stage. The best method is to use a knife and slice the main stem in half. Growth stage 30 is when the ear is at 1 cm (Fig. 13.11). Distinguishing the nodes can also be difficult. They should only be counted when there are more than 2 cm between each (Fig. 13.12). They, too, are best studied by splitting the stem with a knife; it can be very misleading just to feel the outside of the stem.

Table 13.2   The Decimal Code for Growth Stages (GS) of small cereal grains

Code 0   Germination. Subdivided – 00 dry seed to 09 when first leaf reaches tip of coleoptile.

Code 1   Seedling growth

  10 1st leaf through coleoptile

  11 1st leaf unfolded

  12 2 leaves unfolded

  13 3 leaves unfolded

  14 4 leaves unfolded

  15 5 leaves unfolded

  16 6 leaves unfolded

  17 7 leaves unfolded

  18 8 leaves unfolded

  19 9 or more leaves unfolded

Code 2   Tillering

  20 main shoot only

  21 main shoot and 1 tiller

  22 main shoot and 2 tillers

  23 main shoot and 3 tillers

  24 main shoot and 4 tillers

  25 main shoot and 5 tillers

  26 main shoot and 6 tillers

  27 main shoot and 7 tillers

  28 main shoot and 8 tillers

  29 main shoot and 9 or more tillers

Code 3 Stem elongation

  30 ear at 1 cm (pseudostem erect) – visible only if the growing point is dissected

  31 1st node detectable

  32 2nd node detectable

  33 3rd node detectable

  34 4th node detectable

  35 5th node detectable

  37 flag leaf just visible

  39 flag leaf ligule/collar just visible

Code 4   40–49 Booting stages in development of ear in leaf sheath to when awns just visible – GS 49.

Code 5   50–59 stages in the emergence of the inflorescence or ear. Ear fully emerged – GS 59

Code 6   60–69 stages of anthesis (flowering). Flowering complete – GS 69

Code 7   Milk development stages in grain. Late milk – GS 77

Code 8   Dough development stages in grain. Hard dough – GS 87

Code 9   Ripening stages in grain. Grain hard and not dented by thumb nail – GS 92

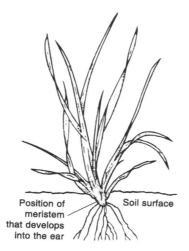

Fig. 13.9 Leafy winter wheat plant with four tillers.

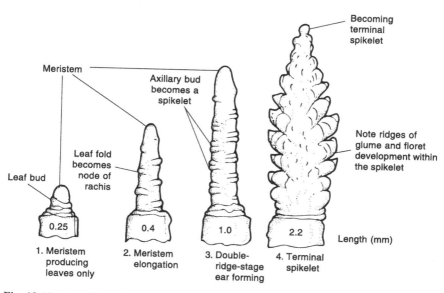

Fig. 13.10 Simplified diagram of an ear forming from the tiller apex of winter wheat (not to scale). (Visible under a microscope or possibly with a good pocket lens.) The apical meristem changes from leaf production to ear formation following cold treatment (vernalisation) and increasing day length (photoperiod). Normally these stages occur in February, March and April and their exact occurrence can be used for efficient timing of pesticides and nitrogen application.

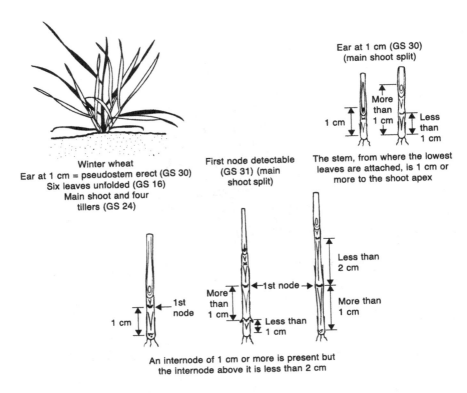

Fig. 13.11   Stem elongation stages in wheat: early stages.

13.4.2   Components of yield

The yield of a cereal crop is determined by the contributions made by the three components of yield:

1. number of ears,
2. number of grains per ear,
3. weight (size) of the grains.

These components are inter-related. By increasing the number of ears (e.g. by denser plant populations or more tillering), the number of grains per ear may be reduced and also the size of the grains. Opinions differ on the ideal number or size of each component and it will vary according to the type and variety of cereal, as well as soil and climatic conditions, time of sowing and seed rate, and the occurrence of weeds, diseases and pests. Ear number is influenced by growth during tillering up to flag leaf emergence. Growth between the flag leaf emerging and flowering affects numbers of grains per ear and grain size is affected by conditions post flowering.

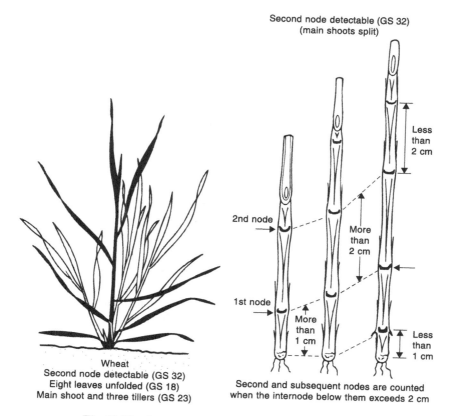

Fig. 13.12   Stem elongation stages in wheat: later stages.

Winter wheat and 6-row winter barley have the least number of ears/m^2 (450–600/m^2), whereas 2-row winter barley has the highest number of ears (750–1000 ears/m^2). The weight of grain is normally heaviest in wheat and lowest in oats and rye.

13.4.3   Seed rates

Accurate spacing and uniform depth of planting of seed in well-prepared seedbeds are very important if optimum yields are to be obtained. This usually involves doing seed counts/kg and carefully setting the drill. Narrow (10–12 cm) rows are preferable. Broadcasting seed can be successful provided the seed is distributed uniformly and properly covered. Stony soils, wet and cloddy conditions can give crops a poor start.

Seed rates should be chosen with the object of establishing the desired plant population. They can vary between 125 kg/ha and 250 kg/ha. The factors to be taken into account are:

- Crop – there are different optimum plant populations for the various cereal crops.
- Seed size, e.g. increase the rate for large seed (low seed counts/kg) and decrease the rate for small seed (high seed counts/kg).
- Tillering capacity – some varieties tiller more freely than others and so their seed rate may be reduced. Winter cereals have more time to tiller than spring cereals and so fewer plants need to be established.
- Seedbed conditions – the seed rate should be increased in cloddy and stony conditions where seedling growth may be poor.
- Time of sowing – for late autumn sowings and very early spring sowings the seed rate should be increased.
- The possibility of seedling losses by pests such as wheat bulb fly, frit fly or slugs; the greater the risk, the higher the seed rate.
- High seed rates can help reduce weed competition.
- Price and quality of seed; very expensive seed will often mean a reduced seed rate.

The most desirable plant population to aim for is debatable; it must be related to the potential yielding capacity of the field. Very low plant populations can ripen unevenly and be prone to weed competition, whereas high populations can lead to smaller ears, low grain weight and increased risk of lodging.

13.4.4   Varieties
There are very many cereal varieties now on the market and new ones are introduced every year. However, there are only a few outstanding varieties of each cereal and these are described in the annual HGCA *Recommended Lists* for cereals.

Improvements in varieties over the last 30 years of the twentieth century greatly contributed to the increasing yield of cereals. Currently, yields have plateaued despite the introduction of new varieties. Choice of cereal variety can affect yield, quality, input requirements such as fungicide and growth regulators, and subsequent returns.

Varieties are recorded for their susceptibility to the main diseases in field trials. Varieties with a rating of 1, 2 or 3 in the HGCA recommended lists are very susceptible compared with those rated 8 or 9 for varieties that are resistant and where the disease is unlikely to reduce yields significantly.

13.4.5   Time of drilling
Time of drilling will be affected by many factors including crop, variety, possible pest, disease and weed problems, soil type, machinery availability, weather conditions and previous cropping. In winter cereals, sowing in late September will normally give a higher yield than drilling at the end of October. Drilling starts, in most parts of the country, with winter barley and first wheats from the

middle of September (occasionally earlier in Scotland). Second and third wheats which could be affected by take-all, or fields with serious grass weed problems, should be drilled last. Cereal sowing will normally be delayed following root crops. The latest safe date for drilling winter cereals will depend on the variety and its vernalisation requirement. The HGCA gives recommendations on the latest safe date for sowing each variety. Spring cereals can be sown during the winter; however, the majority of spring cereals are sown in the spring from January onwards. Time of drilling will depend more on suitable soil conditions than on calendar date.

13.4.6   Seed type and dressing

The most commonly grown category of seed sown is C2 – second generation commercial seed. First generation C1 seed is only sown if it is being grown for seed on contract, or for home-saved seed, or if a very new variety is being tried. Most seed is of Higher Voluntary Standard (HVS) rather than the lower EU Minimum Standard.

Unless growing organic crops, seed (purchased or home-saved) may be treated against fungal diseases such as bunt, leaf stripe or seedling blights (Table 6.2). Depending on the previous crop, pests such as wheat bulb fly and wireworm may be a potential threat (Table 7.1). Insecticides can be included in a seed dressing (a dual-purpose dressing).

13.4.7   Fertilisers

Nitrogen is the main fertiliser required. Rates depend on the type of cereal, market, SNS index and the use of organic manures. In wheat, researchers are looking at canopy management as an aid to achieving optimum yields. In this case nitrogen rates are adjusted according to the size of the crop canopy. Average nitrogen fertiliser use in the UK varies according to the crop, i.e. winter wheat 195 kg N/ha, winter barley 145 kg/ha and spring cereals 105 kg/ha. To avoid leaching, all nitrogen is now applied in the spring as split dressings between February and early May, depending on crop and quality requirements.

Phosphorus and potassium index levels are assessed by soil analysis and can be high on the many arable farms. On fields with high reserves, only maintenance dressings are required or, indeed, none at all. Applications can be made at any convenient time during the growing season. If soil levels are very low (index 0 or 1), the fertiliser should either be applied to the seedbed or combine drilled. Care must be taken with potassium applications if straw is removed from the field rather than incorporated. Barley straw, especially, contains fairly high levels of potassium and removal could well mean that subsequent applications of potassium will need to be increased (Table 13.3).

With the reduction in amount of atmospheric sulphur being deposited many cereal crops are now at risk from sulphur deficiency. In the spring, 20–50 kg SO_3/ha should be applied on these soils. Of the trace elements manganese is the

Table 13.3 All cereals – phosphate and potash recommendations (kg/ha)

	P or K index			
	0	1	2	3 and higher
Straw ploughed in/incorporated				
Winter wheat, winter barley (8 t/ha)				
Phosphate (P_2O_5)	120	90	60	Nil
Potash (K_2O)	105	75	45 (2−)	Nil
			20 (2+)	
Spring wheat, spring barley, rye, triticale, winter and spring oats (6 t/ha)				
Phosphate (P_2O_5)	105	75	45	Nil
Potash (K_2O)	95	65	35 (2−)	Nil
			20 (2+)	
Straw removed				
Winter wheat, winter barley (8 t/ha)				
Phosphate (P_2O_5)	125	95	65	Nil
Potash (K_2O)	145	115	85 (2−)	Nil
			55 (2+)	
Spring wheat, spring barley, rye, triticale (6 t/ha)				
Phosphate (P_2O_5)	110	80	50	Nil
Potash (K_2O)	130	100	70 (2−)	Nil
			40 (2+)	
Winter and spring oats (6 t/ha)				
Phosphate (P_2O_5)	115	85	55	Nil
Potash (K_2O)	165	135	105 (2−)	Nil
			75 (2+)	

Notes:
Some clay soils can release 50 kg/ha of potash annually. On these soils potassium applications should be reduced accordingly.
The recommendations for P and K at index 0 and 1 are for crop requirements as well as building up soil reserves.

most common deficiency and 15–20% of cereals usually require treatment with manganese each year. In Scotland, up to 30% of cereals needs treating with copper, although in England and Wales the area treated is nearer 5%.

13.4.8   Crop protection
There is usually a yield response in cereals from controlling weeds, pests and diseases. The response will depend on the problem and infestation level. On average in the UK, three herbicide, three fungicide, an insecticide and two growth regulator applications are made to winter wheat crops during the growing season (Table 13.4). Pesticide inputs are normally much higher in winter cereals than in spring cereals. Prophylactic/routine treatments are commonly used. Rates are

Table 13.4   Winter cereals – chemical calendar

Month	Crop growth stage*	Herbicides/growth regulators	Insecticides/ molluscicides†	Fungicides
September	Crop sowing	Couch and weed control including volunteers, post-harvest pre-drilling	Seed dressing Control of slugs	Seed dressing
October	Crop emergence Leaf emergence	Grass weed control – black-grass, wild oats, etc. and broadleaved weeds	Control of aphids (BYDV)	
November	Tillering			
December				
January			Control of wheat bulb fly	
February				
March	Stem extension	Broadleaved weed control		T0 – if foliar diseases such as yellow rust are present
April		Growth regulators and control of wild oats and cleavers		Eyespot and foliar diseases – T1
May	Flag leaf emerging			Foliar diseases – T2
June	Ear emergence			Foliar and ear diseases – T3
July	Flowering Ripening		Orange blossom midge (wheat) Aphids (wheat)	
August	Harvest	Couch control pre-harvest		

* Will be affected by sowing date, crop (wheat or barley) and climate.

† Tend to treat when necessary or high-risk situations, not on routine basis.

often reduced depending on the risk of damage. Accuracy of application is best achieved by following the same wheelings each time. These 'tramlines' are normally introduced at drilling. The loss of land caused by tramlines is small and is more than compensated for by the ease and accuracy of spraying. Tramlines also allow easier crop inspection.

With the numbers of different pesticides and formulations, most farmers rely on independent BASIS registered advisers/consultants and distributor representatives to help with their crop walking and spray programmes. Many organisations run cereal/arable crop trial centres where it is possible to see new chemicals and husbandry techniques in practice. Organisations such as NIAB TAG are farmer funded. Information from these centres is normally only available to members. All farmers selling grain pay a small levy to the HGCA to fund cereal research such as the recommended list trials.

13.4.9 Diseases

There are many diseases that can affect cereals from those which are seed-borne (e.g. loose smut), to stem-based problems (e.g. eyespot), to foliar diseases (e.g. mildew and rust). The majority of diseases are caused by fungi, although there are a number of important virus diseases such as barley yellow dwarf virus (BYDV) and the mosaic viruses. Some diseases are specific to particular cereals; others can attack most cereals (see Table 6.2). Routine fungicide programmes are normally used in most crops. Varietal resistance, weather conditions, disease incidence and husbandry, will affect the fungicide programme used. The most cost effective timing for fungicides will vary with the cereal crop. In wheat, for instance, it is essential to keep the flag leaf and ear clean, whereas in winter barley, treatment at early stem extension (GS 31–32) gives the greatest economic return. Care must be taken with fungicide choice as there are diseases which have now developed resistance to some commonly used fungicides (e.g. eyespot resistant to MBCs, mildew to the azoles and strobilurins, septoria tritici resistant to strobilurins and *Rhynchosporium* to the MBCs and some azoles). Table 13.5 summarises the main groups of fungicides used in cereals. To avoid resistance, fungicide mixtures should be used with different modes of action. Any one fungicide should not be applied more than twice in a season.

13.4.10 Weeds

Weed problems in cereals vary, depending on rotations (both past and present), soil types, cultivation systems and area of the country. Continuous autumn sowing of cereals has encouraged autumn-germinating weeds such as black-grass and cleavers, especially on the heavier soils. Mixed rotations with autumn and spring sown crops have a more varied weed flora. Non-ploughing techniques and early drilling of autumn cereals have favoured the bromes.

The most important grass weed problems in winter cereals are wild oats, common couch, black-grass (not the far south-west of England), meadow-grasses (particularly in grass cereal rotations) and the bromes. Common broad-leaved weeds in winter cereals include common chickweed, cleavers, mayweed, speedwell and field pansy. In spring cereals, the polygonums (e.g. knotgrass and redshank) and common hemp nettle (in some areas) are more important.

Grass weed control can prove more difficult in cereals than broad-leaved weeds. The cost of grass weed herbicides, especially wild oat chemicals, is usually higher than herbicides for broad-leaved weeds. A limited range of herbicides are available, some residual and the others foliar acting. Many of the herbicides, other than the main wild oat herbicides, control some broad-leaved weeds (see Table 13.6).

Table 13.5 Examples of fungicide groups for disease control in wheat and barley

Chemical group	Examples of active ingredient	Diseases controlled						
		Eyespot *Wheat*	Mildew *Wheat and barley*	Septoria spp. *Wheat*	Yellow rust *Wheat*	Brown rust *Wheat and barley*	Rhynchosporium *Barley*	Net blotch *Barley*
Azole	Epoxiconazole	(*)	(*)	*	*	*	(*)	(*)
Strobilurin (QoI)	Pyraclostrobin	(*)	(*) w	(*)	*	*	*	*
Quinolinone	Proquinazid		*					
Spiroketalamine	Spiroxamine		(*) w *b	(*)	(*)	(*)	*	
Morpholine	Fenpropimorph		(*)	(*)	(*)	*w (*) b	(*)	
Bezophenone	Metrafenone	*	*	*				
Chloronitrile	Chlorothalonil		(*) w	*	(*)	(*) w	(*)	
Imidazole	Prochloraz	*	(*) w	*	(*)	(*) w		(*)
Anilino-pyrimidine	Cyprodinil	*	(*)				(*)	(*)
Amidoxine	Cyflufenamid		*					
SDHI mixture with azole	Bixafen + Prothioconazole	*	*	*	*	*	*	*

Note: Not all active ingredients in the same chemical group have exactly the same disease control profile. Always check the manufacturer's recommendations. The effectiveness of some of these chemicals may vary in the future depending on development of fungicide resistance.
* Will give control (from good to excellent depending on chemical and timing).
(*) Partial control.
w = wheat and b = barley.

Table 13.6 Examples of herbicide control of grass and broad-leaved weeds in autumn sown cereals

Crops: w = wheat, b = barley, o = oats, r = rye, d = durum, t = triticale

Active ingredients: 1 = Pre-emergence, 2 = Post-emergence (Control at the highest recommended rate)

Trade names – (Always check product literature for rates, timing, safe use and restrictions)

Crops	Loose silky-bent	Awned canary-grass	Seedling ryegrass*	Rough meadow-grass	Annual meadow-grass	Barren brome	Black-grass*	Wild oats*	Active ingredients	Trade names	Charlock	Common chickweed	Cleavers	Crane's-bill	Dead-nettle	Forget-me-not	Field pansy	Common fumitory	Mayweed	Corn poppy	Shepherd's-purse	Speedwell
w b r d t	–	–	–	s	s	s	s	S	Tri-allate 1.(2)	Avadex Excel	r	r	r	r	r	r	r	r	r	r	R	r
w b	–	S	–	S	S	R	S	S	Fenoxaprop-P-ethyl 2	Various	R	R	R	R	R	R	R	R	R	R	R	R
w r d t	–	–	–	S	R	–	S	S	Clodinafop-propargl 2	Various	R	R	R	R	R	R	R	R	R	R	R	R
w b	–	–	S	–	–	–	S	S	Pinoxaden 2	Various	R	R	R	R	R	R	R	R	R	R	R	R
w b r d t	S	s	–	s	–	–	–	–	Pendimethalin 1,2	Various	–	S	–	–	S	S	S	s	s	S	s	S
w b	–	–	S	S	S	s	S	–	Flufenacet + Pendimethalin 1,2	Various	–	S	s	–	S	s	s	–	s	S	S	S
w b o r t	S	–	–	–	–	–	S	–	Flupyrsulfuron-methyl 1,2	Various	S	S	r	S	S	S	r	–	S	S	S	r
w o r t	S	–	–	–	–	–	S	–	Carfentrazone-ethyl + flupyrsulfuron-methyl 2	Lexus Class	S	S	S	S	S	S	s	s	S	S	S	S
w b (d t)	S	–	s	S	S	–	–	–	Prosulfocarb 1,2	Various	–	S	S	S	S	S	s	–	–	–	–	S

Active ingredient	Group	Product	w (r d t)	w r t	w b	w b r d t (o)	w b (o r d t)	w b (r, t, o)
Iodosulfuron-methyl + mesosulfuron-methyl	2	Various	–	S	–	S	–	S
Florasulam + pyroxsulam	2	Broadway star	S	S	S	S	S	S
Picolinafen + pendimethalin	2	Chronicle	S	S	S	S	S	S
Diflufenican	1,2	Various	S	s	–	S	–	S
Diflufenican + flufenacet	1,2	Various	–	S	s	S	S	S
Diflufenican + flurtamone	1,2	Bacara	–	S	s	S	s	S

*Note: Strains of some of these grass weeds have developed resistance to some herbicides included in this table (e.g. fops, dims and sulfonyl ureas)

Symbols: S = susceptible, s = moderately susceptible, r = moderately resistant, R = resistant, – = no information on label.

There are now many cases of resistance of black-grass, wild oats and ryegrass to some commonly used grass weed herbicides. The build-up of resistance is fastest with the ACCase herbicide group (including products such as fenoxaprop-P-ethyl) and the ALS-inhibitor herbicide group (including sulfonyurea herbicides such as iodosulfuron-methyl + mesosulfuron-methyl). Where resistance has been confirmed it is very important to use a mixture of control measures rather than just relying on herbicides. Factors such as cultivations, rotations, stubble hygiene, delayed drilling, crop competition and hand roguing can all help to reduce the risk. Herbicide programmes that include '*stacks*' or mixtures of chemicals with different active ingredients and *sequences* of chemicals with different modes of action applied in close succession should be used. Good control of moderate populations of resistant black-grass has been obtained using stacks and sequences including at least three different active ingredients including products such as triallate, prosulfocarb, and flufenacet. Where possible it is better to treat black-grass pre- and early post-emergence. Repeat treatments with the same herbicide should be avoided. ALS-inhibitor products should not be mixed with other ALS-inhibitor herbicides or applied more than once in a season.

Weed growth stages in relation to herbicide efficiency are explained on the product labels. Timing of control of broad-leaved weeds in cereals is not as critical as for annual grasses, although again better control is achieved when the weeds are small. If dealing with grass weeds in the autumn in the winter cereal crop, then broad-leaved weeds are normally also controlled (Table 13.6). Autumn weed control in a competitive winter barley crop can often mean no requirement for further treatment in the spring except for wild oats or cleavers.

Some broad-leaved weeds, such as field pansy, are more easily controlled by a number of the autumn herbicides, e.g. those including diflufenican (DFF). If an autumn treatment is used, it may be possible to reduce chemical rates in the spring. Some of the broad-leaved herbicides can be applied over a long period without crop loss, unlike the 'hormone' weed killers such as MCPA and mecoprop-P. Metsulfuron-methyl (broad spectrum) and fluroxypyr (mainly for cleavers), depending on the crop, can be applied between early crop emergence and late stem extension. (Note that metsulfuron-methyl is not recommended for application before 1 February) (Table 13.7). A limited number of cases of herbicide resistance to the ALS-inhibitor herbicides (sulfonylureas) have been found in chickweed, poppy and scentless mayweed.

Perennial broad-leaved weeds can be more difficult to control than annual broad-leaved weeds. This is especially the case with thistles, field bindweed, wild onion and docks, mainly because the foliage usually appears after the normal time for spraying annuals. Due to the mass of green foliage late season they can be very troublesome, causing lodging and difficulty when combining. However, spraying glyphosate on the nearly mature crop at least one week before harvest can effectively control all perennial broad-leaved and grass weeds. The weeds must be green and actively growing (grain moisture less than 30%) and sprayer booms set to give good weed coverage. Tramlines, high clearance wheels and a sheet under the tractor minimise crop damage. This technique is approved because of

Table 13.7 General guide to spring broad-leaved weed control in cereals

Crops	Herbicides	Broad-leaved weeds																		
w = wheat b = barley o = oats r = rye d = durum t = triticale	**Active ingredients** – all post-emergence (Control at the highest recommended rates)	Black-bindweed	Charlock	Common chickweed	Cleavers	Crane's-bill	Dead-nettle (red)	Fat hen	Forget-me-not	Common fumitory	Hemp-nettle	Knotgrass	Mayweed	Small nettle	Field pansy	Corn poppy	Redshank	Shepherd's purse	Sow-thistle	Speedwell common
w b o r	MCPA	r	S	–	–	–	–	S	–	s	s	r	r	s	r	s	r	S	r	–
w b o	Mecoprop–P	r	S	S	S	r	s	S	R	S	r	r	r	S	R	s	r	S	s	s
w b o (r d t)	Dicamba+mecoprop–P	S	S	S	s	–	R	S	–	s	s	S	s	S	R	r	S	S	r	s
w b o r d t	Fluroxypyr	S	R	S	S	–	s	S	S	S	–	s	r	–	–	–	r	–	–	r
w b o r t	Ioxynil+bromoxynil (HBN)	S	S	–	R	–	–	S	S	S	S	S	S	S	S	S	S	S	–	S
w b o r d t	Amidosulfuron	–	S	–	S	–	–	–	S	–	–	–	–	–	–	–	–	–	–	–
w b o t	Metsulfuron-methyl	–	–	S	–	S	S	–	s	–	S	s	S	S	s	S	S	S	s	S
w spring b	Metsulfuron-methyl+thifensulfuron-methyl	S	S	S	s	S	S	s	s	–	S	S	S	S	s	S	S	S	–	S
w b o r t	Thifensulfuron-methyl+tribenuron-methyl	–	S	S	s	–	s	–	–	–	–	–	S	–	–	S	S	S	–	–
w b o t (r d)	Metsulfuron-methyl+tribenuron-methyl	s	S	S	–	S	S	s	s	–	S	s	S	S	s	S	S	S	s	S
w b o	Florasulam	–	–	S	S	–	–	–	–	–	–	–	S	–	–	–	–	S	–	–
w b o (r d t)	Florasulam+fluroxypyr	s	S	S	S	–	s	–	S	–	S	s	S	s	–	s	s	S	–	–

Symbols: S = susceptible, s = moderately susceptible, r = moderately resistant, R = resistant, – = no information on label.
Always check the product label for details of timing, application rates, restrictions, tank mixes and safe use.

the very low mammalian toxicity of glyphosate, but the straw should not be used as a horticultural growing medium or mulch. Stubble treatment is often less effective against these weeds due to the limited amount of re-growth.

13.4.11 Pests
Cereals can be attacked by several pests (Table 7.1), some of which cause little damage (e.g. leaf miners); others can cause total crop loss if not controlled. The main pests in cereals include aphids, wireworm, slugs, leatherjackets, wheat bulb fly, frit fly, blossom midge, birds and rabbits. Many pests are specific to some, but not all, cereals; wheat bulb fly does not affect winter oats or spring cereals sown after the middle of March. The previous crop can also have a significant effect on the cereal. Pests following grass include wireworm, leatherjacket and frit fly. Some pest problems only occur in certain areas of the country or on particular soil types, e.g. wheat bulb fly is an eastern counties problem, and slugs are associated with heavier soils especially after oilseed rape. Treatment for pests is not routine every year. Research organisations are trying to improve pest forecasting systems. Cultural control methods such as time of drilling, seedbed conditions and varietal resistance should be used before resorting to chemical control.

13.4.12 Plant growth regulators
Plant growth regulators (PGRs) are mainly used in winter cereals to reduce plant height and increase straw strength and so reduce lodging and brackling; leaning is not a problem.

Lodging occurs when the crop is between being flat and being up to 45 degrees to the ground. Leaning occurs when the crop is standing upright but leaning no more than a 45 degree angle from the vertical. Brackling occurs when the straw in barley breaks just below the ear.

There are two types of lodging. Stem lodging is due to the straw buckling and is affected by stem diameter, straw wall thickness and strength. Root lodging is due to the plant actually being uprooted. The amount of soil moisture, soil texture and rooting depth and spread will affect the amount of root lodging.

Lodging is also affected by a number of husbandry factors:

- Varieties. The HGCA *Recommended Lists* show that there is quite a difference in standing power between crops and varieties. Length of straw is also important. Short-strawed varieties are not as affected by damage caused by the force of high winds.
- Nitrogen use. Excess nitrogen applications, excess soil nitrogen mineralisation and/or too early applications of nitrogen can produce lush growth, dense canopies and weak straw. These crops are prone to stem lodging.
- Stem-based diseases. Fungal diseases such as eyespot and those caused by *Fusarium spp.* can reduce stem strength and increase the likelihood of stem lodging.

- Weeds. Weed competition can weaken straw strength.
- Time of sowing and seed rate. Early drilling of winter wheat and late drilling of winter barley and winter rye produce taller-strawed crops that are more prone to lodging. High seed rates produce crops with a narrow spread of roots that are more liable to root lodging.
- Soil type. Cereal crops grown on shallow, droughty soils can be less affected by lodging than on deeper, more fertile soils. On these soil types care needs to be taken with PGRs as crop damage has been observed.
- Weather. The amount, timing and intensity of wind and rain, situation of field and exposure affect the amount lodging.

Lodging at early ear emergence causes the greatest yield loss (up to 50% reduction has been recorded). On average, yield losses from lodging in wheat are 2.5 t/ha. Other problems induced by lodging can include the production of secondary tillers, uneven ripening and poorer grain quality, especially if the ears start sprouting, there is increased weed competition, a delayed harvest, or increased combining time and drying requirements. By looking at the plant density and canopy size farmers can assess the risk of lodging. Use of growth regulators, rolling in the spring and nitrogen timing can help reduce the lodging risk. Growth regulators tend to be used routinely on fertile soils where high-yielding quality cereal crops are being grown, and where there is a history of lodging.

13.5   Harvesting

Threshing is the separation of the grains from the ears and straw. In wheat and rye the chaff is easily removed from the grain. In barley, only the awns are removed from the grain; the husk remains firmly attached to the kernel. In oats, each grain kernel is surrounded by a husk that is fairly easily removed by a rolling process – as in the production of oatmeal; the chaff enclosing the grains in each spikelet threshes off. Varieties of naked or huskless oats have now been bred; they thresh free from their husks. The majority of cereal crops are harvested using a combine. Only a few crops are traditionally harvested with a binder where the straw is required for thatching, and some crops are foraged for whole crop silage.

Cereal harvesting can be carried out more efficiently if it is spread out by growing early and late varieties of barley and wheat, starting with winter barley in July, followed by winter wheat and spring barley in August/September. Spring wheat is the last cereal crop to harvest. Good weed control, to minimise green material at harvest and avoidance of lodging, is also very important; combining is easier, and less or no drying of grain is required. The combine capacity should be adequate to harvest the crops as they ripen. Delay can result in poor-quality grain (worse if it sprouts in wet weather) and shedding losses. It is preferable to have to dry early-harvested crops than to salvage damaged crops later. Grain losses of 40–80 kg/ha are reasonable (the latter in a difficult season). Slow combining to reduce losses to 10–20 kg can be false economy; more can be lost by shedding due to a delay in harvesting. Timing of harvesting will be affected by area and crops grown as well as quality requirements, combine and drying capacity, labour, use of desiccants and weather conditions.

Grain monitors, if properly used, can be helpful in checking losses. The average rate of working (t/hour) of a combine is about half the maker's rating, which is usually based on harvesting a heavy wheat crop in ideal conditions. Standard combines vary in drum width, straw walker area (the main limiting factor) and header width. Wide headers are desirable for farms where the crops are normally light and the fields reasonably level. Pick-up reels are very useful when harvesting laid crops. The standard combine can work up and down slopes reasonably well, provided the speed is adjusted as necessary. However, combines do not work satisfactorily going across slopes as the grain moves to the low side of the sieves, and little separation takes place because most of the wind escapes on the top side. Some combines are designed to keep the threshing and sieving mechanisms level on sloping ground. The hillside type has a self-levelling mechanism, which can adjust for going across slopes as well as up and down.

The axial-flow combine differs from the conventional machines by having the threshing and separation of grain from straw carried out simultaneously as the material spirals round a large rotor from front to back of the machine. The straw walkers in conventional combines are not the most satisfactory way of separating grain and straw; on some machines these are replaced by a series of drums. This is done in order to increase the output of the combine without increasing the physical size, which is important for access and road transport reasons. These combines, known as 'rotary' combines, require more power and tend to damage the straw more than conventional combines.

The conventional header can be replaced by a stripper header which removes the ears only, and the straw is left standing. This can be an advantage for chopping or ploughing-in the straw. The header will be more expensive, but the work rate can be very high and laid crops do not present serious problems.

Whatever type of machine is used, the manufacturer's recommendations should be followed for the various settings, e.g. drum speed and clearance and fan speeds. On modern machines adjustments can be made easily.

Combines can be fitted with straw spreaders or choppers, which are helpful in dealing with the problem of straw disposal, although the mounted chopper can reduce combining speed as it requires more power. As combines get heavier, they are increasingly being fitted with tracks, to aid traction and reduce soil damage, as well as automatic steering systems.

Cleaning combines to remove weeds, e.g. wild oats, black-grass, barren brome, is very important before moving to a clean field. It is essential that the combine is clean before moving into a seed crop.

13.6   Grain-drying methods

These days most grain is bulk-handled from the combine into and out of store. In some years, when all the cereals are combined at below 15% mc, no drying will be necessary, only possibly cooling.

13.6.1   The options available

Below are listed the most important methods of drying grain:

- *Continuous-flow drier.* The principle is that hot air removes the excess moisture and ambient air and then cools the grain to 10–15 °C. However, in very warm weather this may not be possible and night air may have to be used to cool the heap after drying. These driers are rated as 'x' t/hour taking out 5% moisture in wheat.

  Wet grain (over 22% mc) may require two passes through the drier, taking out about 5% moisture each time.
- *Batch driers.* The drying principle is similar to that of the continuous-flow drier, but the grain is held in batches in special containers during the drying process (extraction rate 6% per hour in small types; 6% per day in silos).
- *Ventilated silos or bins.* Cold or slightly heated air is blown through the grain in the silo. This can be a slow process, especially in damp weather (extraction rate 0.33–1% per day).
- *Bulk storage/on floor drying.* A large volume of cold or slightly heated air is blown through the grain to remove excess moisture. The air may enter the grain in several ways, e.g.:
  (a) from ducts about a metre apart on, or in, the floor;
  (b) from a single duct in the centre of a large heap;
  (c) through a perforated floor which may also be used to blow the grain to an outlet conveyor when emptying.

Floor drying is very popular because it can be carried out in a general purpose building. It is a cheap method and requires very little labour when filling. The rate of drying is 0.33–1% per day. A common mistake with floor drying is to use heat when drying very wet grain. This over-dries the lower layers around the ducts, and the moisture settles out in the cooler upper layers where it forms a crust which impedes the air flow. The heat should be saved for later stages of drying down to about 14% mc. Problems can arise where rubbish is allowed to form in 'cones' as it is loaded into the store. This could prevent air flowing freely past the grain. The wet and dirty grain should be spread as evenly as possible. Wet grain is more spherical than drier grain; the air spaces are usually larger in the wet grain heap, and so air flows more freely through it than through dry grain. This is a useful self-adjusting phenomenon. Grain stirrers can be installed which overcome the problems of uneven drying. They move automatically through the heap of grain on a framework and thoroughly mix the grain as it dries. The relative humidity (RH) of the air being blown through grain determines the final moisture content of the grain. A 1 °C temperature rise from heaters reduces the RH by 4.5%.

- *Mobile grain driers.* These have become popular in recent years. Most are of the recirculating batch type, but portable versions of well-established static batch or continuous flow machines are also in use.

- *Membership of co-operatives.* The other option which farmers have is to dry and store grain with a co-operative group. This store will provide drying and sometimes grain wetting facilities (in a dry year) as well as gravity separators for improving grain quality.

13.6.2   Precautions to be taken in grain drying

Whilst using high-temperature driers, it is important not to overheat the grain as this may result in loss of quality premiums. Seed, malting and milling samples need to be dried at relatively low temperatures as the seed must remain viable. Feed samples can be dried at higher temperatures and therefore the output of the dryer will be increased. Maximum drying temperatures will depend on the type of dryer used and even the individual make of machine. It is important to follow manufacturers' recommendations.

Grain can be stored in bulk for up to one month at 16–17% mc, but for longer term storage it should be dried down to 14% mc. Only fully ripe grain in a dry period is likely to be harvested in the UK at 14% moisture. In a wet season, the moisture content may be over 20% and the grain may have to be dried in two or three stages if a continuous-flow drier is used.

Damp grain will heat and may become useless. This heating is mainly due to the growth of moulds, mites and respiration of the grain. Moulds, beetles and weevils may damage grain that is stored at a high temperature, e.g. grain not cooled properly after drying or grain from the combine on a very hot day. Ideally, the grain should be cooled to 15 °C within two weeks to prevent saw-tooth beetles breeding. This is difficult, if not impossible, in hot weather. For long-term storage the grain should be cooled to below 5 °C to prevent mites and grain weevil breeding. Before storing grain, it is essential to clean and dry the store thoroughly. When clean and waterproof, the building should be fumigated with a suitable approved insecticide to kill any remaining pests.

The temperature of stored grain should be checked regularly. Any rise in temperature can indicate an insect infestation. Rodents and birds must also be kept out of the store. When grain is sold, any chemical treatments must be declared. The Pesticide Notification Scheme is now mandatory. Any kind of grain stored for human consumption must be kept under conditions that satisfy the Food Safety Act 1990. This means that the farmer must exercise 'due diligence' in keeping stores rodent- and bird-proof; these are aspects covered in the various assurance schemes. Other commodities, such as fertiliser, must not be stored in the same building.

13.7   Moist grain storage

Most grain is stored dry. There are, however, alternative methods of preserving grain that can be simpler and cheaper than more conventional systems. An established method is the storage of damp grain, straight from the combine, in sealed silos. Fungi, grain respiration and insects use up the oxygen in the air spaces and give out carbon dioxide, and the activity ceases when the oxygen is used up. The grain dies but the feeding value does not deteriorate whilst it remains in the silo. This method is best for damp grain of 18–24% mc, but grain up to 30% or more may also be stored in this way, although it is more likely to cause trouble when removing it from the silo, e.g. 'bridging' above an auger. The damp grain is taken out of the silo as required for feeding. This method cannot be used for seed corn, malting barley, or wheat for flour milling.

Harvesting grain when moist and then ensiling is a technique that is gaining popularity on some mixed farms. The cereal is combined three to four weeks earlier than normal when the grain dry matter is at around 65%. The grain is then crushed in a machine called a *crimper* and ensiled. A preservative or organic additive is normally used to reduce clamp spoilage. Yield of crimped grain is usually higher than grain harvested at the normal time.

A further method of storing damp grain safely and economically is by sterilising it with a slightly volatile acid such as propionic acid. The acid is sprayed on to the grain from a special applicator as it passes into the auger conveying it to the storage heap. Grain stored in this way is not suitable for milling for human consumption or for seed, but it is very satisfactory for animal feeding and, after rolling or crushing, it remains in a fresh condition for a long time because the acid continues to have a preservative effect.

13.8   Cereal straw

Recoverable straw produced on farms in the United Kingdom is in the range of 2.5–5 t/ha. Virtually all the barley and oat straw is baled for livestock bedding and feed. The feed value for wheat straw is lower than for barley and oats. The use of wheat straw will depend on the area where it is grown and the proximity of livestock enterprises. Each year about 40% of wheat straw is not baled, but is incorporated instead. Other uses include straw for farm and household fuel, mushroom compost, covering overwintered carrots, potato and sugar beet storage. It is also used in horticulture, as well as for thatching and insulation board.

When incorporating straw into the soil it should be chopped, preferably using a combine-mounted chopper. Straw can be incorporated by ploughing to at least 15 cm or, by a non-ploughing method, e.g. heavy discs, to a 10 cm depth. Non-ploughing techniques will tend to encourage annual grass weeds, especially the bromes and black-grass, as well as volunteer cereals. Light cultivations pre-ploughing can encourage weed seeds to germinate, but are not essential for successful straw incorporation. With the introduction of the stripper header there could be more problems incorporating the long un-chopped straw. Rolling the straw prior to ploughing can aid incorporation.

Applying extra fertiliser or additives has little effect on straw breakdown. Where straw had been incorporated for several years, there appears to be no problem with its decomposition. In practice, there should be a small increase in soil organic matter.

13.9   Wheat

Nearly half of the cereal area in the EU is wheat. France, Germany and the UK are the major producers. In the UK just less than two million ha are grown and it is the most widely grown arable crop. The main markets (Table 13.1) are for human consumption (mainly milling), animal feed and for export. Of the remainder, some is used for seed, breakfast foods and distilling whisky. There has been increasing amounts of wheat used for biofuel or bioethanol production. There is an EU intervention scheme for common wheat and barley which gives a minimum support price for grain of the correct quality. World cereal prices are much higher than the EU support price so there are currently no offers for sale of UK grain into intervention.

The majority of wheat is autumn sown winter wheat with only a small amount of spring sown wheat.

13.9.1   Some important qualities of wheat

Approximately five million tonnes of wheat a year are milled in the UK; of this just over 80% is UK grown. Imported wheat makes up for any shortfall in quality requirements. In the production and use of wheat flour, quality requirements can be divided into two categories, for milling and for baking. All milling wheat should satisfy the following general standards:

1. Be free of pest infestation, discoloured grains, objectionable smells, ergot and other injurious materials.
2. Not overheated during drying or storage.
3. Moisture content of 15% or less.
4. Maximum impurities less than 2% by weight.
5. Pesticide residues within limits prescribed by legislation. A 'passport' describing pesticide application and vehicle cleanliness must accompany all loads taken to the mill.
6. Specific weight – usually at least 76 kg/hl required.

Milling quality
This refers to the ease of separation of white flour from the germ and bran (pericarp and outer layers of the seed). It is a varietal characteristic and can be improved by breeding. In the milling process, the grain passes between fluted rollers which expose the endosperm and scrape off the bran before sieving. The endosperm is ground into flour by smooth rollers and in this process the good quality endosperm (hard wheat) breaks along the cell wall, across cells and even the starch granules may be broken. High starch damage is required by bakers as it means that larger

amounts of water can be absorbed into the dough which leads to more water remaining in the final loaf of bread. However, the endosperms of soft wheat break into fine particles with little damage to the starch granules. 'Milling value' is a measure of the yield, grade and colour of flour obtained from sound wheat. Varieties favoured for bread making normally have a hard endosperm. Soft wheat is used for biscuit making. Grain hardness can be assessed using the Single Kernel Characterisation System (SKCS).

Baking/bread making quality
Wheat is very suitable for bread making because the dough produced has elastic properties. This is due to the gluten (hydrated insoluble protein) present. The amount and quality of the gluten is a varietal characteristic, but it can also be affected by soil fertility and climatic conditions.

About 20% of the protein is in the wheat germ and is not important in the baking process. However, the other 80% (in the endosperm) is very important. It can be measured in various ways. For example, protein quantity is assessed by the Dumas test and by the near infra-red reflectance (NIR) technique which is a very rapid dry method. Note in addition, NIR analysis can measure moisture content as well as grain hardness.

Protein quality is best assessed by a mini-baking test, but this takes at least two hours and so is mainly used for testing varieties for bread, biscuit or cake making. Some other tests that can be used include the Zeleny test. A white flour is mixed with lactic acid and isopropyl alcohol and the resulting sedimentation is measured – the greater the amount of sediment the better the quality. The SDS test is similar but faster. Another method is to wash the gluten out of the ground grain and assess it for colour, elasticity and toughness (strength); this takes about 20 minutes. For the export milling market the UK uses the Chopin Alveograph method. In this test the flour is assessed for baking strength. Strong flour for bread making will have a high strength (W value >200). Weak flour is required for biscuit making with a W value <100. In this test air is pumped into a sample of milled grain made into dough and the amount of elasticity (L value) and dough resistance to pressure (P value) are recorded. For bread making, the flour needs to produce a strong elastic dough with a P/L ratio of 0.5–0.9. For biscuit making, the flour needs to form an extensible dough with a P/L ratio of 0.2–0.4. Grain for animal feed tends to produce a very inelastic dough. If grain is overheated during drying (more than 60 °C) then the protein can be spoiled (denatured) and not be suitable for bread making.

To provide large, soft, finely textured loaves, the baker requires flour and dough with a large amount of good quality gluten. Good dough will produce a loaf about twice its volume. The small amount of alpha-amylase enzyme present in sound wheat is desirable for changing some starch to sugars for feeding the yeast in bread making. However, wet and germinated wheat contains excessive amounts of alpha-amylase and so excessive amounts of sugars and dextrines are produced. This results in loaves with a very sticky texture and dark brown crusts. Alpha-amylase activity can be measured by the Hagberg FN (falling number) test. The number is the time in seconds for a plunger to fall through a slurry of ground grain and water, plus 60 seconds for heating time before the plunger is dropped. A high FN indicates a low (and desirable) alpha-amylase level.

If the amino acid cystine is not present in the flour, as may happen if the crop suffered from sulphur deficiency, then the quality of the bread would be poor, even though the grain passed all the usual tests. The amount of water absorption by flour is also important in bread making. It is a varietal characteristic and depends on the protein content and amount of starch granules damaged by milling. A high water uptake by damaged granules of hard wheat means more loaves, which keep fresher longer, from each sack of flour.

Wheat varieties are grouped in the UK by the National Association of British and Irish Millers (NABIM) according to their milling and baking characteristics:

NABIM Group 1 varieties are the best for bread making; they have a hard grain and produce a strong elastic dough.

NABIM Group 2 varieties mostly have bread making potential, they have a hard endosperm and are often included in bread making grists.

NABIM Group 3 varieties have a soft endosperm; produce an extensible but not elastic dough. Grain from these varieties are commonly used in biscuit and cake flour. Some grain is also used in the starch and distilling industries.

NABIM Group 4 varieties are only suitable for animal feed; they can have hard or soft grain and are usually the highest yielding varieties.

Wheat for export

The UK now exports approximately 20% of the wheat produced. British Cereal Exports have produced two brands to aid export of UK wheat; ukp is premium wheat for bread making and uks is soft wheat for a variety of markets. Grain is analysed using the Chopin Alveograph Test when selling into this market as it gives a good indication of end use.

Table 13.8 shows typical wheat quality requirements for the different markets.

Table 13.8   Typical wheat quality requirements

Market	Max moisture content (%)	Min specific weight (kg/hl)	Max impurities (% by wt)	Protein content (% (100 % DM))	Min Hagberg FN (sec)	W value	P/L value
Bread making	15*	76	2	13	250	–	–
Biscuit making	15*	74	2	11.5	220	–	–
Export ukp	15	76	2	11–13	250	>170	<0.9
Export uks	15	75	2	10.5–11.5	220	<120	<0.55
Feed	15	72	2	–	–	–	–

*If grain is to be stored for more than four months, maximum moisture content should be 14%.
W and P/L values are from Chopin Alveograph Tests. Hagberg FN, Hagberg falling number.

13.9.2 Wheat husbandry

Soils and climate

Wheat is a deep-rooted plant that grows well on heavy soils and in the drier eastern and southern parts of this country. Winter wheat will withstand most frosty conditions, but it can be killed by waterlogged soils. The soil pH should be higher than 6.0.

Place in rotation

When the soil fertility is good, wheat is the best cereal to grow as its yields are higher and its returns generally better than those of the other cereals. It is commonly taken for one or two years after grassland, potatoes, sugar beet, beans, peas, linseed or oilseed rape. It has also been grown continuously on many farms. After four or five years, the so-called 'take-all barrier' is passed and yields remain fairly constant. However, many farmers now grow as many first wheats as possible (rather than continuous) because of their higher yields. The increasing problem with herbicide resistant black-grass is making many farmers increase the number of break crops in the rotation.

Seedbeds

A fairly rough autumn seedbed is adequate for winter wheat and helps prevent 'soil-capping' in a mild, wet winter. When soil-acting herbicides are used a fine seedbed is required. In a difficult autumn, winter wheat may be successfully planted in a wet sticky seedbed and usually it still produces a satisfactory crop. Spring wheat should only be planted in a good seedbed.

Time of sowing

Winter wheat can be sown from September to March. Early sowings from the middle of September are usually preferable, but they may not be better than late sowings in some favourable autumns. Occasionally, first wheats are sown at the beginning of September at very low seed rates. These crops normally require higher pesticide inputs and yields can be disappointing. Second wheats should be drilled in October to reduce the risk of take-all. Where grass weeds are a problem, later drilling will give time for some stubble/seedbed cultivations to control the first flush of the seedling weeds. Slow-developing varieties should be chosen for early drilling, and faster-developing varieties for later sowing. However, sowing late may be impossible in some years. The HGCA *Recommended Lists* can be checked for 'latest safe sowing dates'. Spring wheat can be sown from late autumn to April the following year.

Method of sowing

Ideally, the crop should be sown at a uniform depth of 2.5 cm, depending on the moisture level in the soil. The methods used are:

- Combine drill with fertiliser in 15–18 cm rows.
- Grain-only drill in 10–18 cm rows.

- Broadcasting using the fertiliser spinner. This can be very satisfactory (especially if drilling is impossible) provided a good covering of the seed is achieved.

Varieties

These are detailed in the HGCA *Recommended Lists*. Winter wheat normally yields very much better than spring wheat, but the latter is usually of very good milling and baking quality. About 98% of wheat drilled is of winter varieties. Most winter varieties have to pass through periods of cold weather (vernalisation) and increasing day length (photoperiod) before the ears will develop normally hence why winter varieties have a latest safe sowing date recommendation.

Varieties differ in their susceptibility to disease, e.g. yellow rust and mildew, and it is advisable to grow at least three varieties selected to reduce the risk of spread of these diseases (see Diversification Scheme in the HGCA *Recommended Lists*). It is well established that varieties differ considerably in their management requirements such as the best time to sow; fertiliser timing and treatments for yield and quality; disease susceptibility and control; straw stiffness and response to growth regulators; risk of sprouting in the ear; ease of combining and saleability. Examples of recommended winter varieties are shown in Table 13.9.

New varieties are continually being developed for higher yields, quality, disease and pest resistance and standing ability.

Table 13.9   Examples of recommended wheat variety

NABIM group 1 Good-quality milling and bread making	*Sky fall* *Crusoe* *Gallant*
NABIM group 2 Fair bread making	*Cubanita* *Chilton* *Panorama*
NABIM Group 3 Biscuit quality and others	*Zulu* *Invicta*
NABIM Group 4 Animal feed	*Horatio* *Conqueror* *KWS Santiago* *JB Diego*

Seed rates

The usual range is 100–250 kg/ha for winter wheat, and 170–220 kg/ha for spring wheat. A number of factors will affect the number of seeds sown and hence the seed rate:

1. Seedbed conditions (higher rates in very dry, cloddy and stony soils).
2. Weather and time of sowing. The rate could be 30–50% higher in November than in September. The lowest seed rates are used for the early September drillings.
3. Plant population. A target of 250–350 plants/m^2 in the autumn may be reduced to 150–300 plants/m^2 in early spring. In very good conditions fewer than 100 plants/m^2 can give very satisfactory yields, with fewer lodging and disease problems than in thick crops (e.g. 400 plants/m^2). Yield/ha is determined by the number of plants/ha × number of ears/plant × number of grains/ear × average grain weight (this last can be influenced by variety, growing conditions, disease control and possibly by growth regulators). Low plant populations tend to produce more ears/plant (average about 2) and possibly more grains/ear (about 30–50, with a few to over 90) and a heavier grain weight. These low populations are not very competitive against weeds.

The following is a simple way to calculate seed rate, knowing the thousand grain weight (g) – TGW.

$$\text{Seed rate (kg/ha)} = \frac{\text{no. plants/m}^2 \text{ required} \times \text{TGW}}{\% \text{ establishment expected}}$$

Fertilisers

To return those nutrients removed from the soil, 7.8 kg of P_2O_5 and 5.6 kg K_2O should be applied for each tonne of grain/ha expected yield (e.g. at 8 t/ha, 62 kg/ha of phosphate and 45 kg/ha of potash are needed) if the straw is incorporated. If the straw is removed, the potassium rate should be increased to 8.4 kg of P_2O_5 and 10.4 kg of K_2O for each tonne of grain harvested (e.g. at 8 t/ha, 67 kg/ha of phosphate and 83 kg/ha potash is needed). Off-takes are higher in spring wheat when the straw is removed. See Table 13.3 for standard recommendations.

The nitrogen fertiliser rates depend on the SNS index and the use of organic manures. Nitrogen recommendations should be varied according to changes in grain and fertiliser prices (see RB209). The standard nitrogen recommendations for winter wheat in kg/ha are shown in Table 13.10.

Table 13.10 Standard nitrogen recommendations for winter wheat (in kg/ha)

Soil type	SNS Index 1	SNS Index 2
Light sandy soils	130	100
Shallow soils	240	210
Medium and deep clay soils	220	190
Deep silty soils	190	160

Spring wheat, sown in the spring, requires similar nitrogen rates on sandy soils but 180 kg/ha on other mineral soils at SNS Index 1. Grain nitrogen content is a good indication that optimum rates have been used. Nitrogen rates should be increased or reduced by 30 kg/ha for every 0.1% variation from the optimum of 1.9% grain nitrogen for feed varieties and 2.1% for bread making varieties. Another system currently being studied is canopy management in connection with optimum nitrogen applications. The target crop canopy at ear emergence is about 6 units of green material per area of ground (GAI 6). As it is known that there are 30 kg/ha of nitrogen for every unit of green area, by finding out the available soil nitrogen level and fertiliser recovery, optimum nitrogen rates can be calculated.

It is important that adequate nitrogen is available when the crop growth rate is rapid at the beginning of stem extension – GS 30–32. If more than 120 kg/ha is to be applied, it will normally be split either two or three times. The first application of 40 kg should be applied during tillering, no earlier than March. Too early an application of nitrogen will encourage tillering and is also more liable to leach. If the crop is late-drilled, backward, thin or suffering from pest attack, then the first application should be in February to encourage tillering. The main dressing of nitrogen should be applied at GS 31, in April and no later than early May. Timing for spring wheat, drilled in the autumn, should be as for winter wheat. Spring-sown spring wheat should have the nitrogen split between the seedbed and the three-leaf stage. The earlier the drilling, the more nitrogen should be applied post-emergence and vice versa.

To increase protein by up to 1%, for milling wheat, extra nitrogen (30–60 kg/ha) is usually applied with the main dressing or later. A liquid application of urea is often applied by the milky-ripe stage. Late foliar applications of nitrogen usually have no effect on yield.

Spring grazing of winter wheat is now rarely practised, but if there is a need for it and the crop is forward but before GS 31 and the soil is dry, it can provide useful grazing for sheep or cattle in late March. Grazing should cease if the plants are being pulled out of the ground. The crop should be only grazed once and as uniformly as possible in blocks and then top-dressed with an extra 50 kg/ha nitrogen (unless a reason for grazing was to reduce the risk of lodging). Yields are likely to be reduced by this grazing. Grazed crops have shorter straw at harvest, and sometimes less disease.

Seed dressings

A range of fungicide seed dressings are available to control seed-borne diseases such as bunt, fusarium, septoria and loose smut (see the *UK Pesticide Guide*). There are now some fungicide seed dressings that can also reduce the impact of take-all, e.g. silthiofam and fluquinconazole.

After grass where wireworm can be a serious problem, an insecticidal dressing can be included in the seed dressing (dual purpose). Chemicals such as tefluthrin or clothianidin are currently approved. This last is very effective against aphids and seed treatment may be preferable to an autumn aphicide.

Where there is a high risk of wheat bulb fly tefluthrin-treated seed should be used.

Pests

In the autumn, slugs can be a serious problem in trashy, cloddy seedbeds, especially when the crop is late drilled. The slugs can hollow out the seed and shred the young seedling leaves. Slug pellets should be applied if there is a high risk (Table 7.1). Early-drilled wheat can be affected by aphid-transmitted barley yellow dwarf virus (BYDV). The optimum time for control is late October/early November when an aphicide can be applied. Note, some seed dressing can be used to effectively control aphids. In the winter, in susceptible fields, wheat bulb fly may require controlling either in January at egg hatch or later, when the first 'dead heart' symptoms are seen – limited chemicals are available. Aphids and blossom midges can cause direct damage to wheat, at grain fill. If threshold numbers are present, treatment is usually advisable (Table 7.1). Some varieties are resistant to attack by orange wheat blossom midge.

Diseases

Yield and quality of wheat can be very affected by stem-based and foliar diseases. Of the stem base/root diseases take-all and eyespot are the most common. Take-all disease can be a serious problem when growing wheat continuously until the 'take-all barrier' has been passed. On some soil types/rotations second wheats can be affected. On susceptible fields tolerant varieties should be grown, late drilled and/or a seed dressing should be used, e.g. silthiofam or fluquinconazole.

The main foliar diseases in wheat are *septoria tritici* – leaf blotch, yellow rust, brown rust; *septoria nodorum* – leaf and glume blotch, and powdery mildew. The timing of a fungicide spray that gives the largest economic response is at T2 (GS 39). This treatment protects the flag leaf and can eradicate any disease on leaf 2. In practice there is usually a yield response from the TI (GS 31/32) timing, which protects leaf 3 and is the best time to control eyespot. A pre-T1 treatment (T0) timing may sometimes be required if yellow rust is present. The T3 timing (GS 59) is important to control any disease still on the flag leaf and protects the ear against fusarium especially on milling varieties.

Care must be taken with chemical choice especially with the increase in fungicide resistance particularly against septoria, eyespot and mildew. Strobilurins (Qols) should not be applied more than twice to any one crop. Most fungicide programmes use mixtures of different active ingredients and alternate fungicides with different modes of action to reduce the risk of disease resistance developing.

Fungicide programmes usually give a yield response even on disease resistant varieties in a low disease year. For example fungicide control of disease on susceptible varieties may give around 2 t/ha yield response compared with only half that on fairly resistant varieties. Rates and types of fungicides should be varied depending on the risk and presence of disease. Years with low disease pressure and varieties with good disease resistance will only require very low rates of fungicides. Most fungicide programmes contain mixtures of fungicides with different modes of action to give both eradicant and protectant control. For example a good mixture for control of septoria would include an azole, plus an SDHI, plus a multi-site fungicide. Other fungicides will be included for control of specific diseases such as eyespot and mildew (see Table 13.5).

The recognition of other cereal pests and diseases and their control are shown in Tables 6.2 and 7.1.

Weeds

In winter cereals, autumn-germinating grass weeds such as black-grass, the bromes and meadow-grass should be controlled in the autumn with residual herbicides. Care must be taken with chemical choice when resistant weeds are present, especially black-grass. If broad-leaved weeds are controlled in the autumn with chemicals such as difluflenican (DFF) they may only require a reduced dose of herbicide in the spring. Spring herbicide treatments are normally applied by the first node detectable stage GS 31. Control of the very competitive weeds wild oats and cleavers can be later in the spring depending on herbicide choice.

Plant growth regulators

It has been estimated in wheat that there is a 1% yield loss for every 2% of wheat crop lodged at flowering. There is a wide variation in varietie's susceptibility to lodging and requirement for a plant growth regulator (PGR). Growth regulators shorten the internodes and strengthen the stem walls. The reduction in internode length varies (7–20 cm) between varieties and with chemicals. To prevent lodging in winter wheat some of the available PGRs such as chlormequat are applied at, and/or before, stem extension (GS 30/31) so that the lower internodes are shortened and strengthened. The more recently introduced growth regulators, such as trinexapac-ethyl and products containing 2-chloroethyl phosphonic acid offer more flexibility in timing (from GS 30 to at least GS 39).

Harvesting

Winter wheat ripens before spring wheat. The crop is normally harvested in August and September.

Yield

Wheat yields have plateaued over the last ten years (Table 13.11). Yields have also fluctuated across the country mainly due to variations in rainfall.

Table 13.11   UK wheat yields (in t/ha)

	Average	Very good
Grain	8	10
Straw	3.5	5

13.10   Durum wheat

13.10.1   Qualities of durum wheat

Durum wheat is a tetraploid species of wheat (*Triticum turgidum* spp. *durum*) compared with hexaploid common or bread wheat (*Triticum aestivum*). In Europe it is mainly grown in the more southerly countries. Most of the durum wheat grown is for the internal market; only a small amount is exported. If properly grown, durum wheat has a high protein and hard, amber-coloured vitreous endosperm. On milling it breaks into fairly uniform large fragments called semolina, which is made into a range of high quality pasta products by extruding a stiff dough (semolina mixed with warm water) through dies of various shapes. This high quality market (up to a 50% premium) requires grain which is a light amber colour and meets minimum standards, i.e. 12.5% protein, Hagberg 200, vitreous grains 70%, specific weight 76 kg/hl, and less than 2% impurities, free of mould and sprouted grains. Some durum wheat is used for breakfast cereals. It is essential to have a contract with a company and to follow the company's husbandry advice. Currently less than 1000 ha are grown in the UK.

13.10.2   Durum wheat husbandry

Soils and climate
Durum wheat will grow on a wide range of arable soils, but is best suited to medium textured soil. It tolerates dry conditions better than ordinary wheat. As it is a Mediterranean crop it tends to yield better in the eastern wheat growing areas of England. The problem when growing this crop is achieving the required quality and it yields poorly compared with conventional wheat.

Seedbeds and sowing
Seedbed requirements and methods of sowing are similar to those for wheat crops.

Time of sowing
As durum wheat does not require vernalisation, it can be sown in either October or March. Autumn-sown crops are susceptible to frost damage.

Varieties
Varieties all have awns on the ear – so are bearded. Current varieties include *Lloyd* and *Pescadou*.

Seed rates
180–200 kg/ha. The target should be 300 plants/m^2.

Fertilisers
Recommendations as for milling wheat.

Pests
Pests as for wheat.

Diseases
Durum wheat is very susceptible to eyespot and ergot.

Weed control
There are only a limited number of products that are approved for use in durum wheat, there are a few chemicals with Extensions of Authorisation for Minor Uses (EAMUs) in durum wheat.

Harvesting
An autumn-sown crop is usually combined in August. If possible, it should be given priority (about 20–22% mc for combining) to obtain good vitreous grains and a high Hagberg number. If it is left too late, the endosperm is white and there is a risk of sprouting. Grain should be stored at 15% mc.

Spring-sown Durum wheat is normally harvested in September.

Yield
Durum is not as high yielding as ordinary wheat, averaging about 20% less. The range is 3–7 t/ha.

13.11 Barley

13.11.1 Uses of barley
After wheat, barley is the next most important cereal crop grown in Europe. About a third of the world trade in barley is from the EU; France, Germany and Spain and are the main producers followed by the UK. Approximately one million ha of barley is grown in the UK of which over 60% is now spring sown. The grain is used mainly for:

- Feeding to pigs (ground), cows and intensively fed beef (rolled). Some feed barley is exported.
- Malting. Over a third of barley sold in the UK goes for malting/distilling. The UK is one of the main countries producing malt in the world and is one of the major malt exporting countries. Grain for malting should be plump and sound with a high germination percentage (about 98%) but a low nitrogen percentage – less than 1.85% nitrogen. Some varieties are more suitable than others for malting (HGCA *Recommended Lists*). The final use for the malt will affect the grain nitrogen requirements. A premium, which varies from year to year, is paid for good malting barley (Table 13.12).

In the malting process the grain is soaked (steeped), and then the grain is allowed to sprout. During this germination period the starchy endosperm is converted to sugars. Once all germinated the grain is dried in a kiln at high temperatures and

Table 13.12   Typical barley quality requirements

Market	Maximum moisture content (%)	Minimum specific weight (kg/hl)	Maximum admixture (%)	Maximum nitrogen range (%)	Minimum germination (%)
Malting	15	–	2	1.55–1.75	98
Export malting	13.5–15	64	2	1.70–1.85	95
Export feed	13.5–15	64	2	–	–
Feed	15	63	2	–	–

finally the rootlets (culms) removed to leave the sweet malt grains. In the brewing process the malt is crushed and then soaked in warm water in mash tuns. The sugars then dissolve in the liquor before being drained off. The remainder of the malt (brewer's grains) is a valuable cattle food. The sugary liquor (wort) is boiled with hops to give it a bitter flavour and keeping quality, and to destroy the enzymes. The strained wort is then fermented with yeast which converts the sugars to alcohol and so produces beer. To make whisky, hops are not added; the fermented wort is distilled to produce a more concentrated alcoholic liquid (malt whisky).

Some malt is used in other products, e.g. malt vinegar and breakfast cereals. Most barley straw is baled for livestock bedding and feed.

13.11.2   Barley husbandry

Soils and climate
Barley can be grown on arable land throughout the UK, provided the pH of the soil is about 6.5; it is more affected by a low pH than is wheat. Barley grows better than other cereals on thin chalk and limestone soils. However, it will grow on a wide range of soils provided they are well drained. On organic and very fertile soils, it may lodge, especially in a wet season, and the grain is unlikely to be of malting quality.

The crop is not exacting as regards climate, but little winter barley is grown in the colder parts of the UK.

Place in rotation
Barley, unlike wheat, is usually grown when the fertility is not very high. On many farms, provided that the soil conditions are right, it has been grown continuously on the same fields producing reasonable yields. However, this was before the mosaic viruses became a serious problem. Reducing the number of winter barley crops in the rotation is now important to slow down the build-up of the disease. Spring barley is unaffected. As an early harvested crop, winter barley is commonly grown before oilseed rape.

Seedbeds
Generally a finer seedbed than wheat is required. Shallow sowing at 3–5 cm is important.

Time of sowing
Winter barley can be sown from mid-September to early November. It is important to have the plants well-established before the winter. Too forward crops can be susceptible to frost damage as well as foliar disease and BYDV. Spring barley can be sown from January to early April. A good seedbed is more important than early drilling. Yields are usually reduced from drilling after the end of March. The sowing of barley follows the same lines as for wheat.

Varieties
See HGCA *Recommended Lists.* Variety is very important if growing barley for malting. The Institute of Brewing and Distilling (IBD) publish an annual approved list for varieties for brewing, malt distilling and grain distilling use. If growing barley for feed, then disease resistance and resistance to lodging should be considered as well as yield. Six row varieties are normally higher yielding than two-row varieties (especially the hybrid varieties) though grain quality is usually poorer.

Examples of recommended varieties including IBD approved varieties are shown in Table 13.13.

Modern two-row varieties of winter barley yield better than spring barley – especially in areas which suffer from drought in the summer. However, many spring varieties have very good malting quality compared with the winter varieties. To obtain best results with winter barley, the crop must be sown early enough to develop a good root system and become well-tillered before winter. Winter barley has declined in popularity in recent years partly due to higher input costs as well as some very variable harvests.

Table 13.13   Examples of recommended varieties of barley

	Winter	Spring
Malting quality for brewing (two-row)		
IBD approved	SY Venture	Quench, Concerto
	Cassata	Propino,
Malting quality for distilling		
IBD approved	Flagon	Shuffle
	Cassata	Belgravia
Feeding quality		Waggon,
(two-row)	KWS Cassia, KWS Glacier	Garner
(six-row)	KWS Meridian	
	Volume (hybrid)	
BaYMV resistant	KWS Tower, KWS Cassia	
	SY Venture, Volume	

Seed rates

Winter barley 150–180 kg/ha; spring barley 125–180 kg/ha. The target is to establish 250–300 plants/m^2, which should result from sowing 300–400 seeds/m^2.

Pests

The main pest problem which requires control, particularly in early-drilled winter barley, is to reduce barley yellow dwarf virus transmitted by aphids (Table 7.1). In very susceptible areas the seed should be treated with clothianidin or an aphicide should be applied at the end of October to the emerged crop. Early-drilled crops may require two applications.

Diseases

The barley crop can be affected by a large number of seed-borne diseases such as leaf stripe, *Fusarium*, smuts and snow rot. Foliar diseases can also be prevalent in both winter and spring barley, such as *Rhynchosporium*, mildew, brown rust, net blotch and *Ramularia*. The mosaic viruses (barley yellow mosaic virus (BaYMV) and barley mild mosaic virus (BaMMV)) have become a problem where winter barley has been grown too frequently. Control is only by rotation and use of resistant varieties.

Many of the seed treatments are similar to those applied to wheat although usually other chemicals are included such as imazalil to control leaf stripe. In winter barley leaves 2 and 3 contribute more to yield than in wheat. It is the T1 (GS 30–32) fungicide timing in winter barley that gives the greatest yield response compared with the T2 application. Occasionally a pre-T1 treatment is required if there is early infection with powdery mildew or brown rust. The T3 (GS 59) treatment gives the lowest response in barley, it is only necessary if the T2 treatment was inadequate or not applied – see Table 13.5 for the range of products and diseases controlled. In spring barley the T2 timing gives the largest yield response compared with the T1 timing at GS 26–31. Normal practice in barley is for at least a two-spray programme at T1 and T2. For effective control and to reduce the risk of resistance, fungicides with different modes of action should be used. Low rates of fungicides can be used when growing disease resistant varieties when the amount and risk of disease is low.

Weeds

Autumn herbicide treatment is desirable in winter barley for both annual broad-leaved and grass weeds, using suitable herbicides (Table 13.6). Long runs of winter cereals tend to encourage grass weeds such as black-grass and the bromes, and control can be expensive as well as difficult if there are herbicide resistant weeds present. There is a very good opportunity to control these difficult autumn germinating grass weeds before drilling spring barley. Usually only a broad-leaved weed herbicide plus possibly a wild oat herbicide are required in spring barley (Table 13.7).

The very early ripening time of winter barley makes it easier to control perennial weeds using pre-harvest glyphosate. This is especially the case with onion couch which senesces earlier than common couch (Chapter 5).

Fertiliser requirements

Nitrogen fertiliser recommendations are lower than for wheat. If growing crops for malting where a low grain nitrogen is required then less N should be applied than on a crop grown for feed. Remember to adjust rates depending on fertiliser and grain prices as well as if organic manures have been applied. The fertiliser requirements are shown in Table 13.14.

Early application of nitrogen to winter barley often encourages foliar diseases. The nitrogen should be applied to the seedbed with late-sown spring crops (other than on sands), whereas with the early-sown crop it should be split to avoid leaching.

The phosphate and potash recommendations for winter barley are similar to those for winter wheat and slightly lower for spring barley (Table 13.3). Remember the extra requirements for phosphate and potash if straw is removed.

Plant growth regulators

Winter barley particularly can be susceptible to lodging. Stiff-strawed varieties should be chosen. To control lodging in winter barley, chlormequat is not as effective in barley as in wheat; products containing 2-chloroethylphosphonic acid or trinexapac-ethyl can be used. The later applications will control necking and brackling.

Results in spring barley are variable. Highest risk of lodging in spring barley is on thick crops growing in fertile soil, especially if there is wet and windy weather conditions after flowering.

Harvesting

Winter barley is ready for combining from mid-July to early August. Spring barley combining normally starts between mid-August and early September.

In a crop containing late tillers, harvesting should start when most of the crop is ready. Malting crops are usually left to become as ripe as possible before harvesting. Harvesting of feed barley is often started before the ideal stage. This is especially the case if a large area has to be harvested and the weather is uncertain. The ears of some varieties and over-ripe crops can break off very easily, resulting in serious losses.

Table 13.14   Fertiliser requirements of barley

	Feed	Malting
Winter barley	60–210 kg/ha depending on SNS index. If applying more than 100 kg then split: 40 kg/ha in February and the rest in April.	40–160 kg depending on SNS index. Apply by end of March.
Spring barley	Up to 160 kg applied by early May.	Up to 140 kg applied by end of March.

Table 13.15   UK winter barley yields (in t/ha)

	Average	Good
Grain	6.3	7.5
Straw	2.75	4

Yield

Yields of barley can be more variable than yields of winter wheat. Spring barley grain yields are usually about 1 t/ha lower than these for winter barley. Barley yields are shown in Table 13.15.

13.12   Oats

13.12.1   Uses of oats

Oats are now a minor cereal crop in Europe though the EU is the second largest producer worldwide. The higher yields of wheat, as well as no intervention support for oats, have been the main factors affecting the declining oat area. Two-thirds of the UK crop is milled for human consumption.

The best quality oats may be sold for making oatmeal which is used for bread making, oatcakes, porridge and breakfast foods. Oats for human consumption should have a high specific weight, 50 kg/hl, a moisture content of 14% or less and less than 10% screenings through a 2 mm sieve. Quality oats should also have a high kernel content. There are no standards for oil and protein content. Non-food consumption of oats includes use in cosmetics and shampoos. A small amount of UK oats are exported onto the world market.

Oats are particularly good for horses and are also valuable for cattle and sheep. However, they are not very suitable for pigs because of their high husk (fibre) content. Naked oats species have a very high feeding value (better than maize) because the husk threshes off the grain are much more suitable for pig and poultry feed. The normal oat grain contains about 20% by weight of husk (the lemma and palea) but it is easily separated from the valuable groat in naked varieties. Varieties of naked oats have been developed, but they are about 25% lower yielding than conventional oats. Growers should ensure that they have a contract before growing naked oats.

The major part of oat straw is baled and used for bedding/feed. Quality of the straw is usually better than for wheat.

13.12.2   Oat husbandry

Soils and climate

Oats will grow on most types of soil; they can withstand moderately acid conditions where wheat and barley would fail. If too much lime is present, manganese deficiency (grey-speck) may reduce yields. Oats do best in the cooler and wetter northern and western parts of the country, but even in these areas have been replaced by barley on many farms.

Place in rotation
Oats can be taken at almost any stage in a rotation of crops. They are a useful take-all break crop and are often a lower input crop than other cereals.

Seedbed and methods of sowing
These are similar to wheat and barley.

Time of sowing
Winter oats, late September–October. Spring oats, February–March. Most naked oats are winter sown.

Varieties
These are detailed in the HGCA *Recommended Lists*. Winter varieties are not so frost-hardy as winter wheat or barley. They usually yield better, especially in the drier districts, and are less likely to be damaged by frit fly than spring oats. Winter varieties are mainly grown in England and spring varieties further north where frost can be serious. Examples of winter varieties: *Balado, Gerald, Rhapsody, Mascani* and *Fusion* (naked variety). Examples of spring varieties: *Canyon, Ascot, Firth* and *Lennon* (naked variety).

Seed rates
190–250 kg/ha for both winter and spring oats. The target plant population is 300 plants/m^2 for spring oats and 250–300 plants/m^2 for winter oats. Seed rates are lower for naked varieties.

Fertilisers
120 kg/ha nitrogen is usually considered the optimum amount for winter oats on most mineral soils at SNS Index 1 and 110 kg/ha for spring oats. In winter oats nitrogen should be applied at stem extension stage. In spring oats application as for spring barley. (See Table 13.3 for phosphate and potash requirements.) Remember to adjust fertiliser rates if any organic manure is applied.

Spring grazing
Very occasionally winter crops are grazed in the spring. This may be desirable if there is a risk of lodging, because grazing results in a shorter straw.

Pests
The main pest problem is that of aphids transmitting BYDV. Oats are the most susceptible crop. An insecticidal seed dressing should be used, or treat with an aphicide in late October. The cereal cyst nematode is encouraged by close cropping

with oats. Nowadays, very little damage is found. Stem nematodes can be a problem in rotations where several susceptible crops are grown. This is controlled by widening rotation and growing resistant varieties.

Diseases

Mildew and crown rust are the major diseases of oats. Oat mosaic virus is not commonly seen; the current winter varieties are resistant. Winter oats may require a fungicide at GS 31 against mildew and at GS 39 against crown rust and mildew. Oats are not so susceptible as wheat and barley to *Fusarium* sp. in the ear. Mycotoxin levels are usually very low in the grain.

Weeds

Grass weeds are very difficult to control in the oat crop. Few chemicals are recommended for grass weed control. It is preferable to grow oats where there is no black-grass or wild oat problem. Oats are, however, a very competitive crop against weeds (see Tables 13.6 and 13.7).

Growth regulators

Oats can be very susceptible to lodging. Choose a stiff-strawed variety and/or apply a PGR in high risk situations. Chlormequat formulations are useful in preventing lodging if applied at GS 32, as is trinexapac-ethyl at GS 30–31.

Harvesting

Winter oats are usually harvested just before winter wheat. Likewise, spring oats (depending on the district) are normally harvested before spring-sown spring wheat. Oats, like barley, are liable to shedding when ripe. The grain is not very dense and requires 40% more storage space than wheat.

Yield

Winter oats yield similar to winter barley, spring oats yield up to 1 t/ha less. Yields are shown in Table 13.16.

Table 13.16   UK winter oats yields (in t/ha)

	Average	Good
Grain	6.3	7.7
Straw	3.5	5

13.13 Rye

13.13.1 Uses of rye

Only 4% of the cereals grown in the EU is rye and Germany and Poland are the main producers. Rye is grown on a small scale (5000–6000 ha) in the UK for grain or very early grazing. When milled the grain is used mainly for making rye crisp-bread and there has been an increased demand for rye in soft grain bread. It is not widely used in the UK for feeding to livestock. Rye should only be grown on contract for milling. The long, tough straw is very good for thatching and bedding, but not for feeding. Usually about half the straw is baled and the rest is incorporated into the soil.

13.13.2 Rye husbandry

Soils and climate
Rye will grow on poor, light, acid soils and in dry districts where other cereals may fail. It is mainly grown in such conditions for grain because, on good soils, although the output may be higher, it does not yield or sell so well as other cereals. It is extremely frost-hardy and withstands much colder conditions than the other cereals.

Place in rotation
Rye can replace other cereals in a rotation, especially where the fertility is low. It can be grown continuously on poor soils with occasional breaks. It is more resistant to take-all and so it can be grown as a third or fourth cereal.

Seedbeds and methods of sowing
The seedbed requirements and methods of sowing are similar to those for wheat.

Time of sowing
For grain production, the crop should preferably be sown in September; there are at present no spring rye varieties grown in the UK.

Seed rate
150–190 kg/ha. 30% lower seed rates are used for the hybrid ryes.

Varieties
Rye, unlike the other cereals, is cross-fertilised and so varieties are difficult to maintain true to type and new seed should be bought in each year. Most of the varieties now grown are hybrids. Seed is expensive but yields can be up to 15% higher than conventional varieties.

Examples of hybrid rye varieties include *Askari; Capitan* is a conventional variety.

Fertilisers
Depending on the SNS index and soil texture, the nitrogen requirements are in the range of 20–150 kg/ha. This should be applied as a top-dressing in April. For phosphate and potash requirements see Table 13.3.

Spring grazing

The grain crop can provide useful grazing in late February/March, depending on soil conditions, and before stem extension. Grain yield will be lower.

Crop protection

Rye is a very competitive crop with good disease resistance so inputs tend to be lower than for other cereals. A limited range of grass and broad-leaved herbicides is available (Table 13.6 and Table 13.7); there are a few chemicals with EAMUs in rye. Because rye is tall and often weak-strawed, a growth regulator is required, e.g. chlormequat (GS 30–31) or trinexapac-ethyl at GS 30–32 or 2-chloroethyl phosphonic acid at late stem extension.

Rye is less susceptible to take-all than wheat or barley. It can, however, be affected by eyespot, mildew and brown rust. Hybrid varieties tend to be more susceptible to brown rust than conventional varieties. It is usually worthwhile to apply a fungicide at T1. As rye is open-flowering, it is more susceptible than other cereals to ergot. No very effective fungicide control is currently available, but crop rotation and deep ploughing will reduce the risk.

Harvesting

The crop is normally harvested before winter wheat. It is combined when the grain is hard and dry and the straw is turning from a greyish to a white colour. The grain sprouts very readily in a wet harvest season, so it is usually harvested at 20% moisture content.

Yield

Rye yields are given in Table 13.17.

Table 13.17 UK rye yields (in t/ha)

	Average	Good
Grain	6.0	7.5
Straw	4	5

13.14 Triticale

13.14.1 Characteristics of triticale

This is produced by crossing tetraploid durum wheat with diploid rye and treating seedlings of the sterile F1 plants so that their chromosome number is doubled and they become reasonably fertile. It is bearded and intermediate between wheat and rye in most of its characteristics. Mainly winter varieties are being grown in the UK. There is no EU support. Triticale is in direct competition with other feed grains. The area grown in the UK is small, only approximately 20 000 ha.

13.14.2 Triticale husbandry

Soil and climate
Triticale is probably best suited to marginal and lighter soils for use as a feed grain of similar value as wheat. Triticale can often yield better than other cereals on these marginal soils. It is very winter hardy and tolerant of drought.

Seedbeds and methods of sowing
The seedbed requirements and methods of sowing are similar to those for wheat.

Seed rate
270–330 seeds/m^2 (125–160 kg/ha); minimum plants after winter 150/m^2.

Time of sowing
Mid-September to October. Winter triticale has a requirement for vernalisation so it should not be sown after the middle of February.

Fertilisers
Nitrogen requirements are the same as those for oats and rye. 120 kg/ha of N are recommended for a mineral soil at SNS 1. Remember nutrient contributions from organic manures. Nitrogen should be top-dressed at the end of March/early April depending on the SNS indices. For phosphorus and potassium requirements see Table 13.3.

Varieties
Examples include *KWS Fido, Tribeca, Agostino, Tulus* and *Grenado*.

Crop protection
Triticale is a much lower input crop than winter wheat and some varieties of triticale can be quite weak-strawed, particularly on less droughty soil. The growth regulators such as chlormequat, 2-chloro-phosphonic acid and trinexapac-ethyl will reduce lodging.

It is a very competitive crop, though herbicides are still required. There are more restrictions on the use of some grass weed herbicides in triticale than with the wheat crop. Several broad-leaved weed herbicides are recommended and some have EAMUs in triticale (see Tables 13.6 and 13.7).

Triticale is susceptible to ergot and eyespot, and fairly resistant to many cereal foliar diseases such as brown rust. A fungicide at T1 for control of eyespot and, where necessary, foliar diseases is recommended. *Rhynchosporium* is common in triticale but usually is not a serious problem. Take-all does not affect triticale to any extent.

Harvesting
The crop is harvested in August, as for wheat. Drying and storage advice is also the same as for wheat.

Yield
The average yield for triticale is 5–6 t/ha.

13.15   Maize for grain

Maize is a tall annual grass plant with a strong, solid stem carrying large narrow leaves. The male flowers are produced on a tassel at the top of the plant, and the female some distance away on one or more spikes in the axils of the leaves. This separation simplifies the production of hybrid seed. After wind pollination of the filament-like styles (silks), the grain develops in rows on the female spike (cob) to produce the maize ear in its surrounding husk leaves.

Climate limits the production of grain maize in the UK. It is a sub-tropical plant (C4) and in this country it is confined to southern England where the yields can be low (5–7 t/ha). There is an increasing amount of grain maize being grown for crimping. In more marginal areas a few growers are growing the crop under plastic so that it can be sown and harvested earlier. It is common practice not to combine the crop until October or November. If the crop is not being crimped then it usually requires quite a lot of drying. The situation is likely to remain this way until, or if, more cold-tolerant varieties are produced. Currently the UK imports around one million tonnes of maize for animal feed.

Grain maize grows well in many southern countries in Europe, especially in France and Italy. Yields are very high averaging between 8 and 9 t/ha of grain. With this yield, grain maize is the third most important cereal in the EU after wheat and barley. Most of the maize crop is used for animal feed with a smaller amount milled for industrial use (starch) and for human consumption. There is an increasing market for maize for production of bioethanol. For crop husbandry see Chapter 18.

There is a limited market for 'corn-on-the-cob' or sweet corn as a vegetable. This latter is a special type of maize in which some of the sugar produced is not converted into starch and is harvested when the grain is in the milky stage.

13.16   Sources of further information and advice

Further reading
Alford D V, *Pest and Disease Management Handbook*, Blackwell Science, 2000.
Bayer CropScience, *Pest Spotter*, Bayer CropScience Ltd, 2011.
Defra, *Fertiliser Manual* (RB209), TSO, 2010.
Francis, S A, *British Field Crops*, NHBS, 2009.
Gooding M J and Davies W P, *Wheat Production and Utilisation*, CABI, 1997.
Hay R K M and Porter J R, *The Physiology of Crop Yield*, Wiley-Blackwell, 2006.
Henry R J and Kettlewell P S, *Cereal Grain Quality*, Chapman & Hall, 1996.
HGCA, *Grain Storage Guide for Cereals and Oilseed Rape*, HGCA, 2011.
HGCA, *The Wheat Disease Management Guide*, HGCA.
HGCA, *The Barley Disease Management Guide*, HGCA.
HGCA, *The Wheat Growth Guide*, HGCA, 2008.

HGCA, *The Barley Growth Guide*, HGCA, 2006.
HGCA, *Avoiding Lodging in Winter Wheat – Practical Guidelines*, HGCA, 2005.
HGCA, *Managing Weeds in Arable Rotations – a Guide*, HGCA, 2010.
HGCA, *Recommended Lists for Cereals and Oilseeds*, HGCA, annual.
HGCA and BASF, *The Encyclopaedia of Cereal Diseases*, HGCA and BASF plc, 2008.
Lainsbury M A, *The U.K. Pesticide Guide*, BCPC, updated annually.
McLean K A, *Drying and Storing Combinable Crops*, Farming Press, 1989.
Murray T D, Parry D W and Catlin N D, *Diseases of Small Grain Crops*, Manson Publishing, 2009.
Senova, *Oats from Breeder to Market*, Senova, 2011.
Ullrich, S E, *Barley: Production, Improvement and Uses*, Wiley-Blackwell, 2011.
Welsh W, *The Oat Crop*, Chapman & Hall, 1995.
Wrigley C W and Batey I L, *Cereal Grains*, Woodhead Publishing Ltd, 2010.

Websites
www.ukagriculture.com
www.hgca.com
www.niab.com
www.defra.gov.uk
www.bayercropscience.co.uk
www.ukmalt.com
www.nabim.org.uk
www.rothamsted.ac.uk
www.quoats.org

14

Combinable break crops

DOI: 10.1533/9781782423928.3.337

Abstract: This chapter opens with an introduction to the concept of world markets for oilseeds and proteins and how this affects farmers' decisions to grow break crops. It then tackles the agronomy of the main combinable break crops including oilseed rape, linseed, field peas and field beans. It briefly covers more esoteric crops such as flax, lupins, navy beans and sunflowers. Each crop is dealt with by describing the variety choice, the establishment, crop nutrition, crop protection, and harvesting and storage, as well as a brief discussion of their markets and quality aspects required.

Key words: oilseeds, proteins, oilseed rape, peas, beans, markets, quality.

14.1 Introduction

As a group the fortunes of combinable break crops tend to fluctuate, sometimes on an annual basis because of adverse weather and changes in the world supply situation. They are non-cereal crops that can be combine harvested. Generally, they should not be grown more often than one year in five nor within four years of each other, although many farmers are risking growing oilseed rape more frequently because of the good margins it produces. For some of the crops it is essential to ensure market availability before growing the crop and buy-back contracts are sometimes available. Industrial crops with a contract for non-food use may be grown and in all parts of Europe, and most of the rest of the world, bio-diesel production from oil sources such as oilseed rape is being encouraged.

The range of crops, often with autumn- and spring-sown varieties, offers an option for most soil types, pH and crop rotations. Seasonal weather may adversely influence crop establishment, harvesting and sample quality. Home-saved seed has been an option for some of the crops but with specific quality requirements for the harvested product, and a trend to lower seed rates/plant populations, high quality seed with an appropriate treatment is advisable.

On a world-wide basis a large range of crops produce edible oil with soya, palm, rape (canola) and sunflower currently supplying over 75% of the total. However, the market for vegetable oils cannot be separated from the market for oilseed meals which are in great demand for intensive livestock systems and therefore influence the popularity and the price of oilseed crops. Soya beans grown on a large scale in both North and South America are a major influence on world trade and price of oils and proteins, but their importance has been slightly reduced in northern Europe by the use of Double Low oilseed rapes, which may be used as direct substitutes for both soya oil and meal. There is also the threat of imported soya contaminated with GM varieties, which has led to an increase in demand for other, home-grown oilseeds, particularly on the European market.

The burgeoning economies of countries such as China and India, and their rapidly increasing populations, are influencing the market. The issue of GM crops is yet to be resolved and has the potential to impact even further on both the quantity and quality of oil produced. In the EU 27, the major oilseed crop is rape followed by sunflower and a smaller area of soya. The UK is the fourth largest producer of rape after France, Germany and Poland, whereas Romania is the largest sunflower producer with Italy and Romania being the main soya producers.

14.2   Oilseed rape

Oilseed rape is now firmly established as the most important and profitable break crop in the United Kingdom. The small black seed contains 38–40% oil which is extracted by crushing and used for the manufacture of spreads and cooking oils. Oilseed processors will only accept seed which is low in both erucic acid and glucosinolate for human or animal consumption (the Double Low varieties). The protein-rich residue left after the oil has been removed can be included in animal rations at a higher rate now that the glucosinolate levels are below 25 micromoles per gramme of seed but still limit the amount of meal which may be fed to non-ruminants. All recommended varieties for human consumption are now Double Lows. Oilseed rape for non-food use can be high in erucic acid rape (HEAR) varieties. Care must be taken to avoid cross-contamination, e.g. by volunteer plants. There are now new restored hybrid varieties which grow bigger through hybrid vigour, require more room to grow and are usually sown at lower seed rates than conventional varieties. There are also high oleic, low linolenic (HOLL) varieties which are grown under contract specifically for deep frying chips, mainly in fast food outlets. Biodiesel can be made from any variety.

Soils and climate
The crop will grow in a wide range of soil and climatic conditions provided the land is well drained, pH over 6, and the soil and subsoil structure is good.

Rotation

Ideally, rape and other *Brassica* crops should not be grown more than one year in five, in order to avoid a build-up of diseases such as clubroot and also of pests. It is an alternative host for sugar beet nematode and this could affect the place taken by sugar beet in a rotation.

Varieties

These are detailed in the HGCA UK *Recommended Lists for Oilseeds*. The winter rape lists are regionalised between North and East/West.

Winter and spring swede rape (*Brassica napus*) are used for most of the UK's rape crop with winter varieties being the most important. There is now some interest, particularly in the north of the country, in spring and winter turnip rape (*Brassica rapa*) varieties and some spring varieties are being grown for their earlier maturity. Varieties are replaced on a regular basis and this has allowed for improvements in disease resistance, plant types and the introduction of hybrid varieties which are now filling a significant market share. Varieties vary in their yield, oil and glucosinolate content, resistance to disease and lodging. Examples of recommended varieties are shown in Table 14.1.

There are also varieties being developed for Clearfields® technology, developed by BASF, which provides tolerance to certain imidazolinone herbicides.

Time of sowing

Winter crops are sown in early August to early September, usually following winter barley but can be grown after early-maturing wheat. For spring crops, sow in early March, if possible.

Seedbeds

The seed is very small and so a quality seedbed with fine, moist soil conditions is required. Economic and time pressure on growers has increased the popularity of direct drilling into, or broadcasting onto, stubble which can be very successful on soils with good structure. It is also a way of retaining moisture in the seedbed during the height of summer. Whichever method is used, seedbed condition and crop establishment must be good and the seed needs to be about 2–3 cm

Table 14.1   Recommended varieties of oilseed rape (2014/15)

	Winter	Spring
Conventional	*Charger (E/W)*	*Shelley*
	Trinity (E/W)	*Amulet*
	Anastasia (N)	
Hybrids	*Incentive (UK)*	*Dodger*
	PT211 (UK)	*Dokrin*
	Compass (UK)	*Makro*

deep. Compacted pans and large flat stones restrict the growth of the deep taproots. The risk of slug activity should be reduced as much as possible, especially in a wet summer.

Seed rates and plant population

Seed size varies considerably between varieties and seed types and should always be taken into account. For conventional and hybrid winter varieties, in ideal conditions, 30–40 seeds/m² should be sown to give 25–35 plants/m² in the spring resulting in a seed rate of 2–3 kg/ha. In practice, most farmers err on the side of caution and assume that spring conditions will not be ideal, so use slightly higher seed rates. For spring varieties the conventional varieties should be sown at 120 seeds/m² and the hybrids at 100 seeds/m² to give the required higher plant populations of between 60–80 plants/m², and this results in seed rates of 5–6 kg per hectare as the seeds are smaller in size. The ultimate aim is to produce the optimum number of pods and seeds per square metre so that the available light is used efficiently for seed fill.

Fertilisers

The recommendations for oilseed rape in the *Fertiliser Manual* (RB209) are based on yields of 3.5 t/ha of winter OSR and 2 t/ha of spring OSR. To return those nutrients removed from the soil 14 kg of P_2O_5 and 11.0 kg of K_2O should be applied for each tonne of seed/ha expected over and above these yields. This only need apply when indices are below the target index of 2 for each element.

The nitrogen rates depend on the SNS index (page 68) and the possible use of organic manures. Unlike winter wheat, a 30 kg/ha application of seedbed nitrogen may be justified for winter oilseed rape on soils with 0–2 SNS index. Spring nitrogen rates may vary from 0–220 kg/ha according to the SNS index. For rates less than 100 kg/ha the whole dressing may be applied in late February or early March. If more than 100 kg/ha are to be applied split dressing is recommended, with half in late February or early March and the remainder by late March or early April.

For spring oilseed rape rates vary from 0–120 kg/ha according to the SNS index. All the nitrogen can go in the seedbed except on light sand soils with rates greater than 80 kg/ha where 50 kg/ha should be applied in the seedbed and the remainder by early May.

If sown too early or at too high a seed number, or excessive nitrogen is available, winter rape may develop too much leaf and this may result in a yield reduction. Current research advocates a lower seed number than used conventionally and assessing spring fertiliser nitrogen requirement from crop size (Green Area Index, GAI) which indicates the amount of nitrogen already in the crop (about 40 kg/ha per unit of GAI) and future soil supply predicted by measuring the Soil Mineral Nitrogen (SMN) in the top 90 cm of soil in February. Most crops need a total nitrogen supply of 150–200 kg/ha for optimum yield. However, in deciding rate of fertiliser nitrogen required to make up any shortfall a recovery of 60% should be assumed.

In some crops, e.g. on light soils with a high pH, boron deficiency may occur (stunted growth, curled leaves with rough mid-ribs and some stem cracking) so 20 kg of borax or 10 kg of 'Solubor' per hectare can be applied to the seedbed (if a deficiency has been diagnosed) or sprayed on the leaves in early spring. Oilseed rape has a high sulphur requirement and much of the UK is now at high risk of sulphur deficiency because of the reduction in deposition from the atmosphere. Where a deficiency has been diagnosed 50–75 kg/ha SO_3 (20–30 kg of elemental sulphur) should be applied in the early spring as a sulphate-containing fertiliser.

At Mg indices 0 and 1, between 50 and 100 kg/ha of MgO (30–60 kg Mg/ha) should be applied every three or four years.

Seed treatments

Iprodione/thiram protect against *Alternaria* and damping-off diseases and tend to be used as standard.

The EU ban on neonicotinoids from December 2013 means that seed dressings containing clothianadin, imidacloprid and thiomethoxam can no longer be used for the control of cabbage stem flea beetle (CSFB) or the peach-potato aphid, a vector of turnip yellows virus. Foliar applied pyrethroids are now the only control for CSFB, but resistance is now widespread.

A pyrethroid spray may be required in October/November if scarring by the larvae is found on more than 70% of leaf petioles. (Scarring may be seen as brown or brownish-purplish pitting on the upper surface of the leaf petiole.)

Pests

Slugs may be a serious problem of winter oilseed rape, particularly in trashy or cloddy seedbeds or when late established. Metaldehyde or equivalent should be applied where there is a high risk, and method of establishment may influence the timing of this application. The peach-potato aphid may need a pyrethroid or pirimicarb spray in October/November, but populations of the aphid are highly resistant to both.

In the spring, seed weevil and bladder pod midge may require control and in spring rape, or late winter rape, pollen beetle may also be a problem (Table 7.1).

Pigeon damage

This can be very serious in some years, especially in areas with a high pigeon population, near woods, where only a small area of rape is grown and where there is little else for the pigeons to eat in the winter.

Although the crop is often eaten in late autumn and early winter, the greatest damage is caused by grazing in late winter and early spring when the new shoots and buds are developing. This leads to uneven flowering and ripening and creates problems for timing of spraying and harvesting, as well as loss of yield and a lower oil percentage. Many sound and visual scaring devices such as bangers, kites, balloons and scarecrows are available and when used they should be moved around from time to time. Shooting is also helpful, particularly when the weather is severe and in the more vulnerable periods. A full crop is always likely to be less damaged than a thin patchy crop and direct drilling into a long cereal stubble can put them off landing in the field during early autumn.

Rabbit damage
This is an increasing problem during winter where rabbit populations are getting larger. Electric fencing and gassing are possible solutions.

Diseases
Light leaf spot and *Phoma* may be a problem in both autumn and spring time and varietal resistance should be taken into account. A Crop Monitor service is available on the HGCA website for a number of important crop diseases, including phoma and sclerotinia, and this is now available as an App for a number of smart phone systems, as well as being on Twitter. Light leaf spot should be sprayed if bleached spots (no black picnidia) are seen in the autumn/winter. *Phoma* has leaf spots with black picnidia and early attacks, before the end of October, are most damaging and should be treated with, for example, tebuconazole when 10–20% of plants are affected, with a follow-up spray 4–6 weeks later. An attack at later growth stages may be treated with the same group of fungicides which may also give useful control of *Alternaria* and *Sclerotinia* (Table 6.2).

Weeds
Winter oilseed rape is a very competitive crop. A fairly high weed population, e.g. 50–100 weeds/m^2, can have little effect on the yield of a vigorous crop. Many farmers grow oilseed rape as a cleaning crop. This is because there is a wide range of herbicides available which will control weeds that are otherwise difficult to control elsewhere in the rotation, e.g. black-grass and brome. Other brassicae are difficult to control and should be dealt with in other crops if possible. Cleavers can be controlled pre-emergence with clomazone alone or in mixtures, or post emergence with quinmerac in mixtures. Volunteer cereals are one of the main problems, especially if the rape is established using minimum cultivations after winter barley. Several graminicides, e.g. fluazifop-P-butyl (a fop) and cycloxydim (a dim) will control the cereals and other grass weeds, but grass weed resistance to these products is increasing and may limit choice of chemical. Currently there are no grass weed resistance problems with herbicides such as metazachlor and propyzamide.

In well-established crops, the time of weed removal is not as important as in backward thin crops. Most broad-leaved weed herbicides are applied in the autumn. A number are very persistent and the manufacturer's recommendations should be followed concerning the cultivations before the next crop. There is only a limited range of products which can be used in spring rape.

An early-drilled vigorous crop may smother many weeds; however, if managing the crop to optimise canopy size then weed control may require extra attention. Treatment should be targeted to difficult weeds like cleavers (Table 14.2).

Table 14.2 General table of weed susceptibility to some commonly used herbicides in oilseed rape, combinable peas and field beans

Crop	Herbicides	Timing	Annual broad-leaved weeds																Grass weeds					
			Black bindweed	Charlock	Chickweed	Cleavers	Fat hen	Hemp-nettle	Knotgrass	Marigold corn	Mayweeds	Nettle small	Nightshade black	Pennycress	Poppy common	Redshank	Shepherd's purse	Speedwell	Thistle creeping	Black-grass	Brome barren	Meadow grass annual	Cereal volunteers	Wild oats
Oilseed rape	Metazachlor	Pre or post-em	s	r	S	r	s	r	R	–	S	s	S	R	s	s	S	S	–	S	–	S	–	–
	Propyzamide	Post-em	S	R	S	s	s	–	s	–	R	S	S	–	R	S	R	s	–	S	S	S	S	S
	Clopyralid	Post-em	s	–	–	–	–	–	r	S	S	S	–	–	–	r	–	–	S	R	R	R	R	R
	Graminicides, e.g. quizalofop-p-ethyl	Post-em	R	R	R	R	R	R	R	R	R	R	R	R	R	R	R	R	R	(S)	S	S	S	(S)
Peas and field beans	Pendimethalin	Pre-em	S	–	S	s	S	S	S	S	s	S	S	S	–	S	S	S	–	s	–	S	S	–
	Clomazone	Pre-em	–	–	S	S	s	–	–	–	–	–	–	–	S	–	S	–	–	–	R	s	R	–
	Imazamox + pendimethalin	Pre-em	S	S	S	–	S	s	S	s	s	–	s	–	S	S	r	S	–	R	S	S	R	–
	Bentazone	Post-em	S	S	S	s	S	s	R	S	S	S	s	S	–	S	S	S	–	s	–	R	s	–

Symbols: S – susceptible, s – moderately susceptible, r – moderately resistant, R – resistant, ( ) high risk of resistance.
See product label for timing, application rates, restrictions and safe use.

Growth stages
A knowledge of the growth stages of the oilseed rape plant will be helpful in deciding the best time to treat the crop for a specific problem (Table 14.3).

Table 14.3   Growth stages in oilseed rape

	Definition	Code
0	Germination and emergence	
1	Leaf production	
	Both cotyledons unfolded and green	1,0
	First true leaf	1,1
	Second true leaf	1,2
	Third true leaf	1,3
	Fourth true leaf	1,4
	Fifth true leaf	1,5
	About tenth true leaf	1,10
	About fifteenth true leaf	1,15
2	Stem extension	
	No internodes ('rosette')	2,0
	About five internodes	2,5
3	Flower bud development	
	Only leaf buds present	3,0
	Flower buds present but enclosed by leaves	3,1
	Flower buds visible from above ('green bud')	3,3
	Flower buds level with leaves	3,4
	Flower buds raised above leaves	3,5
	First flower stalks extending	3,6
	First flower buds yellow ('yellow bud')	3,7
4	Flowering	
	First flower opened	4,0
	10% all buds opened	4,1
	30% all buds opened	4,3
	50% all buds opened	4,5
5	Pod development	
	30% potential pods	5,3
	50% potential pods	5,5
	70% potential pods	5,7
	All potential pods	5,9
6	Seed development	
	Seeds expanding	6,1
	Most seeds translucent but full size	6,2
	Most seeds green	6,3
	Most seeds green-brown mottled	6,4
	Most seeds brown	6,5
	Most seeds dark brown	6,6
	Most seeds black but soft	6,7

(Continued)

Table 14.3   Continued

	Definition	Code
	Most seeds black and hard	6,8
	All seeds black and hard	6,9
7	Leaf senescence	
8	Stem senescence	
	Most stem green	8,1
	Half stem green	8,5
	Little stem green	8,9
9	Pod senescence	
	Most pods green	9.1
	Half pods green	9,5
	Few pods green	9,9

Note: To estimate later leaf stages judge the number of lost leaves by their scars. Note that stages from bud to seed development should normally apply to the main stem and seed development stages should normally apply to the lowest third of the main inflorescence. Otherwise, branch position on the inflorescence should be stated. Senescence stages apply to the whole plant.

Harvesting

Shedding losses may become significant if harvesting of oilseed rape is delayed. Consequently, whilst some growers combine winter rape direct, most prefer to either desiccate or swath the crop. The drying-out period for desiccated crops is about 10 days compared to 15–20 days for swathed crops. However, it is important that swathed crops be combined when fit or shedding losses may again be significant. Combining early or late in the day or during dull weather may reduce shatter losses. Tall crops on exposed sites are best swathed as the pods become brittle, while desiccation, especially with glyphosate, is better when there are perennial weeds present in the crop. Spring crops are often more determinate and can be more readily direct cut.

Drying and storage

The moisture content of the seed at harvest time will be in the range 8–15% for swathed crops and 10–25% for direct-combined crops. The contract price is usually based on 9% moisture content and so it should be dried to a slightly lower percentage. Damp rape seed must be dried as soon as possible, either by a continuous drier (not above 60 °C, and then cooled quickly) or by bulk drying on the floor or in bins. The undried seed should not be piled more than 1.25 m deep, and the drying ducts covered with hessian.

Normally, dry cold air is used for drying, but some heat may be required to lower the relative humidity to 70% so that the seed dries to 8% moisture content. Seed coming in hot from the combine on a fine summer's day should be cooled rapidly to avoid moulds and mites and maintain oil quality. When rape seed is being conveyed in trailers or lorries, great care should be taken to block all holes through which it might escape.

Yield

Winter oilseed rape has a yield range of 2–5 t/ha with 3.5 t/ha being average. For spring oilseed rape the yield range is 1–3 t/ha with 2 t/ha being average.

14.3   Linseed and flax

Linseed and flax are different varieties of the same plant which has been grown in this country since Roman times. Figure 14.1 shows the main difference between the two varieties. Linseed is short-strawed with capsules (bolls) to give a higher yield of seed than flax which is long-strawed and grown for its fibre.

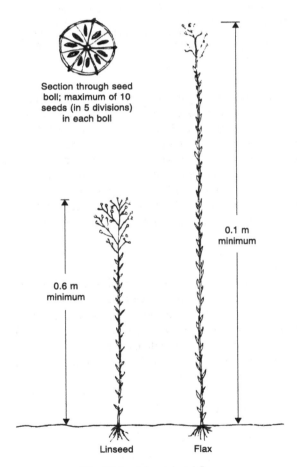

Fig. 14.1   Linseed and flax.

14.3.1 Linseed

The seed contains about 40% oil and the residue is a high protein animal foodstuff. The drying oil produced from linseed is mainly used for making paints, putty, varnishes, oil cloth, linoleum, printer's ink, etc. Over the last 30 years the introduction of plastics and latex paints and the use of other oils have reduced its importance. However, with a move towards more 'natural' products the use of linseed oil has increased. Better dyeing techniques have meant that linoleum can be made bright and attractive to compete with modern PVC floor coverings. Unlike PVC linoleum is biodegradable and it does not give off poisonous gases when burnt. The majority of varieties are spring sown but developments in plant breeding have resulted in the introduction of stiff-stemmed, autumn-sown EarlyCut varieties and of edible oil (linola) types which are usually grown on buy-back contracts. There is also a small but expanding market for quality (high Omega 3) organically grown seed to be used for human consumption for which traditional varieties are suitable.

Soils and climate
Linseed can be grown in any of the arable areas of the UK but moisture-retentive soils are preferable because of the small root system of the plant. The pH should be about 6.5.

Rotation
Because of the comparatively small area grown, pests and diseases are not generally a serious problem. Nevertheless, linseed should not be grown more often than one year in five. It is not related to any other break crop so it can be grown in all rotations. It leaves a good seedbed for a following crop of wheat with a fine tilth and few slugs.

Varieties
Linseed varieties show continuous improvement with regard to yield, shorter and less fibrous straw, earlier and more uniform ripening, less shattering of seed and disease resistance. Current varieties include *Brighton, Juliet, Batsman* and *Festival*. They are all brown seeded varieties as opposed to yellow (HGCA UK *Descriptive List for Spring Linseed*).

Seed and sowing
The seed should be treated to control damping-off and drilled 1–2 cm deep into a fine, firm and moist seedbed. For autumn-sown crops 800–950 seeds/m^2 should be sown in late September/early October. For spring-sown crops 600–700 seeds/m^2 should be sown in late March/early April when soil temperatures are 6–8 °C to obtain an established population of about 400 plants/m^2. Seed rates will range from 44–60 kg/ha because of variation in seed size between varieties.

Fertilisers
On fertile soils there is little response to nitrogen, phosphate and potash and at SNS index 1 the crop has a nitrogen requirement of 50–100 kg/ha, phosphate 30 kg/ha and potash 10 kg/ha at index 2.

Weed, disease and pest control
A well-grown crop is essential to reduce the likelihood of problems. Linseed competes poorly with weeds. Any weed problem not controlled during the growing season can be desiccated at harvest using diquat, glyphosate or glufosinate-ammonium (as in oilseed rape, dried peas and field beans). Glyphosate can also be used pre-harvest for the control of perennial weeds. As linseed is a relatively minor crop, there are at present only a small number of chemicals which are approved for use. These include amidosulfuron, iodosulfuron, bentazone, bromoxynil + clopyralid for broad-leaved weed control. Several graminicides can be used to control non-resistant grass weeds.

Crop protection products are available but generally with 'off-label' approval. Flea beetle control may be a problem because neonicotinoid seed treatments have been withdrawn (Table 7.1).

Flowering
Most varieties have pale blue (a few have white) flowers which appear in early July. The plant flowers early in the morning, but only for a few hours. The flowering period for the crop may last several weeks. It is normally self-pollinated.

Harvesting
Linseed is ready for harvesting when the top of the plant is dry, the stems are yellow-brown and the leaves have fallen from the base. The stem can still be green, indeed it flows better through the combine when it is green than when it is completely ripe and the fibres are breaking out of it. The seeds should rattle in the bolls, be plump and brown and, for direct combining, show the first signs of shattering. Linseed straw is very tough and wiry and so the knife must be kept sharp. The crop should usually be desiccated before combining, especially where there are weeds. The secret of successful desiccation is achieving penetration below the canopy of capsules so that the chemical can dry out the top of the stems. Higher than normal volumes of water should be used and 'crop tilters' are useful as long as they do not damage the crop.

Plant breeders are being successful in producing varieties which have their capsules near the top of the crop. The cutter bar of the combine may then be set high to keep as much of the fibrous stem as possible out of the machine – this will reduce blockages and tangles. The long straw can then be ploughed under after harvest and allowed to rot. Combining should only be carried out when the crop is dry and a stripper header is advantageous. The straw is exempt from the burning ban and can be burnt in the swath or baled and used in on-farm boilers to provide heat.

Yield
Linseed has a yield range of 1–2.5 t/ha with an average yield of about 1.5 t/ha.

Drying and storage
The seed must be carefully dried to about 8% moisture content for safe storage. It is harvested in late August/early September at about 12% moisture. Linseed is a

small slippery seed which can easily fall through tiny holes; this must be watched carefully at all stages, i.e. at sowing, combining, in trailers and in the drier and storage buildings.

14.3.2   Flax
This is grown mainly for the fibres in the stem which are present as long bundles around a woody core. Traditionally, flax has been grown for the linen market (long fibre) whereas UK-produced fibre is classified as short fibre and is used for textile blends and for industrial markets. There has been an increased interest in both long and short fibre as a natural product with a low carbon footprint compared with imported material.

The growing of the crop is similar to linseed but in the UK the crop may be harvested early by mowing or later, after desiccation, by combine therefore also obtaining some seed. After cutting it is important for the crop to 'ret' which is a natural process which results in the woody part of the stem being easily removed. This retting in the field is very important and in-field conditioners are available to speed up the process.

14.4   Sunflowers
Sunflower plants grow successfully throughout the UK in gardens and coverts for game birds, but the agricultural crop is restricted mainly to the south-eastern part of England, mainly south of a line drawn from the Wash to Exeter. The crop is restricted in area because of its need for warmth to reach maturity and its dislike of wet conditions during seed ripening when botrytis head rot can be a problem.

Modern sunflower varieties are potentially high-yielding, semi-dwarf varieties with a requirement from drilling to harvest of 1300 to 1400 accumulated day degrees (over a base of 6 °C). They are produced in a similar way to maize hybrids, by using male sterility and restoring gene technique.

The area sown in the UK is very small, approximately 800 hectares, yielding up to 2.5 t/ha. The decorticated (dehusked) seed contains over 40% of valuable, high quality edible oils which can be used in food manufacture, margarine and cooking oils, but the major part of UK production goes into pet foods because the soft shells of the seed make it suitable for native birds.

A well-drained soil with a pH of over 6 is essential and the crop is best sown after cereals or fallow. A rotation of one year in five or six is recommended to reduce the risk of infection from *Sclerotinia sclerotiorum*. Fertilisers used are 25–50 kg/ha of nitrogen, and 40–60 kg/ha of phosphate and potash which should be applied to the seedbed as a compound. Too much nitrogen can cause lodging, can favour disease development, delay maturity and lower seed oil content. Sunflowers are sensitive to boron deficiency.

The crop is sown in late April/early May when soil temperature in the top 10 cm is 10 °C. Sowing rate depends on seed size and the aim should be to achieve

8–11 plants/m^2 using either a precision or air drill at a depth of 2.5–5 cm in 34 cm rows. Herbicides such as pendimethalin are available for broad-leaved weed control, although inter-row hoeing can be successful.

Slugs can be a problem during establishment, as well as pigeons and rabbits.

The crop is ready for harvesting in mid-September/October when seed has reached 30% moisture content or less. At this stage the back of the head is deep yellow and the bracts have begun to turn brown. Desiccation is not used. A conventional cereal combine can be used with similar settings as for harvesting beans. Seed is dried on a drying floor at a maximum depth of one metre using cold air until the moisture content reaches 15%. After cleaning, seed should be dried to 9% using heat.

14.5   Soya beans

These are potentially attractive to replace imports particularly with the concern about using products from genetically modified crops. New varieties from Eastern European breeding programmes are now available and have potential for production in southern England in good years and the seed should be inoculated with *Rhizobium* bacteria.

14.6   Evening primrose

The seed contains about 20% oil which is a very rich source of gammalinolenic acid (GLA) and this is highly valued for certain pharmaceutical uses. The market for evening primrose oil is very small and is dominated by Chinese and Eastern European production, although the plant is still grown on a small scale in the UK. It is recommended that this crop should only be grown on contract for a reputable end-user.

14.7   Borage

Borage is a naturalised annual plant which has grown wild or in gardens for many centuries. It was developed as a crop plant for its seed, with an oil content of about 30% which is very rich in GLA. It therefore competes with evening primrose oil and again, the crop should only be grown on contract for a reputable end-user.

14.8   Combinable pulses

More than 40 species of grain legumes belonging to the Leguminosae family are grown throughout the world and unlike other crops do not need nitrogen fertiliser because the *Rhizobium* bacteria found in nodules on the roots are able to fix atmospheric nitrogen. They are excellent break crops providing a yield advantage in the following crops which benefit from the residual nitrogen. The nutritional value of the grain legume seeds is of major importance in both human and animal nutrition providing protein, starch and fibre. The grains of these crops are traded as pulses but this term excludes leguminous oilseeds such as soya beans and under EU regulations this crop is treated as an oilseed.

The EU imports about two-thirds of its vegetable protein requirement and the area sown to pulse crops is less than 4% of the arable area.

The EU divides pulses into two categories:

1. Protein crops – peas, field beans, lupins.
2. Other grain legumes – chickpeas, lentils, vetches.

In the 1990s the EU became the world's major producer of dry peas with Spain and France providing 75% of total production. China is the major producer of field beans but Australia and the UK, the major EU producer, are the main exporters. Australia is the major producer and exporter of lupins whilst India is the major producer of pulses, mainly dry beans, chickpeas and lentils but has to import from countries such as Australia, Canada and Turkey to try and meet demand. In the EU Spain is the main producer of other grain legumes.

In the UK, the main pulses grown are the various types of beans and peas with a developing interest in lupins. The grain from these crops is primarily used for animal feed but premium markets are available for human consumption (mainly peas) and for pigeon feeding (mainly maple peas and small seeded beans). The current interest in traceabililty of food products, the aversion to GMOs and the residual interest in organic farming may encourage the production of these crops. There is also now an interest in using these crops, mainly peas, for whole-crop silage.

14.8.1   Field beans

Soils and climate
Winter field beans do well on the heavier soils provided they have a good structure and are well drained. In the north of the country there is an increased risk of frost damage and later harvest. Spring beans are more suitable on medium soils but may then suffer from drought.

Place in rotation
Field beans are a good break from cereals and are often followed by winter wheat which benefits from the residual nitrogen. To reduce the risk of building up persistent soil-borne diseases such as footrots, field beans and related crops should not be grown within five years of themselves or each other. For spring beans, in addition, stem rot (*Sclerotinia sclerotiorum*) may be a problem

and host crops such as linseed and oilseed rape should therefore have the same restrictions.

Seedbeds

Requirements are similar to those for cereals but ideally a little deeper. As a result of soil type and establishment method, the autumn seedbeds are fairly rough and may impair the activity of soil-acting herbicides.

Time of sowing

Winter beans are sown mid-October to mid-November. If sown too early in a mild autumn, the resultant soft growth is easily damaged by hard frosts and may be attacked by an early infestation of chocolate spot disease. This early spread of the disease can subsequently be difficult to control. Spring beans may be sown as soon as soil conditions permit after early February. This early sowing may reduce the effect of potential drought and may result in earlier harvesting.

Method of sowing

Winter beans may be ploughed-in at 10–15 cm deep or may be broadcast on the ploughed surface and covered by harrowing, but more than half the area of winter beans is now drilled. The spring crop is usually drilled up to a depth of 7.5 cm, the required depth being influenced by choice of herbicide.

Varieties

These are detailed on the Processors and Growers Research Organisation (PGRO) website. Winter beans are generally large seeded and are marketed for animal feed. Spring beans are usually smaller seeded and also used for animal feed but some varieties may attract a premium if suitable for specialist markets. There are pale and black hilum varieties and tic beans which are small and round. *Maris Bead* (a spring variety) has small rounded seed (tic type) suitable for the pigeon trade. The newer white-flowered tannin-free varieties are more attractive to some feed compounders and for on-farm use for non-ruminants but may be lower yielding. There are no white–flowered currently on the recommended list. Examples of recommended varieties are given in Table 14.4.

Table 14.4   Recommended varieties of field beans (2014/15)

	Winter	Spring
Coloured flower	*Tundro*	*Vertigo*
(high tannin)	*Wizard*	*Fanfare*
	Clipper	*Boxer*

*Clipper is a 'black' hylum winter variety.

Seed rates
The usual range for winter beans is 160–200 kg/ha and 160–250 kg/ha for spring beans. A number of factors will affect the number of seed sown and therefore the seed rate:

- Thousand seed weight (TSW) (range 330–570 g).
- Seedbed conditions and time of sowing, e.g. higher seed rate when sowing late in the autumn or early in the spring.
- Plant population. Winter beans require an established plant population of 18–20 plants/m^2 post-winter and the chosen seed rate should allow for 20–25% field loss. For spring beans the required population is about 40 plants/m^2 with a field loss of 0–5%. Dense populations tend to increase disease problems whereas thin crops are less competitive to weeds. In some areas rooks may reduce plant populations.

Seed rate may be calculated as follows:

$$\frac{\text{TSWg} \times \text{target population/m}^2}{\% \text{ germination}} \times \frac{100}{(100 - \text{field loss})} = \text{seed rate kg/ha}$$

Seed treatment
Certified seed is advised but if using home-saved seed it should be tested for germination and seed-borne leaf and pod spot (*Ascochyta fabae*) infection. Low levels of this disease may be treated with thiram. It is recommended that all field bean seed is tested for stem and bulb nematode (*Ditylenchus dipsaci*) and should not be used for seed if nematodes are present. PGRO offer a testing service.

Fertilisers
At index 2, 40 kg/ha of P and 30 kg/ha of K are required in the seedbed assuming a 3.5 t/ha yield. 11 kg of extra P and 12 kg of extra K will be needed for every tonne of yield above 3.5. However, allowance should be made for nutrients applied as organic manures. No nitrogen fertiliser will be required, but a magnesium fertiliser will be required at soil index 0 and 1.

Weeds
Initially, field beans are susceptible to weed competition, although later they are a very competitive crop. Most growers rely on herbicides for weed control, although in-crop harrowing can be used as an aid to keep the weeds under control. Most herbicides used are residual and are applied pre-emergence. Simazine is no longer approved so farmers look to more expensive materials such as triallate, propyzamide and carbetamide. Bentazone is the only post-emergence herbicide approved for broad-leaved weed control. Post emergence fops and dims can be used for grass weed control, allowing for resistance. In spring beans, several of the pea herbicides are approved pre-emergence. This chemical only controls a limited range of weeds.

A sound rotation with good weed control in the previous crops particularly of grass weeds and perennial broad-leaved weeds will reduce the need and cost of weed control in field beans.

Glyphosate can be used pre-harvest for desiccation and will also control perennial weeds.

Pests
Slugs may be a serious problem for winter beans especially where the crop is late sown in a cloddy seedbed and metaldehyde or equivalent should be used. Rooks and pigeons may also damage the crop and bird scarers or other methods of control may be required.

Bean seed beetle (Bruchid beetle) may be a particular problem in crops for seed or human consumption and may be controlled by a pyrethroid insecticide. This may also be used for the control of pea and bean weevil which tends to be more of a problem in spring beans and a monitoring system is available to forecast spray timing. Black bean aphid is also mainly a problem of spring beans and may be controlled by pirimicarb (Table 7.1). However, if insecticides are to be used when the crop is flowering then consideration should be given to protecting the bees which are useful in cross-pollinating the flowers.

Diseases
Wide rotations and the use of healthy seed will reduce the risk of some diseases whereas others are more influenced by weather conditions and for chocolate spot, mainly a problem in winter beans, early drilling and high populations may predispose the crop to infection. A range of approved products are available such as metconazole, chlorothalonil plus cyproconazole, azoxystrobin and tebuconazole for chocolate spot control and early infection by the disease will require treatment. There are no products approved for the specific control of leaf and pod spot but some will give partial control (Table 6.1). Resistant varieties are available.

Harvesting
Winter beans are usually ready for harvesting in August or September; spring beans from late August on. They ripen unevenly, lower pods first, and although the leaves wither and fall off early combining has to be delayed until the stems lose their green colour. If required, a desiccant such as diquat (Reglone) may be used when at least 90% of pods are dry and black, particularly in weedy crops. Loss of shed seed when combining may be reduced by combining in dull weather or in the morning or evening. The crop is at its easiest to combine when the seed is at 18% moisture content. Beans for seed and specialist markets such as human consumption or pigeons must be carefully harvested and handled.

Drying and storing
Field bean seeds are large and must be dried carefully to ensure the moisture is removed from the centre of the seed and to avoid cracking. In continuous dryers,

Table 14.5 Field beans indicative yields (t/ha)

	Average	High
Winter beans	3.5	5.0
Spring beans	3.25	4.5

seed of over 20% moisture should be dried in two stages. On-floor dryers may be successfully used but if the ducts are widely spaced some beans near the floor may not dry properly as the air escapes easily. For long-term storage, beans should be dried to 14% moisture content and the marketing standard is normally 14% moisture and 2% impurities or a combination of the two up to 16%.

Yield
The indicative yields in t/ha are shown in Table 14.5.

14.8.2   Dry harvested peas

Soils and climate
Peas grow best on well-drained loams and lighter soils. The pH should be about 6–6.5; if it is too high, manganese deficiency is likely and may reduce the quality of the harvested sample. Lodging may be a problem where fertility is high and so suitable varieties should be chosen.

Compared to the other pulses, peas are more tolerant of drought stress and high rainfall may make combining difficult and result in stained produce which may be less marketable. In dry situations irrigation may be of benefit at flowering and again at the pod-swelling stage but this should not be done at the expense of higher value crops such as potatoes.

Place in rotation
Peas should not be grown more often than one year in five and should be kept four years apart from other pulses, oilseed rape and linseed to avoid the build-up of soil-borne diseases and pests.

Seedbed
Ploughing in the autumn and allowing weathering to occur will allow the preparation of a seedbed with a minimum number of cultivations because peas are sensitive to over-compaction and do not require too fine a tilth. On very light land, spring ploughing will work which allows over-wintered stubbles.

Time of sowing
As soon as possible after mid-February but seedbed condition and soil temperature are important considerations.

Method of sowing
Narrow rows are advisable as this tends to result in better competition with weeds, higher yields and easier combining. Seed should be covered with about 3 cm of settled soil after rolling.

Varieties
The main market for combining peas is for animal feed and high yield is important. For the specialist markets such as human consumption, pigeons and micronising for pet food, choice of an appropriate variety is the main consideration and the premium will need to compensate for the lower yield. Disease resistance, standing ability and ease of combining are particularly important in a wet season and the newer semi-leafless and tare-leaved varieties tend to be better for these characteristics. Varieties are detailed on the PGRO website.

Types of peas

- *Marrowfat*. This is the main type used for human consumption, the seeds are large, dimpled with a blue-green seed coat. A good colour and freedom from waste and stains are required quality factors so timeliness of harvest is important.
- *White peas*. These have round, smooth seeds with a white/yellow seed coat. They are mainly used for animal feed but are also sold for use in soups and as split peas.
- *Small blues*. These have small, round, smooth, seeds with a blue-green seed coat. There is limited use for canning as small processed peas.
- *Large blues*. These have large, round, smooth seeds with a blue-green seed coat. Samples with good colour may also be used for micronising to be used for pet food or in packets for human consumption.
- *Maple peas*. These have purple flowers and round brown-speckled seeds. These varieties are grown as whole-crop forage or for the seed to be sold for pigeons.

Examples of recommended varieties of dry harvested peas are shown in Table 14.6.

Seed rates
The usual range is 200–250 kg/ha or a little higher. However, an adequate plant population is important and TSW and field losses should be taken into account.
Suggested target populations for dry harvested peas are shown in Table 14.7.

Table 14.6   Examples of recommended varieties of dry harvested peas (2014/15)

White peas	*Salamanca*	*Mascara*
Large blue peas	*Crackerjack*	*Campus*
Marrowfat	*Sakura*	*Neon*
Small blue peas	*None on the recommended list*	
Maple pea	*Mantara*	*Rose*

Table 14.7   Suggested target population for dry harvested peas

	Plants/m^2
Some marrowfats	65
Others	70

Field losses may vary from 20% down to 5% depending on sowing time and seedbed conditions.

The seed rate may be calculated as follows:

$$\frac{\text{TSWg} \times \text{target population/m}^2}{\% \text{ germination}} \times \frac{100}{(100 - \text{field loss})} = \text{seed rate kg/ha}$$

Thousand seed weight and germination percentage may be obtained from the seed merchant.

Fertilisers
At index 2, 40 kg/ha of P and 30 kg/ha of K are required in the seedbed assuming a 4 t/ha yield. 8.8 kg of extra P and 10 kg of extra K will be needed for every tonne of yield above four tonnes. However, allowance should be made for nutrients applied as organic manures. No nitrogen fertiliser will be required, but a magnesium fertiliser will be required at soil index 0 and 1. In areas of sulphur deficiency 25 kg/ha SO_3 (10 kg of elemental sulphur) should be applied pre-drilling.

If manganese deficiency occurs (sometimes called Marsh Spot), symptoms being yellowing of leaves between veins which remain green and a brown spot in the centre of the pea seed, 5 kg/ha of manganese sulphate should be applied, repeated at full flower and again 10–14 days later.

Seed treatment
Certified seed is advisable but there is no certification standard for *Ascochyta* and disease-free seed or thiram should be used. Pre-emergence damping-off is prevented by using thiram which should be used on all pea seed. Seed-borne *Ascochyta* can be controlled by cymoxanil + fludioxonil + metalaxyl M which will also give control of downy mildew in susceptible varieties.

Pests
Pea weevil may cause the characteristic 'U' shaped notches around the edges of the leaves but the main damage is as a result of the larvae feeding on the root nodules. A monitoring system is available to predict the need for sprays. Pea aphids if present in large numbers may result in a severe reduction in yield. A predictive model is available and an approved pyrethroid should be applied when 15% or more of plants are affected.

Pea moth larvae feed upon the developing seeds reducing the yield and value of the crop particularly for use as seed or for human consumption. A pea moth monitoring service is available and growers may use pheromone traps to aid spray decision making. One or more sprays of an approved pyrethroid will be required (Table 7.1).

Diseases
Crop rotation, varietal resistance and clean or treated seed are all vital for the pea crop particularly for diseases such as *Ascochyta* spp. Modern triazoles and strobiluron fungicides can give useful control of *Ascochyta*. Some varieties are resistant to pea wilt and some have good resistance to downy mildew, otherwise an appropriate seed treatment should be used. Diseases such as *Botrytis* and *Mycosphaerella* may be a problem during flowering particularly if the weather is wet and may be treated with an appropriate fungicide such as cyprodinil, pyrimethanil or azoxystrobin with or without chlorothalonil (Table 6.1).

Weeds
Good weed control is essential in peas which are not a very competitive crop and as well as increasing yield will aid combining and drying of the grain. A range of pre-emergence and foliar-applied herbicides is available but a few varieties are sensitive to herbicides and this should be checked. Where soil type, seed bed conditions and soil moisture are appropriate a pre-emergence herbicide such as isoxaben + terbuthylazine or pendimethalin is the preferred option. Once the crop is at the recommended growth stage post-emergence herbicides, for example, bentazone plus MCPB should be applied as soon as possible but it is advisable firstly to ensure the pea leaves are well-waxed by using the crystal violet dye test. Volunteer cereal, or grass weeds such as wild oats may be controlled with herbicides such as cycloxydim or propaquizafop, but if the grass weeds may be herbicide resistant then a specialist should be consulted (Table 14.2).

Growth stages
A knowledge of the growth stages of the pea plant will be helpful in deciding the best time to treat the crop for a specific problem (Table 14.8).

Harvesting
The price of combining peas for human consumption is largely dependent on quality and great care should be taken at harvest. A clean crop may be combined direct whereas if weedy a desiccant may be helpful but first consult the processor if the crop is for human consumption. Peas for animal feed may be combined a little later and drier than for human consumption and if the crop is lodged the combine must be fitted with efficient lifters and it may be necessary to combine one-way.

Drying and storage
As with beans, peas must be dried and stored with care. The relatively large size of the seed makes drying more difficult and low temperature dryers are safer. Peas

Table 14.8 Growth stages in peas

Code	Definition	Description
Germination and emergence		
000	Dry seed	
001	Imbibed seed	
002	Radicle apparent	
003	Plumule and radicle apparent	
004	Emergence	

Vegetative stage refers to main stem and recorded node. Two small-scale leaves appear first and the nodes where these occur are not recorded.

Code	Definition	Description
101	First node	(leaf fully unfolded, with one pair of leaflets, no tendrils present)
102	Second node	(leaf fully unfolded, with one pair of leaflets, simple tendril)
103	Third node	(leaf fully unfolded, with one pair of leaflets, complex tendril)
10×	*x* node	(leaf fully unfolded, with more than one pair of leaflets, complex tendril found on later nodes)
	Last recorded node	(any number of nodes on the main stem with fully unfolded leaves according to cultivar)

Reproductive stage refers to main stem, and first flowers or pods apparent.

Code	Definition	Description
201	Enclosed buds	Small flower buds enclosed in terminal shoot
202	Visible buds	Flower buds visible outside terminal shoot
203	First open flower	
204	Pod set	A small immature pod
205	Fat pod	
206	Pod swell	Pods swollen, but still with small immature seeds
207	Pod fill	Green seeds fill the pod cavity
208	Green wrinkled pod	
209	Yellow wrinkled pod	Seed 'rubbery'
210	Dry seed	Pods dry and brown, seed dry and hard

Senescence stage refers to lower, middle and upper pods on whole plant.

Code	Definition	Description
301	Desiccant application stage. Lower pods dry and brown, seed dry, middle pods yellow and wrinkled, seed 'rubbery', upper pods green and wrinkled	
302	Pre-harvest stage. Lower and middle pods dry and brown, seed dry, upper pods yellow and wrinkled, seed 'rubbery'	
303	Dry harvest stage. All pods dry and brown, seed dry	

These definitions and codes refer to the main stem of an individual plant. Definitions and codes for nodal development refer to all cultivars; only nodes where a stipule and leaf stalk develop are recorded; descriptions in brackets refer to conventional leaved cultivars only.

for human consumption are dried at a lower temperature than peas for animal feed and if moisture content is over 24% a lower temperature should be used and two passes may be more appropriate. For long-term storage, peas should be dried to 14% moisture content and the normal marketing standard is 14% moisture and 2% impurities or, a combination of the two up to 16%.

Yield
The indicative yields in t/ha are as follows: average, 3.75; high, 4.75.

14.8.3   Lupins
Lupin seed has a protein content of 30–40% with about 12% oil and therefore if the crop could be successfully grown in the UK it would compete with imported soya. Modern varieties have been bred for low alkaloid content (sweet lupins) and new, more determinate autumn- and spring-sown varieties are available with a yield potential of up to 4 t/ha.

The main commercial interest is in three species:

- White lupin (*Lupinus albus*) with winter and spring varieties.
- Blue lupin (*Lupinus angustifolius*) with spring varieties.
- Yellow lupin (*Lupinus luteus*) with spring-sown varieties.

Lupins require a pH below 7 and the seed should be inoculated with *Rhizobium* bacteria to encourage nodulation. It is important to establish a good plant population and time of sowing is critical for winter varieties. The crop does not require nitrogen and on fertile soils has little requirement for phosphate and potash. Crop protection products are available but generally with Extensions of Authorisation for Minor Uses (EAMUs) approval and advice should be obtained.

14.8.4   Dried 'navy' beans
These are the white seeded beans used for canned baked beans and if successful would have the potential to replace substantial imports. BBSRC sponsor a breeding programme to try and produce varieties which are well adapted to UK conditions. However, seasonal weather has caused problems and commercial arrangements have not yet been satisfactory.

14.9   Sources of further information and advice

HGCA, *Oilseed Rape Guide*. Available online at: http://www.hgca.com/content.output/ 5121/5121/Crop%20Management/Disease/Oilseed%20disease%20control.mspx
Defra, *The Fertiliser Manual RB 209*, The Stationery Office (TSO), 2010.
HGCA, *UK Recommended Lists for Oilseeds*. Available online at: http://www.hgca.com/ content.template/0/0/Home/Home/Home.mspx

PGRO, Information and advisory service for pulse growers and users.
PGRO, Pulse Agronomy Guide. Online through membership.
CABI, *The UK Pesticide Guide* (updated annually).
Weiss E A (ed), *Oilseed Crops*, 2nd edn, Blackwell Science, Oxford, 2000.

15

Root crops

DOI: 10.1533/9781782423928.3.362

Abstract: This chapter deals mainly with potatoes and sugar beet and, for each, covers markets, the industry and quotas in the case of sugar beet, quality aspects required, variety choice, establishment and soil management, crop nutrition, crop protection and harvesting and storage. There is a section on irrigation management for the potato crop and, for the sugar beet crop, a look forward to EU sugar regime change, discussing what might happen.

Key words: potatoes, sugar beet, quota, soil management, weed control, blight, rhizomania.

15.1   Introduction

Cash root crops are an important part of the cropping systems of a significant number of farmers throughout the main arable areas of the UK. They form an important element of the rotation, acting as a break from combinable crops and, if grown well, provide higher gross margins than cereals, oilseeds and legumes. Both sugar beet and potatoes require capital investment in specialist machinery, or the ability to use contractors, and most potato enterprises need investment in buildings suitable for long-term storage of the produce as well as grading/handling systems.

15.2   Potatoes

The production of potatoes in the United Kingdom was, until 1997, strictly controlled by a quota system administered by the Potato Marketing Board (PMB). A heavy excess levy and possibly a fine were imposed if the area grown by a farmer exceeded the quota. Quotas could be purchased or leased from other growers. The Board also controlled the marketing of the crop annually by setting and supervising quality standards and riddle sizes. Additionally, it organised a support buying scheme to help maintain reasonable prices in difficult seasons, as well as promoting the product, carrying out market research and funding research and development.

On 1 July 1997 a new organisation, the Potato Industry Development Council, was formed to take over some of the activities of the old PMB. It is now called the Potato Council, is part of the Agricultural and Horticultural Development Board and its objectives are to:

1. Deliver value for money for levy payers in everything it does.
2. Improve efficiency and productivity in the industry to help levy payers have thriving businesses.
3. Improve marketing in the industry to help profitability and customer awareness.
4. Improve services that the industry provides to the community.
5. Improve ways in which the industry contributes to sustainable development.

It does this by funding research and development, helping to transfer technology into the industry, collecting and disseminating market information and promoting British potatoes to consumers in the UK and overseas.

It is a levy-funded body with levies coming from growers with over 2 ha of potatoes (c£43/ha) and purchasers each time potatoes are moved (c19 pence/tonne). In 2011 there were approximately 2500 growers with 146 000 ha registered. The levy is compulsory and the PC is robust in its attempts to collect outstanding debts.

The most important change to the industry was the cessation of quotas and intervention buying. Both of these factors had a stabilising effect on potato prices, although the weather played a part in seasonal price fluctuations even with quotas in place. The years immediately following 1997 saw a virtually free market place with prices high one year and low the next as growers increased and decreased acreages. The trend was for the bigger and more efficient growers to expand at the expense of the smaller producers, who could make use of 'economies of scale'. The number of growers today continues to shrink but the level of production remains fairly stable because of better yields and better storage techniques.

The main areas of development for the Council and the industry are to increase the environmentally sustainable production of ware and seed potatoes and to increase potato quality to meet end-user requirements. Projects currently funded by them include the improvement of input precision and the reduction of production costs, the improvement of understanding of the factors affecting eating quality and the Fight Against Blight (FAB) campaign.

Potatoes are tubers in which starch is stored and, depending mainly on the variety, they vary in shape (e.g. round, oval, kidney, irregular); skin colour (mainly cream and/or red); flesh colour (mainly cream or lemon); and depth of eyes (where sprouts develop).

In this country, potatoes are used mainly for human consumption, but in a glut year some subsidised lots may be used for stock feeding. In several EC countries, some of the crop is used for producing starch, glue, alcohol, etc. The consumption of potatoes in the United Kingdom, at 125 kg/head, is fairly constant from year to year and so the prices obtained may vary considerably according to the supply. An increasing proportion of the crop is processed as frozen chips and crisps or other potato snack products.

Good quality ware potato tubers:

- Are not shrivelled.
- Are not damaged, frosted or diseased.
- Are free of greening (an indication that a toxic substance, solanine, is present); secondary-growth irregularities such as knobs, cracks and glassiness; sprout growth which spoils quality and increases preparation costs.
- Are of good shape and size (40–80 mm).
- Have clean skins with shallow eyes for easy peeling.
- Have not been damaged by pests such as wireworms, slugs and cutworms.
- Have flesh which does not turn brown or blacken before or after cooking.
- Do not break down when being boiled or fried.

Potatoes must have special qualities to be suitable for processing:

- For crisps, chips and dehydration the tubers must have a high dry matter (starch) content and a low reducing sugar content. Fry colour in the potato snack market is all-important and too much sugar produces dark brown crisps and chips which are not acceptable to the UK consumer.
- For canning, small (20–40 mm), waxy-fleshed, low, dry matter tubers are required which do not break down in the cans.

In the United Kingdom, varieties are classified as first and second earlies and maincrops. The difference in times of maturity between varieties is mainly because the earlies are 'long day' potatoes, i.e. they can produce a crop during the long days of early and mid-summer, whereas the late varieties are neutral or 'short day' types and will not produce full crops unless allowed to grow on until the shorter days in early autumn. Early and some second early potatoes can be considered as a 'high risk/high profit' crop. Yields are lower than maincrops, but prices are often high, especially very early in the season. Hitting the market at the right time before the price begins to fall is the secret of profitable early potato production. It is no longer the case that old maincrop potatoes coming out of store in the spring are poor quality, soft and 'tired' tubers. Better technology and a better understanding of storage requirements has meant that consumers can buy firm, clear-skinned potatoes throughout the year. This means that the novelty of the new crop soon wears off with prices of first earlies falling rapidly by mid-May. There is also strong competition at this time of the year from imported potatoes from southern Europe and the Middle East. Technology and better climates mean that countries such as Cyprus, Egypt and Israel can produce new potatoes and export into the UK virtually all year round, so seasonality has all but disappeared.

Another important change in the UK market is the shift in customer preference. The big supermarkets have pushed 'babies and bakers' with specific varieties being required for each type, currently *Charlotte* for babies and *Marfona* for bakers. The traditional 'dirty' early scraper, which needed washing and scraping in the kitchen, has been replaced by small, clear and set skinned baby potatoes which can be dropped into the pan straight from the pack, and which suit our more hectic lifestyle because of the greater convenience.

Many of the maincrop growers enter into contracts with processors or supermarkets. Some growers have entered into agreements with local packhouses, while others have set up their own on-farm pack lines to provide supermarkets with pre-packed, labelled and bar-coded produce. Maincrop growers have less opportunity to 'play the market'. Older maincrop varieties such as *Russet Burbank* are back in favour and grown on contract for the French fries trade.

15.2.1 Varieties

The following are the more important (or promising) varieties as at 2013:

First Earlies: *Maris Bard, Premiere, Accord, Amora.*
Ware Second Earlies: *Marfona, Estima, Saxon, Charlotte, Nadine, Osprey, Wilja, Shepody.*
Ware Maincrop: *Desiree, King Edward, Maris Piper* (versatile variety suitable for ware and chipping), *Markies, Melody, Harmony, Pentland Dell, Fontane.*
French Fries Maincrop: *Russet Burbank.*
Crisps Maincrop: *Lady Rosetta, Saturna, Hermes.*
Set Skin Small: *Maris Peer.*

It should be noted that all healthy plants of the same variety are alike because they are reproduced vegetatively.

15.2.2 Seed rates

The decision on the most profitable seed rates for potatoes is complicated by factors such as cost of the seed and expected price for the crop; variety and size of the seed must also be considered carefully. The figures given in Table 15.1 are for normal-sized seed (30–60 mm), but if healthy small seed (20–30 mm) is used the rate can be reduced by at least 25%. In some countries it is common practice to cut larger seeds into pieces by hand or machine before planting. The cut pieces are more likely to rot than whole seed, but they may be treated with a fungicide for fusarium control.

Table 15.1 Yield and seed rate for potatoes

	Yield of tubers (t/ha)	Seed rate (t/ha)	Time of planting	Time of harvesting
Earlies	10–30	4–5	Feb–Mar	Late May/July
Maincrop	35–65	3–4	April	Aug/Oct
Seed	25–40	4–5	April	Sept/Oct
Canning	10–20	5–7	Mar–June	June/Oct

Only basic or certified seed can be bought or sold. Growers can plant uncertified home-grown seed, but this is risky if the crop is from poor stock and is not protected by suitable insecticides against aphids, which spread leaf roll and mosaic virus diseases. Some varieties have a high resistance to virus diseases and so are better suited to growing-on as home-grown seed. A reliable tuber-testing service should be used to check any doubtful stocks of seed before planting.

UK seed crops of potatoes are grown in Northern Ireland, Scotland or at a high altitude in England. Disease and aphid levels are lower in these areas.

15.2.3   Sprouting (chitting) tubers before planting

Fewer than half of the potato crops in the United Kingdom are grown from sprouted seed. It is a necessity with earlies to obtain high early market prices and sprouting should be started in the winter before planting so that single sprouts develop on each tuber (an apical dominance effect). It is desirable for maincrops because, on average, it increases yield by 3–5 t/ha. It also allows more flexibility of planting time, and reduces the risk of yield losses if blight occurs early. However, many maincrop growers have given up the technique of chitting and physiological ageing because of mechanised planting regimes. Chitted potatoes are more difficult to plant and the chits often get knocked off.

Seed crops also benefit from sprouting. Rogues may be removed before planting; the crop bulks earlier and so the haulms can be destroyed early to check the spread of virus diseases by aphids. To obtain high yields from maincrops and seed crops, several sprouts should be encouraged to grow on each tuber (multi-sprouting).

The physiological age of seed tubers can have a marked effect on how most varieties develop. Physiologically old seed is produced from early planted seed crops which are lifted early in warm conditions and, when in store, have heat units added to them by keeping the temperature at levels above 4 °C. Such seed produces earlier crops with fewer tubers, but the crop is more susceptible to drought. Physiologically young seed, produced and stored in cooler conditions, is better for maincrops which grow on to high yields.

The rate of growth of the sprouts is controlled by temperature and is usually quickest at about 16 °C, but varieties differ considerably. The strength of the sprout is controlled by light. Short, sturdy sprouts are formed in well-lit buildings such as glasshouses or barns fitted with warm-white fluorescent tubes. Storage buildings must be frost-proof and well ventilated to avoid high humidity and condensation. Chitting trays, each holding about 15 kg (60–80 per tonne), are still the most popular containers for sprouting, especially for earlies. Higher capacity white plastic trays, or wire crates and pallet boxes holding at least 500 kg, can speed up the loading of high output maincrop planters.

Well-developed sprouts (2–3 cm) give maximum yield advantage. However, they are easily damaged by mechanical planting and so mini-chitted seed, with sprouts less than 5 mm, is used with automatic machines on at least one-third of sprouted maincrops. Mini-chitting is achieved by temperature manipulation. If well-sprouted tubers are taken from a warm store and planted into cold ground, some varieties are likely to emerge very late due to 'coiled sprout' development, or 'little potato' may develop when no shoots appear and the old tuber is converted into a few new tubers.

15.2.4   Soils and climate

Earlies do best on light, well drained soils in areas free of late frosts, but some frost protection is possible with irrigation. These areas are usually coastal, where the sea acts as a storage heater and a buffer against cold temperatures. Covering with polythene sheeting can hasten crop establishment. This is removed when the crops grow bigger.

Maincrops are best suited to deep, fertile, loam soils because high yields are very important, especially if prices are low. Potatoes will grow in acid soils (under pH 5.5); common scab can cause problems on high pH soils in a dry season. Irrigation, if it is possible, should be considered in order to suppress the disease.

Place in rotation

Potatoes can be taken at various stages, but usually after cereals. To control cyst nematodes (PCN), maincrops should not be grown more often than two years in eight, seed crops one year in five to seven years, but earlies may be grown continuously if lifted before the nematodes develop viable cysts (i.e. before 21 June). There is some varietal tolerance of PCN species.

15.2.5   Seedbed preparation

Except on light soils, early ploughing is necessary to allow for frost action. In spring deep cultivations, harrowing, discing and/or power-driven rotary cultivators are used, as required, to produce a fine deep tilth without losing too much moisture. The land is then ridged up using only the weathered soil available. Bringing up unweathered soil will result in smearing of the ridge sides.

Some soils are very stony, which is likely to cause harvesting problems. The stones can be collected mechanically and removed, or mechanically separated into windrows between the ridges by a 'destoner'. If sufficient tilth has not been formed over winter by frost action, it may be necessary to form mini-ridges to start with (see section 15.2.6 below) and, after planting, build these up to cover the potatoes as the soil dries out. Building up the ridges in several stages also allows the soil to warm up faster and, in addition, this method can hasten the growth of earlies.

15.2.6   Planting

Nearly all potato crops in the United Kingdom are planted mechanically with 2–7 row machines which are either fully or semi-automatic. Hand planting is still practised on a few farms growing earlies. The ridges must be well-formed to protect the new tubers from blight spores and light, which would cause greening. Some farmers who grow on lighter soils are now planting potatoes in beds. There are usually three rows in beds 150–180 cm wide, i.e. the width of two ridges. This is preferable for irrigation (there is less run-off), and it produces more tubers in the 40 mm size and less greening. Modern machinery can deal with the extra throughput of soil (it may, in fact, reduce damage), but it is only feasible in fairly dry conditions on light soils.

Spacing of tubers (sets)

Row width for earlies and seed potatoes is 60–70 cm; for maincrop, the commonest width is 90 cm. There is less yield loss and fewer problems in producing good ridges, keeping them free of clods, and fewer 'greened' and blighted tubers with the wider rows. The spacing between the sets will depend on the seed rate and the size of the seed. Table 15.2 gives an indication of optimum seed rate and set numbers per hectare. This assumes that the seed size is 35–55 mm, the rows are 90 cm apart and the seed price about twice that likely to be obtained for the ware. Some land needs to be destoned before planting. The destoner passes the ridge or bed soil over a moving web and the stones and clods can be windrowed alongside the row. The spaces on the webs of the destoner should ideally match those of the harvester. This should mean that, at harvest, all the soil drops way leaving just potatoes on the harvester webs. This makes harvesting easier, causes less damage to the tubers and prevents stones in the soil from restricting growth.

On highly fertile soils, especially if irrigated, and where higher yields (over 70 t/ha) are expected, some of the tubers may be too large (over 80 mm). In these conditions, the seed rate should be increased by 10% to give more tubers of desirable size. Some varieties such as *Cara* produce fewer tubers than average and so they are often too large, hence the higher seed rates. Other varieties such as *King Edward* and *Maris Piper* produce a large number of tubers and they may not grow big enough (under 40 mm) and so a lower seed rate is acceptable.

Table 15.2   Optimum populations and seed rates

Sets/50 kg	Tuber spacing 90 cm row	Sets/ha	t/ha
630	31 cm	36 000	2.86
650	25 cm	45 000	3.46
600	32 cm	35 000	2.96
700	30 cm	35 500	2.70
800	31 cm	36 000	2.25
750	33 cm	34 000	2.26
750 (*Cara*)	17 cm	65 000	4.30

15.2.7   Manures and fertilisers

Potato crops benefit from organic manures such as farmyard manure and slurry. Manure can be liberally applied when available at approximately 30–50 t/ha of farmyard manure or slurry equivalent. Normally this is ploughed-in during the autumn. Fertilisers are either broadcast during the spring cultivations or, more efficiently, placed in the ridges by the planter. Too much nitrogen applied to the seedbed usually delays tuber development which can reduce yields in some seasons, but is not usually a factor taken into consideration. Too much nitrogen can also delay haulm senescence and skin set. This can lead to more harvest damage and to lower dry matters. The tuber size is increased by nitrogen and potash, whilst phosphate increases the number of tubers (Table 15.3). Dry matter can be decreased by high levels of potash.

Nitrogen rates for potatoes are not actually as simple as Table 15.3 suggests. Growers must take into account the soil nitrogen supply, the length of growing season and the longevity of the haulm. The last factor is a varietal one and varieties fall into one of four categories:

1. Short haulm longevity (determinate varieties).
2. Medium haulm longevity (partially determinate varieties).
3. Long haulm longevity (indeterminate varieties).
4. Very long haulm longevity.

In general, more nitrogen is needed for the determinate varieties and, the longer the growing season, the more nitrogen is needed. This gives nitrogen rates ranging from 0 kg/ha up to 270 kg/ha depending on the above factors. Recommendations are that, unless the soil is prone to leaching, e.g. very light with irrigation, all nitrogen should be applied to the seedbed. Where there is a risk of leaching, two-thirds should be applied in the seedbed and the remainder just after emergence. No reductions should be made for placement of fertiliser, nor should more or less be used when using beds as opposed to individual ridges.

Table 15.3   Effect of fertilisers on growth of potatoes

	N kg/ha (assumed SNS index 0–1)	P kg/ha (assumed index 2)	K kg/ha (assumed index 2)
Earlies	0–140	170	300
Second earlies and maincrop	40–270	170	300
Seed (burnt off early)	40–210	170	300

Notes:
(i) Recommendations for phosphate and potash are for incorporation into the seedbed. A small reduction can be made for placement alongside seed. Where more than 300 kg/ha of potash (K) is required, e.g. at lower indices, half should be applied in the late autumn and the other half in the spring seedbed.

(ii) Allowances should be made for N, P and K applied as organic manure.

15.2.8   Weed control
Potatoes are a very competitive crop once they meet across the rows, but early weed emergence, which is not controlled, can reduce yields. Weeds can also affect potato quality and ease of harvesting.

Weeds, by tradition, have been controlled in good weather by cultivations, including ridging after planting. However, cultivations may be impossible in wet seasons and there can be considerable loss of valuable moisture and damage to crop roots in dry seasons. They can also produce clods in some soil conditions. Consequently, in most potato crops, weeds are now controlled by herbicides; often a mixture of contact-acting and residual products.

Pre-planting
Perennial weeds should, if possible, be controlled by glyphosate applied preharvest in the previous cereal, oilseed rape, pea, or bean crop (the weeds must be green and actively growing), or by treatment in the autumn.

Pre-emergence of potato shoots
Diquat, carfentrazone-ethyl, rimsulfuron or glufosinate-ammonium will kill most emerged seedling weeds. These chemicals work by contact action. Other residual herbicides (singly or in mixtures) can also be used, e.g. clomazone, linuron, flufenacet, or pendimethalin. Metribuzin is more persistent, but there are varietal restrictions. On organic soils it can be incorporated into the ridge to improve the control of weeds.

Post-emergence of potato shoots
There are very few products which can be used post-emergence in potatoes. Metribuzin can be applied to some crops and varieties, as can the foliar-acting herbicides, bentazone and rimsulfuron. Several graminicides are approved for grass weed control, e.g. cycloxydim and propaquizafop.

Haulm destruction
Chemicals are normally applied to destroy the haulm; this encourages skin set and reduces the spread of blight to the tubers. The following may be used: diquat (but not in dry conditions), carfentrazone-ethyl, pyraflufen-ethyl or glufosinate-ammonium.

Volunteer potatoes from tubers and seed are an increasing problem, particularly in close rotations. Glyphosate, pre-harvest in other arable crops, can aid control.

15.2.9   Disease control

Blight
This is the worst fungal disease which attacks the potato crop. It can seriously reduce yield by killing the foliage early; during periods of heavy rain the spores of the fungus can be washed into the soil and onto the tubers, so causing them to rot in the ground or during storage. The disease spreads rapidly in warm, moist

conditions which, when they occur, are known as 'Blight Periods'. These are recorded by the Meteorological Office, and information can be obtained by farmers by telephone or through the Internet. Growers can then respond by an application of fungicides; a wide range is available. Most fungicides have a protectant and systemic action and are effective if sprayed regularly with the leaves being well covered. A common programme is to use propamocarb mixtures early in the season with follow-up sprays of newer materials such as cymoxanil, cyazofamid, amisulbrom, ametoctradin, fenamidone and zoxamide. Spray intervals range from 7 to 14 days depending on material used and pressure of disease. The use of irrigation in the summer usually triggers a Blight Period and a spray programme should be started soon after.

Tin compounds used to be available for killing spores which fell on the soil before they reached the tubers, but are now no longer approved. To reduce the risk of spores spreading from the leaves to the tubers, the haulms should be destroyed when about 70% have been killed by blight; this is especially important if heavy rain is expected (Table 6.2). It is also important to think about preventative measures to stop the spread of blight. Potato dumps and volunteers in neighbouring fields are important sources. Reject and waste potatoes should be deeply buried and volunteers should be killed off in other crops before they become a source of infection.

Other diseases of potatoes are mainly seed-borne and can be controlled by treating seed with fungicides, e.g. imazalil, flutolanil, and pencycuron. They include gangrene, silver scurf, dry rot, skin spot and stem canker (Table 6.2).

Aphids
Increased physical damage to the crop and the spread of virus by aphids in some hot summers have necessitated the use of soil-acting granular insecticides such as oxamyl or sprays such as lambda cyhalothrin, on more than half the potato crops in the country (Table 7.1). However, because of the historical widespread use of organophosphates growers had to rely increasingly on sprays containing pyrethroids and carbamates. Aphids are now showing resistance to these materials as well, so a careful programme of mixtures of active ingredients (some containing controversial neonicotinoids others exhibiting antifeeding properties, e.g. pymetrozine) is advised, especially when treating seed crops.

15.2.10 Irrigation
(See also Chapter 8.) In dry seasons, very profitable returns can be obtained from irrigating the potato crop. Quality may be improved by the tubers being more uniform in size and having less common scab, but the dry matter percentage may be lowered if an excessive amount of water is used late in the crop's life. It is important to aim to stop irrigation four to five weeks before the intended lifting date. If water is applied to potatoes which have almost died due to drought, secondary growth may develop as knobs and cracks; this will obviously spoil the crop.

It is vital that irrigation water is used economically and effectively. Restrictions on licences for water abstraction are likely to increase in the future and the cost of water will rise. Accurate irrigation scheduling is vital and over-watering should be avoided. Many growers are investigating the use of trickle irrigation systems and methods of preventing run-off from conventional irrigators.

Earlies may be protected from frost damage by keeping the soil moist on those warm sunny days which are followed by radiation frosts at night. In more severe cases, they can be protected by spraying with about 2–3 mm of water per hour during the frosty period; the latent heat given out, as icicles form on the crop, prevents damage to the leaves. Irrigation may also be used to assist in breaking down clods when preparing spring seedbeds in a dry time.

15.2.11 Harvesting

Earlies are harvested when the crop is still growing. To make lifting and later cultivations easier, the tops should be destroyed with a flail-type machine. Earlies must be treated gently because the skins are very soft. Maincrops, especially if they are to be stored, should not be harvested until the tuber skins have hardened. This is usually about three weeks after the tops have died, or have been desiccated.

Historically, much of the desiccation on maincrops was carried out using acid but this is no longer allowed. Glufosinate-ammonium is an alternative cleared for use on maincrop ware potatoes and is widely used. Newer developments include the manufacture of large propane tractor-mounted burners which can straddle three beds at a time and have a high work rate.

Early crops are often picked by hand, but most maincrops are lifted mechanically. One-row, two-row and bed harvesters of various designs and capabilities are available to lift and load the crop into trailers or into bags. Harvesters can be manned or unmanned. Very high rates of working are possible with some machines in stone-free and clod-free conditions, when it is reasonably dry.

Damage to tubers by harvesting operations can result in serious losses due to bad design and/or faulty operation of the machines. Damage can be extremely serious in very dry soil conditions when the tubers are somewhat dehydrated and more susceptible to damage, and the clods are very hard. Machinery and handling equipment is now designed to limit the number and intensity of impacts that the potato suffers during the harvesting, storing and grading process.

15.2.12 Storage

A high proportion of the maincrop has to be stored. Storage may only be for a short period for some of the crop, but in other cases the potatoes may have to be stored until May or June when the home-grown earlies come on to the market again. Obviously, badly damaged, diseased and rain-wet tubers will not store satisfactorily and should be sold at harvest time.

Most crops are now stored indoors, either in on-floor bulk stores or in buildings designed to hold large 1000 or 1500 kg boxes which are stacked on top of each other from floor to ceiling. The pallet boxes are expensive (£50–60 each), but ventilation is usually very satisfactory and the tubers retain a good appearance. Care is required when filling bulk stores to ensure an even distribution over the floor and to avoid 'soil cones' which can prevent air circulation. If the heap is more than 2 m deep, some form of duct ventilation will be required to prevent overheating. In the more sophisticated buildings, very accurate temperature control is possible by using refrigeration and recirculation ventilation and most growers of contracted maincrop potatoes would have invested in this type. In most ordinary stores it is usual to put a deep (300 cm) layer of straw or nylon quilt on top of the heap to prevent greening and frost damage, and to collect condensation moisture, so keeping the tubers dry. Good ventilation, and allowing the heap to warm up to 15 °C for 7 to 10 days after filling, helps to dry the tubers and to heal wounds. Some farmers apply fungicides to prevent rotting in store.

For late storage (by whatever method), sprout growth has to be prevented in the spring. This can be done by keeping the heap cool (4–7 °C) by cold air ventilation. This method is not satisfactory for crisping and chipping potatoes, because some starch changes to sugars (some reversal of this process is possible by warming the tubers before sale). Alternatively, sprout suppressants such as chlorpropham on processing potatoes can be applied when the crop is going into store.

15.2.13 Grading
The stored tubers are riddled (graded or sorted) during the winter or early spring when the best potatoes (ware) are separated from chats (small tubers), diseased, damaged, over-sized and misshapen tubers. They are usually sold off either in 25 kg paper bags, in bulk to processors or are packed for supermarkets either on-farm or at centralised packing stations.

15.2.14 Yield
Average yields for potato crops are shown in Table 15.1.

15.3   Sugar beet

The sugar which is extracted from the crop supplies the United Kingdom with approximately 60% of its total sugar requirements. About 113 000 hectares are grown on contract by 3600 growers for British Sugar Plc (a private monopoly) which has four factories in England, all of which are situated on the eastern side of the country. A contract price per tonne of clean beet containing a standard percentage of sugar (usually 16%) is determined annually by the EU. Weight of raw beet, sugar percentage, dirt tare, transport allowance, early and late delivery allowances, a crown tare adjustment (now fixed at 6.77%) and purity bonuses all determine the price which the grower will receive.

There is a world surplus of sugar and, to restrict output, a quota system operates for the production of sugar in the EU and this affects the price the grower receives. The annual beet contract for individual UK growers is expressed as a contracted tonnage, which is the amount of quota beet which receives the top price for the production of, at present, approximately 1.4 million tonnes of sugar.

The quota is adjusted to a 16% sugar level. Any sugar surplus to quota is 'non-quota' beet and is sold (usually at a lower price) on the world market after carrying forward about 70 000 tonnes to the next year's production. This is an insurance against the following year's crop being a poor one for some reason. Nationally the quota should be met, if not, it could mean a lower quota in subsequent years. There is also an industrial beet contract of about 500 000 tonnes which is used for biofuel production.

Each grower is allocated a quota and is expected to fulfill it. There are good reasons for meeting the contracted tonnage each year:

1. To realise the potential enterprise gross margin of the crop.
2. To avoid a reduction of quota in subsequent years for the grower. The advice is to endeavour to exceed the quota by up to 10% to compensate for yield fluctuations outside the grower's control and which may occur in any one year. Growers accept the fact that some beet will receive the non-quota price.

The average sugar content in the UK is about 18% but this can also differ from year to year depending on the growing season and, in particular, the amount of sunlight the crop receives. Sugar percentages can often be 20% or higher in southern Europe because of the extra solar radiation, but the yields will be lower.

There are several useful by-products from the sugar beet industry:

- Beet tops – a very succulent food, but which must be fed wilted; they are not always easy to utilise but can be ensiled or fed *in situ* in the field.
- Beet pulp – the residue of the roots after the sugar has been extracted; an excellent feed for stock. This can be sold wet for inclusion in cattle or sheep rations or as a bagged dry feed, very popular for horses, as a source of energy.
- Factory Waste Lime (LimeX) – used by the factories for purification and sold as a liming product back to the growers. It has a useful nitrogen content as well as containing some trace elements.
- A soil conditioner called TOPSOIL, produced from the 300 000 tonnes of soil received by the factories each year.
- Tomatoes grown in 18 hectares of glasshouses heated by factory waste water.

There is also an opportunity to use surplus beet to produce bioethanol and this is now done at British Sugar's Wissington factory which produces 70 million litres per year.

15.3.1   Soils and climate

Sugar beet can be grown on most soils, except heavy clays which are usually too wet and sticky; thin chalk soils which do not retain sufficient moisture and on which harvesting is difficult, and very stony soils on which drilling and harvesting can be very difficult. Moderately stony soils can be cultivated with newly developed machines which bury the stones below the rooting zone in a one-pass operation.

Sugar beet is a sun-loving crop which will not grow well when there is too much rain and cloud. Sugar is produced by the conversion of the energy of sunlight into the energy of the plant – sucrose in the case of sugar beet. Consequently sunlight, and the amount of crop leaf area which can absorb it, is an important factor in determining the yield of sugar produced by the crop. The grower should see that as much sunlight as possible is intercepted and utilised by a well-developed crop (irrigation may help) in the long days of May and June. This emphasises the importance of early drilling and establishment of the crop as well as a uniform distribution of the plants. However, drilling too early will expose the young plants to periods of low temperature. This vernalisation will make the plants prone to bolting with the subsequent production of a seed head and a dramatic loss of sugar in the root.

15.3.2   Place in rotation

In order to prevent the build-up of *Rhizomania* (Table 6.2), as a condition of the contract sugar beet may not be sown in a field which has grown any Beta species, e.g. fodder beet, mangels, red beet, in either of the two preceding years. This restriction also helps to prevent the build-up of the beet cyst nematode (Table 7.1) and weed beet, although most *Brassica* crops as host crops can perpetuate the nematode. A three or four year interval is, in fact, preferable between sugar beet and any other closely related crops, as well as *Brassicae*.

Ideally, sugar beet will follow a cereal. This allows an opportunity for stubble cleaning but still allows time for careful October/November ploughing with subsequent essential weathering of the topsoil. Care must be taken to avoid certain cereal herbicides which can remain in the soil in sufficient quantities to affect the germination of the beet crop.

15.3.3   Seedbed preparation

The importance of a good seedbed for sugar beet cannot be over-emphasised. The success or failure of the crop can, to a great extent, depend on the seedbed. It must be deep yet firm, fine and level (see also Section 8.6.2).

In most seasons, except on heavy soils, the less work done on the seedbed in the spring the better the conservation of moisture and the better the chance of avoiding compaction. The one-pass cultivation/drilling technique with front-mounted equipment and rear-mounted drill has much to commend it. An alternative method is to use a controlled wheeling system where the widths of machines match up and tractor wheelings are confined to specific tramlines across the field. This is made much easier these days with the advent of precision farming technology, where GPS systems can provide accuracy within the field to within a few centimetres. Good early winter ploughing followed by a correctly timed winter cultivation (where possible) is, as always, a prerequisite on the majority of soils. The exceptions are the light peat and sandy soils; they can be ploughed and furrow pressed in the spring, as weathering is unnecessary.

A well-worked under-structure by deep loosening equipment with no soil inversion is, in some situations, preferable to the plough. The latter should be avoided when blowing (wind erosion) could be a problem. Non-ploughing techniques may also help to prevent weed beet germinating in the beet crop. The weed can more easily be dealt with in other crops in the rotation.

15.3.4   Varieties

The British Beet Research Organisation (BBRO) recommended list and the BS agronomist can help the grower to decide which varieties to use. There is a selection of varieties and they can be grouped according to the yield of roots, the sugar percentage, their resistance to rust and powdery mildew and their resistance to bolting. Apart from perpetuating weed beet, bolting is highly undesirable as the roots become woody with a low sugar content. If there are more than 10% bolters in the crop, there is a risk that some of the bolted beet may finish up as part of the sample taken for determining the sugar percentage at the factory gate. Varieties with a high resistance to bolting should be used for early sowing and in colder districts. There are also assessments of resistance to Beet Cyst Nematode and AYPR, an aggressive form of *Rhizomania*.

All varieties are purchased from British Sugar under a seed supply scheme.

15.3.5   Manures and fertilisers

The following recommendations can be made:

1. Farmyard manure – 25 tonnes of well-rotted manure per hectare should, ideally, be applied in the autumn. Only about 30% of the sugar beet grown in the country actually receives FYM.
2. Lime – a soil pH of between 6.5 and 7.0 is necessary. Alkaline soils can suppress some plant foods and, on peat soils, which may have a low content of phosphorus, magnesium and manganese, a pH 6.0 to 6.5 is acceptable.
3. Salt – 200 kg/ha should be broadcast a few weeks before sowing where the K index is less than 2. 100 kg of salt should be applied when the K index is 2 and the level of Na is below 25 mg/kg of soil. No salt or potassium is needed when the K index is 3 or above. Dressings of 200 kg of salt will have no effect on soil structure, even on those with a poor structural stability.
4. Magnesium – magnesium deficiency has become more evident in recent years, particularly on light, sandy soils. If necessary 500 kg/ha Kieserite (magnesium sulphate) should be applied. Magnesium limestone can also be used if available. This will help to maintain magnesium levels, as well as remedying any lime deficiency.
5. Boron – may also be necessary, particularly (but not always) on well-limed sandy soils where it could be in short supply (Table 6.2). A boronated fertiliser is often recommended in this situation.

Table 15.4 Other plant foods for sugar beet

	N (SNS index 0) kg/ha	P (index 2) kg/ha	K (index 2) kg/ha
With FYM	100	40	75
Without FYM	120	50	100

6. Manganese – a deficiency of this trace element is found where the pH on peaty soils is above 6 and on sandy soils above 6.5, and also on old ploughed-up grassland. Manganous oxide incorporated with the pelleted seed material is a useful insurance against manganese deficiency, but it may also be necessary to spray manganese sulphate as soon as deficiency symptoms show on the leaves (Table 6.2). A severe deficiency can reduce sugar yield by up to 30%.

Other plant foods required can be summarised as in Table 15.4, assuming phosphate and potash indices at 2 to 3. Kainit can be used at 500–600 kg/ha instead of potash, salt and magnesium (except in severe cases of magnesium deficiency where Kieserite is still necessary). It should be applied some weeks before sowing the seed. The phosphate and potash can be broadcast and worked into the seedbed at any time over the preceding winter months.

Except after a very wet autumn and winter which may leave nitrogen reserves on the low side, nitrogen in excess of 120 kg/ha is unnecessary. Excessive nitrogen produces more leaf area with subsequent shading of lower leaves and the possibility of reduced sugar yields. Ammonium nitrate (not urea) must be applied in the spring 30–40 kg/ha broadcast straight after drilling, and the balance after the beet has emerged but before the beet seedling forms a rosette of leaves, which may trap nitrogen prills or granules leading to scorch.

Excess nitrogen, potassium and sodium also leads to higher levels of impurities in the sugar. This makes the sugar more difficult and expensive to extract in the factory. This is now one of the measured factors determining price so it is important that growers tailor their fertiliser applications to the crop requirements and take into account the nutrient contribution from any organic fertiliser applied in the form of slurry or manure.

15.3.6 Seeds and sowing
Genetical monogerm seed has now replaced the hitherto used 'multigerm natural seed' – a cluster of seeds fused together which usually produced more than one plant when it germinated and meant that the crop required thinning by hand (labour intensive and expensive). Monogerm seed contains a single embryo from which only one plant will grow.

All monogerm seed is pelleted, i.e. it is coated with clay to produce pellets of a uniform size and density – 3.50–4.75 mm. Unpelleted monogerm seed is rather lens-shaped and, as such, cannot easily be handled in the precision drill. The pelleting of seed allows the use of seed dressings. All seed (apart from that sold for organic production) is now sold treated with an insecticide against aphids and more and more seed is being sold as 'primed', such as 'Advantage' seed where part of the germination process has been completed before pelleting. Seed is also colour-coded according to the seed dressing.

Seed rate
This depends on seed spacing and row width. The recommendation is that pelleted seed should be sown at between 1.00 and 1.43 units/ha (each unit has 100 000 seeds) to give a final optimum plant population of around 90 000 per hectare, allowing for field losses.

Time of sowing
Ideally the crop should be sown as soon as possible after 1 March and aiming to finish within 7 to 10 days. It is best completed by early April at the very latest, after which the yield penalty gradually increases until, in May, about 0.6% sugar yield is lost for each day's delay in sowing.

Drilling-to-a-stand
All of the national sugar beet crop is now drilled-to-a-stand, i.e. the individual seeds are placed separately in the position required for the plant, using a precision drill. This aims to give each plant an equal growing area so that the crop produces uniformly sized roots (making harvesting more efficient and effective) and minimises shading of one plant by another. The number of plants which make up a crop of sugar beet is one of the most important factors in determining not only the yield, but also the sugar content of the roots. The main aim of drilling-to-a-stand is to have a reasonably spaced, yet adequate, plant population. Ideally, this should average 90 000 plants/ha; when it is less than 80 000 there can be a serious reduction in final yield.

The seed spacing should be decided on the basis of the germination percentage of the seed (obtainable from the factory which supplies the seed but a minimum of 90%), the possible losses before and after plant emergence (soil conditions, spray damage, inter-row cultivation, wireworm and other pests) and the row width. As far as is consistent with any inter-row work and the harvesting of the crop, the row width should be as narrow as possible, but at present it is unlikely to be less than 45 cm because of the restrictions imposed by harvester design. If, as is more usually the case, a 50 cm row width is being used, the seed is spaced between 14 and 21 cm apart depending on the desired plant population.

Shallow-sowing will aid germination; 25–30 mm is ideal with a well-prepared seedbed. A press wheel fitted immediately behind the seed coulter effectively compresses the soil over the seed and in dry weather this will help germination.

The bed system
The development of the bed system for growing beet has logically followed the now common practice of overall spraying of the crop for weed control and the use

of multi-row harvesters. One of the most important advantages of the bed system is that all tractor wheelings are avoided where the beet is actually grown. The wheelings are concentrated on, at the most, 15% of the land surface (i.e. the area between the beds – suggested width of 60 cm) and, if necessary, any compaction there can be dealt with by soil-loosening tines following the tractor wheels.

A reduction in compaction should lead to better establishment and a consequent quicker crop cover and earlier interception of essential sunlight. To compensate for the larger beet growing in the rows adjacent to the 60 cm gap (less competition), the plants should be closer spaced than those in the inner rows. In this way a more uniform sized crop should be obtained. Harvesting of the beds (at least four rows at a time) could be a problem in some seasons. With the right conditions, beet grown in beds will almost always do better than conventionally grown crops, both in terms of yield and sugar percentage.

The prevention of wind damage

A survey by British Sugar showed that about 40 000 hectares of the national sugar beet crop was at risk from wind damage. However, only about 32 000 ha was actually protected. Unless preventive measures are taken, it is not unusual for crops in vulnerable situations to be completely destroyed or damaged beyond an economic recovery. This necessitates a late re-drilling with a consequent serious loss of yield, quite apart from the extra cost involved. It is on the lighter and/or organic soils that the beet is at greatest risk from the wind, from sowing until the 6-leaf stage.

Straw cover, as a wind barrier, can be provided to reduce the wind speed at ground level by creating a straw planted 'hedge' about 110–120 cm apart between or across the beet rows, using a straw planting machine. Another type of barrier is a spring cereal, normally barley, drilled at about 60 kg/ha some three weeks before the beet is sown (the latter being drilled between the cereal rows) or, using an ordinary corn drill, the cereal can be sown at right angles to the intended direction of the beet rows. Before there is any competition, and when the beet has established itself sufficiently to withstand any damage, the cereal is sprayed out using, for example, cycloxydim.

Instead of a straw cover the beet can be drilled direct into ploughed and furrow-pressed land. Compaction is avoided and the press helps to stabilise the top few centimetres of the soil. Direct drilling into the undisturbed soil, usually stubble, using precision drills modified for the purpose, is sometimes used as a technique to prevent blowing.

Chemicals such as polyvinyl acetate, which bind the surface particles together, are commercially available. They are, however, expensive (more than £100/ha) and are not widely used. A barley cover crop is much cheaper.

15.3.7   Weed control

Band spraying with herbicides at the time of drilling is no longer the usual method of controlling weeds within the row. When it was introduced it was cheaper than conventional overall spraying and also it had the advantage of leaving weeds for a time between the rows (to be removed later). These provided alternative food for pests and on light soils the weeds, acting as a barrier, helped to reduce wind damage. However, band spraying is a slow operation and it has to be followed up by inter-row cultivations. Consequently, most growers now use a low-volume low-dose technique for overall application of the herbicide. Normally, a sequence of sprays is necessary, but application should always be carried out when the weeds are in the cotyledon to first true-leaf stage.

A method of weed control brought across from the continent is the FAR technique. This involves spraying low rates of a mixture of chemicals onto flushes of cotyledon weeds. The components of the mixture are:

F = phenmedipham
A = activator (usually ethofumesate)
R = residual component (e.g. metamitron, chloridazon or lenacil).

Typical costs of this programme are £72–90/ha.

Where overall application has not been carried out, steerage hoeing (in the early stages of plant establishment) will be necessary, perhaps twice, to assist in weed control. Further hoeing, after the beet has reached the 5–7-leaf stage, should be undertaken, but only as much as is necessary. Weeds have to be kept in check, but any inter-row cultivation does mean a loss of moisture. It is another point in favour of overall herbicide application.

Couch and other perennial weeds should be controlled in the year prior to planting the crop. Weeds can seriously reduce beet yields if not controlled, particularly in the first eight weeks following the sowing of the crop.

It is important that the correct type and amount of herbicide are applied (according to the weed growth stage) and the soil conditions are suitable, i.e. fine and moist. This technique relies on treatment when the weeds are at the cotyledon stage. On average, a sugar beet crop may have four applications of herbicides. Some inter-row cultivations using a steerage hoe are still carried out mainly where there is a problem with weed beet.

Table 15.5 shows weed susceptibility to some commonly used herbicides.

Pre-sowing
Pre-sowing treatments, using contact-acting products like glyphosate, may be required to kill weeds that have not been controlled during seedbed cultivations.

Pre-emergence
The majority of crops receive a pre-emergence treatment. Choice of chemical will depend on potential weed problems and soil type. Chloridazon +/− ethofumesate, quinmerac or metamitron are commonly used.

Post-emergence
It may be satisfactory or necessary to leave all annual weed control to this time. However, bad weather conditions may delay spraying and many weeds could

Table 15.5 General table of weed susceptibility to some commonly used herbicides in potatoes and sugar beet

Crop	Herbicides	Timing	Black bindweed	Charlock	Chickweed	Cleavers	Fat hen	Hemp-nettle	Knotgrass	Marigold corn	Mayweeds	Nettle small	Nightshade black	Pennycress	Poppy common	Redshank	Shepherd's purse	Speedwell	Thistle creeping	Black-grass	Brome barren	Meadow grass annual	Cereal volunteers	Wild oats
			Annual broadleaved weeds →																Grass weeds →				
Potatoes	Metribuzin	Pre- and post-em	s	S	S	R	S	S	S	s	S	S	r	s	S	S	S	S	R	S	–	S	–	r
	Linuron	Pre-em	S	S	S	S	r	–	r	S	S	S	–	–	S	S	S	S	–	R	–	–	–	R
	Rimsulfuron	Post-em	r	S	S	S	S	S	S	–	S	S	r	–	–	S	–	–	–	S	–	–	S	S
	Carfentrazone-ethyl	Early post-em	S	–	–	S	S	–	S	–	–	–	–	–	–	S	S	S	–	–	–	–	–	–
Sugar beet	Chloridazon	Pre-em	S	S	S	r	S	S	S	S	S	S	S	S	S	S	S	S	–	–	–	S	–	R
	Metamitron	Pre- or post-em	r	s	S	R	S	s	S	S	S	S	r	S	S	s	s	S	r	r	–	S	–	R
	Phenmedipham	Post-em	S	S	S	r	S	s	s	s	r	S	r	s	S	s	S	S	R	R	R	R	R	R
	Ethofumesate	Pre- or post-em	S	s	s	S	S	S	S	–	S	s	s	s	s	s	S	s	R	S	–	S	s	s

Symbols: S – susceptible, s – moderately susceptible, r – moderately resistant, R – resistant.
See product label for timing, application rates, restrictions and safe use.

grow past the susceptible stages. Using the low-dose technique, there is less restriction on timing of the post-emergence herbicides in relation to crop growth stage. Mixtures of herbicides with different modes of action are commonly used, e.g. the contact herbicide, phenmedipham with or without metamitron or other residual types such as ethofumesate or the foliar acting triflusulfuron-methyl. A clopyralid spray can be used to give better control of established thistles and mayweeds, as well as suppressing volunteer potatoes. Graminicides can be used to control wild oats and some grasses and severely check common couch although it is preferable not to use these products if resistant grass weeds are present. Glyphosate can be applied with a rope-wick or roller machine to control weed beet.

15.3.8   Disease control

(See also Chapter 6.) *Virus yellows*. It may be necessary to spray the crop to kill the aphids which spread virus yellows (Table 6.2), or to apply the insecticide as a seed dressing, e.g. clothianidin + beta-cyfluthrin. Viruses which cause the leaves to turn yellow are particularly damaging to sugar beet because they restrict the conversion of solar energy into sucrose. Warnings are issued by British Sugar about the likelihood of infected aphids appearing in crops.

Powdery mildew. A fungicide should be used if the disease becomes prevalent in late July or August (Table 6.2). Some cereal triazole and strobiluron fungicides are approved for use in sugar beet for the control of mildew as well as rusts and *Ramularia* leaf spot. Sulphur can also be used for powdery mildew control.

Rhizomania. (See also Table 6.2.) This viral disease was spreading through the UK having been imported into the UK in 1987. It is spread by a soil fungus and there is no chemical control. Its effects on beet are severe. Roots are shrivelled, fanged and covered in fine hairs and cankers. Sugar percentage is drastically reduced. It can be spread in soil on machines, vehicles and any root crop grown in the field. A field quarantine system used to be in place but new, tolerant varieties are now used and the disease has become much less of a problem.

15.3.9   Irrigation

Although it is a deep-rooted crop, the beet plant will respond to supplementary water in the majority of seasons. This will obviously depend on summer rainfall as well as soil type.

If, by irrigation, the plant can develop a reasonable canopy of leaves early on in its life, it will be able to make more efficient use of sunlight for growth and sugar production. But the crop will also need water during June and July. In hot, dry and bright weather it may require up to 350 mm during these two months. Many soils (notably the sands), particularly in eastern England, are unable to supply this need. With irrigation on the farm and, as part of an overall irrigation plan, a water-balance sheet will have to be kept to determine the soil moisture deficit and when it will be necessary to irrigate the sugar beet crop (see Section 8.2.3). However, irrigation should not be necessary after August, even if there is a water shortage; by then the deep root should easily be able to cope.

15.3.10 Weed beet

Weed beet is a big problem for sugar beet growers. Weed beet is any unwanted beet within and between rows of sown beet and, of course, other crops. Unlike true beet, which is a biennial, it behaves like an annual and produces seed in one year. It originates from several sources, but mainly from naturally-occurring bolters in the commercial root crop, from contamination of seed by cross-pollination from rogue plants, from beet groundkeepers either growing in crops following beet, or from old clamp and loading sites, and from the seed shed from weed beet itself.

Once weed beet becomes established it is self-perpetuating (although it can remain dormant for many years) and on average it produces 2000 seeds in the year. However, only about 50% survive, but nevertheless a light infestation (perhaps 1000/ha) can mean one million weed beet/ha the following year.

Control measures. It is very important to prevent any weed beet seed returning to the soil and so, to start with, inter-row work should be carried out to clear any seedlings which are between the rows (at present there are no selective herbicides to control weed beet in sugar beet). Following this, as far as possible, all bolters should be removed from the beet field, no later than the middle of July, i.e. before the seed has set. Rogueing, i.e. by hand pulling, is the most economical method if there are fewer than 1000 weed beet/ha. If rogueing is not possible, the bolters can be cut down mechanically. This is done usually twice, the first time 7–10 cm above crop leaf height and then, 14–20 days later (about the end of July), to prevent lateral shoots and late-flowering plants from producing seed, cutting is repeated – this time at crop leaf height.

Instead of mechanical cutting, the 'weed wiper' fitted to the front or rear of the tractor can be used. This consists of a boom containing a herbicide (at present glyphosate is recommended) which permeates through a nylon rope 'wick' attached to the boom. As it goes through the crop the wick wipes the taller (than the beet plants) weed beet and bolters and so deposits the herbicide. Usually more than two treatments are necessary, starting at the beginning of July and then repeating treatment at 10–14 day intervals until the middle of August.

In addition to preventing the propagation of seed from bolters or weed beet which have already contaminated the crop, it is important that weed beet in crops other than sugar beet should also be controlled. Weed beet found growing on old clamp and loading sites, headlands and waste land must also be destroyed.

The effect of cultivations on weed beet is being examined to try and find out the ideal sequence of cultivations necessary to reduce the weed beet population more rapidly. Direct drilling, tine cultivation and ploughing are all being studied.

Finally, with a bad infestation in a field, it will be necessary for the grower to widen their rotation to six to seven years, rather than the more usual three to four year interval.

15.3.11 Harvesting

In theory, early to mid-November is the right time to harvest the beet crop. In most years it is still slightly increasing in weight, with the sugar percentage reaching a maximum at that time before beginning to fall off.

However, apart from the fact that, in most years, conditions for harvesting deteriorate as autumn proceeds, if all the beet were delivered at this time there would be tremendous congestion at the factory. Therefore a permit delivery system is used, whereby each grower is given dated loading permits which operate from the end of September until about the end of January (although this date could be extended into February or even March in late seasons). It means that growers must be prepared to have their beet delivered to the factory at intervals throughout the period. It is important that the permits are used and are not lost through failure to have sufficient beet out of the ground for delivery to the factory. In the early part of the harvesting period (up to about mid-October), when the beet is still growing in the field, the aim should be to keep just ahead of the delivery permit, in case bad weather stops lifting. As far as possible, the poorest part of the crop should be harvested first. It will never be able to respond to open weather at the end of the year in the way that a full crop can. More beet should be clamped in November and, depending on soil conditions, all lifting should be completed by mid-December.

To minimise dirt tares, direct delivery from field to factory should be avoided, especially when the beet is harvested in wet conditions. If the beet can be clamped at least a week before delivery, and then reloaded using an elevator with a cleaning mechanism, dirt tare will be considerably reduced; it should not exceed 10%. Very dirty loads may not be accepted by the factory. Factories are also discouraging farmers delivering their beet in small loads by tractor and trailer.

Topping. 'Defoliation' is now the preferred method rather than the old fashioned 'scalping'. It gives a 3–5% yield increase by leaving a rounded top, rather than one that has been cut with a knife. British Sugar now make a standard 6.77% crown adjustment to the price received so the setting up of the scalping blade on the harvesters to ensure optimum top tares has become much less important.

Harvesters. There are a number of different types of harvesters and harvesting systems. These range from the single-row trailed tanker harvester to the multi row (2–8 rows) self-propelled tanker, as well as the single and/or multi-row side-loading trailed machine. Depending on the distance to the clamp, the tanker harvester can be operated as a one- or two-man system. This can fit in well with the steady supply of beet needed for the permit delivery arrangements without seriously interrupting other work on the farm. A survey by British Sugar indicated an average 8% loss of yield by machine harvesting, i.e. about 3.5 t/ha per 45 t yield of crop per hectare. Over topping and breakage of roots below the ground and leaving whole roots on the ground are the main reasons.

15.3.12 Storage

Beet stored on the farm should be done carefully, with the clamp adjacent to a hard road and accessible at both ends so that beet which has been clamped longest can be delivered first. Plenty of air circulation should be allowed. Ideally, use a ridge-type clamp for beet which is clamped early whilst the temperature is still high and there is a risk of overheating in the pile. However, as it gets colder this is less important and large square clamps, or heaps against a wall, can be built.

It is important to prevent the beet being damaged by frost, either in the ground or in store, as this has a very quick and deleterious effect on the sugar percentage. Badly frosted beet will be rejected by the factory as it is useless for sugar extraction. Depending on the district and incidence of frost, the clamp may have to be covered. Plastic sheeting, rather than straw, is now being used. It has the advantage that it can be removed quickly, with a rise in ambient temperature and easily replaced when necessary. It is also important to avoid clamping damaged or diseased beet for long periods. Although the beet root is more robust than the potato, it is still possible to break the skin by careless handling. Sucrose will then begin to leak out with a consequent reduction in sugar percentage.

15.3.13 Yield

- Roots (washed), 50 t/ha.
- Sugar at 18%, 9 t/ha.
- Tops, 25–35 t/ha depending on the variety. A stock farmer may prefer a large topped variety, although fewer are grown now.

15.4   Future trends

BBRO has introduced its 4×4 plan – a 4% yield increase in 4 years between 2012 and 2015. It sets targets for various aspects of agronomy including achieving 100% of seed dressed with an insecticide, bringing drilling dates forward to maximize growing season length, using full rate fungicides on 100% of the crop, doubling the level of weed beet control, trying to achieve 100% of control of other tall weeds in the crop, sowing 80% of crop with the top six recommended varieties and increasing the area of land tested for BCN, amongst other things.

The EU sugar regime is due to be revised in 2015, the current one having been in place since 2006. It is as yet unclear about what exactly will happen, but there are clear signs that quotas will disappear. This is likely to bring much more volatility to the EU market, with bigger price fluctuations and less certainty about sugar beet profitability.

15.5   Sources of further information and advice

Harris P. M. *The Potato Crop: The Scientific Basis for Improvement*, Chapman & Hall, 1991.
Draycott, A. P. (ed) *Sugar Beet. World Agriculture Series*. Blackwell, 2006.
http://www.beetthebest.co.uk/downloads/BBRO%20Spring%20Crop%20Management%20
   Bulletin.pdf
http://www.beetthebest.co.uk/downloads/Harvesting%20and%20Storage%20Tech%20
   Guide.pdf

16

Industrial crops

DOI: 10.1533/9781782423928.3.387

Abstract: This chapter discusses the global pressure on finite resources and investigates the crops that might be used as sources of renewables and grown as an alternative to food crops. It describes specific crops such as Miscanthus, short rotation coppice, oil crops for bio-diesel and carbohydrate crops for bio-ethanol production. It also look at crops grown for renewable fibre, pharmaceuticals, essential oils and neutraceuticals. It discusses the use of crops for anaerobic digestion (AD) and the wider problems of growing on land that might otherwise be used for much needed food crops in countries suffering from food shortages, or under environmental threat.

Key words: renewables, bio-diesel, bio-ethanol, energy, anaerobic digestion.

16.1 Introduction

16.1.1 Global pressure for renewables

There are finite supplies of fossil fuels and an increasing (but still debated) risk of climate change because of greenhouse gas emissions. While the discovery of shale gas, and the controversial process of fracking, may alleviate the supply problem for a while, the world looks to invest in renewable industrial crop technology and research. Globally the demand is increasing, particularly by the developing Asian economies. As the price of fossil fuels continue to rise, and production and transport costs rise with it, world economies are focusing their attention on what can be obtained from farming and by natural mechanisms such as wind, solar and water power. Because of the poor photosynthetic efficiency of crops it is predicted that, long term, industrial crops for energy will not be as important as directly harnessing solar power or the use of wind and water turbines.

However, crops remain an important source for other industrial uses such as lubricant oils and chemicals used for plastic manufacture for example. They have the potential to substitute for fossil hydrocarbons in these areas, releasing them for energy use and delaying the time when the world runs out. Natural fibres from tropical and temperate crops are replacing artificial fibres like polypropylene and nylon derivatives.

Another reason for not looking for long-term provision of energy from crops is the effect it has on food supply. Attempting to supply only a fifth of the world's energy demand through biofuels would mean virtually no land left to satisfy human nutritional demands.

Already, biofuel production in Brazil, the United States, the Far East and Europe is a controversial issue, both from a food supply point of view and from an environmental damage one. However, at the current time the various renewable fuel obligations signed up to by most developed countries through the United Nations Kyoto Protocol to the United Nations Framework Convention on Climate Change (UNFCCC) agreement of 1997 (in force in 2005) provide farmers in individual countries with the opportunity to exploit these demands for renewable industrial crops.

16.1.2   UK policy
As signatories to the Kyoto Protocol the UK is committed, amongst other things, to investigate renewable energy sources, reduce GHG emissions and become more energy efficient. Legislation applicable to the UK includes a climate change act, an EU Renewable Energy Directive, the Renewables Obligation and the Renewable Transport Fuels Obligation. The Department for Environment, Food and Rural Affairs (Defra) produce substantial guidance for farmers looking to diversify into the industrial crop markets, and point them in the direction of the numerous schemes and grants available to make a start. A number of planting grants were available from Natural England for biomass crops such as willow, poplar and Miscanthus. These covered 50% of the actual or on-farm planting costs and commited farmers to growing the crops for at least five years, but the scheme finished in 2013. Annual crops grown for biofuels, e.g. cereals or oilseed rape, are eligible to be included in the Single Payment Scheme (SPS) and there are processing aid payments for primary processors of fibre crops.

A National Non-Food Crops Centre (NNFCC) was established by the government in 2003. It has evolved into a consultancy company which offers services to producers and consumers on bioenergy, biofuels and bio-based products.

16.1.3   General areas of use
Apart from biofuels for transport, bio-liquids for heating and electricity, and biomass crops, there are a number of other important uses for industrial crops. These include:

- Lubricants and waxes
- Surfactants (wetting agents)

- Surface/paper coating
- Printing inks
- Pharmaceuticals and nutraceuticals
- Cosmetics
- Essential oils
- Fragrances
- Flavourings
- Dyes
- Packaging/compostable plastics
- Building materials
- Bio-composites
- Textiles
- Adhesives.

Starch and sugar crops like wheat and sugar beet, as well as bioethanol, can be used to produce carbohydrates for use in the manufacture of plastics, adhesives and surface coatings. Oilseed crops such as oilseed rape and sunflowers can, as well as biodiesel, produce lubricants, surfactants and slip agents. The main fibre crops such as flax, hemp and cereal straw can be used for making paper, textiles, insulation material and bio-composites for the car industry.

The main biomass crops are Short Rotation Coppice (mainly willow) and Miscanthus grown for pellets, billets or chips for use mainly in co-fired power stations where they are burnt with coal, or in domestic or community heating systems. Crop wastes such as straw can be burnt in on-farm boilers to provide heating for grain drying.

16.2   Specific crops: Miscanthus

Miscanthus species (sometimes erroneously called elephant grasses) are fast growing and high yielding tropical or sub-tropical perennial grasses which can survive and grow well in temperate climates. They are propagated commercially using rhizomes which are planted in rows and allowed to grow up each year before harvesting. They can produce commercial yields for 15–20 years before they need replanting. The species used in Europe is a hybrid *Miscanthus giganteus*.

16.2.1   Growth cycle
New shoots emerge in March. These grow to maximum height (2.5 to 3.5 m in a mature crop) by late August and produce bamboo-like stems. The crop dies off during the autumn and loses most of its leaves leaving only the cane. This is then harvested in February, usually with a modified forage harvester and, once dry enough, baled. It should yield between 10 and 15 oven-dried tonnes (ODTs) per hectare but has the potential to yield more if conditions are optimal.

Where can it be grown?

Miscanthus can be grown on a wide range of soil types, including reclaimed land where the annual leaf fall will add organic matter to the topsoil. It will grow happily in a wide range of temperatures but will yield best in areas of medium to high rainfall. To receive the planting aid the site must not compromise any other conservation area, nor block any footpaths.

Planting

The rhizomes can be planted with a specialist rhizome planter or a modified potato planter after the site has been prepared. Glyphosate is often used pre-planting to get rid of perennial weeds and the site may also need sub-soiling in the autumn prior to planting to remove compaction. Rhizomes should be bought from a specialist producer in the UK to ensure they are disease free. Each piece should have at least three buds and should be planted at about 7 cm deep at a rate of 10 000 to 15 000 per hectare, depending on field conditions. The aim is to establish 10 000 plants per hectare but emergence can vary between 50 and 95%. The ideal time to plant is in March or April which gives a longer first year growing season and better development of the root systems which will have to support the plant for the next 15 or 20 years.

Nutrition

The offtake of a crop yielding 14 ODTs will be around 85 kg N/ha, 15 kg P and 120 kg K, so these should be replaced. It is a very efficient user of soil nutrients and usually does not respond to more than these maintenance dressings. Timing of fertiliser is not critical although the logical time to apply the nitrogen would be in May, before rapid growth occurs.

16.2.2   Crop protection

Weed control in Miscanthus is relatively simple. A herbicide may be needed in year two, while the crop is establishing, for the control of grass and broad-leaved weeds and this must be done during the dormant period between harvest and new growth. After that, the leaf litter keeps most weeds well suppressed.

There are no serious diseases or pests that will economically affect the yield of the crop. With an increasing area of the crop there may be potential for aphid infestation carrying viruses such as BYDV.

16.2.3   Harvest and storage

The driest samples are achieved by late harvesting in early summer, but this compromises the following year's growth. The first year's growth is not usually harvested, so in the second year the crop is cut and swathed with a forage harvester, allowed to dry and then baled. Bales can be stored short-term outside if stacked carefully. If the crop has been chipped this will need well ventilated storage undercover if kept for any length of time and may have to be dried.

16.3   Short Rotation Coppice

Short Rotation Coppice (SRC) mostly uses willow (osier) or poplar, with willow being most popular. Unlike Miscanthus it is harvested every three years instead of annually, to allow the crop to regrow into a small tree. However, it should last 10 years longer than Miscanthus if properly managed and newer varieties have the potential to give higher ODTs per hectare on an annual basis (around 18 ODTs/ha/yr or 54 ODTs per ha per harvest). It needs specialist equipment for planting and harvesting and is usually used in the form of chips or billets (small rods 5–15 cm long). It is used in power stations or in on-farm or in-building specially adapted boilers.

16.3.1   Growth cycle

The crop is harvested in the winter after leaf fall. New shoots emerge the following spring from the 'stool' or stump and are then allowed to grow for three years before harvesting again.

Where can it be grown?

It can be grown on a wide range of soils but needs sufficient moisture within a metre of the surface to grow to its fullest potential and will survive seasonal waterlogging. Its impact on the environment and landscape must be taken into account as it could be in the ground for 30 years and grow up to eight metres tall. There is also a lot of bare ground left after harvests and during the establishment phase so soil erosion risks must be considered.

Planting

Like Miscanthus, the land should be carefully prepared for good establishment with sub-soiling if needed and perennial weed control with glyphosate. Bulky organic manure with low nitrogen is beneficial and rabbits should be kept out of the field by fencing for at least two years after planting. Unlike Miscanthus, there are several varieties available which have been tested by the Forestry Commission. Rods or cuttings are used which are then pushed into the ground by a step planter so they are left sticking vertically out of the soil in rows. A population of 15 000 cuttings per hectare is needed although lower populations will mean thicker stems and larger billets. They are usually planted in twin rows with enough room between each set of twin rows to allow a tractor to pass through the crop. Planting can take place as early as the end of February but is more likely to be in March/April. The cuttings will throw out three or four shoots which are allowed to grow through the summer and then cut back in the winter to encourage more shoots to form (similar to encouraging tillering in cereal crops). The field headlands and the 1.5 m tractor pathways can be grassed to prevent soil erosion.

Nutrition

No fertiliser is needed during the establishment year and it is almost impossible to spread fertiliser in years two and three of growth, so all must be applied during year one after a harvest. A crop yielding 30 ODTs will remove 90 kg/ha N, 55 kg/ha P and 72 kg/ha K. It can be a useful crop to apply treated sewage sludge and the nutrient value of this should be taken into account before applying artificial fertilisers.

16.3.2   Crop protection

A residual herbicide can be applied after planting and again at any time during dormancy if required. Pendimethalin is often used. In years two and three the crop is very competitive with weeds and no herbicides should be required, indeed they would be impossible to apply.

Melampsora rust is the most important disease and willows are highly susceptible. Varietal resistance is being developed and it is recommended that four or five varieties should be planted randomly within a field to slow down the spread of the disease. Fungicides are not generally used.

Willow beetles are the main pests of willows and poplars and should be spot treated with a knapsack sprayer if in large colonies. An overall spray of insecticide is not practical and could be environmentally damaging.

16.3.3   Harvest and storage

These are small trees that are being harvested and so require much more robust equipment than Miscanthus. Forage harvesters are used with specialist headers, usually with two circular blades to slice through the bottom of the stems and a feeder mechanism to pull the cut shoots into the machine. The chips are blown into a silage trailer and carted off the field. Billets are produced using converted sugar cane harvesters which produce pieces of stem about 20 cm in length and 9 mm thick. Because they are thicker than chips, billets allow natural ventilation and will dry quite happily down to 30% MC in a heap outside. Chips may need to be stored on a drying floor to remove moisture by blowing cool air through them. They can heat up rapidly if left too wet.

16.4   Wetland crops

These are by-products of wetland management where areas are cleared of mainly excess reeds and rushes. This can be dried, compressed into briquettes and burnt as a raw material but other uses are being researched, including pyrolysation into a sustainable form of charcoal.

16.5   Crops for anaerobic digestion (AD)

Methane digester systems can use almost any wet organic substance to produce gas as long as the right conditions exist within the digester. Livestock farmers typically use a mixture of slurry with straw or maize. The agronomy of maize will be the same as for forage maize grown for silage (see Section 18.3.1) but if it is being grown on an all-arable farm, artificial nitrogen rates will need to be higher than that used on dairy farms, because of the lack of manure or slurry applications. Maize grown for AD should also be grown to maximize dry matter per hectare, rather than starch, so narrower rows are often used to encourage tillering and increase height. Some varieties are better than others at this and should be used.

16.6   Woodland biomass

Two main products can be considered. The first is waste from forestry and sawmill activities and the second is biomass produced from woodland or forests grown specifically for this purpose.

Forestry waste (trimmings) can be used for wood chips while sawdust from the mills can be compressed into pellets and burnt in boilers.

Stemwood produced from managed forestry or woodland is often better quality than that from short rotation coppice and no specialist equipment is needed for harvest. It is usually chipped or cut into manageable sized logs.

16.7   Liquid biofuel crops

The two main markets are for biodiesel and bioethanol as a petrol substitute. In the UK the main crops grown for these are oilseed rape, cereals, sugar beet and fodder beet. They have the potential to cut carbon dioxide emissions by 50–60% as well as reducing other pollutants such as smoke particles.

There are no special agronomic requirements of these crops grown for biofuels. Any oilseed rape variety can be used for biodiesel and wheat is the favoured cereal for bioethanol as its starch yield is higher than barley. British Sugar have a bioethanol plant at their Wissington factory which can also use wheat.

16.8   Lubricant oil crops

Oil crops should be able to produce erucamide so High Erucic Acid Rape (HEAR) varieties and Crambe are the two that are most common. HEAR is grown the same as conventional double-low rape (see Chapter 14) and Crambe is similar to spring oilseed rape in its requirements. Crambe (*Crambe abyssinica*) is a member of the mustard family. The oil contains about 58% erucic acid and the seed coat is high in glucosinolates so must be removed before the meal is suitable for animal feed, and can then be used as a fibre source for low grade paper production.

The crop is planted at the same time as spring rape, flowers within 35 days and can be harvested 90 days after planting by the same techniques used for rape, i.e. combined direct, desiccated or swathed.

16.9   Fibre crops

There are two main crops of commercial interest to UK farmers, flax and hemp. Miscanthus, linseed straw and cereal straw can also be converted into useable, low grade fibre. Hemp core (shiv) gives a short length fibre for use in packaging, insulation and biocomposites whereas flax, if grown and harvested properly can give high quality, long fibre for use in linen manufacture. Flax or hemp grown on eligible land attracts a Fibre Processing Aid payment which is made to the processor but often shared with the grower as part of the contract. Hemp must be licensed through the Home Office as it is classed as a narcotic.

16.9.1   Hemp (*Cannabis sativa*)

A low input crop which can be grown on a wide range of soils but grows best in a pH of 6.5 or above. It can substitute for broad leaved break crops in a rotation and is very competitive against weeds. It is sown in the spring with a target population of 150 stems/m². Moderate amounts of nitrogen (80–120 kg/ha) can be applied onto the seedbed and, apart from a possible pre-emergence herbicide, no other crop protection is usually needed.

The crop should be cut in August and left in the field to ret (fibres are separated from the stem by the action of wetting and drying). Then it is rowed up and baled. Dual-hemp crops can be grown for fibre and oilseed, in which case they are combined and the straw left to ret and then baled normally.

16.9.2   Flax (*Linum usitatissimum*)

Commercial flax-processing plants to produce long fibre suitable for linen do not exist in the UK at the moment, but it is widely grown in the rest of Europe. The agronomy is similar to linseed (see Chapter 14). The crop is sown in the spring at a plant density of 2000 per m² (much higher than linseed) to encourage long straight stems, receives maintenance dressings of P and K and often no nitrogen at all. The higher plant densities may encourage more disease, especially botrytis, so fungicides may have to be considered. Weed control is the same as for linseed.

Crops grown for long fibre are pulled out of the ground by special machines. This gives the maximum length of fibre which goes down into the main root. The crop is then retted like hemp, although in Europe there are factories which can artificially ret in tanks. Short-fibre crops can be cut and baled after retting, similar to the hemp crop.

16.10   Pharmaceuticals, neutraceuticals, essential oils and cosmetics

These are small-scale crops usually grown on contract with a buyer such as Springdale Crop Synergies, Norfolk Essential Oils or Statfold Seed Oil, and the acreages are often tiny. Commercial flower and fruit production is outside the scope of this book, and of course, many oils are imported from the tropics or sub-tropics, but the main potential crops in the UK are species such as lavender, mint, chamomile and a number of other herbs. Pharmaceuticals and neutraceuticals (body and skin care products) are mostly based on oilseed crops such as borage, evening primrose, Gold of Pleasure (Camelina), Echium, Calendula and Lunaria.

Many of these crops are very low yielding, e.g. borage averages around 0.4 t/ha and requires a high level of attention to detail for their agronomy, so farmers will expect good returns based on a realistic contract price. Unless margins match those of the mainstream arable crops, growers will not bother with these lesser-known ones.

16.11   Carbohydrate crops

The main product from these crops is industrial starch used in many different ways. These include packaging, coatings for textiles and paper, adhesives, explosives, cosmetics, plastics and, of course, bioethanol which has already been covered elsewhere in this chapter.

All crops have a carbohydrate content but the ones of most interest for commercially extractable starch or other products include maize, wheat, sugar beet, potatoes and barley. There may be the potential to develop more esoteric crops such as quinoa, sorghum and chicory for starch production.

The agronomy of these crops will be mostly the same as if they were being grown for food production.

16.12   Sources of further information and advice

Defra and BABFO *Guide to Renewable Biofuels for Transport*. Available at: http://adlib. everysite.co.uk/resources/000/015/083/Biodiesel_and_bioethanol_the_facts.pdf

Defra, *Growing Short Rotation Coppice*, 2004. Available at: http://www.naturalengland. org.uk/Images/short-rotation-coppice_tcm6-4262.pdf

Energy Crops Scheme: Establishment Grants Handbook: 3rd Edition (version 3.1) January 2013 (NE125), Natural England. Available at: http://publications.naturalengland.org.uk/publication/46003?category=43017

HGCA, *Industrial Uses of Arable Crops*. Available at: http://www.hgca.com/publications/documents/Industrialusesarablecrops1.pdf

The Bioeconomy Consultants, formerly the National Non-Food Crops Centre (NNFCC). Available at: http://www.nnfcc.co.uk/

Interactive European Network for Industrial Crops and their Applications (IENICA). Available at: http://www.ienica.net/

17

Fresh produce crops

DOI: 10.1533/9781782423928.3.396

Abstract: The fresh produce sector is technically challenging: the range of crops is large and different production systems are employed for many of them. This chapter gives an overview of the production of field-grown fresh produce crops and includes root, bulb, leaf, legume and soft fruit examples. The general agronomy of each crop is presented with details on production from soil preparation, nutrition and planting through to harvesting and postharvest management.

Key words: fresh produce, field-grown, vegetables, soft fruit.

17.1   Growing fresh produce crops

Fresh produce is a term used to describe edible horticulture crops. The fresh produce sector is technically challenging; the range of crops is large and different production systems are employed for many of them. The crops are harvested fresh and, in contrast to arable crops, may be harvested regularly for a number of months. Some of the key differences between fresh produce and broad-acre arable crops are:

- Very diverse range of crops.
- High value crops.
- Intensive crop inputs.
- Labour use is high.
- Mainly marketed fresh (some crops can be stored for months).
- Cold chain is important in distribution.
- Close interactions along supply chain.

There have been major technological advances and changes in the fresh produce sector that have affected how fruit and vegetables are grown in the UK. Many large farms now specialise in a small range of fresh produce crops, growing produce both in the UK and abroad to supply multiple retailers for 12 months of the year. Crop protocols and codes of practice have been drawn up between growers and retailers to promote good practice, enable crop traceability and give assurance to the consumer. Over 80% of produce is now sold through multiple retailers and an increasing proportion of that is sold ready prepared. The demand for organic produce has increased but sales are sensitive to the general economy.

It is very important for the grower to know the requirements of the market, to produce only what is needed when it is needed, in suitable quality and quantity and following crop protocols. It is usually advisable to arrange contracts with buyers, such as processors, freezers, pre-packers, multiple retailers or cooperatives in advance of growing the crop. Alternatively, all or part of the produce may be sold direct to the public at the farm gate, through pick-your-own or door-to-door delivery such as the 'box scheme' used by some organic growers.

There have been many developments in the farm-scale production of fresh produce that have resulted in a high level of mechanisation from seed to supermarket shelf. Seed choice and sowing and planting systems combine to result in a uniform crop that comes to maturity evenly and which often can be harvested at a single pass. Husbandry techniques aim to optimise crop growth while reducing inputs to the benefit of the farmer, the environment and the consumer. Harvesting and postharvest methods aim to transfer the crop from field to customer as efficiently as possible and in the best condition. These developments may be summarised as follows:

- F1 hybrids give uniformity and vigour.
- Seed is available graded or pelleted to help with drilling, and treated or primed to aid and speed germination.
- In a number of crops young plants can be raised under protection in small modules or larger blocks allowing efficient transplanting and promoting quick establishment in the field.
- Bed systems make best use of the available land and give good access to the crop for all tractor operations.
- Plastic mulches and fleece crop covers extend the growing season in the field.
- Integrated crop and pest management reduces inputs by applying fertiliser and water only as needed for crop growth and targeting pesticides at those weeds, pests and diseases actually present in the crop.
- Specialised harvesting rigs ensure fast and efficient removal from the field; mobile packhouses and modern grading, washing and packaging facilities ensure the crop is then transferred from field on to distribution quickly while maintaining quality.

Limiting factors on the individual farm may be:

- Need for specialist machinery.
- Unsuitable soils, poor drainage or lack of irrigation.
- Difficulty obtaining suitable casual staff and supervision.
- Lack of a suitable water supply for irrigation.

17.1.1   Value and volume

The UK is not self-sufficient in fresh produce. The proportion of crops marketed in the UK has generally declined in recent years due to increased imports and in 2011 UK-produced crops accounted for 58% of vegetables and only 12% of fruit marketed in the UK (Table 17.1). Fresh produce crops are high value and UK fresh produce production is similar in value to the UK wheat crop. Defra produces a summary of basic horticultural statistics each year where the area, volume and value of the main crops are summarised.

17.1.2   Assurance and food safety

A major challenge for growers of some fresh produce crops such as leafy salads or soft fruit is that they may be consumed fresh without being cooked. Any microbial contamination on the crop at harvest may be eaten by a consumer and there have been cases of food-borne illness linked to contaminated fresh produce crops around the world. The only way to prevent this happening is to prevent the contamination of the crop. The main risks to the crop come from the use of manures as soil conditioners, contamination by livestock or wild animal manure, contaminated water sources, and equipment and worker hygiene. Growers are required by many customers to follow strict standards of production to reduce the risks of contamination. The most widely followed scheme in the UK is the Red Tractor Fresh Produce Scheme which has general standards for all crops and specific crop protocols.

17.1.3   Use of integrated pest management (IPM)

Many fresh produce crops are classified as minor or specialist crops. The recent harmonisation of EU pesticide registration has led to the revoking of approval for

Table 17.1   Area, tonnage, proportion of the UK market supplied and value of UK production of fruit and vegetables in 2011

	Planted area (Ha)	Value of UK production (£1000)	UK supply to the market (%)	Production (1000 t)
Vegetables				
Field	120 219	909 630	–	–
Protected*	688	303 488	–	–
Total	120 907	1 213 119	58	2570
Fruit				
Field/orchard	29 157	598 146	–	–
Glasshouse†	180	39 311	–	–
Total	150 244	637 457	12	427

* Including glasshouse crops and mushrooms.
† Mainly soft fruit.
Source: Defra.

use of a significant number of pesticides used by fresh produce growers. As a consequence growers are increasingly relying on a combination of IPM techniques (see Chapter 7). For example, the loss of suitable herbicides has led to mechanical weeding/hoeing being used in combination with herbicides in many crops. Where the number of available insecticides has been reduced the repeated use of the remaining products needs to be avoided to prevent development of resistance in the pest. The alternatives to pesticides are limited; one option is to use netting/mesh during risk periods to minimise damage, for example, to control cabbage root fly in brassica crops.

17.2   Fresh peas

There are two main types of peas produced in the UK: *vining peas* are used for canning fresh (sometimes called 'garden' peas), freezing or drying and are normally grown under contract, or by cooperating groups. The contracting companies usually supply the variety they require and decide on times of sowing based on a 'heat unit' system. This is in order to spread the harvesting period to ensure a regular and constant supply to the processing factories. *Fresh peas (market pick or pulling peas)* make up ~5% of the UK crop and are grown to be sold fresh in the pod to be home-shelled or eaten whole (e.g. mange tout or sugar snap peas) or as shelled fresh peas.

Vining peas are best suited to the main arable areas and where there are local processing plants or markets (most pea growers are concentrated in East Anglia and Humberside), while most vining pea growers are members of cooperatives, often linked to a specific processing factory. The factory must ensure a continuous and reliable supply of produce so that it can operate efficiently. Growers must ensure, therefore, that the crop comes ready for harvesting over several weeks. This is achieved by: (a) growing different varieties with different maturity dates, or (b) staggering the drilling of the peas based on the calculation of accumulated heat units (AHUs). Thus, if the difference between one day and the next in East Anglia at harvest time is 11.5 AHUs, and the vining capacity of the cooperative is 40 hectares per day, each block of 40 ha must be drilled 11.5 AHUs apart back in the spring. This could mean delays in drilling of several days if the weather is cold. This should result in the shared harvesters moving around the pea crops just at the right stage of ripeness. If there is a delay because of weather or breakdown, then the peas can often get too tough for freezing or canning. They can then be allowed to grow on for harvesting dry.

Soils and climate
Peas suit deep, free-working loams. Pea roots are very sensitive to the physical condition of the soil and compaction and waterlogging are very damaging so light sands and heavy clays are not suitable. Peas do not grow at temperatures below 4.4 °C so overwintered fresh peas are only suited to frost-free coastal areas in the south of England.

Cultivations
Peas are sensitive to compaction and an adequate tilth is required for planting. Land can be autumn ploughed and weathered over winter or spring ploughed on lighter soils.

Drilling/planting and timings
Vining peas are drilled from February to May. Sequential plantings give a steady supply of crops over the harvest window. Fresh peas can be autumn planted and overwintered for an early crop or spring planted for later harvests.

Rotation
Peas can be a valuable break crop in cereal-based rotations. Peas are N-fixers and the crop debris contains high levels of N. Legumes (peas and beans) should be grown no more than one year in five to prevent the build-up of pests and diseases.

Pests and diseases
Peas can be attacked by a range of pests including pea and bean weevil, pea cyst nematode, aphids, moths, thrips, pea midge, silver Y moth and slugs. Damping off, foot and root rot, leaf and pod spot, downy mildew, powdery mildew, botrytis, virus and bacterial blight can affect the pea crop and may need controlling.

Weed control
Peas do not compete well with weeds and efficient weed control is essential to avoid yield loss. Perennial weeds should be dealt with before planting and thistles, oilseed rape and groundkeeper potatoes can be a particular problem. Nettles and thistles can interfere with hand picking and, in machine-harvested crops, weeds can lead to crop rejection from the processor. Inter-row cultivations may be effective when the peas are between the second and fifth node growth stage.

Nutrition
Nitrogen-fixing *Rhizobia* on the roots provide the crop with nitrogen. Fertiliser N or FYM has no benefit and can delay or suppress nodulation. For a soil with an index of 2, the fertiliser recommendations for fresh peas would be 85 kg/ha of P_2O_5 and 40–90 kg/ha of K_2O and for vining peas would be 40/ha of P_2O_5 and 20–40 kg/ha of K_2O.

Irrigation
Peas are sensitive to soil moisture and irrigation may be beneficial from the beginning of flowering and during pod swelling.

Harvest and storage
Vining peas are harvested in June and July and are ready for harvesting when the crop is just starting to lose its green colour and the peas are still soft. The firmness

of the peas is tested daily (near harvest time) with a tenderometer. The reading for freezing peas is about 100 and, for canning peas at 120. The crop must be cut as soon as possible. The pods are harvested using self-propelled vining machines that separate the peas from the pods. The shelled peas are rapidly transported to the processing plant for freezing or canning.

Fresh peas are harvested by removing the pods when the peas are at a required maturity. The peas can be either hand-harvested or mechanically harvested. The use of overwintered crops, crop covers and variety choice means that fresh peas can be harvested from May to September.

Fresh peas do not store well and deteriorate quickly in warm temperatures with much of their sugar turning to starch within a few hours. Peas should be cooled and distributed rapidly to retailers using the cold chain.

17.3   Broad beans

Broad beans are grown for freezing, canning and for the fresh market. There are two groups of beans based on flower colour, which is important for the processed crops, particularly canned broad beans. Coloured flower varieties give a pink-brown tinge to the beans and water during the canning process. White flowered varieties are used for both canning and freezing, varieties with coloured flowers are only used for freezing.

Soils and climate
Broad beans have a strong taproot and can cope with a wide range of soil types but require a pH of at least 6.0 to 6.5. If the pH is below 5.5 growth will suffer and an application of lime will be needed.

Drilling/planting and timings
The large seeds are likely to be damaged in some drills. Vacuum drills and some types of belt-fed drills are more suitable. For small areas the crops may be hand-sown.

Retail crops can be sown November/December and overwintered to be harvested late May/June. The main crops are sown from February to May, for harvest from June to August.

Processing crops can be sown from early February to May as directed by the processors. Successional sowings give a regular supply of beans of the correct maturity over the harvest window from mid July to the end of August.

The normal row width is 45 cm with 13 cm spacing along the row.

Rotation
Broad beans can be a valuable break crop in cereal-based rotations. Beans are N-fixers and the crop debris contains high levels of N. Legumes (peas and beans) should be grown no more than one year in five to prevent the build-up of pests and diseases.

Pests and diseases
Pea and bean weevil, stem and bulb nematode, black bean aphids, bruchid beetle, chocolate spot, foot rot and rust are usually the main problems in broad beans.

Weed control
Broad beans do not compete well with weeds and efficient weed control is essential to avoid yield loss. Perennial weeds should be dealt with before planting and wild oats, thistles and groundkeeper potatoes can be a particular problem. Inter-row cultivations may be used if the crop is grown in wide rows.

Nutrition
Nitrogen-fixing *Rhizobia* on the roots provide the crop with nitrogen. Fertiliser N or FYM has no benefit and can delay or suppress nodulation. For a soil with an index of 2, the fertiliser recommendations for fresh broad beans would be 100 kg/ha of P_2O_5 and 50–100 kg/ha of K_2O. The recommendation for processed broad beans would be 40 kg/ha of P_2O_5 and 20–40 kg/ha of K_2O.

Irrigation
The crop is sensitive to soil moisture deficit and higher yields are obtained from irrigating at early flowering and pod fill.

Harvest and storage
Hand-harvested crops are normally picked two or three times with sequential plantings picked over from May to August. Machine-harvested crops for processing are harvested from mid-July to the end of August to fit in with pea harvesting. Processed crops are harvested according to tenderometer readings (TR) and crops for freezing have a TR of 110–140, for canning the TR should read 130 and over. Machine-harvested crops can be cut and left to wilt before using a mobile viner. Single pass combine harvesters are also available.

Broad beans deteriorate quickly at warm temperatures and pre-pack/retail crops should be cooled rapidly after harvest, while crops destined for canning or freezing should be transported quickly to the processor to prevent deterioration.

17.4   Green beans

Green beans (Dwarf French beans) are grown mainly for processing but there is a small retail market for fresh green beans. Green beans are usually classified according to pod width: Fine beans (8.0–9.0 mm pod width); Medium fine beans (9.0–10.5 mm pod width); and Medium fine/large beans (>10.5 mm pod width).

There are many varieties on the market and, where growing for processors, advice should be sought from them regarding their requirements. Consideration should be given to yield, quality and ease of harvesting, together with the growing environment on each individual farm.

Soils and climate

Green beans need to be grown in loams and the lighter types of soil. It is a late-sown, short-season crop and so requires a deep, free-draining, moisture retentive soil. Soils that cap, and those of a high organic matter content (peat soils), which will produce excessive vegetative growth should be avoided. A site offering shelter and a southerly aspect is ideal. Most of the crop is grown in the eastern counties south of the Wash.

Cultivations

The soil surface should be prepared so that following drilling it will be as level as possible to prevent problems when machine harvesting.

Drilling/planting and timings

Green beans are sown between mid-May and late-June, as early drillings can be damaged by late frosts. The seed testa is easily damaged and seed must be handled with care. Pneumatic precision or belt feed-drills are used with treated seed. The final plant population for processing is 30–40 plants/m^2 and for the fresh market 20–25 plants/m^2. The normal row width for mechanically harvested crops is 30–40 cm with 6–7 cm spacing along the row.

Rotation

Legumes (peas and beans) should be grown no more than one year in five to prevent the build-up of pests and diseases.

Pests and diseases

Green beans are subject to attack by a number of pests especially caterpillars of Silver Y moth, bean seed flies, aphids, slugs and cutworms. Halo blight, botrytis, rust and sclerotinia can be a problem in some seasons.

Weed control

Green beans do not compete well with weeds and efficient weed control is essential to avoid difficulties with machine harvesting. Perennial weeds should be dealt with before planting and thistles and groundkeeper potatoes can be a particular problem. If weeds are not controlled, the crop may be unsaleable. Inter-row cultivators/weeders can lead to irregular soil conditions and cause problems with machine harvesting. Weeds are controlled mainly through herbicides.

Nutrition

Green beans respond very well to nitrogen. In contrast to broad beans, the nitrogen fixing, *Rhizobium* bacteria that nodulate green beans are not found in many UK soils so the seed/soil can be either inoculated when drilling or nitrogen fertiliser can be applied. For a soil with an index of 2, the fertiliser recommendations for green beans would be 120 kg/ha N, 100 kg/ha of P$_2$O$_5$ and 50–100 kg/ha of K$_2$O. To prevent fertiliser leaching and reduced germination a maximum of 100 kg N/ha should be applied at drilling. The remainder should be applied as required by the crop.

Irrigation

Green bean yield responds well to irrigation during early pod development but little benefit is observed when irrigating before the flowering stage. Overhead irrigation should be avoided when flowers are present as the wet petals stick to developing pods leading to fungal infections of the pods.

Harvest and storage

Green beans are harvested from July to October. Processing crops are harvested using one-pass mechanical harvesters. Retail crops can be machine or hand harvested. The harvested bean pods must be whole, undamaged, separated (not in clusters) and free of stems, leaves, soil and stones. As the crop matures, the pods lengthen rapidly and they then enlarge as the seeds develop. Maturity should therefore be judged according to pod width and not length. Flat-podded beans must be harvested before the development of seeds becomes obvious externally. The beans should not be allowed to over-mature or quality will be reduced. This is assessed by taking a random sample of 10 plants, taking the most mature seeds from the most mature pods on each plant and measuring the length of 10 seeds. As a guide, for all beans the length of one seed should be no more than the pod width.

Green beans deteriorate quickly at warm temperatures and pre-pack/retail crops should be cooled rapidly after harvest, crops destined for canning or freezing should be transported quickly to the processor to prevent deterioration.

17.5   Lettuce

Lettuces are grown for sale as a whole head or for further processing for foods such as sandwiches and salads. Most types are field grown with the main types being Iceberg/Crisphead, Cos/Romaine, Speciality/Continental and Little Gem. The traditional flat or butterhead lettuces are mainly grown under protection.

Soils and climate

Lettuce can be grown on a wide variety of soil types. An ideal soil has good water retention properties but is not too heavy as this may limit field access by machinery for planting, crop applications or harvest if wet. Very light soils may require frequent irrigation and additional fertiliser applications.

Cultivations

On heavy land it may be necessary to subsoil prior to ploughing and power harrowing before transplanting. Lettuce can be grown in beds or on the flat.

Drilling/planting and timings

All commercial lettuce in the UK is grown from transplants (Fig. 17.1) although it is possible to grow lettuce from drilled seed. Lettuce is susceptible to frost and can be grown without protection from the last frost in the spring to the first frost in the autumn. Early crops may be transplanted using blocks (Fig. 17.2) from February

Fig. 17.1   Lettuce block.

Fig. 17.2   Transplanting lettuces.

and will be covered in fleece or polythene for the first 2–4 weeks to reduce the risk of frost damage. Lettuce takes 6–8 weeks to mature depending on lettuce type and time of year with summer transplantings maturing most quickly. Plants are spaced at approximately 30–35 cm but closer plant spacing may be used for Little Gem.

Rotation
Due to the short production time, lettuce can have two or three crops grown in one field in a season. One year breaks are possible but longer breaks are beneficial.

Pests and diseases
Lettuce are attractive to a wide range of pests especially aphids, caterpillars and slugs. The diseases downy mildew, botrytis and sclerotinia are the main problems when growing the crop.

Weed control
The reduced number of active ingredients means that the control of weeds in lettuce has become difficult. Mechanical or hand weeding may be the only option where sites are badly affected with weeds. As a consequence there is an increase in the use of vision-guided mechanical hoes which can remove weeds between plants within the row.

Nutrition
For a soil with an index of 2, the fertiliser recommendations based on Iceberg lettuce would be 160 kg/ha N, 150 kg/ha of P_2O_5 and 100–150 kg/ha of K_2O. Lettuce are not responsive to high levels of N and excess levels may reduce shelf life. Due to potential risk to health from having too much nitrate in a diet, there is a maximum level of nitrate allowed in lettuce leaves governed by EU legislation. The maximum value depends on season, method of production and lettuce type. Growers need to manage N to avoid excess applications. Trace element deficiencies should be identified by tissue analysis.

Irrigation
Lettuce has a very high water content (>95%) and irrigation has an important role in crop production. Sufficient water is needed after transplanting to help rapid crop establishment. Subsequent irrigation requirements will vary according to soil type, crop growth stage and rainfall. Growers can use irrigation-scheduling decision support systems to decide when and how much water to apply. Care is needed not to over irrigate crops, due to nutrient leaching and reduced postharvest quality. Under-irrigation will lead to reduced yield and may increase tip scorch in some crops. The water source used needs to be free of human pathogens that would contaminate the crop.

Harvest and storage
Lettuce are harvested by hand. Many are harvested in field rigs (Fig. 17.3) that pack and label the product in the field. The heads are selected for size and cut

Fig. 17.3   Harvesting lettuces.

at the base of the plant. The outer leaves may be left on (as protection in transport for processing heads) or removed and the head wrapped in a plastic bag before labelling in the field and dispatching to the multiple retailer. Lettuces are rapidly cooled then stored and transported at low temperatures in refrigerated vehicles.

17.6   Baby leaves and herbs

Baby leaf crops are leaf crops grown at high seed rates and harvested with 3–5 true leaves. They are grown mainly for further processing and sold washed or unwashed in bags although baby leaves can be sold loose or bunched. The greatest volume crop is spinach, grown both for salads and cooking, but a wide range of other crop species and types are also grown including brassicas, such as mizuna or mustard; red and green lettuce, including green oak leaf and lollo rosso; and herbs such as flat leaf parsley and coriander.

Soils and climate
Baby leaf crops will grow on most soils but frequent access to the crop is needed by machinery. Light free-draining soils are best suited to intensive production but attention to nutrition is needed as light soils may have low nutrient retention.

Cultivations
Fine seedbeds are required for uniform emergence of baby leaf crops. This can be achieved using power harrows following discing. Baby leaves are commonly grown in beds to minimise compaction and aid harvesting.

Drilling/planting and timings
Baby leaf crops are grown at high seed rates and seed cost is a significant production cost. For example spinach will be drilled at 120–140 seeds/m^2. Crops develop quickly and can take 21–28 days from emergence to harvest. The size of the crop at harvest will dictate the market with crops grown for micro, baby and teen leaf specifications. Crops may be covered by fleece to hasten emergence at the start of the season.

Rotation
Ideally crops will be grown in a field one year in three. It is common for multiple crops to be grown in the same field within a year. Consideration should be taken of mixed fields, i.e. where leaves of different plant families such as brassicas and lettuces are being grown.

Pests and diseases
Baby leaves and herbs are attractive to a wide range of pests especially aphids, leaf miners, caterpillars and slugs. The contamination of the crop with insects is also a commercial issue. Seedling diseases can affect the crops such as *Pythium* and *Rhizoctonia* but the main disease is downy mildew which can lead to complete crop loss if untreated. Resistant lines are available for some of the larger volume crops like spinach but downy mildew evolves new strains rapidly and plant breeders are always playing 'catch up'.

Weed control
Due to the small plants and short production cycle soil sterilisation is sometimes carried out prior to the production of baby leaves. Sterilisation is usually carried out using metam-sodium or dazomet (Basamid) but steam sterilisation is also used. Some growers have also experimented with mulching to suppress weeds and this can work in some situations.

Nutrition
For a soil with an index of 2, the fertiliser recommendations based on spinach would be 75 kg/ha N, 150 kg/ha of P_2O_5 and 125–175 kg/ha of K_2O. Baby leaf crops will require less nitrogen than a mature full leaf crop but nitrogen needs to be adequate to avoid pale leaves. Due to potential risk to health from having too much nitrate in a diet, there is a maximum level of nitrate allowed in baby leaf leaves including spinach, lettuce and rocket, governed by EU legislation. The maximum value depends on season, method of production and crop. Growers need to manage N to avoid excess applications. Trace element deficiencies should be identified by tissue analysis.

Irrigation

Water is very important for germination and crop uniformity. Micro-sprinklers are commonly used as they can apply small amounts of water frequently. Large droplet sizes can damage the foliage so fine mists are often used. The short crop cycle means that baby leaves and herbs have shallow roots 20–50 cm deep. The water source used needs to be free of human pathogens that would contaminate the crop.

Harvest and storage

Baby leaves are rarely hand cut (except for bunched herbs) and machine harvesting is common. The crop is cut 1–3 cm above the soil surface using a trailed or self-propelled harvester. It is possible to reharvest a crop more than once following regrowth but this gives a poorer quality crop with variable leaf sizes. The leaves are then cooled rapidly and stored at high relative humidity often with misting or fogging.

17.7   Cabbages

Different types of cabbage are grown for year round availability. There are six types, defined by appearance and time of maturity.

1. *Spring greens* (Collards) are sold as loose leaf or open hearted heads. They are traditionally cut from overwintered crops from March to May but production is increasing through the year with winter crops grown on early sites in Cornwall and Kent.
2. *Early summer cabbage* (Primo) is sold as whole heads. They are produced from overwintered crops or spring planted crops and harvested from May.
3. *Late summer/autumn cabbage* is sold as whole heads. They are produced from late spring planted crops and harvested from July to October.
4. *Winter cabbage* (January King and green types) is sold as whole heads. They are produced from late spring planted crops and harvested from November to March.
5. *Winter white/red storage cabbage* is grown for retail as a whole head or for processing. Retail heads are approximately 0.5–1 kg whereas processing heads may be >3.5 kg. They are produced from late spring planted crops and harvested October onwards. The ability to store them for 6–9 months gives all year round availability.
6. *Savoy cabbage* is sold as whole heads. They are the most frost tolerant cabbage type with distinctive 'bubbled' leaves. They are produced from spring/summer planted crops and harvested from June to April.

Soils and climate

Cabbages can be grown throughout the UK but problems with ring spot are greater in areas with heavier rainfall. Cabbage can be grown on a wide range of soil types and grow well on moisture retentive soils. Lighter, sandier soil types may require irrigation. Heavy soils should be avoided with winter cabbage where machinery access will be needed in the winter months. Care is needed in light fenland soils where direct drilling can be affected by loss of soil through wind erosion. A soil pH level of 7.0 to 7.5 is required, particularly where club root may be a problem. Lime should be applied well before planting/drilling. Capping soils should be avoided where crops are direct drilled.

Cultivations
Cabbages have moderate to deep roots. A firm soil with a good tilth is required. Compacted soils may require sub-soiling before ploughing and harrowing/discing. Cabbages can be grown in beds or on the flat with tramlines.

Drilling/planting and timings
Cabbages are predominantly produced from transplanted modules. Spring greens are the exception, being mainly direct drilled at high densities. Transplanting gives a better crop production programme and more even crop maturity. Where whole head crops are direct drilled, two seeds are sown together in each location at the final spacing for the crop, and the weaker plant removed by hoeing.

- *Spring greens* are mainly drilled direct in August, transplanted in September or direct drilled in April to June at its final crop spacing 38 × 13 cm.
- *Early summer cabbage* can be autumn drilled and overwintered, drilled in February or transplanted April/May at a final spacing of 46 × 23 cm for harvesting from May. Plastic covers can be used on early crops which advance harvests by up to two weeks.
- *Summer cabbage* is drilled in March/April or transplanted May/June, at a spacing of 46 × 23–46 cm, for harvesting July to October.
- *Winter cabbage* varieties are drilled/transplanted in May/early June for harvesting from mid-November onwards. The target spacing is 46 × 23–46 cm.
- *Winter white/red storage cabbage* is drilled April/May or transplanted May/June and harvested from October onwards. The target spacing is 46 × 36 cm.
- *Savoy cabbage* is drilled March to May, or transplanted April to July and harvested September onwards. The target spacing is 46 × 30–46 cm.

Rotation
Cabbage can fit well into a rotation leaving a high amount of residual nitrogen in the soil as the waste leaves are incorporated into the soil. Where club root is a problem, particularly on acid soils, brassicas should be grown no more than one year in four or five with a planned liming policy before planting. It is possible to grow brassicas continuously on a high pH soil.

Pests and diseases
Cabbage root fly can be a problem, particularly with summer and winter cabbage, and treatments will be required from April onwards. Aphids, caterpillars, cutworms, pollen beetle, cabbage stem weevil and slugs all attack cabbages. Wood pigeon is a particular problem on overwintered crops. Damping off and

wirestem can be a problem at the seedling stage. Club root, downy mildew, dark leaf spot, ring spot, white blister and spear rot can all be a problem during the growing season.

Weed control

Cabbages can compete well with weeds due to their large leaves but established perennial weeds should be controlled prior to transplanting or drilling. With the limited number of approved herbicides available many growers rely on mechanical methods of weed control (Fig. 17.4). Oilseed rape volunteers and potato groundkeepers can cause a problem in some crops.

Nutrition

For a soil with an index of 2, the fertiliser recommendations for most cabbage would be 180–260 kg/ha N, 100 kg/ha of P_2O_5 and 150–200 kg/ha of K_2O. Storage cabbage would require 280 kg/ha N and spring greens between 180 and 270 kg/ha N but the same P and K recommendations would apply. Cabbage growing over the winter months may require split application of nitrogen fertiliser with no more than 100 kg/ha N at transplanting and the remainder to support later growth.

Irrigation

Irrigation can improve evenness of maturity but it is only economical to irrigate summer crops particularly on lighter soil types. Irrigation can be applied by overhead booms or trickle tapes.

Fig. 17.4   Inter-row weeding cabbages (Garford Farm Machinery Ltd).

Fig. 17.5   Cabbage harvesting rig.

Harvest and storage
Mechanical harvesting is becoming available for processing crops but nearly all
other crops are hand harvested (Fig. 17.5). Cabbages can be packed on rigs or
harvested in to crates. Loose outer leaves are stripped back and the head cut
leaving approximately 10–15 mm of stem. Variable maturity of cabbage means
that fields may be harvested a number of times with usually three or more passes.
In the summer the field may be harvested every three days and in autumn/winter
every seven days. Cabbages can deteriorate in warm conditions and should be
cooled rapidly after harvest. Long-term storage requires humidification equipment.

17.8   Broccoli

Broccoli (also known as Calabrese) is grown for freezing, processing and retail
markets. The harvested head (spear) consists of closed flower buds which will
open if the crop is left in the field too long after maturity. Green broccoli is the
main type with a single terminal head. Secondary spears are usually more
succulent and a there are number of sprouting broccolis where secondary spears
are harvested including 'Tenderstem' types and Purple Sprouting Broccoli.

Soils and climate
Broccoli is grown around the UK from the East of Scotland to the SW of England
and can be grown on a wide range of soil types and grows well on moisture

retentive soils. Lighter, sandier soil types may require irrigation. Heavy soils should be avoided with later crops where machinery access will be needed in wet conditions. A soil pH level of 7.0 to 7.5 is required, particularly where club root may be a problem. Lime should be applied well before planting/drilling. Capping soils should be avoided where crops are direct drilled.

Cultivations
Broccoli does not root deeply. A firm soil with a good tilth is required. Compacted soils may require sub-soiling before ploughing and harrowing/discing. Broccoli can be grown in beds or on the flat with tramlines.

Drilling/planting and timings
Broccoli is planted from March to July and harvested from June to November/December and is produced predominantly from transplanted modules. Transplanting gives a better crop production programme and more even crop maturity. Transplants are planted at spacings of 60×40 cm. Direct drilling is more common in Scotland. High-density crops can be sown direct to a stand. Modules are commonly raised under glass at commercial propagators from September to June and are ready for planting from March. Early crops are produced using plastic or fleece crop covers and can bring crops to maturity two weeks earlier. Use of covers needs managing as high temperatures under the covers can cause physiological damage and disease problems may develop in humid conditions. Crop covers need to be removed when heads are 10–15 mm in diameter, 4–6 weeks before harvest.

Rotation
Where club root is a problem, particularly on acid soils, brassicas should be grown no more than one year in four or five with a planned liming policy before planting. It is possible to grow brassicas continuously on a high pH soil.

Pests and diseases
Cabbage root fly can be a problem and treatments will be required from April onwards. Aphids, caterpillars, cutworms, pollen beetle, cabbage stem weevil and slugs all attack broccoli. Damping off and wirestem can be a problem at the seedling stage. Club root, downy mildew, dark leaf spot, ring spot, white blister and spear rot can all be a problem in the field.

Weed control
Established perennial weeds should be controlled prior to transplanting or drilling. Oilseed rape volunteers and potato groundkeepers can cause a problem in some crops.

Nutrition
For a soil with an index of 2, the fertiliser recommendations for broccoli would be 165 kg/ha N, 100 kg/ha of P_2O_5 and 125–175 kg/ha of K_2O.

Irrigation

Irrigation is needed on light soils and often with later crops. Irrigation can damage the wax layers on the head increasing the risk of spear rot and should end twenty days before harvest.

Harvest and storage

Machine harvesters are being developed but currently crops are harvested with field rigs. The head is cut and trimmed by hand before packing into trays. Side shoots will be packed into field crates and taken to a packhouse for grading and final packing. Broccoli is cooled quickly after harvest to prevent deterioration. Care is needed as broccoli can be damaged at temperatures below 2 °C. Broccoli does not store well and is dispatched into the supply chain within a few days of harvest.

17.9   Cauliflowers

Cauliflower is grown for freezing, processing and retail markets. The harvested head is known as a curd. A creamy white curd is the most common cauliflower but green, purple and yellow colours are available along with baby and romanesco cauliflowers, the latter having a distinct geometric pattern to the curd.

Different types of cauliflower are grown for year round availability. There are four types, defined by time of maturity.

1. *Early summer cauliflower* is produced from overwintered or January sown modules for harvesting from late May to early July.
2. *Late summer and autumn cauliflower* is produced from modules transplanted from April to July and harvested from July to mid-November.
3. *Winter cauliflower* is transplanted in June and July and harvested from December through to early March. These cauliflowers are susceptible to frost and are grown on frost-free coastal land in Kent and the south-west of England and Wales.
4. *Winter hardy cauliflower* (spring heading) is transplanted in July, overwintered and harvested from March to June.

Soils and climate

Cauliflower can be grown on a wide range of soil types, growing well on moisture retentive soils. Lighter, sandier soil types may require irrigation. Heavy soils should be avoided with later crops where machinery access will be needed in wet conditions. A soil pH level of 7.0 to 7.5 is required, particularly where club root may be a problem and lime should be applied well before transplanting. Winter cauliflowers harvested from December to March are grown in frost-free coastal areas.

Cultivations

Cauliflower does not root deeply. A firm soil with a good tilth is required. Compacted soils may require sub-soiling before ploughing and harrowing/discing. Cauliflower can be grown in beds or on the flat with tramlines.

Drilling/planting and timings

The crop is mainly produced from 14 ml modules. Early summer production can be brought forward by using larger module/blocks giving a larger plant at transplanting or by using crop covers.

- *Early summer cauliflower* modules are either sown in October and overwintered in unheated glasshouses, or sown in January and grown in heated glasshouses. The modules are transplanted in March for harvesting from late May to early July. The target spacing is 50×50 cm
- *Late summer* modules are sown February to April, transplanted April to May at a spacing of 60×40 cm and harvested from late July to early September.
- *Autumn cauliflower* modules are sown during late April to mid-June, transplanted June to July at target spacing of 60×40 cm and harvested from mid-September to mid-November (or the first frost).
- *Winter cauliflower* modules are sown May, transplanted June to July at a spacing of 60×50 cm and harvested from December to early March.
- *Winter hardy cauliflower* modules are sown in June, transplanted July at a spacing of 60×50 cm and harvested from early March until early June.

Use of covers needs managing as high temperatures under the covers can cause physiological damage and disease problems may develop in the humid conditions. Crop covers need to be removed when heads are 10–15 mm in diameter.

Rotation

Where club root is a problem, particularly on acid soils, brassicas should be grown no more than one year in four or five with a planned liming policy before planting. It is possible to grow brassicas continuously on a high pH soil.

Pests and diseases

Cabbage root fly can be a problem and treatments will be required from April onwards. Aphids, caterpillars, cutworms, pollen beetle, cabbage stem weevil and slugs all attack cauliflower. Damping off and wirestem can be a problem in propagation. Club root, downy mildew, dark leaf spot, ring spot, white blister, bacterial soft rot and black rot can all be a problem in the field.

Weed control

Established perennial weeds should be controlled prior to transplanting or drilling. With the limited availability of herbicides, weed control programmes usually use an approved herbicide followed by inter-row cultivations. Oil seed rape volunteers and potato groundkeepers can cause a problem in some crops.

Nutrition

For a soil with an index of 2, the fertiliser recommendations for cauliflower would be 235 kg/ha N, 100 kg/ha of P_2O_5 and 125–175 kg/ha of K_2O. The winter hardy types growing over the winter months may require split application of nitrogen fertiliser with no more than 100 kg/ha N at transplanting and the remainder to support later growth.

Irrigation
Irrigation can improve evenness of maturity and it is important to irrigate autumn crops, transplanted in June and July, particularly on lighter soil types. Irrigation can be applied by overhead booms or trickle tapes.

Harvest and storage
The head is cut and trimmed by hand before packing into trays. The crop will be cut a number of times taking heads at the correct size for specifications but leaving the smaller heads to grow on. Cauliflower is cooled quickly after harvest to prevent deterioration. Care is needed as the heads can be damaged at temperatures below 2 °C. Cauliflower does not store well and is dispatched into the supply chain within a few days of harvest.

17.10 Brussels sprouts

Brussels sprouts are the axillary buds growing in the leaf axils along the stem. They are grown for the fresh market and quick freezing, they can be sold loose, pre-packed or on the stem and the tops can also be sold rather like a small cabbage. Most varieties are now F1 hybrids offering a more uniform plant with an even distribution of buttons along the stem length and evenness of maturity (ideal for machine harvesting), compared with traditional open pollinated types. By planting 6–7 different varieties which mature at different rates Brussels sprouts can be harvested from August to April.

Soils and climate
Brussels sprouts can be grown on a wide range of soil types and grows well on moisture retentive soils. Lighter, sandier soil types may require irrigation. Heavy soils should be avoided with later crops where machinery access will be needed in wet conditions. A soil pH level of 7.0 to 7.5 is required, particularly where club root may be a problem. Lime should be applied well before planting/drilling.

Cultivations
A firm soil with a good tilth is required. Compacted soils may require sub-soiling before ploughing and harrowing/discing.

Drilling/planting and timings
Processed (freezing) crops may be direct drilled from March to April at close spacings. Transplanting of modules starts in April, until the first week in June at a target spacing of 60 × 38 cm. Later planting will lead to reduced yield even with late maturing varieties. Early varieties mature in 5–6 months, whereas late varieties mature in 10–11 months.

Rotation

The crop fits well into a cereal rotation on land that is clean and fertile. The earliest crops are cleared by the end of September and so it is possible to follow with winter wheat. Where club root is a problem, particularly on acid soils, brassicas should be grown no more than one year in four or five with a planned liming policy before planting. It is possible to grow brassicas continuously on a high pH soil.

Pests and diseases

Cabbage root fly can be a problem and treatments will be required from April onwards. Aphids, caterpillars, cutworms, pollen beetle, cabbage stem weevil and slugs all attack Brussels sprouts. Damping off and wirestem can be a problem at the seedling stage. Club root, downy mildew, powdery mildew, dark and light leaf spot, ring spot, white blister and canker can all be a problem in the field.

Weed control

Established perennial weeds should be controlled prior to transplanting or drilling. Oilseed rape volunteers and potato groundkeepers can cause a problem in some crops.

Nutrition

For a soil with an index of 2, the fertiliser recommendations for Brussels sprouts would be 270 kg/ha N, 100 kg/ha of P_2O_5 and 150–200 kg/ha of K_2O. Nitrogen top-dressing may be required when recommended rates are high. The top-dressing is generally applied within two months of the base-dressing application.

Irrigation

Irrigation may be applied before drilling or after transplanting to aid establishment. Irrigating early and mid-season sprouts may increase yield and quality.

Harvest and storage

In earlier-maturing varieties due to be single-harvested in August and early September the crop is often 'stopped' by removing the growing point (or terminal bud) from the plant, or by destroying the bud with a sharp tap using a rubber hammer. This removes apical dominance and stimulates sprout growth producing a 5–10% higher yield. Some varieties tend to be more cylindrical in set and stopping is not needed. Stopping may be done 4–10 weeks before harvest when half of the sprouts are 12 mm in diameter.

The tops of the plant, which resembles a small open hearted cabbage can be removed and sold in October and November. The major part of the crop is machine harvested, usually beginning in August and extending until March, according to the variety and season. Occasionally, the earliest maturing varieties are picked over by hand before the whole plant is cut for stripping. Where hand harvested, the crop is usually picked over three or four times during the season. In warm conditions sprouts can become discoloured and deteriorate and need to be cooled rapidly after harvest.

17.11   Bulb onions

Bulb onions are grown widely and fit well into arable rotations. The dominant type of onion grown is a Rijnsburger type. These are globe-shaped bulbs needing long day lengths to initiate bulbing. Brown onions are the most common type and are grown for cooking and smaller sizes are used for pickling. Red onions are increasing in popularity.

Soils and climate
Bulb onions can be grown on a wide range of mineral and peat soils provided that they are well drained, have a good available-water capacity, a pH of 6.5 or more and can provide a good seedbed. Bulb onions can be grown in many parts of the UK, but do best in the eastern and south-eastern counties where it is often drier at harvesting time, which helps promote skin quality.

Cultivations
Onions develop best in loose, well-structured soil. Ideally the land should be deep ploughed and disced before planting. Capping soils should be avoided. In some fields destoning may be needed to remove large stones that could affect onion development and harvesting.

Drilling/planting and timings
Onions can be grown from seed or sets. Onion seed is prilled to aid precision drilling. Sets are produced from high-density crops producing small 'bulbs' 1–3 cm in size and stored over winter at moderate temperatures (10–15 °C) to prevent bolting. Rijnsburger types initiate bulbing in mid-July regardless of planting date but larger plants in spring will be ready to harvest earlier in the summer.

Onions can be planted in autumn and over-wintered for an earlier harvest. Seed can be direct-drilled in August for harvest the following June/July or sets planted in October for harvest the following June/July. The majority of the UK crop is spring planted; some are direct drilled in February–March to harvest late August/September; and a significant proportion is grown from sets planted in January–March and harvested in late July/August. Onions can also be grown from multi-seeded modules propagated under glass in February for planting in April, but this is no longer common.

Onions are planted by precision drill in shallow double rows, 1–2 cm deep. A bed system is used with the seed spacing dictating bulb size. A target of 5–7 cm bulbs would require 50–100 plants/m^2. Sets are planted at similar spacings using specialist planters.

Rotation

Onions (and other alliums) should not be grown more often than one year in four, and ideally one year in six in the same field, because of the risk of nematode, onion fly and white rot disease.

Pests and diseases

The following can affect bulb onions: leaf blotch, downy mildew, onion white rot, neck rot, onion fly, cutworms, bean seed fly, thrips and bulb and stem nematode.

Weed control

Perennial weeds should, wherever possible, be controlled in the previous year. Bulb onions are not very competitive and grass weeds can be a problem.

Nutrition

Onions are responsive to P and K but response to N is poor. High K can increase keeping quality but excess N reduces keeping quality with soft tissues prone to rots. For a soil with an index of 2, the fertiliser recommendations for onions would be 110 kg/ha N, 100 kg/ha of P_2O_5 and 125–175 kg/ha of K_2O.

Irrigation

Onions have a shallow, poorly-developed root system with most root activity in the top 40 cm of soil. Adequate soil moisture is important for growth of new adventitious roots at the base of the bulb. An even supply of water to the bulbs is needed while they are developing. A period of drought followed by heavy applications of water can lead to the outer skins splitting, reducing quality. Late irrigation should also be avoided as it delays harvesting and gives rise to softer bulbs with more bacterial rots and irrigation is usually stopped when bulbs begin to mature approximately three weeks before harvest.

Harvest and storage

For long-term storage the crop is usually sprayed with maleic hydrazide, applied preharvest as a growth depressant at 10% leaf fall. Some stores are using low levels of ethylene to suppress sprouting. Bulb onions are ready for harvesting when about 80% of the tops have fallen over. This is usually from June to September, depending on the time and method of planting. If left later, yields might increase slightly, but the outer skins are more likely to crack which can lead to disease loss in store.

Onion tops are removed in the field with a flail harvester at harvest time and may be left in the field for 1–2 days to dry in windrows or immediately lifted with adapted potato harvesters and either box stored or bulk-loaded into store.

Preparation for storage is completed in three separate stages: drying, curing and cooling. Throughout the process, temperature and humidity control are crucial for the control of quality. In the initial drying stage, warm air (25–30 °C) is blown through ventilation ducts until the moisture is removed from the leaves and tops of the stems (approximately three days), to bring the onions to 'rustle dry' condition. In the curing stage, in the next 2–4 weeks, air at about 25 °C is blown intermittently through the stack and the relative humidity is reduced to 65%. This removes the moisture from the bulb neck tissue and the skins should now be a golden colour. Once the bulbs are fully dried, in the cooling stage, the stack temperature is lowered slowly to prevent sprouting.

Stores can be cooled by refrigeration systems or by ambient cooling – which is where fans automatically draw in outside air when temperature is 3 °C or more below crop temperature. Minimum temperature in ambient stores should be 5–8 °C. Refrigeration at 0–1 °C is essential for bulbs scheduled to be marketed from February to the end of May/early June. A controlled atmosphere is necessary for storage beyond this period. When the onions are removed from long-term storage, air of 2.5 °C greater than the stack temperature is blown through the stack to prevent condensation forming on the onions. Onions are graded for quality and size before sale and marketable yield varies with conditions and timings from 30 to 50 t/ha.

17.12 Leeks

Leeks are alliums but do not bulb and are less pungent than onions. The crop can grow well around the UK and is available from July to May. Leeks are sold with the leaves trimmed into an inverted V, straight or stripped back. Leeks sold as pre-pack are usually smaller than those sold loose. Leeks can be harvested from the field over winter as they are tolerant of frosts.

The varieties change over the season with more rapid maturing varieties providing the early crops. Varieties that can hold well in the field are suited to overwintering.

Soils and climate
Leeks can be grown on a wide range of soil types, but the most suitable are sandy loam to sandy clay loam, silts and some peat based soils. It is best to avoid very light soils as these can spoil the crop if soil gets blown down into the leaves. Heavy soils may restrict access to the field in winter.

Cultivations
The land should be ploughed, bed formed, destoned or declodded before planting. Capping soils should be avoided with drilled crops.

Drilling/planting and timings
Leeks can be raised from direct drilled seed or blocks/modules. Leeks are planted from January until July and harvested from July to May. The new season starts with spring planted transplants harvested in July and August. Spring drilled crops are then harvested from August to December. Late spring drilled crops are overwintered and harvested from January until May. In general, crops drilled from January to April give crops before Christmas and May drillings give crops

after Christmas. Early crops can be grown under plastic/fleece covers until mid-May to bring them forward two weeks.

Modules or blocks are machine planted and gapped up by hand. Blocks are bigger, giving an earlier crop but are more expensive. Seed is drilled using a precision drill in 4 rows across a bed 45 cm apart. Spacings between plants are usually 7.5–15 cm for modules and 4–7.5 cm for seed. Leeks are smaller at maturity at the higher density, being better suited to pre-pack crops.

Rotation

Leeks (and other alliums) should not be grown more often than one year in four, and ideally one year in six in the same field. Consideration should also be made of other non-allium crops in the rotation that are susceptible to bean seed fly, i.e peas, beans, cucurbits.

Pests and diseases

Cutworms, bean seed fly, onion fly, leek moth and thrips can be a problem in the crop. Rust, foot rot, leaf blotch, white tip and white rot are the main disease problems in leeks.

Weed control

Perennial weeds should, wherever possible, be controlled in the previous year. Leeks compete weakly with weeds. Too many weeds can reduce yields and cause difficulty at harvest, particularly when harvesting mechanically. Mechanical or hand weeding may be required where sites are badly affected with weeds.

Nutrition

Leeks are responsive to nitrogen but applications should match crop growth as over-application of nitrogen can result in very soft crops that are very prone to disease. High levels of N at planting can reduce germination. Where crops are overwintered a split application of nitrogen can be applied as a top dressing in spring. Top dressing of a growing crop can lead to granules lodging in the leaves and damaging them, so care is needed to avoid this. For a soil with an index of 2, the fertiliser recommendations for leeks would be 170 kg/ha N, 100 kg/ha of P_2O_5 and 125–175 kg/ha of K_2O.

Irrigation

Leeks require adequate soil moisture for yield and a lack of soil moisture during crop development can lead to bolting in the crop. Care is needed with maintaining adequate soil moisture following drilling as the seeds and seedlings are sensitive to dry conditions. Irrigating newly planted modules can help wash soil around the modules improving contact with the soil.

Harvest and storage

Leeks can be harvested by hand or machine. Specialist leek harvesters are available that undercut the leeks and trim the leaves before placing them in bulk boxes. Leeks for pre-pack can be manually trimmed and the outer leaves stripped back. Roots need trimming back but care is needed not to damage the base plate of the leek. The leeks should be cooled and held between 2° and 10 °C prior to despatch. Leeks should not be harvested when frozen as they can be damaged by handling. The gap in harvest in May and June can be supplied by stored leeks. These require refrigerated, controlled atmosphere stores.

17.13 Carrots

Carrot production for human consumption in this country is now a large and specialised enterprise for most growers. Carrots are grown for many different markets – fresh, frozen and processing – and quality requirements of the buyers can be exacting. Uniformity and freedom from damage and disease are very important. Carrot out-grades and rejects may be fed to livestock.

There are six main types of carrot, based on shape. The cylindrical stump rooted *Nantes* types are popular for pre-packing and some processing; small finger carrots for freezing are usually *Amsterdam Forcing* types; small *Chantenay* types are used for pre-pack and canning; where large, uniformly-coloured carrots are required for slicing and dicing, *Autumn King* and *Berlicum* types are suitable. Long thin *imperator* types are used for pre-pack or cut and polished to make a number of 'baby' carrots per root.

Varieties are also grouped by date of harvest: first earlies end of May, second earlies end of June and maincrop from July.

Soils and climate
The climate in most arable areas of the country is suitable for carrots; the main limiting factor is soil type, which should not restrict root growth, usually 50–75 mm deeper than the required length of carrot. Sandy loams, loamy sands and fen peats are ideal for carrots. Soils should be well drained but moisture retentive, stone free (to avoid 'fanging') and non-capping. Crops are grown around the UK from the north-east of Scotland to the south of England to provide continuity of supply.

Cultivation
To produce more than 70% of carrots of a desired size is very difficult, and every effort has to be made to prepare a really good, uniform seedbed. The ground will be destoned and beds raised by a bed former. Carrots can be grown on the flat or in beds. Beds are commonly used to prevent compaction and aid lifting.

Drilling/planting and timings
Coated seed is drilled with a precision drill, i.e. Stanhay or Mini Air (Fig. 17.6) into a moist seed bed. Seed rate will depend on the germination percentage and field conditions at sowing and the desired size of carrot at harvest. Seed rate

Fig. 17.6   Precision drilling carrots (Stanhay Webb Ltd).

ranges from approximately 60 seeds/m^2 for first earlies to 175 seeds/m^2 for maincrops. Chantenay types are drilled at very high rates of approximately 900 seeds/m^2. To prevent the carrots at the edge of a bed being oversized, seeds may be drilled at differential rates across the bed with more seed placed on the outside rows in a 60/40 split.

First earlies are sown under polythene in October. Second earlies are also sown under polythene from December to February (Fig. 17.7). Maincrop carrots are sown in the open from March to early July.

Rotation
A rotation of one crop in five years is preferable to avoid disease. Sugar beet and potatoes host violet root rot.

Pests and diseases
Carrot fly, cutworms, willow-carrot aphids and nematodes, and the diseases violet root rot, black rot, leaf blight, cavity spot, scab and *Sclerotinia* rot are the main problems in the growing of the crop. Some varieties are more susceptible to cavity spot than others.

Weed control
The loss of active ingredients suited to carrot production means that increasingly machine and hand weeding may be needed to control weeds such as fool's parsley, hemlock and wild carrot, which are difficult to control with herbicides. Volunteer potatoes also pose a problem in carrots. Particular care is needed in managing weeds where crops are grown under plastic.

Fig. 17.7   Crop covers – carrots.

Nutrition
For a soil with an index of 2, the fertiliser recommendations would be 40 kg/ha N, 100 kg/ha of P_2O_5 and 125–175 kg/ha of K_2O. Leaf analysis should be taken for manganese and copper.

Irrigation
Irrigation is essential in most of the carrot growing regions, particularly on sandy soils and for early crops. Irrigation should be scheduled to avoid wet–dry cycles which can cause root splits and cracking.

Harvest and storage
The crop is harvested from June onwards. The roots are very easily damaged and careful handling is required at all stages. Early crops can be lifted using top-lifters, which pull the roots up by the leaves, and then top the roots. Top-lifters work well during the summer until the end of October, when the tops become too weak. From late October onwards, share-lifters (Fig. 17.8) are used which require the tops to be flailed-off first. To maintain freshness, carrots are often lifted at night, when it is cooler and the road network is clear allowing rapid transport to the packhouse.

Carrots can be lifted and held in long-term cold stores, but they lack the fresh appearance associated with newly lifted carrots. Most stored crop is left in the ground during the winter and lifted as required. Crops are either earthed-up or more commonly covered with black polythene sheets with a

Fig. 17.8   Share-lifter.

layer of straw 30 cm deep (approximately 100 t/ha). This protects the crop from frost and helps to stop regrowth in the spring. Lifting may continue until the crop becomes too woody or otherwise unsaleable, usually by early May. Yield can vary from 20 t/ha for early crops to well over 60 t/ha for maincrops, however rejects in grading can be 30% or more.

Maximum utilisation of the crop is important. Carrots are often washed and polished (using brushes to scrub the surface) to remove light blemishes of scab and cavity spot, increasing the proportion of class 1 crop.

17.14   Edible swede and turnips

Swedes and turnips are grown for sale as a whole root or for further processing as prepared vegetables, for example for soups.

Soils and climate
Swedes and turnips require uncompacted well-drained loams, silts and light clay loams. The crops can be grown around the UK but are mainly grown in areas with cool summers, such as the south-west, west and north of England, Wales and Scotland.

Cultivations
A fine, firm and moist seedbed is needed for uniform crop establishment.

Drilling/planting and timings
Directly drilled in beds for mechanical lifting at a target spacing of 20–25 plants/m². Early crops are drilled in April and harvested from July to September. The main crop is drilled from April to May in the north of the UK and May to July in the south for harvesting from September to March.

Rotation
Where club root is a problem, particularly on acid soils, swedes and turnips should be grown no more than one year in four or five after another brassica crop with a planned liming policy before planting.

Pests and diseases
Control measures may be necessary against cabbage root fly, flea beetle, aphids, caterpillars, cutworms, turnip sawfly and slugs. Hares, pigeons and rabbits can severely damage crops. Use of covers can be very effective at reducing pest damage, especially as there are very few recommended pesticides for use in these crops. Club root, powdery mildew, downy mildew, *Phoma*, *Alternaria* and *Rhizoctonia* can all be a problem.

Weed control
Established perennial weeds should be controlled prior to transplanting or drilling. Oilseed rape volunteers and potato groundkeepers can cause a problem in some crops. Mechanical methods with tractor hoes may be used where practical, which reduces the need for chemical control.

Nutrition
For a soil with an index of 2, the fertiliser recommendations for swede would be 70 kg/ha N, 100 kg/ha of P_2O_5 and 150–200 kg/ha of K_2O. Small edible turnips require lower rates of N.

Irrigation
In dry periods, irrigation will ensure even growth and quality of crop. Large variations in soil moisture can lead to growth cracks affecting quality.

Harvest and storage
Swedes and turnips can be hand pulled or lifted mechanically using a modified potato lifter and transported to the packhouse in crates or bulk boxes. Swedes are harvested from July to the following March. Late crops can be stored for several months in cold stores. Turnips store less well.

17.15 Strawberries

Strawberry production is widely distributed around the UK and has expanded partly due to the use of protection. Strawberries can be produced as field-grown crops covered with temporary structures such as Spanish or French tunnels, or under fixed structures such as glasshouses or fixed polythene clad tunnels. Open field-grown crops may also be covered with floating fleece to advance growth or insulated with straw, whilst they are still dormant in late winter, to delay harvest.

There are two main types of strawberry plant: June bearers and Everbearers. June bearers naturally initiate flower buds in autumn, become dormant overwinter and then produce flowers in late spring and produce fruit over 3–4 weeks in summer with the plants being used for 2–3 years. Everbearers naturally initiate flower buds in spring and summer and flower and produce fruit over 10–12 weeks during the summer. Everbearers may be cropped for one year and then pulled out.

By using cold stores, a range of planting dates, a mix of types/varieties and unheated polytunnels, strawberries can be produced from mid-May to late October/early November. Heated tunnels or glasshouses can extend production by a month at the start and end of the season but incur additional costs for heating. Early crops can be 'forced' with a double layer of floating fleece in the open or under polytunnels. At growing temperatures above 28 °C strawberry growth may stop (thermodormancy) and fruit quality is reduced. The maximum growing temperature is 25 °C. Crops can be cooled by removing covers or venting structures.

Soils and climate
Strawberries can be grown on a wide range of soils but very heavy clay soils can lead to poor aeration and waterlogging. Care is needed to avoid waterlogging as this can increase the spread of soil-borne diseases in winter and cause significant root death if soil becomes anaerobic. This can be avoided through field drainage and the use of raised beds.

Strawberries can also be produced using soil-less substrates, commonly peat or coir mixes in bags or troughs. Bags can be placed on the ground or on raised 'table top' systems to ease picking (Fig. 17.9).

Drilling/planting and timings
Young plants are planted into the soil when dormant. A range of planting material can be used with the crown size influencing fruit yield. Larger crowns or waiting bed plants do not store well and are ideal for spring forcing. Misted tips or rooted runners cold-store well and are suitable for summer planting.

Dormant June bearers can be lifted when dormant in winter and held in a cold store at −1 °C before being sequentially planted in spring and summer. Fruit is ready to harvest approximately two months later and these are known as '60 day plants'. Everbearer varieties are usually planted in early spring.

Plant spacing depends on the size of starting material and time of planting but is generally 10 plants per metre in a double row.

Rotation
A wide rotation of five or more years will benefit yield. The minimum rotational break should be three years following a three year crop of strawberries, or two

Fig. 17.9   Polytunnel grown tabletop strawberries.

years following a two year crop of strawberries. These can be successive one year crops or multi-year production from one crop.

Rotations including linseed, lucerne, flax, hops, peas, runner beans or potatoes should be avoided as these may increase levels of *Verticillium*. Care is also needed in the use of residual herbicides in preceding crops.

Pests and diseases
The main pests are vine weevil, blossom weevil, spider mites, tarsonemid mite, caterpillars, capsids and aphids. Slugs can be a problem under plastic mulching.

Powdery mildew, Botrytis, red core, crown rot and *Verticillium* wilt need to be controlled during the season.

Care is needed when applying pesticides under protection as some plastics will reduce the amount of UV light slowing the degradation of some active ingredients. Growing crops under protection helps to keep fruit dry and can reduce the occurrence of Botrytis and other fungal diseases. Protected structures are also suited to the use of biological control using insect predators. *Verticillium* wilt may be reduced through soil sterilisation prior to raising beds using steam or chemical fumigants.

Weed control
Perennial weeds should be controlled prior to mulching and planting. Polythene mulching of raised beds can help to control weeds. Hand weeding of weeds from planting holes may be needed during crop production.

Nutrition
As strawberry is a long-season crop, nutrition can be fine-tuned using leaf analysis through the growing season. Nutrition in soil-less growing media can be applied through the trickle irrigation system, i.e. fertigation, but controlled release fertilisers can be used. The optimum levels of nutrients required, pH and conductivity are reported in RB209. In field production, with soil with an index of 2, the fertiliser recommendations would be 0–60 kg/ha nitrogen, depending on soil type and type/variety with Everbearers requiring more nitrogen, 40 kg/ha of P_2O_5 and 80 kg/ha of K_2O.

Irrigation
Fruit size and yield are affected by irrigation and this is important under protected structures and should be scheduled according to crop requirement. Trickle tape/ drip irrigation is successful as it avoids wetting the fruit. Irrigation may be applied using overhead misting if crops are too hot during early growth. Overhead irrigation systems should be avoided during fruiting as this damages fruit and increases fungal disease, especially Botrytis.

Harvest and storage
Strawberries are hand harvested into punnets and removed quickly from the field and cooled to remove field heat. They are normally packed and labelled in a packhouse before distribution, using the cold chain, to retailers and markets.

17.16   Sources of further information and advice

Further reading
Alford D V, *Pest and Disease Management Handbook*, Blackwell Science, 2000.
Anon, *Vegetable Yearbook and Buyers Guide*, ACT Publishing, annual publication.
Anon, *The Berry Year Book*, ACT Publishing, annual publication.

Bayer CropScience, *Pest Spotter*, Bayer CropScience Ltd, 2011.

Biddle A J and Cattlin N, *Pests, Diseases and Disorders of Peas and Beans*, Academic Press, 2007.

Brewster J L, *Onions and Other Vegetable Alliums*, CABI publishing, 2008.

Davies G and Lennartsson M, *Organic Vegetable Production: A Complete Guide*, The Crowood Press Ltd, 2012.

Davies G, Sumption P and Rosenfeld A, *Pest and Disease Management for Organic Farmers, Growers and Smallholders: A Complete Guide*, The Crowood Press Ltd, 2010.

Davies G, Turner B and Bond B, *Weed Management for Organic Farmers, Growers and Smallholders: A Complete Guide*, The Crowood Press Ltd, 2008.

Defra, *Fertiliser Manual* (RB209), TSO, 2010.

Dixon G R, *Vegetable Brassicas and Related Crucifers*, CABI publishing, 2007.

Fordham F and Biggs A G, *Principles of Vegetable Crop Production*, Collins, 1985.

Gratwick M, *Crop Pests in the UK*, Chapman and Hall, 1992.

Hancock J F, *Strawberries*, CABI publishing, 1999.

Koike S T, Gladders P and Paulus A O, *Vegetable Diseases – A Colour Atlas*, Manson Publishing Ltd, 2007.

Lainsbury M A, *The UK Pesticide Guide*, BCPC and www.cabi.org, published annually.

Maynard D N and Hochmuth G J, *Knott's Handbook for Vegetable Growers*, Wiley-Blackwell, 2006.

McKinlay R G, *Vegetable Crop Pests*, Macmillan Press, 1992.

Rubatzky V E, Quiros C F and Simon P W, *Carrots and Related Vegetable Umbelliferae*, CABI publishing, 1999.

Ryder E J, *Lettuce, Endive and Chicory*, CABI publishing, 1998.

Sherf A F and Macnab A A, *Vegetable Diseases and their Control*, John Wiley & Sons, 1986.

Wein H C, *The Physiology of Vegetable Crops*, CABI publishing, 1997.

Websites

http://assurance.redtractor.org.uk/rtassurance/global/home.eb

www.britishgrowers.org

www.defra.gov.uk/statistics/foodfarm/landuselivestock/bhs/

www.hdc.org.uk

www.pgro.org

Part IV

Grassland and forage crops

18

Arable forage crops

DOI: 10.1533/9781782423928.4.433

Abstract: This chapter deals with forage crops other than grasses and perennial forage legumes. A full account of their husbandry is included and their suitability for different circumstances. These are annual crops grown for feeding to livestock, either for their yields of roots as in the case of fodder beet, mangels and swedes, or, for their yields of leaves for grazing. The subject of 'catch' cropping with forage brassicas is also covered. Further sections deal with the use of annual forage crops for ensiling, by far the most important of which is forage maize.

Key words: forage roots, forage brassicas, catch crop, forage maize, Ontario heat units.

18.1 Crops grown for their yield of roots

Fodder beet and, now to a much lesser extent mangels, are grown for feeding to cattle and sheep in the south of the UK where they are better croppers than swedes and maincrop turnips which are mainly grown in the north. Occasionally fodder beet is grown as a cash crop, for sale, and there has been some increase in interest in it in areas where sugar beet growing is no longer possible due to lack of processing facilities. With its lower dry matter content, the mangel has now almost died out as a forage crop. The financial difficulties facing the livestock sector, the absence of machinery suitable for root growing on stock farms and the development of contractor operations focused on forage maize, all have caused a major decline in the use of most of these crops.

18.1.1 Fodder beet

This has been bred from selections from sugar beet and mangels. At one time it was quite popular as a feed for pigs. Fodder beet should not be grown on heavy and/or poorly drained soils, nor on stony soils. There could well be establishment and harvesting problems in these conditions.

Varieties

An important consideration, apart from dry matter yield, is the cleanliness of the roots after harvest. Dirty roots may be less palatable and cause digestive upsets when fed to livestock. Another important factor is the dry matter percentage. *Kyros* has been a popular variety for many years and gives average yields of clean roots at about 16% dry matter. *Magnum* is an example of a variety with very high yields and a higher dry matter of about 19%. *Blizzard* is a variety well suited to harvesting with sugar beet machinery. Sheep and young cattle, if they are to be fed on fodder beet, do better on the lower dry matter varieties. *Feldherr* at about 14% dry matter is suitable for hand lifting or grazing *in situ* by younger animals, but has a lower yield than the others. *Ribondo* is a variety with *Rhizomania* resistance. Herbicide tolerant varieties are also being developed but have not, so far, been approved for farm use in this country.

Many aspects of the growing of fodder beet are similar to those of sugar beet, but there are some important differences and the chief points to remember are described below.

Seed rate

Seed is usually sold in acre packs of about 50 000 seeds (125 000 seeds or 2.5 packs/ha). Graded (3.50–4.75 mm) pelleted seed is used, which should be precision drilled if possible.

Time of sowing

Early to mid-April is best; sowing earlier will cause bolting (some varieties are more susceptible). Later sowing will reduce yields.

Most crops are now drilled to a stand with monogerm seed which precludes the need for thinning out. However, a small proportion of monogerm fodder beet seed will still produce more than one seedling and so the crop may need 'rough singling'. The drill width will depend on the harvesting system; it is normally 50 cm with a 15–17 cm spacing in the row. The aim is to achieve at least a 65 000/ha plant population, but because of inevitable plant losses through pests and diseases, post-emergence herbicides and/or accidents with inter-row work, if carried out, an initial target figure of 85 000 plants/ha is not unrealistic.

Manures and fertilisers

As with sugar beet, fodder beet responds to sodium, and agricultural salt is often used as a basal fertiliser. About 400 kg/ha (supplying 200 kg/ha Na_2O) should be applied and worked into the soil at least a month before the crop is drilled. The phosphate and potash can also be applied at this time. If salt is not applied potash fertiliser should be increased by 100 kg/ha K_2O. Additional magnesium may also be needed on some soils with Mg indices of 0 or 1. Normally half of the nitrogen will be applied on the seedbed and the balance at the 3–7 leaf stage. General fertiliser recommendations are shown in Table 18.1.

Table 18.1   Fertiliser recommendations (kg/ha) for fodder beet and mangels

	SNS*, P or K index						
	0	1	2	3	4	5	6
Nitrogen (N)	130	120	110	90	60	0–40	0
Phosphate (P$_2$O$_5$)	110	50	50	0	0	0	0
Potash (K$_2$O)	170	140	110(2−)	40	0	0	0
			80(2+)				

*Soil Nitrogen Supply Index – see Defra RB209.

Agricultural salt is recommended for all soils except fen silts and peats. About 400 kg/ha of salt (200 kg/ha Na$_2$O) should be worked into the seedbed well before drilling. If salt is not applied potash should be increased by 100 kg/ha. Where FYM and slurry have been used, the above recommendations should be reduced to take account of the available nutrients so supplied. These recommendations have been based on Defra RB209 (8th edition) issued in June 2010.

A pH of 6.5–7 is necessary. Acid conditions cause stunted and misshapen roots. However, as with sugar beet, over-liming can induce boron and manganese deficiencies. A soil test for boron availability is a useful guide.

Crop protection

It is important to keep the crop clean for at least the first six weeks and details of suitable pre- and post-emergence herbicide programmes suitable for fodder beet (usually very similar to those used for sugar beet) are given in section 15.3.2.

Weed beet can be a problem. In fodder beet it tends not to be treated as seriously as in the sugar beet crop. Apart from the fact that the problem is perpetuated if the weed is allowed to grow unchecked (obviously a serious problem if sugar beet is also grown on the farm), it does compete with the fodder beet itself and with the following crops in the field. Hand rougueing is the best form of control.

Fodder beet seed should be dressed with thiram against footrot. It is also possible to guard against the wide range of soil-borne insect pests (including flea beetle) which attack both fodder beet and sugar beet by having the seed dressed with imidacloprid. Mangel fly and the aphid vectors of beet yellows virus can also be controlled by this treatment. Where fodder beet is being grown after ploughed-out grass, leatherjackets may be a serious problem.

Virus yellows can be very serious in some years, particularly if an early attack by aphids occurs (Table 6.2). In the absence of imidacloprid seed treatment a spray of cypermethrin may be necessary. Powdery mildew, appearing as a white powdery growth on the leaves in early autumn, is not usually considered serious enough to treat with a fungicide.

Harvesting

This usually takes place in October and November, although, depending on the season, the crop can continue to grow on through the autumn. However, late harvesting generally increases the risk of more difficult conditions obtaining, particularly on heavier soils. Many growers will also harvest earlier at the expense of yield in order to follow with winter cereals (usually wheat) and to achieve a clean root sample.

The crop is lifted either by specialist harvesters or modified (or outdated, cheap) sugar beet harvesters. Top savers can be used when necessary or a forage harvester can be used for topping the beet (an inaccurate technique, unless the crop is very evenly grown). Topping must be carried out correctly when it is intended to store the beet for any length of time. The roots should be topped at the base of the leaf petioles. Overtopping soon results in mould growth; leaving too much green material causes high losses through increased respiration in the clamp.

Fodder beet can be fed fresh to stock and even strip fed in the field, but normally it has to be stored. This must be done carefully and, for outside storage, a temperature of 3–6 °C should be maintained in the clamp. Heating in the clamp should be avoided, although the beet should be adequately protected against frost. A straw covering or plastic sheet can be used, but a strip along the top of the clamp should be left uncovered to aid ventilation, except in freezing conditions. The storage capacity needed will vary according to the size of the beet but is usually in the range 700–750 kg/m^3.

Like sugar beet, fodder beet can produce a substantial yield (about 10 t/ha) of tops which can also be fed to livestock. These have a low energy content but much higher crude protein (<15%) than the roots. Some sheep farmers feed them to lambs *before* harvesting the roots. Alternatively they can be fed in the field after harvesting is complete and, normally, after wilting to reduce the soluble oxalate content which may cause scouring.

Yield

- Fresh yield: 80–100+ t/ha.
- Dry matter yield: 11–16 t DM/ha – up to 20 t DM/ha in favourable conditions.
- Dry matter yields of tops may also be 2–3 t DM/ha.

Wholecrop fodder beet for silage
In the 1990s a technique was evolved for ensiling chopped fodder beet with ammonia treated straw to give an acceptable silage product. Early harvesting in dry conditions was advised and the selection of a variety yielding clean roots was also important since the amount of soil entering the clamp should obviously be minimised. The lack of availability of suitable machinery for harvesting and chopping beet appears to have precluded any advancement of this system.

18.1.2 Mangels (mangolds)
As shown in Table 18.2, there is a significant difference in dry matter yield and energy content when comparing mangels with fodder beet. Very few mangel crops are now grown.

Table 18.2 Typical yields and nutritional values of forage crops compared with grass

Crop	Typical fresh annual yield (t/ha)	DM (%)	Dry matter yield (t/ha)	ME* (MJ/kg DM)	Crude protein (%)
Fodder beet	90	17.0	15.3	12.5	7
Mangels	120	11.0	13.2	12.4	9.2
Yellow turnips	70	8.5	6	12.7	11.2
Forage turnips	40	10	4	12.5	11.5
Swedes	90	11.0	9.9	13.5	9.5
Kale	55	15.0	8.3	11.4	20
Forage rape	45	11.0	5	10.4	19
Fodder radish	50	10	5	9	2.3
Forage maize	40	30.0	12	11.5	9
Wilted grass for silage†	40–60	25.0	10.0–15.0	11	13.7

*ME – metabolisable energy

†Grass yields in particular may vary substantially according to growing conditions.

Published figures regarding the yields and compostion of all these crops vary substantially and farm experience too can be extremely variable. The figures presented represent the potentials in average conditions with near optimum inputs.

The husbandry of mangels is similar to that of fodder beet. Most of the varieties such as *Wintergold* (a low dry matter traditional variety) are multigerm, with the seed producing more than one plant when sown (*Peramono*, a monogerm variety, is an exception). The seed, although graded and pelleted, will produce a relatively full row which should be thinned. The target population is 65 000 plants/ha.

Although the mangel grows with much of its root out of the ground, harvesting, except by hand, is difficult. The root 'bleeds' very easily and so the tops (not the crowns) are either cut or twisted off. The tops are small and are rarely fed. The roots are not trimmed. Ideally, they should be left for a period in the field to 'sweat' out in small heaps covered with leaves to protect them from frost damage. Following this they can be clamped in the same way as fodder beet.

18.1.3 Swedes and turnips

Although their popularity has declined in the past 30 years, swedes (especially) and maincrop turnips are still grown, particularly in the northern and western parts of the United Kingdom. In the north of England and Scotland these crops are also known as 'neeps'. In appearance, the difference between the two crops is that swedes have smooth, ash green leaves which grow out from an extended stem or 'neck' whilst turnips have hairy, grass green leaves which arise almost directly from the root itself. Swedes are normally the higher yielding crop with dry matter levels between 9–13%. Turnips are usually grown in less fertile soils at higher elevations and are lower yielding with slightly lower dry matter contents.

Both crops are valuable for cattle and sheep and, depending on the variety grown, they can also be used as table vegetables. There has been a small scale revival in their popularity for this purpose. A limited market has also developed for growing turnips on contract, for freezing.

The growing of swedes and main crop turnips is similar.

Varieties

There are limited numbers of swede varieties available; some, such as *Ruta Otofte*, *Marian* and *Airlie* have been available for many years. More recent introductions are *Siskin* and *Invitation* (which have good resistance to club root disease), *Brora* and *Gowrie*. *Massif* is a recent introduction of a maincrop turnip and is higher yielding than the traditional *Aberdeen Green Top Scotch*.

Climate and soil

The crops like a cool, moist climate without too much sunshine. Powdery mildew can be a problem in the warmer and drier parts of south and east England. With the exception of heavy clays, most soils are suitable.

Seedbed

A fine, firm and moist seedbed is necessary to get the plants quickly established. In very wet districts the crops can, with advantage, be sown on a ridge. This traditional practice is mainly carried out in northern areas and in Scotland. It is not widespread now, mainly because of the use of precision seeding and improved chemical weed control.

Manures and fertilisers

The average fertiliser requirement is summarised in Table 18.3. If possible, 25–40 t/ha farmyard manure should be applied in the autumn. This is especially important for improving the water holding capacity of lighter soils. Slurry can be used instead, in the spring, at about 35 000 l/ha.

Less nitrogen is needed in wetter areas, although exactly how much is used will depend on the soil nitrogen supply status. More phosphate may be needed on heavier soils, but this will again depend on the soil phosphate index; if the fertiliser is placed 5–10 cm below the seed, the phosphate can be reduced. Potash is important, but savings can be made by using organic manures. As with fodder

Table 18.3 Fertiliser recommendations (kg/ha) for maincrop forage swedes and turnips

	SNS*, P or K index						
	0	1	2	3	4	5	6
Nitrogen (N)	100	80	60	40	0–40	0	0
Phosphate (P$_2$O$_5$)	105	75	45	0	0	0	0
Potash (K$_2$O)	215	185	155(2−) 125(2+)	80	0	0	0

*Soil Nitrogen Supply Index – see Defra RB209. Where FYM and slurry have been used the above recommendations should be reduced to take account of the available nutrients so supplied. These recommendations have been based on Defra RB209 (8th edition) issued in June 2010.

beet, boron deficiency (also known as brown heart or 'raan' in swedes) may occur in some soils and, particularly if the pH is high, a soil sample to determine boron availability is advisable.

Lime is important as swedes and turnips are often grown on potentially acid soils. Soil pH should be above 6. Club root can be prevalent under acid conditions. However, over-liming is equally serious as it can induce boron deficiency symptoms (Table 6.2).

Time of sowing

Mid-April until the end of June is ideal. Swedes can now be sown earlier than has been the case in the past, thus increasing the yield potential. Powdery mildew, which is more likely with earlier sowing, can be reduced by chemical treatment, e.g. with tebuconazole, however, an increasing number of swede varieties now show reasonable resistance to this disease.

The yellow-fleshed turnip is normally sown in May and June.

Seed rate

Rates vary between 0.5 and 4.5 kg/ha. The lower amount is used with precision drilling. Pelleted seed is sown 7–15 cm apart in the row at row widths of 17–35 cm. This is particularly applicable where overall chemical weed control is carried out and/or the crop is to be folded. The aim is for a high plant population of up to 100 000/ha.

For inter-row cultivation, or if the crop is to be lifted, the plants are spaced (or rough-singled) at 20–22 cm apart with row widths of 50–75 cm. A plant population of 60 000/ha is acceptable, although, if the roots are rather large and dry matter content is low, they may not keep well. Both swedes and turnips can also be broadcast at 3.5–5 kg/ha.

Seed treatments containing thiram for protection against soil-borne fungi should normally be used. An approved treatment for flea beetle control is a pre-sowing application of deltamethrin. Leatherjackets are likely to be a problem in fields recently ploughed from grass.

Weed control

The stale seedbed technique, using cultivation or a contact herbicide to kill weeds prior to drilling is frequently used. An alternative is a pre-emergence application of metazachlor herbicide. Inter-row cultivations or harrow combing are also carried out by some growers, particularly on organic farms.

Harvesting swedes

Swedes, like turnips, are very often grazed *in situ* in the field, but both crops can deteriorate under wet and frosty conditions. Varieties with globe-shaped roots, rather than tankard shaped, are less prone to damage when harvested by machine. In most districts they are lifted in late autumn, before they are fully matured. It is advisable to allow the roots to ripen in a clamp to minimise scouring when fed to stock. In mild districts, swedes may be left to mature in the field; for this purpose, high dry matter, more winter hardy types are recommended and the swedes would only be stored for two to three weeks, with the grower endeavouring to keep harvesting just ahead of any possible bad weather. Swedes are vulnerable to rotting under poor storage conditions. They should be handled carefully and not clamped in wet conditions. Clamps should be no higher than two metres and adequately protected against frost. Because of the low dry matter content and probable soil contamination, the tops are usually wasted. However, they are quite high in protein.

Harvesting turnips
The main crop (yellow fleshed) turnips are normally harvested in October when the outer leaves begin to decay. They can also be grazed *in situ*, but are often lifted and stored in one main clamp. Machines which merely top, tail and windrow the roots for subsequent collection are now being replaced by complete harvesters. On mainly livestock farms harvesting is often contracted out.

Yield

Swedes, 65–110 t/ha (7–12 t DM/ha).
Turnips (yellow-fleshed) 60–80 t/ha (5–7 t DM/ha).

18.2   Crops grown for grazing

18.2.1   Kale
Kale is grown for feeding to livestock, usually in the autumn and winter months. It can either be grazed in the field or cut (normally with a forage harvester) and carted off for feeding green. In this case a heavier yielding crop is needed. After wilting, kale has also been successfully made into silage both in clamps and big bales. It is mostly grown in the south and south-west of England. The area sown to kale, cabbage (for stock feeding) and forage rape has declined substantially in recent years and it would probably be true to say that the majority of kale crops at present are grown for game cover.

Climate and soil
Kale is an adaptable crop, although under very dry conditions it may be difficult to establish. For grazing it is essential that it is grown on well drained fields.

Place in rotation
On livestock farms kale is often direct drilled into an old ley after glyphosate spraying. This may be after an early grazing or a silage cut has been taken. Kale can also follow a catch crop sown after cereals in the previous autumn. Club root disease (Table 6.2) can be a problem and kale should not be grown in the same field more often than one year in three.

Seedbed

A fine, firm, and clean seedbed is required. Kale is very suitable for direct drilling and this has the important advantages of conserving moisture at sowing time, leaving a much firmer surface for grazing and good annual weed control. After sowing in conventional seedbeds good consolidation is important, for moisture conservation and good seed–soil contact. Rolling is also important immediately after direct drilling has taken place in order to close the slits where the seed has been placed to retain moisture.

Manures and fertilisers

Up to 50 t/ha of farmyard manure can be applied in the autumn. This is especially important when a heavy yielding crop is desired. Slurry at about 35 000 l/ha can also be used in the spring, prior to cultivation. A light dressing of slurry applied to a direct drilled field after drilling can also be beneficial.

Total plant nutrients required are summarised in Table 18.4. Less nitrogen is required if kale follows grass. Fertiliser is usually applied during final seedbed preparations. The nitrogen application can be split and partly top dressed when the young plants have up to five leaves. A pH of 6–6.5 is recommended.

Varieties

Maris Kestrel and *Hereford* give good yields of dry matter and are highly digestible. *Camaro* is one of the highest yielding varieties and *Keeper* (as its name suggests) is favoured for game cover. *Caledonian* and *Bitterne* are club root resistant varieties.

Time of sowing

Kale can be sown from the end of March until mid-July. The heaviest yielding crops are normally associated with early sowing.

Seed rate

Precision drilled: 0.5–2 kg/ha, the seed spaced at 2.5–10 cm apart in 35–75 cm rows. Drilled: 3–5 kg/ha in 35–75 cm rows. Broadcast: 5–7.5 kg/ha.

Table 18.4 Fertiliser recommendations (kg/ha) for kale

	SNS*, P or K index						
	0	1	2	3	4	5	6
Nitrogen (N)	130	120	110	90	60	0–40	0
Phosphate (P_2O_5)	110	80	50	0	0	0	0
Potash (K_2O)	260	230	200(2−) 170(2+)	130	0	0	0

*Soil Nitrogen Supply Index – see Defra RB209. Where FYM and slurry have been used the above recommendations should be reduced to take account of the available nutrients so supplied. These recommendations have been based on Defra RB209 (8th edition) issued in June 2010.

Weed and pest control
Residual herbicides such as metazachlor and propyzamide are among those approved for use on kale crops. 'Stale seedbed' cultivation, steerage hoeing or harrow combing of drilled crops are possibilities on organic farms. Deltamethrin spray is an example of an approved post-emergence treatment against flea beetle. Leatherjacket damage is likely when the crop is spring sown after grass.

Utilising the crop
Strip grazing behind an electric fence reduces the need to handle the crop. Light or well drained soils are essential, otherwise both stock and soil suffer. Direct drilled crops are particularly suited to strip grazing. Wastage can be high at 15–30% and up to 70%. Starting to graze kale in the early autumn and mowing before grazing behind an electric fence can both help to improve utilisation. If strip grazing is not possible a forage harvester may be used to chop the crop coarsely and blow it into a trailer for feeding in the yard.

Kale contains an imbalance of trace elements such as iodine and manganese and a chemical called S-methyl cysteine sulphoxide (SMCO); very high intakes should be avoided as they can lead to haemolytic anaemia and fertility problems.

Yield
35–75 t/ha at about 15% dry matter (5–11 t DM/ha).

18.2.2   Forage catch cropping

Forage catch cropping is the practice of taking a quick growing crop between two main crops. The term is usually applied to crops of quick growing white turnips, or rape, or mixtures of the two, sown for autumn or early winter grazing after an early harvested cereal (usually winter barley) or any crop taken early for wholecrop silage.

Forage catch crops are usually direct drilled since this helps to conserve soil moisture. It is also a valuable technique for weed control since the grass weeds and cereal volunteers, which almost invariably germinate alongside the forage crops, can be effectively controlled either by a glyphosate spray pre-drilling, or a post-emergence application of an effective approved graminicide. Apart from weed control and the probable need to control slugs, the only other important treatment for these crops is nitrogen fertiliser, so they have the additional advantage of being very cheap to grow.

Quick growing white turnips (stubble turnips)
The quick growing white turnips originated in The Netherlands and are therefore also known sometimes as Dutch turnips. The name stubble turnips has arisen from the practice of growing them as a catch crop on the stubble following an early harvested cereal crop. However, in the north of England and in Scotland, because of later cereal harvests, it is seldom possible to obtain a reasonable yield by stubble growing. In colder regions they are sown earlier and, indeed, in England,

these turnips can be sown as early as April to provide a useful mid-summer feed in July. There is nothing against establishing a second crop of turnips after the first crop from a spring sowing has been utilised. However, repeated sowing in the same field year on year, would inevitably lead to a problem with club root disease.

Sowing should be completed ideally by mid-August. Later sowing produces much reduced yields and may not be worthwhile. Direct drilling is now the norm; it helps to conserve moisture at what is usually a dry time of the year. Broadcasting seed into a cereal crop before harvesting using a pneumatic fertiliser spreader or 'autocasting' seed when the cereal is combined are time saving establishment techniques which have been tried in recent years. Success depends very much on the effectiveness with which the straw is chopped and spread, on rolling soon after sowing, and on the incidence of rainfall. In dry conditions these techniques are unlikely to be very successful.

Direct drilling and grazing a crop of forage turnips is a valuable way in which to obtain additional dry matter production from a poor grass ley prior to reseeding in July or August. It is also a useful technique in that it enables the old sward to be completely eliminated prior to establishing the new seeds.

Seed rate
2–5 kg/ha drilled; 6–9 kg/ha broadcast. A common practice is to mix with forage rape for autumn sheep feeding.

Varieties
Popular varieties are the tetraploids *Vollenda*, *Whitestart* and *Frisia*. *Samson* and *Delilah* are diploid varieties. *Appin* and *Tyfon* are varieties with low root yields, mainly leaves. These varieties, however, are unsuitable for spring sowing as they are likely to bolt. There are some varietal differences in susceptibilities to club root, powdery mildew and *Alternaria* diseases.

Fertiliser
The main need is for nitrogen; turnips will respond to between 75 and 100 kg N/ha. This should be broadcast at the time of sowing or top dressed shortly after the crop has emerged. Phosphate and potash are not usually needed although this will depend on soil indices and should be supplemented at index 2 and below.

Weed control
Spraying with glyphosate prior to direct drilling into an old grass ley is normal. A useful technique is to spray glyphosate prior to the final utilisation of grass. This will not harm the stock and ensures a complete kill of the old sward. The use of an approved graminicide such as fluazifop-P-butyl can be helpful for controlling grass weeds and cereal volunteers when sowing after a cereal crop.

Utilisation
Crops are usually ready for grazing about three months after drilling. Whereas most crops are utilised by sheep, cattle, including dairy cows, can also benefit considerably.

Some dairy farmers regularly sow turnips for summer and early autumn grazing to supplement grass growth and both milk yield and quality can benefit.

Yield
30–40 t/ha of fresh yield is the norm for autumn sowings at about 10% dry matter. Much higher, but quite variable, yields are possible when sowing takes place in favourable conditions in spring or in July.

Forage rape
This is a quick growing palatable leafy crop which is ready for grazing about 12 weeks after sowing. It generally lacks winter hardiness and should therefore be used before the end of the year.

Varieties
Relatively little plant breeding has been undertaken in forage rapes in recent years and so *Emerald*, which was first introduced in the 1970s, is still in use today. *Hobson* is a popular recent introduction. A common failing of forage rape varieties is susceptibility to club root. *Sparta* and *Dinas* both have moderate resistance to this disease.

Seed and sowing
The seed is either drilled at 4.5 kg/ha or broadcast at 10 kg/ha from the end of April until mid-August, usually following, in this last case, an early harvested grain crop. Direct drilling is a suitable method of establishment. Rape is often sown as a mixture with quick growing turnips for feeding to sheep.

Fertiliser
Up to 100 kg N/ha can be used. Phosphate and potash are not always necessary, but, as with stubble turnips, this will depend on soil indices.

Yield
30–50 t/ha at 11–12% dry matter.

Rape kale and Hungry Gap kale
These are not true kales but are related to, and similar to, rape. Both are particularly susceptible to club root and mildew. They are grown on a small scale only in the southern part of the United Kingdom, and can be used for a limited period after Christmas. They can produce useful regrowth for grazing in March and April – the so called 'hungry gap' period before spring grass is available.

Seeding and growing requirements are similar to those for normal forage rape. *Swift* and *Redstart* are two example varieties.

Fodder radish
This crop is grown for forage, green manuring (see section 4.7.8) and game cover. By producing quick ground cover it can also be used as a protection against wheat

bulb fly. It is very quick growing (some varieties grow a metre high) and is suited to most soil conditions. *Ikarus*, *Apoll* and *Radical* are varieties currently available. Fodder radish is normally sown in July for forage at 8 kg/ha drilled, or 13 kg/ha broadcast. For green manuring it may be sown earlier, with a heavier seed rate of 17 kg/ha. Up to 75 kg N/ha can be applied with the seed. Phosphate and potash should not be necessary unless soil analysis indicates a deficiency.

Fodder radish should be grazed-off before flowering and whilst it is still palatable, being normally utilised within 8–12 weeks of sowing. This crop is very susceptible to frost but resistant to club root and mildew. It can also provide a good rotational break on farms where sugar beet is grown and there is limited evidence that fodder radish can reduce soil populations of sugar beet cyst nematode.

18.2.3   Winter cereals as grazing crops

Rye and triticale are often sown specifically as grazing crops. Early sowing in August or September can result in a worthwhile crop for grazing as early as November. The seed rate for rye should be about 200–250 kg/ha and for triticale 180 kg/ha.

Grazing in March or April following an early application of about 50 kg N/ha is the normal time for utilising these crops. On many stock farms this period often coincides with a shortage of winter feed and occurs prior to the availability of spring grass for grazing. When an early grazing takes place it may be possible also to graze a regrowth three to four weeks later or even to take this regrowth for a silage cut. Neither crop is particularly palatable to stock and it is common practice also to sow about 10 kg/ha of an Italian ryegrass variety to improve this.

Average dry matter yields of around 3.5 t/ha are possible from rye and triticale. After the last grazing the fields would be suitable for cropping with kale or forage maize, following an application of FYM or slurry.

Well-grown crops of winter wheat are also occasionally grazed and the practice is widespread in other countries. Care should be taken to ensure that poaching does not occur and that grazing does not take place after stem extension (Zadoks growth stage 31, see Section 13.4.1) or grain yields may be seriously affected.

18.3   Crops grown for ensiling

18.3.1   Forage maize

Apart from grass, forage maize has become the most important arable forage crop grown in the UK. In an emergency, maize can be cut early for feeding green to stock and there is a small area grown for grain, but most of the crop is now used for ensiling for livestock feeding or for biogas production. At the turn of the millennium the area sown was just over 100 000 ha, most of it in the Midlands and southern England. This had expanded to about 164 000 ha in 2011 with the crop increasing in popularity in lowland areas of Wales, Northern Ireland and Scotland.

Maize was introduced into this country at the beginning of the twentieth century and early trials on its suitability as a crop for ensiling were carried out at Wye College in Kent. Although high yields were shown to be possible, two main factors were responsible for the relatively slow uptake of the crop. The first was that the varieties available at that time were mainly of American origin and did not mature early enough for British conditions. In particular, the development and yield of cobs (the most valuable part of the plant, rich in cereal starch) was not good. The second reason was the absence of suitable and reliable mechanical systems for harvesting, ensiling and feeding-out.

The rapid development of the crop from the early 1960s has taken place in response to two main stimuli. One has been the introduction of high yielding, early maturing hybrid varieties, capable of developing good yields of cobs and high starch percentages in many parts of the UK. The other has been the advent of high capacity, contractor-based growing and harvesting machinery systems. The increasing use of contract harvesting systems, not only for maize but also for grass silage, has developed concurrently with the decline in the labour forces on many livestock farms.

A further reason for the rise in the maize area has been that it satisfies the need for a high energy forage for winter feeding. Good maize silage has an ME value of 11.2–11.3 MJ/kgDM and a D value of between 70–75%. Cereal starch normally accounts for between 25% and 30% of the dry matter. This augers well for the production of high yields of high protein milk, and maize silage fits in well with the move to complete diet feeders which are the basis of such systems. Tower silos have been tried, but now largely abandoned by UK farmers in favour of long narrow bunker clamps, which better suit current handling systems, and which minimise the likelihood of secondary fermentation at feed-out.

Climate, site and soil types

Maize is a C4 plant of sub-tropical origin. This means that is has a different way of photosynthesising compared with most other crops grown in the UK. In particular, it responds well to high temperatures and requires a minimum soil temperature of 10 °C before active growth commences. Crops grown at low altitude in the coastal areas of southern England and Wales will, therefore, have much higher yield potentials, because they experience higher levels of temperature during the growing season.

Maize is best grown on fields with a southerly aspect, and does not normally do well in cold conditions at high altitude. Fields with steep slopes should also be avoided because of the risk of soil erosion. Maize is well suited to light to medium textured soils with good drainage. Growing maize on poorly drained clay soils is not advisable; good crops can be grown, but in a wet autumn substantial harvesting difficulties may be encountered.

Forage maize should form part of a rotation of crops. On some farms this may not be possible and maize may have to be grown for several years in the same fields. Where maize provides a very high (up to 100%) proportion of the winter bulk feed ration there may be no alternative to repeated cropping.

However, such practices are risky and may lead to problems with soil structure and with trash-borne diseases such as maize eyespot (*Kabatiella zeae*) which can have devastating effects on yields (Table 6.2).

Cultivations
Winter ploughing following soil loosening in a reasonably dry autumn is the best preparation. However, where maize fields are to receive large quantities of organic manures, ploughing may be delayed until the spring, to facilitate application and immediate incorporation. In dry areas it is preferable to roll immediately after ploughing, to conserve moisture. Power harrowing to produce a medium tilth then takes place immediately before the maize is precision drilled. Rolling after drilling should then complete the operation.

Choice of variety
Apart from the choice of field and seedbed preparation, as described above, the choice of a suitable variety is the main factor influencing the success of growing maize. A great many varieties (all hybrids) are available to UK farmers. NIAB TAG together with BSPB (The British Society of Plant Breeders) produce an annual descriptive list with data drawn from a range of about ten sites in England and Wales. Varieties are no longer categorised according to maturity classes, as has been the case in recent years, but growers are now given a simplified list of first and second choice varieties for 'favourable' or 'less favourable' sites. Information for Scotland is available from SASA and from DARD in Northern Ireland.

The suitability of a site for forage maize depends on calculations of accumulated temperature which the crop is likely to experience. Farmers can categorise their growing conditions by referring to published climatic data expressed as 'Ontario heat units' (OHUs). ('Maize heat units' in some publications also describe temperature accumulation, but in a slightly different way.) The Meteorological Office has produced a map that depicts the areas of southern England and Wales which achieve suitable levels of Ontario heat units in nine years out of ten (see Fig. 18.1). It should be remembered, however, that these data were compiled some years ago and take no account of more recent climatic changes, but the principles involved are relevant still. It should also be remembered that altitude and aspect, in particular, will affect temperature on a local basis. The most favourable areas for maize are those that experience more than 2500 OHUs in nine years out of ten. These areas have the highest yield potential. The less favourable areas are those experiencing 2300–2500 OHUs or less.

Apart from comparing dry matter yields, the NIAB TAG/BSPB information also gives details of likely levels of ME, the percentage starch content and cell wall digestibility. Not all of the varieties available in the UK are currently listed. For information about the suitability of non-listed varieties for growth in specific farm situations, information should be sought from the seeds company concerned.

It is not sensible to refer to the attributes of specific recommended varieties in this publication since new introductions are made and recommendations change

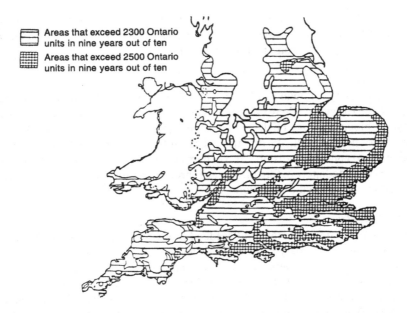

Fig. 18.1   Map of Ontario heat units showing areas most suited to growing maize. *Source*: Meteorological Office, Bracknell, UK.

each year. Good information can also be readily obtained from reputable seed suppliers and from the UK Maize Growers Association Ltd.

Seed rate and sowing date

Almost all maize is precision sown with fungicide dressed seed. Optimum seed rates are between 100 000 and 120 000 seeds/ha with seeds being sown in rows about 75 cm apart. Trials have been carried out to compare the conventional row spacing with closer drilling or drilling in double rows (alternately close drilled and conventional row widths) from which some slight advantages are apparent. Some farmers have adapted cereal drills to sow maize seed in 37.5 cm rows and saving the contractor costs of precision drilling. Maize seed is best sown at between 5 and 10 cm deep, depending on soil moisture content.

Sowing date is important and the majority of maize crops are drilled in the latter half of April in southern England. A more precise guide is soil temperature and the soil at 10 cm should have reached 8–10 °C before drilling commences. When drilling is delayed beyond the first week of May, substantial reductions in yield are inevitable.

Sowing maize under plastic film

There has been some interest in the use of biodegradable plastic film for establishing maize, particularly in marginal areas. Increased dry matter yields and accelerated maturity have been achieved; however, the technique has not become widespread mainly for reasons of cost effectiveness.

Fertilisers

pH is an important consideration for maize growers since the crop is often grown on light or sandy soils which have a predisposition to acidity. If lime is required it is best applied in the autumn or on the ploughed furrow prior to final seedbed preparation in the spring. The field should be limed to a target pH of 6.5.

The response of maize to added fertilisers is variable because the crop is often grown in conditions of very high fertility. Theoretically, an average (say 12 tDM/ha) crop would remove about 120 kg N/ha, 55 kg/ha P_2O_5 and 175 kg/ha K_2O. After deducting the theoretical values of nitrogen, phosphorus and potassium delivered by the large quantities of FYM and slurry applied to maize fields, it is probable that the average crop requires no additional fertiliser apart from about 50–60 kg N/ha. However, it is very common practice for maize to receive at least 125 kg/ha of mono-ammonium phosphate (MAP) worked into the seedbed or placed close to the seed at drilling. This provides readily available nitrogen and phosphorus to enhance the vigour of seedling growth in the early stages of crop development. However, there is evidence to suggest that, on fertile sites (index 3 and above) especially, where large quantities of organic manures are habitually applied, and in spite of an apparent visual response in the early stages, there is seldom a worthwhile final yield response. Extra nitrogen, where required, may be worked into the seedbed or top dressed in the early stages of growth. The current recommendations for maize are summarised in Table 18.5.

Crop protection

Seedling fungal diseases are normally protected against by a thiram seed dressing. Methiocarb seed dressing acts as a seedling bird repellant and also has some activity against frit fly larvae. Effective weed control is important and there are a range of approved pre-emergence herbicides available such as flufenacet and

Table 18.5 Fertiliser recommendations (kg/ha) for forage maize

	SNS*, P or K index						
	0	1	2	3	4	5	6
Nitrogen (N)	150	100	50	20	0	0	0
Phosphate (P_2O_5)	115	85	55	20	0	0	0
Potash (K_2O)	235	205	175(2−)	110	0	0	0
			145(2+)				

*Soil Nitrogen Supply Index – see Defra RB209. Where FYM and slurry have been used the above recommendations should be reduced to take account of the available nutrients so supplied. Serious pollution can be caused by over-application of organic manures. This is particularly important on fields that have grown maize for several years. Where appropriate, the relevant NVZ (Nitrate Vulnerable Zone) Action Programme measures must be complied with and, in all cases, the Codes of Good Agricultural Practice for the protection of water, soil and air. These recommendations have been based on Defra RB209 (8th edition) issued in June 2010.

isoxaflutole, S-metolachlor and pendimethalin. Post-emergence products include terbuthylazine, mesotrione, prosulfuron and bromoxynil. The use of a stale seedbed, coupled with steerage hoeing or harrow combing, is a viable alternative for those who wish to grow maize on organic farms or who wish to reduce herbicide use.

Maize eyespot (*Kabatiella zeae*) has caused severe yield loss in some years in Wales and in the south and south-west of England, and an approved fungicide such as flusilazole may be worth applying. It is important to plough in maize stubble as soon as possible after harvest to reduce carry-over of inoculum to a following crop and, where possible, to avoid following maize with maize in the same fields year on year. Smut (*Ustilago maydis*) may occur in very hot dry conditions and there are some instances too of *Septoria* leaf blotch on maize crops.

Undersowing forage maize
The establishment of Italian ryegrass in maize stubble can provide valuable additional grazing, as well as reducing, to some extent, the risk of soil erosion and nutrient leaching after harvest. Where only a non-residual herbicide, such as bromoxynil has been used for post-emergence weed control, it has also been possible to establish Italian ryegrass undersown in forage maize.

Harvesting
Harvesting usually takes place in the early autumn (weather permitting) and the best time for most crops is in early October. The ideal dry matter level is about 30% which can readily be assessed by the thumbnail test (i.e. when a thumbnail can just be pressed into the grain on the cobs). At this stage the yields of dry matter and starch are likely to be optimum. Earlier harvesting may lead to a very acid fermentation because of high sugar levels and, almost certainly, to a loss of dry matter yield. Later harvesting of very dry material may result in hard, uncracked grains in the silage and to difficulty in achieving the very fine chop required for good fermentation and efficient utilisation.

Details of the ensiling process for making maize silage are given in Section 22.2.3.

Yields
30–50 t/ha fresh yield (9–15 t DM/ha).

Cob-only maize harvesting options
The harvesting of maize cobs with forage harvesters equipped with maize picker-headers, for chopping and ensiling, gives rise to the very high energy ensiled products which have become known as ground ear maize (GEM) or corn cob mix (CCM) depending on the amount of leaf which remains with the cob before it is chopped. The remaining stover is spread on the field prior to being ploughed in.

Environmental concerns and 'cross-compliance' issues
A loss of soil structure and a subsequent loss of yield can follow from repeated harvesting of maize in the same fields in wet autumn conditions. Furthermore,

nutrient enrichment can follow from the repeated heavy dressings of FYM and slurry often applied to maize fields. This, coupled with the erosion of soil which inevitably takes place, both before and after harvest, is leading to quite serious pollution problems in some areas involving not only nitrate leaching, but also substantial losses of phosphate. Where appropriate, the relevant NVZ (Nitrate Vulnerable Zone) Action Programme measures must be complied with.

Farmers have currently to comply with the Codes of Good Agricultural Practice for the protection of water, soil and air and a series of 'Good Agricultural and Environmental Conditions' standards in order to be eligible for the Single Payment Scheme. Factors such as maintaining land drainage systems, using early maturing varieties to facilitate an early harvest and cultivating and ploughing, if possible, soon after harvest to remove wheelings and compaction, are important components of good soil management and particularly relevant to maize growers and should help to mitigate some of the problems associated with the crop. Further details of these requirements and guidance on the Soil Protection Review process can be found in the relevant Defra, SASA, DARD and Environment Agency publications.

18.3.2   Other cereals as silage crops

Good silage can be made from virtually any cereal, harvested as a whole crop. The most popular is winter wheat, for which the husbandry is virtually the same as would be the case if it were destined for grain production. Making whole crop silage provides the farmer with the ultimate in flexibility since the precise tonnage required may be harvested for ensiling and that part of the crop not required left for combining. Whole crop cereals often substitute for maize in the northern part of the UK but in other areas, too, the benefits of earlier harvesting (July or August at the latest) leave the grower with many more post-harvest cropping opportunities.

Yields
15–40 t/ha fresh yield (dependent on the dry matter per cent at harvest), 10–16 t DM/ha.

Spring cereals, too, can be taken for wholecrop silage and there is a tradition in some parts of the UK of sowing mixtures of spring cereals with legume crops as 'arable silage mixtures'. Traditionally, such mixtures contained spring barley or oats mixed with forage peas or vetches. More recently, mixtures of spring wheat or spring triticale with forage lupins have become popular. These crops are normally harvested when the cereal grain is 'cheesy' in July and yields of 6–10 t DM/ha and more in good conditions are possible.

18.3.3   Forage peas

Specific varieties of peas grown for forage purposes (such as *Magnus*) have been introduced in recent years. However, good yields may also be obtained from conventional grain varieties. Peas for ensiling are normally harvested when the pods are fully formed and the grain has started to fill. The yields are not particularly high (6–8 t DM/ha) but the great virtue of peas is their speed of growth. A crop sown in early May with reasonable average rainfall could be harvested in mid-July at about 8 tDM/ha and 20% crude protein. Earlier sown crops could be harvested in June.

Peas are an excellent crop for undersowing and the semi-leafless grain varieties in particular, sown at about 50 seeds/m^2, can offer a good cover for establishing a ryegrass/clover ley or even lucerne. In an emergency, peas can be an excellent crop for grazing and, in view of their high tannin content, there is little fear of bloat.

The husbandry of peas grown for forage is very similar to that for grain crops. The main difference is that since the crop is intended for grazing or ensiling there is little point in spending heavily on weed or disease control. In fact a good stale seedbed coupled with drilling in late April can preclude the need for a herbicide altogether. Pea and bean weevil damage to undersown legumes can, on occasions, be a problem.

18.3.4   Other annual forage crops grown for ensiling

Vetches are a traditional legume crop. Winter and spring varieties are available. Vetches are not normally grown alone but, most commonly, mixed with cereals such as oats or barley and used as an arable silage crop. They are a good smothering crop and suitable for inclusion in organic rotations for reducing weed competition and for increasing the soil nitrogen status.

Forage lupins and forage soyabeans are both crops under development in the UK with dry matter yields and protein levels apparently similar to forage peas. Lupins, in particular, are very sensitive to soil conditions, especially pH, and care should be taken to avoid high pH sites. Soyabeans have been grown for both hay and ensiling in the UK but are extremely sensitive to temperature and to day length, making the choice of a suitable variety very important. Furthermore, failure to nodulate effectively is a problem that will necessitate the use either of an effective *Bradyrhizobium* inoculant or some nitrogen fertiliser.

Other crops grown for ensiling on occasions include linseed and field beans. Little information is available on the yield potential or feeding value of these crops when grown for forage.

18.4   Sources of further information and advice

ADVANTA, *Profit from Forage Crops*, 2007.
Defra, *Fertiliser Recommendations for Agricultural and Horticultural Crops*, RB209 (8th edition), The Stationery Office, 2010.
Environment Agency, Maize Growers Association, *Managing Maize: Environmental Protection with Profit: A Guide to Profitable Maize Growing and Safeguarding the Environment*, Environment Agency, 1997.

Hocknell J, *Fodder Beet–Growers Guide and Variety Comparison*, Kingshay Farming Trust, 2001.

Kingshay Farming Trust, *Forage Costings Report*, 2008.

Lane G P F and Wilkinson J M (eds.), *Alternative Forages for Ruminants*, Chalcombe Publishing, 1998.

Maize Growers Association, Wholecrop Forage Conference: Wednesday 5th February 2003, held at Harper Adams University College, Newport, Shropshire, *Conference Proceedings*, 2003

NIAB TAG, *Livestock Crops* (annual).

Wilkinson J M, Allen D M and Newman G, *Maize: Producing and Feeding Maize Silage*, Chalcombe Publishing, 1998.

Young N E, *The Forages and Protein Crops Directory*, Context Products, 2002.

19

Introduction to grass production/ characteristics of grassland and the important species

DOI: 10.1533/9781782423928.4.454

Abstract: In this chapter the reader is introduced to the grass crop in its different manifestations, either as temporary grassland ('leys') or as permanent pasture. Approximate figures are given for the land areas covered and there is a brief introduction to the subject of the identification of the important grasses and forage legumes. An account is also given of the value of these, namely the ryegrasses and red and white clovers and example varieties of those in current use are cited. The chapter concludes with descriptions and examples of seed mixtures suitable for both conventional and organic, agricultural and equine use.

Key words: permanent pasture, ley, ryegrass, clover, seeds mixture.

19.1 Types of grassland

Grass is the United Kingdom's most important crop and grassland (including rough grazing, heath and moorland) is a valuable natural resource, covering more than 70% of the area utilised by agriculture. Recent Defra statistics show the approximate areas of the main types of grassland. They are summarised in Table 19.1. For comparison, the total area of land in arable crops in the UK in 2011 was about 4 830 000 ha.

Grassland can be broadly classified as described below.

19.1.1 Uncultivated grasslands

These represent about 42% of the total grassland area. They consist of the following types.

Table 19.1   Areas of grassland in the UK

	'000 hectares
Uncultivated grassland	
'Sole right' rough grazing	3891
Commons	1199
Cultivated grassland	
Permanent pasture	5877
Temporary grassland or 'leys'	1278

Rough mountain and hill grazing

The plants making up this type of grassland are not of great agricultural value. They consist mainly of fescues, bents, *Nardus* (mat grass) and *Molinia* (purple moor grass) as well as cotton grass, heather, bracken and gorse. Burning on a regular planned basis at between 10 and 15 year intervals can have beneficial results, both for grazing stock and wildlife. Traditionally, sheep, beef cattle and ponies were farmed in these areas with grouse shooting and deer stalking increasing in importance from the mid-nineteenth century. The payment of beef and sheep premiums, together with 'Less Favoured Area' supplements, such as the Hill Livestock Compensatory Allowances (HLCA), have encouraged overstocking, particularly with sheep, and a consequent decline in sward quality. The HLCA was replaced by the Hill Farm Allowance, in part to help with reducing stocking densities. This, in turn, was replaced in June 2010 by the Uplands Entry Level Scheme, further increasing the incentive not to overstock. In some areas, where the soil is extremely acid, conifer afforestation has been successfully carried out, but this area is now reducing.

Lowland heaths

Heather, bracken and grasses are the main species to be found here. Sheep's fescue is often the dominant grass on the predominantly acid soils. These heaths are to be found in south and east England and some of them have been reseeded. Traditional management involved regular burning and grazing by cattle and ponies, but nowadays many heaths are not grazed at all.

Calcareous downland

This occurs predominantly in southern England. Apart from herbs and broad-leaved flowering plants, grasses such as sheep's fescue and erect brome are found on these chalk and limestone soils. The traditional management of these grasslands involved predominantly grazing by both cattle and sheep. Unimproved areas of calcareous downland are now restricted to the steeper slopes.

Wetland and fen areas in east and south-west England

Any unreclaimed areas are mostly poorly drained and are dominated by water-loving plants such as cotton grass, rushes and sedges. Fens in eastern England, when drained, are associated with intensive arable cropping. In south-west

England areas such as the Somerset Levels have been partly improved by reseeding but are frequently inundated by floodwater.

Maritime swards
These consist of salt marshes and coastal dune areas with marram and cord grasses. The 'machair' areas of coastal Scotland and the Western Isles contain a wide variety of rare plants and offer a unique and valuable habitat for wildlife.

19.1.2   Cultivated grasslands
These represent about 58% of the total grassland area. They consist of the following.

Permanent pasture
The statutory definition of permanent pasture is grassland which is more than five years old. An alternative one would be 'grassland not normally included in an arable rotation'. The agricultural value of a permanent pasture is dependent mainly on its content of perennial ryegrass and white clover. A sward containing more than 30% of its annual dry matter production as perennial ryegrass would be very productive and would not normally be thought to require much improvement. Other, less productive grasses, such as bents (*Agrostis* spp.) and meadow-grasses (*Poa* spp.) and a great many more, make up the balance. Although much less productive than the ryegrasses, these constituents perform a useful function in that they create a 'thatch' over the surface of the soil which, if well developed, has the capacity to carry stock physically and reduces the severity of the well known phenomenon of treading or 'poaching' which can occur in wet conditions. In addition to clover and other legumes, a large number of broad-leaved plants also occur. Many of these, such as buttercup, dock, thistles and ragwort are weeds and ragwort is extremely poisonous. Some others, however, such as yarrow, burnet and ribwort are termed herbs and thought to have beneficial effects on grazing livestock.

Significant areas of species-rich permanent pasture are the subject of statutory protection, e.g. as Sites of Special Scientific Interest (SSSIs) or as EU-designated Special Areas of Conservation. Others have been targeted by one of a number of agri-environmental schemes, such as the classic Countryside Stewardship and Environmentally Sensitive Area schemes or the current Environmental Stewardship scheme in England and their equivalents by the devolved administrations in Scotland, Wales and Northern Ireland. Management strategies for such pastures are normally designed to maintain soil fertility at low or decreasing levels, in order to maintain or increase the species diversity and structure.

Temporary grassland or 'leys'
These are temporary swards which have been sown to grass or grass/clover mixtures for a period of up to five years. A less precise definition is 'grass in an

arable rotation'. Leys represent an important opportunity for the restoration of fertility to arable land. Fields which have been grazed, or which have included significant quantities of forage legumes, will contribute substantial amounts of available nitrogen to the following crops. Opportunities exist also for the reduction of some of the weed, pest and disease problems associated with arable farming. Winter wheat grown after a ley will normally yield very well indeed. In organic systems clover leys are one of the main ways in which soil fertility can be enhanced for subsequent arable cropping.

19.2   The nutritive value of grassland herbage

The great value of grass and indeed of other forages is their potential to provide cheap (often the cheapest) sources of energy and protein for farm livestock. Many calculations have been carried out to demonstrate their value and, in particular, the way in which good grassland management, and the achievement of high yields of digestible nutrients, can bring about worthwhile improvements in livestock performance, and a reduction (or complete elimination) of the need to buy in other feeds. Data published by Kingshay Farming Trust and by the Agriculture and Horticulture Development Board (previously by the Milk Development Council) clearly demonstrate this and, in particular, point to the very low costs associated with grazing.

19.2.1   The effects of grass maturity

Short, leafy grass is rich in protein and highly digestible. As it matures the proportion of yield represented by cell walls increases whereas the proportion of cell contents decreases. This results in an increase in the percentage of fibre and a decrease in the percentage of crude protein with maturity. As a result of the much greater leaf and stem area of the mature plant, and the greater opportunity for photosynthesis, the sugar (water soluble carbohydrate or WSC) percentage also increases with maturity. This explains why it is usually easier to make a well-fermented silage from mature grass and why it is more difficult to do so from young, leafy grass.

19.2.2   Digestibility ('D value')

Digestibility is one of the main characteristics which determines the feeding value of herbage plants, i.e. the amount of the plant which is actually digested by the animal. In the past, the digestibility of forages was only determined by *in vivo* animal feeding trials. With the development of the laboratory *in vitro* technique which simulates rumen digestion, it is possible for the digestibility of any species of herbage plant to be assessed very much more easily and cheaply without using animals directly.

The digestibility of herbage plants is expressed as the 'D value' or DOMD. It is the digestible organic matter expressed as a percentage of the dry matter. High digestibility values are normally desirable as it means that animals are able to obtain large amounts of nutrients from the herbage being fed. It is now accepted that digestibility is a major factor affecting intake. Animals usually eat more of a highly digestible feed. However, other factors, such as sward height and density in a grazing situation and the pH, chop length and dry matter percentage of silages are also important. Dry matter yield increases as the plant grows; the proportion of leaf (the most nutritious and digestible part of the plant) decreases and the proportion of stem (more fibrous, lignified and less digestible) increases. This causes a gradual decline in the D value of the whole plant with maturity.

The rate of decline depends on the grass species and the variety. Cocksfoot, for example, has a lower D value than the ryegrasses. The D value of particular varieties on any date is usually determined by the earliness or lateness of flowering. The D value of first cuts normally declines at a rate of between 3 and 5 units per week. The rates of decline in Timothy and Italian ryegrasses are slower than in the perennial ryegrasses. D values also differ between grass and legume species. For example, white clover, when young, has a D value of about 80 and the rate of decline is only about 0.8 units per week. Red clover and lucerne are substantially less digestible and their rates of decline are between 2.5 and 2.8 units per week.

When cutting for silage, higher yields will result when the crop is cut at a more mature stage of growth. Many farmers aim to cut grass at or before it achieves a D value of 67. When making field-cured hay, a D value of 60 or even less is a more realistic expectation. This reflects the more mature nature of the crop which can be expected by the time the weather becomes suitable for hay making (if it ever does!). Information about the relative D values of different species and varieties of grasses can be found in the *Recommended Grass and Clover Lists* leaflet for England and similar information published by the devolved administrations in Scotland, Wales, and Northern Ireland.

The approximate conversion of D values to ME (metabolisable energy) data for rationing purposes can be accomplished by the application of the following simple equations:

For fresh and dried grass and hay, ME $(MJ/kgDM) = 0.15 \times DOMD$
For grass silage, ME $(MJ/kgDM) = 0.16 \times DOMD$

19.3   Identification of grasses

19.3.1   Grass species of economic value

Over 150 different species of grasses can be found growing in this country, but only a few are of any importance to the farmer. They are: short duration ryegrasses, perennial ryegrass, Timothy, cocksfoot and meadow fescue. With the exception of the short duration ryegrasses, most of the grasses described in Table 19.2 may be found in permanent pastures, either as indigenous types or bred varieties which were sown when the pasture was established. Before it is possible to recognise plants in a grass field, it is necessary to know something about the parts which make up the plant.

Table 19.2   Recognition of grasses in the vegetative stage

	Short duration ryegrasses	Perennial ryegrasses	Meadow fescue	Cocksfoot	Timothy
Leaf sheath	Definitely split	Split or entire	Split	Entire at first, later split	Split
	Pink at base	Pink at base	Rolled in shoot	Folded in shoot	Pale at base
	Rolled in shoot*	Folded in shoot			Rolled in shoot
Blade	Broad	Narrow	Narrow	Broad	Broad
	Margin smooth	Margin smooth	Margin rough	Margin rough	Margin smooth
	Dark green	Dark green	Lighter green	Light green	Light green
Lower side of blade	Shiny	Shiny	Shiny	Dull	Dull
Ligule	Blunt	Short and blunt	Small, blunt, greenish-white	Long and transparent	Prominent and membranous
Auricles	Medium size and spreading	Small, clasping the stem	Small, narrow and spreading	Absent	Absent
General	Veins indistinct when held to light	Not hairy	Veins appear as white lines when held to light	Not hairy	Base of shoots may be swollen
	Not hairy		Not hairy		Not hairy

*Identification of hybrid ryegrass is more difficult; it can show a mixture of vegetative features similar to both Italian and perennial ryegrasses. Round tillers predominate.

19.3.2   Identification of vegetative parts

Stems

There are two types of stems found on grass plants. They are:

- The *flowering stem or culm*. This grows erect and produces the flower. Most stems of annual grasses are culms.
- The *vegetative stem*. This does not produce a flower, and has not such an erect habit of growth as the culm. Perennial grasses have both flowering and vegetative stems.

Leaves

These are arranged in two alternate rows on the stem and are attached to the stem at a node. Each leaf consists of two parts (Fig. 19.1):

- The *sheath* which is attached to the stem.
- The *blade* which diverges from the stem.

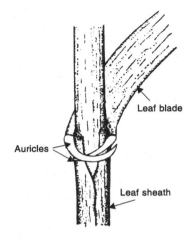

Fig. 19.1   Parts of the grass leaf.

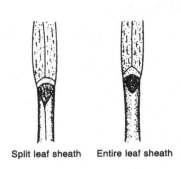

Split leaf sheath   Entire leaf sheath

Fig. 19.2   Parts of the grass leaf: sheath.

Fig. 19.3   Parts of the grass leaf: rolled
leaves leading to round shoots.

Fig. 19.4   Parts of the grass leaf: folded
leaves leading to flattened shoots.

The leaf sheath encloses the buds and younger leaves. Its edges may be joined
(entire) or they may overlap each other (split) (Fig. 19.2). If the leaves are rolled
in the leaf sheath, the shoots will be round (Fig. 19.3), but when folded the shoots
will be flattened (Fig. 19.4).

Other structures
At the junction between the leaf blade and leaf sheath is the *ligule*. This is an
outgrowth from the inner lining of the sheath (Fig. 19.5). *Auricles* may also be
seen on some grasses where the blade joins the sheath. These are a pair of claw-
like outgrowths (see Fig. 19.1).

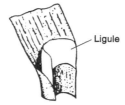

Fig. 19.5   Parts of the grass leaf: ligule.

In some species, the leaf blade will show distinct veins when held against the light and, according to the variety, the underside may be shiny or dull.

19.3.3   The inflorescence or flower head

The inflorescence consists of a number of branches called spikelets which carry the flowers. There are two types of grass inflorescence:

- S*pikes*, where the spikelets are attached to the main stem without a stalk (Fig. 19.6).
- *Panicles*, where the spikelets are attached to the main stem by a stalk (Fig. 19.7).

In some grasses the spikelets are attached to the main stem with very short stalks to form a dense type of inflorescence termed 'spike-like' (Fig. 19.8). The spikelet is normally made up of an *axis*, bearing at its base the *upper* and *lower glumes* (Fig. 19.9). Most grasses have two glumes. Above the glumes, and

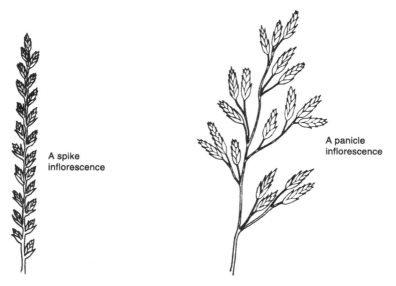

A spike
inflorescence

A panicle
inflorescence

Fig. 19.6   A spike inflorescence.   **Fig. 19.7**   A panicle inflorescence.

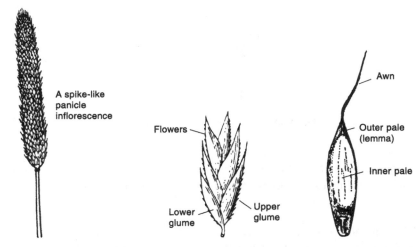

Fig. 19.8 A spike-like panicle inflorescence.

Fig. 19.9 The spikelet.

Fig. 19.10 The pales.

arranged in the same way, are the *outer* and *inner pales*. In some species these pales may carry *awns* which are usually extensions from the pales (Fig. 19.10).

Within the pales is the flower. The flower consists of three parts (Fig. 19.11):

- The male organs – the *stamens*.
- The female organ – the rounded *ovary* from which arise the feathery *stigmas*.
- A pair of *lodicules* – at the base of the ovary. These are indirectly concerned with the fertilisation process, which is basically the same in all species of plants (Fig. 1.22).

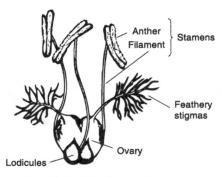

Fig. 19.11 The flower of the grass.

19.4   Identification of legumes

The important perennial forage legumes used by farmers in the UK are red and white clovers and lucerne. Sainfoin, birdsfoot trefoil and alsike are less important. A detailed description of each is given in Table 19.3.

Leaves
With the exception of the first leaf (which may be simple), all leaves are compound. In some species the mid-rib is extended slightly to form a mucronate tip. Other features on the leaf may be serrated margins, the presence or absence of marks, colour and hairiness (Fig. 19.12 and Fig. 19.13). The leaves are arranged alternately on the stem, and they consist of a stalk which bears two or more leaflets according to the species (Fig. 19.14).

Stipules
These sheath-like structures are attached to the base of the leaf stalk. They vary in shape and colour (Fig. 19.12).

Flowers
The flowers are brightly coloured and, being arranged on a central axis, form an indefinite type of inflorescence (see Fig. 19.14).

Table 19.3   How to recognize the important perennial forage legumes

	Leaves, etc.	Stipule	General	Species
Mucronate tip	Centre leaflet with prominent stalk	Broad, serrated and sharply pointed	May be hairy	Lucerne
	Leaflets serrated at tip 6–12 pairs of leaflets, plus a terminal one	Thin, finely pointed	Stems 30–60 cm high Slightly hairy	Sainfoin
	Trifoliate (or pentafoliate if including the stipules)	Leaflet-like	Not hairy	Birdsfoot trefoil (yellow and orange/red flowers)
No mucronate tip	Trifoliate, dark green with white half-moon markings on upper surface	Membranous with greenish-purple veins Pointed	Hairy	Red clover
	Trifoliate, serrated edge with or without markings on upper surface	Small and pointed	Not hairy	White clover
	Trifoliate, serrated edge, no leaf markings	Long tapering point, never red-veined	Not hairy	Alsike clover (pink or white flowers)

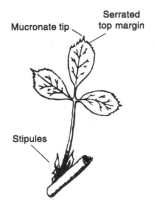

Fig. 19.12   Possible legume leaf features.

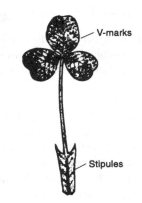

Fig. 19.13   Possible legume leaf feature variations.

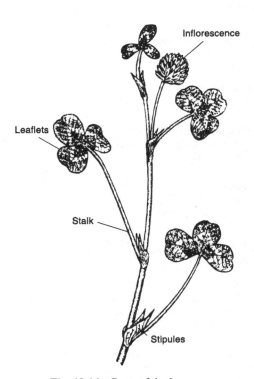

Fig. 19.14   Parts of the legume.

19.5   Grasses of economic importance

19.5.1   Types of grass varieties

Details of good quality, reliable varieties for England and Wales, from UK and European plant breeders, can be found in the *Recommended Grass and Clover Lists* leaflet which is funded by the Grass Levy Scheme and is published annually using data from trials carried out by NIAB TAG and evaluated by a panel of experts. Farmers in Scotland and Northern Ireland have access to information of a similar quality published by SAC and DARD respectively. Seed catalogues and the online information provided by reputable seed merchants can also be valuable sources of information. Details of grasses which have been bred for amenity and sports surfaces are provided by the Sports Turf Research Institute (STRI) also on an annual basis.

Early and late varieties

The *Recommended Grass and Clover Lists* leaflet classifies perennial ryegrass varieties according to date of heading (the date when 50% of the ears in fertile tillers have emerged) as 'early', 'intermediate' and 'late'. Early heading varieties are usually associated with early spring growth and a less densely tillered plant with an erect habit of growth. These varieties are not very persistent. In the past, they were referred to as hay types. Late heading is associated with densely tillered plants which generally means good persistency. They have a prostrate growth habit and are more suitable for grazing. These varieties used to be called pasture or grazing types. Intermediate varieties combine the advantages of both the previous groups of varieties, namely relatively early spring growth but later flowering and good persistency. They were previously referred to as dual-purpose types.

Further information given by NIAB TAG concerns the comparative dry matter yields under cutting or simulated grazing regimes, the approximate heading dates for the first cut, D values, and a 'ground cover index' which gives an indication of persistency. Useful details are also given about the resistance of individual varieties to crown rust and, in the case of Italian and hybrid ryegrasses, to ryegrass mosaic virus and mildew. It should be remembered that these recommended lists are amended and updated with new information each year as plant breeders introduce improved varieties.

Diploids and tetraploids

Tetraploid varieties have bigger cells and twice the number of chromosomes compared with diploids. They also have larger seeds and bigger leaves and establish quicker than diploids. The *Recommended Grass and Clover Lists* leaflet gives details of whether varieties are diploid or tetraploid. Seed merchants' catalogues usually provide similar information. Tetraploid varieties are often included in mixtures to improve palatability because of their high sugar content. The dry matter content of tetraploids are usually lower than those of diploids and their persistency and winter hardiness are also often inferior. As a result of being less 'aggressive' with fewer tillers, tetraploids make excellent companion grasses for legumes, especially white clover. Their high sugar content also makes them ideal for conservation as silage, particularly when the grass is wilted. They are often the first choice, too, for the very high dry matter (<65% DM) 'haylage' which has become so popular as a roughage feed for horses. However, when a ley is being established specifically for making hay, some farmers prefer to stipulate diploid varieties, since tetraploids may prove more difficult to dry. However, tetraploid varieties are sometimes preferred by the buyers of hay for the equine trade.

19.5.2   Details of individual species

Short duration ryegrasses
This group consists of varieties of three species: Westerwolds, Italian and hybrid ryegrasses.

Westerwolds ryegrass
This is an annual and the quickest growing of all grasses (Fig. 19.15 and Table 19.2). A good crop can often be obtained within 12–14 weeks of sowing. It should not be undersown. It performs best when direct sown in the spring and summer, as it is not winter-hardy, except in very mild districts. NIAB does not recommend any varieties but a current one is *Lifloria. Hellen, Pollanum* and *Lemnos* are recent examples of more winter-hardy varieties.

Fig. 19.15   Short duration ryegrass.

Italian ryegrass
This is short-lived (most varieties persist for 18–24 months) and very quick to establish. Sown in the spring, Italian ryegrass can produce good growth in its seeding year and an early grazing the following year. For optimum production it is best sown in summer or early autumn and will then produce a full crop in the following year. It does well under most conditions, but responds best to fertile soils and plenty of nitrogen. Like all ryegrasses, its winter-hardiness is improved when surplus growth is removed in the autumn. Although stemmy, it is palatable with a high digestibility. *Muriello, Alamo* and *Davinci* are examples of current NIAB-recommended diploid varieties. *Gemini, Dorike* and *Danergo* are examples of recommended tetraploids.

Hybrid ryegrass
Some varieties in this group are similar to the less persistent Italian ryegrass varieties and others to the perennial ryegrasses. A feature of hybrid ryegrasses is their increased longevity (3–4 years) compared with Italians, and good resistance to diseases. Most hybrid varieties are tetraploids. However, annual yields are usually lower than for the Italians.

Solid and *AberExcel* are examples of currently recommended varieties. *AberEcho* and *AberEve* are recent introductions with improved yields and, in the case of *AberEcho*, very good early spring growth.

Perennial ryegrass
Although it forms the basis of the majority of long leys and is the most important grass found in good permanent pasture, perennial ryegrass is, depending on the variety, also used in medium term leys (Fig. 19.16 and Table 19.2). It is quick to establish and yields well in the spring, early summer and autumn. It does best under fertile conditions and responds well to nitrogen. Crown rust disease is a problem – particularly in the south-west of England. The choice of a resistant variety is the best method of control (Table 6.2). Examples include:

Early varieties:

- *January* and *Kimber* are examples of recommended diploid varieties.
- *Anaconda* and *AberTorch* are examples of recommended tetraploids.

Intermediate varieties:

- *Butara* and *AberDart* are examples of recommended diploid varieties.
- *Orion* and *AberGlyn* are examples of recommended tetraploids.

Late varieties:

- *Foxtrot, Inoval* and *Cancan* are examples of recommended diploid varieties.
- *Delphin* and *Twymax* are examples of recommended tetraploids;

Details of perennial ryegrass varieties suitable for amenity and for sports turf use can be found in the annual recommended list of the Sports Turf Research Institute.

Fig. 19.16   Perennial ryegrass.   **Fig. 19.17**   Timothy.

Timothy

Timothy is fairly slow to establish and is not particularly early in the spring (Fig. 19.17 and Table 19.2). It is less productive than the other commonly used grasses, but is very palatable, although its digestibility is not as good as that of the ryegrasses. It is often included in grazing leys with perennial ryegrass. Timothy is winter-hardy and does well under a wide range of conditions, except on very light dry soils. Examples of recommended varieties include:

- Early: *Promesse.*
- Intermediate: *Motim.*

Cocksfoot

This is quick to establish and fairly early in the spring (Fig. 19.18 and Table 19.2). It is a moderately high-yielding grass, but unless heavily stocked, most of the existing varieties soon become coarse and unpalatable. Cocksfoot is a deep-rooting grass and therefore often used in dry or drought-prone situations.

Cocksfoot does not need fertile soil conditions and produces well in late summer and autumn. Most varieties show good winter-hardiness. Cocksfoot should be considered as a special-purpose grass on drier, lighter soils in areas of low rainfall. Cocksfoot varieties are not as digestible as the other important grasses.

Example varieties: *Sparta* and *Prairial.*

Meadow fescue

Meadow fescue is not used very much now (Fig. 19.19 and Table 19.2). This is because the once famous Timothy/meadow fescue mixture is no longer popular; it

Fig. 19.18   Cocksfoot.   **Fig. 19.19**   Meadow fescue.

is not so productive as the perennial ryegrass-based swards. Meadow fescue is slow to establish, but fairly early to start growth in the spring and has a high digestibility. NIAB does not recommend any varieties at present. Examples of those currently available are *Rossa* and *Cosmolit*.

Other grasses and weed grasses
Tall fescue is a perennial. Once it is well established, varieties such as *Falcon* and *Starlett* are useful for early grass in the spring, but production after this is mediocre. It is very hardy and it can be grazed in the winter. Because of its long growing season it is sometimes used for green crop drying enterprises. The small fescue grasses, such as creeping red fescue and sheep's fescue, are useful under hill and marginal land conditions, and in some situations they will produce more than perennial ryegrass. They have no practical value under lowland farm conditions. They are typical 'bottom' grasses, and they produce a close and well-knit sward. Chewings fescue is often included in seed mixtures for lawns and playing fields. Strong creeping red fescue, as its name implies, is a very resilient grass and a frequent constituent of grass horse gallops. It is also sometimes included in grass mixtures for horse paddocks because of its hard wearing capability. *Boreal* and *Reverent* are frequently used varieties.

The brome grasses occur in both permanent pasture (e.g. meadow brome) or as weeds of arable land (e.g. sterile or barren brome). Improved varieties of bromes (e.g. *Grasslands Matua*) have been introduced in the past from New Zealand but have proved unsuccessful due to lack of persistency under UK conditions.

Rough stalked meadow-grass is indigenous to the majority of soils, but it prefers more moist conditions. In the later years of a long ley, it can make a useful contribution to the total production of a sward if white clover has been suppressed by heavy nitrogen fertilising. Rough stalked meadow-grass has become an important arable weed on some farms. Smooth stalked meadow-grass (Kentucky blue grass) spreads by rhizomes, and can withstand quite dry conditions. It is very persistent and hard wearing and is sometimes included in mixtures for horse paddocks or gallops. Annual meadow-grass is a very common weed of arable and grassland.

Crested dog's tail is not very palatable because of its wiry inflorescence. These days it is only used in seed mixtures for lawns, playing fields and sometimes for horse paddocks.

The bents (*Agrostis* spp.) are very unproductive and unpalatable grasses and, under most conditions, they can be considered as weeds. Creeping bent is very common both in arable land and permanent pastures where it performs a useful function in creating a 'thatch' which can help to reduce poaching. Black bent and loose silky bent are important arable weeds on light land in south and east England. Browntop bent is often included in mixtures for lawn and amenity turf and Highland bent is a useful constituent of grass gallops, racecourses and amenity areas.

Yorkshire fog is an extremely unpalatable grass except when very young. It is especially prevalent under acid conditions where fields have been repeatedly cut for hay. Barley grass is a common weed of permanent pastures in southern England. It is also extremely unpalatable.

Tall oat grass is another very common grass of permanent pastures and hay meadows, and, as onion couch, it is also an important weed of arable land. Common couch occurs widely in permanent pastures and arable land but is less of a problem since the advent of glyphosate-based herbicides.

Blackgrass (also known as slender foxtail) and wild oat are important arable weeds as are both Italian and perennial ryegrasses, particularly in situations where they have been allowed to set seed. Awned canary grass is an increasing problem on arable land in south-east England.

A guide to the recognition of some of these are given in Table 19.3. More detailed descriptions and illustrations can be found online and in standard reference books such as Hubbard (1984).

19.6   Forage legumes of economic importance

The clovers of greatest agricultural importance are the red and white clovers. Red clovers are used in short leys for conservation. White clovers are used for longer duration grazing leys where they act as 'bottom plants'; with their creeping habit of growth they knit the sward together and help to keep out weeds. Although the majority of clovers are palatable, with a high feeding value, they are not so productive as the grasses and they should not be allowed to dominate the sward (aim for 30% clover).

The digestibility of red clover and most other legumes, except white clover, is not as high as that of the grasses, but their voluntary intake at equal digestibility is higher. Because of the ability of legumes to fix atmospheric nitrogen through the action of the *Rhizobium* bacteria in their root nodules, a balanced mixture of grasses and clover enables farmers to economise on fertiliser nitrogen inputs, and will leave a useful legacy of nitrogen for the following crop. This is particularly important in the case of organic and other low input farms, where clover-based leys are one of the most important ways of naturally enhancing the fertility of the soil prior to arable cropping. Estimates of the quantity of nitrogen supplied to a mixed sward by a strong stand of clover, under UK conditions, vary from 150–200 kg N/ha/year.

19.6.1   Red clovers

Red clovers are usually a constituent of short term leys for cutting for hay or silage (Fig. 19.20 and Table 19.3). Normally the mixture would also contain Italian or hybrid ryegrasses or Timothy although red clover can be sown alone. Although used more for conservation, some of the more persistent improved varieties are useful for aftermath grazing. In recent years red clover leys have

Fig. 19.20   Red clover.

become very popular for fattening store lambs. No breeding work had been undertaken in the UK for many years but the Institute of Biological, Environmental and Rural Sciences (IBERS) (formerly the Institute of Grassland and Environmental Research (IGER)) at Aberystwyth started a programme in the late 1990s. There are two main types:

- *Early red clover.* Merviot is the most commonly used variety and the industry standard. It has good resistance to *Sclerotinia* (clover rot, Table 6.2), but is susceptible to stem eelworm (Table 7.1). *AberRuby, AberClaret* and *AberChianti* are all introductions from the IBERS breeding programme. *Milvus* (not currently a NIAB recommended variety) is an introduction from Switzerland which is very persistent and has good resistance to *Sclerotinia* and stem eelworm.
- *Late red clover.* These varieties are later to start growth in the spring than the early red clovers. *Altaswede* (not currently a NIAB recommended variety) yields about 10% less than the early varieties but is more persistent.

19.6.2   White clovers
These should be regarded as the foundation of the grazing ley (Fig. 19.21 and Table 19.3). They are not so productive as the red clovers but are much more persistent. There are four types, classified according to leaf size.

- *Very large-leaved white clovers.* The only clover in this group is *Aran*. It has become popular in recent years due to its ability to survive in conditions of high fertiliser nitrogen application and is suitable for inclusion in mixtures for dairy cow grazing and cutting. *Aran* is resistant to slug damage but shows poor resistance to *Sclerotinia* infection (Table 6.2).
- *Large-leaved white clovers.* Alice is one of the most popular varieties from this group and it is useful in a similar context to *Aran*.
- *Medium-leaved white clovers.* These are normally included in leys where a variety of different managements are likely. *AberHerald* is a variety which claims earlier spring growth than many others. *Crusader* and *AberDai* are also very popular.
- *Small-leaved white clovers.* These are used in the long ley and are particularly favoured for sheep grazing. They are rather slow to establish but can become dominant. *Kent wild white, AberAce* and *S184* are the most popular varieties in this category.

Clover blends
Since the management applied to leys containing white clovers can vary so much (sheep/cattle grazing, silage/hay cutting) it has become common practice to include blends of varieties of white clover from several of the categories described above, in ley mixtures. This practice is designed to ensure that one or more varieties can contribute well under each of the different management systems imposed.

Fig. 19.21   White clover.

19.6.3   Lucerne (or alfalfa)

This is a very deep-rooted legume and it is therefore useful on dry soils, although it can be grown successfully under a wide range of soil conditions provided drainage is good (Fig. 19.22 and Table 19.3). For best establishment lucerne should be sown in the spring (May is the ideal month). Undersowing or direct sowing after winter barley harvest are less satisfactory. Lucerne can be sown alone or with a companion grass such as cocksfoot or meadow fescue. It yields best, however, as a single species stand. In the UK, lucerne is mainly used for silage or haymaking but it is also very suitable for zero grazing in dry summers. Lucerne also forms the basis of several large-scale green crop drying ventures. The main problems associated with lucerne growing in the UK concern Verticillium wilt, a soil-borne fungal disease (Table 6.2) and stem eelworm. A variety which shows good resistance to both is *Vertus*. There are no NIAB recommendations for Lucerne at the moment. The majority of seed used in the UK is now being imported from France. Resistance to stem eelworm is an important factor.

19.6.4   Sainfoin

The growth characteristics of this most traditional of forage legumes are similar to those of Lucerne (Fig. 19.23 and Table 19.3). Generally, it yields less well than

Fig. 19.22   Lucerne.

Fig. 19.23   Sainfoin.

lucerne or red clover but can do reasonably well on dry calcareous soils. Traditionally it was used for hay (especially valued for horses) and the aftermath growth was used for fattening lambs. Sainfoin is currently the subject of much scientific research as it has two unusual qualities; it is non-bloating which is unusual for a legume and it is also a natural anthelmintic. The most commonly used companion grass is meadow fescue but a late heading tetraploid perennial ryegrass, such as *Condesa*, will also give good results. Spring (April/May) sowing is preferred and sainfoin will establish well both direct sown and undersown in spring cereals.

- *English giant*. This is high yielding but short lived, usually for just one harvest year. There is currently very little seed available.
- *Common sainfoin*. There are just two traditional varieties available in the UK: *Hampshire common* and *Cotswold common*.

19.6.5   Birdsfoot trefoil or lotus
Although indigenous in many lowland pastures Birdsfoot trefoil is becoming increasingly popular as a constituent of grazing mixtures alongside clovers (Table 19.3). In common with sainfoin it is a non-bloating legume and also has some anthelmintic properties. Examples of current varieties are *Leo* and *Bull*.

19.6.6   Alsike
Alsike is sometimes included as a substitute for part of the red clover content of a ley (Table 19.3). It grows better in acidic or wet soils than red clover. It should never be used for horse hay as it can cause photosensitivity and even liver damage.

19.6.7   Other forage legumes
Wild red clover and Yellow suckling clover are indigenous species in many pastures but of low productivity. Crimson clover and Black medick (also known as Trefoil or Yellow trefoil but not to be confused with the perennial Birdsfoot trefoil) are sometimes sown as 'winter annuals' for subsequent hay or silage crops. Sometimes these species are mixed with Italian ryegrass, or even, in an organic situation especially, sown as a green manure crop to increase the soil nitrogen supply for a spring-sown crop.

19.7   Herbs

These are deep-rooting plants which are generally beneficial to pastures. However, to be of any value they should be palatable, in no way harmful to stock, and they should not compete with other species in the sward. They have a high mineral content which may benefit the grazing animal.

Yarrow, chicory, rib grass and burnet (Figs 19.24–19.27) are the most useful of the many herbs which exist. They can be included in a seed mixture for a grazing

Fig. 19.24   Yarrow.

Fig. 19.25   Chicory.

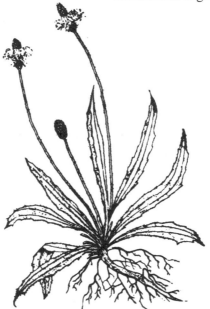

Fig. 19.26   Rib grass.

Fig. 19.27   Burnet.

type of long ley. They are not cheap, however, and one or more of them may well establish in a sward of its own accord. Chicory is occasionally sown as a crop specifically for fattening lambs and a New Zealand variety *Grasslands puna* has proved successful and popular in the UK.

19.8   Grass and legume seed mixtures

19.8.1   Traditional mixtures

Grasses and legumes may be sown as single species or even single variety swards but it has become normal (sometimes for poorly defined reasons) to sow mixtures. It is of course possible to create a rationale for this phenomenon. So, for example, the mixing of species or varieties which exhibit desirable factors such as high yields, early spring growth, palatability, persistency and winter hardiness all in the same field would seem to offer the farmer the combined benefits of all of them. Add to this further factors such as drought tolerance, suitability for cutting or grazing by a variety of different stock, to say nothing of the desirability of clover and herbs in a mixed sward, and it is possible to devise a mixture of such complexity, and with so many constituents, that it is likely that none of the desired characteristics would be demonstrated to any worthwhile degree. However, this was quite common practice for many years and so, for example, the traditions of the 'Cockle Park' and 'Clifton Park' mixtures and those devised by Frank Newman Turner survive even to this day. Indeed, the need to re-establish some of the species rich grasslands so rashly discarded over the last 50 years has created a resurgence of interest in them.

19.8.2   Modern seed mixtures

The following section contains examples of mixtures which would be suitable for mainstream agricultural and organic use. Examples of good quality varieties recommended by NIAB for general use have already been given in Sections 19.5 and 19.6. The reader should not feel restricted by these however, and it is strongly recommended that the current issue of the *Recommended Grass and Clover Lists* leaflet, together with information from reputable seed houses, be consulted for the most up-to-date information. Organic producers should also check for the availability of organically produced seed, and use it if available at the appropriate inclusion rate.

Short-term mixtures

Westerwolds ryegrass is only suitable for a one-year crop and is normally sown in the spring for grazing or cutting in the same year. The majority of short term (one or two year) mixtures incorporate Italian ryegrass varieties because of their very high yield potential. They are suitable for early grazing and for cutting for hay or silage. The inclusion of early perennials or hybrids is often practised in order to extend the potential of the ley into a third year. However, yields will almost certainly suffer in year three as the Italians will have died out. Tetraploid varieties with high sugar contents are favoured for silage and haylage production. Where it

Table 19.4   Examples of short-term leys (1–2) years

Plant	Ley characteristics
35 kg/ha Westerwolds ryegrass or 25 kg/ha Westerwolds ryegrass 10 kg/ha Italian ryegrass 35 kg/ha (14 kg/acre)	A one-year ley ideal for sowing in spring or autumn; not suitable for undersowing; it should produce two good silage cuts within the year. Expensive option but has been excellent where waterlogging has damaged previous grass crop.
32–40 kg/ha Italian ryegrass* or 16–20 kg/ha Italian ryegrass* 16–20 kg/ha hybrid ryegrass* 32–40 kg/ha (13–16 kg/acre)	These leys can be productive for up to two years (some extended to three where conditions suitable); usually autumn-sown after cereals or maize; suitable for early bite grazing and subsequent silage cuts.
16–20 kg/ha hybrid or Italian ryegrass* 10 kg/ha early red clover 26–30 kg/ha (10.5–12 kg/acre)	Suitable for conventional or organic use over two years (hybrid up to three); suitable for silage, hay and aftermath grazing; select diploid ryegrass varieties for haymaking.

*Blends of diploids and tetraploids; where tetraploid varieties are predominant, use the higher seed rate.

is intended to make traditional hay regularly, diploid varieties are sometimes preferred. The most suitable types of clover are the large- or very large-leaved white clovers and/or red clover. Example mixtures are shown in Table 19.4.

Medium-term cutting/grazing mixtures
This type of ley typically will be taken for one or more silage cuts and followed by grazing. Early, intermediate and late perennial ryegrasses will all be suitable but obviously early heading perennials would be favoured for early cutting. These mixtures also contain late perennial ryegrasses which are most suited to grazing. It is normal to include both tetraploid (more palatable and better for making silage) together with diploid (more persistent) varieties. A white clover blend including both large- and very large-leaved and medium-leaved varieties would be suitable. Example mixtures are given in Table 19.5.

Long-term grazing mixtures
Intermediate and late perennial ryegrasses predominate in such mixtures and some tetraploids and Timothy can be included for improved palatability. However, the lower persistency of the tetraploids precludes their large-scale use in these mixtures. A white clover blend with small- medium- and large leaved varieties should be used, especially where grazing may be carried out by sheep as well as cattle. Example mixtures are given in Table 19.6.

Mixtures including lucerne and sainfoin
Lucerne and sainfoin may be sown with a grass companion but there is some evidence that lucerne, in particular, yields better as a single species sward. Red

Table 19.5 Examples of medium-term leys (up to 5 years)

Plant	Ley characteristics
15 kg/ha tetraploid hybrid ryegrass 10 kg/ha intermediate perennial ryegrass* 5 kg/ha late perennial ryegrass* 5 kg/ha Timothy 2.5 kg/ha large or very large-leaved white clover 37.5 kg/ha (15 kg/acre)	A 'dual-purpose' type mixture suitable for one or two cuts of silage followed by aftermath grazing. For later cutting the proportion of early ryegrass varieties could be reduced and replaced with intermediate and late varieties.
12 kg/ha tetraploid hybrid ryegrass 7 kg/ha intermediate perennial ryegrass* 6 kg/ha late perennial ryegrass* 5 kg/ha cocksfoot 4 kg/ha Timothy 2.5 kg/ha white clover blend 5 kg/ha early red clover 41.5 kg/ha (17 kg/acre)	A deeper rooting ley, suitable for inclusion in an organic rotation, with a high proportion of red and white clovers. This ley could be cut for silage and then grazed. Alsike clover, which can grow well in low fertility, wet or acidic situations could be substituted for part of the clover content. Alsike should not be used in leys intended for horse hay.

*Blends of diploids and tetraploids.

Table 19.6 Examples of long-term leys (over 5 years)

Plant	Ley characteristics
10 kg/ha intermediate perennial ryegrass* 20 kg/ha late perennial ryegrass* 5 kg/ha Timothy 2.5 kg/ha medium/large leaved white clover blend 37.5 kg/ha (15 kg/acre)	A long-term mixture suitable for intensive grazing by dairy cows.
30 kg/ha late perennial ryegrass* 3 kg/ha small leaved white clover 33 kg/ha (13.5 kg/acre)	An extreme long-term pasture type ley for grazing by sheep or horses.
5 kg/ha early perennial ryegrass 5 kg/ha intermediate perennial ryegrass 15 kg/ha late perennial ryegrass* 3.75 kg/ha late red clover 3.75 kg/ha white clover blend 2.5 kg/ha birdsfoot trefoil 35 kg/ha (14 kg/acre)	A long-term mixture suitable for organic farming. Birdsfoot trefoil is included to reduce the problem of bloat. It may also have some anthelmintic benefit. May also be suitable as a productive sward for equine use, needing little artificial nitrogen.

*Blends of diploids and tetraploids.

clover, cocksfoot and meadow fescue are suggested as suitable species for inclusion. Examples of suitable mixtures are shown in Table 19.7.

Mixtures suitable for equine use
Table 19.8 gives some examples of mixtures suitable for equine use.

Table 19.7 Examples of mixtures incorporating lucerne and sainfoin

Plant	Characteristics of mixture
90 kg/ha sainfoin (unmilled seed) 6 kg/ha cocksfoot or meadow fescue ————————— 96 kg/ha (39 kg/acre)	Where milled seed is available, a substantial reduction in the weight of sainfoin seed sown would be possible. This mixture could last for about three years.
20 kg/ha lucerne (8 kg/acre)	Timothy, cocksfoot or meadow fescue are suitable companion grasses for lucerne. 5 kg/ha of lucerne could be substituted by 10 kg/ha of grass seed. Lucerne will persist for up to five years in UK conditions.
12 kg/ha lucerne 7 kg/ha early red clover ————————— 19 kg/ha (7.75 kg/acre)	Sowing these two legumes together helps to compensate if pest or disease damage reduces the contribution of either. Red clover will die out after about two years.

All of these mixtures would be suitable for use on organic farms.

Table 19.8 Examples of mixtures suitable for equine use

Plant	Ley characteristics
14 kg/ha perennial ryegrass (intermediate/late) 7 kg/ha meadow fescue 7 kg/ha creeping red fescue 3.5 kg/ha Timothy 3.5 kg/ha smooth stalked meadow-grass (SSMG) ————————— 35 kg/ha (14 kg/acre)	A permanent pasture for use where productivity is required, such as for studs. Provides some 'bottom' to the ley to resist damage from hooves. Timothy can provide some drought resistance, while SSMG also known as 'Kentucky Blue Grass', is another palatable species popular for equine use.

For native ponies, and other situations where productive ryegrass pastures may risk weight issues and laminitis, the PRG can be replaced by increased use of fescues, SSMG and the addition of cocksfoot, rough stalked meadow-grass and crested dogstail in descending order of inclusion

25 kg/ha creeping red fescue 12.5 kg/ha dwarf perennial ryegrass 12.5 kg/ha SSMG ————————— 50 kg/ha	An example for use on gallops to provide a dense turf capable of resisting damage from flying hooves, providing some 'spring' reducing jarring to limbs, and having some ability to recolonise divots from hoof fall. Not suited for polo pitches due to rhizomes (use newer dwarf PRG varieties tolerant to very close mowing).

There are many new varieties of ryegrass with rapid establishment in cool climates, developed for sports turf; they are in use for racecourses and polo grounds to repair, renew and renovate swards that have heavy use. For winter National Hunt racecourses, pre-germination enables divots to be renovated within 2 weeks in temperatures as low as 3 °C. Some of these sports turf varieties will even germinate in soil temperatures as low as 3.5 °C.

19.9   Sources of further information and advice

Andrews J and Rebane M, *Farming and Wildlife: a Practical Management Handbook*, Royal Society for the Protection of Birds, 1994.

Frame J, *Improved Grassland Management*, Farming Press, 2000.

Frame J, *Temperate Forage Legumes*, CAB International, 1998.

Hague J and Hutchinson M, *Forage Costings for Home Grown Crops*, Kingshay Farming Trust, 1999.

Hopkins A (ed.), *Grass: its Production and Utilization*, 3rd edn, Blackwell Science, 2000.

Hubbard C E, *Grasses: a Guide to their Structure, Identification, Uses and Distribution in the British Isles*, 3rd edn (revised by J C E Hubbard), Penguin Science, 1984.

Kingshay Farming Trust, *Grass and Clover Mixtures for Grazing and Silage*, Kingshay Farming Trust, 1998.

Milk Development Council, *Winter Sward Management*, Milk Development Council, 2000.

Monsanto, *Renewed Grass for Improved Milk and Meat Production*, 1998.

NIAB, *Grasses and Herbage Legumes Variety*, leaflet (annual).

Nature Conservancy Council, *Effects of Agricultural Land Use Change on the Flora of Three Grazing Marsh Areas*, 1989.

Palmer J, *Future for Wildlife on Commons. Part I: The report; Part II: The case studies*, 1989.

Sheldrick R D, Newman G and Roberts D J, *Legumes for Meat and Milk*, Chalcombe Publishing, 1995.

Sheldrick R D (ed.), 'Grassland management in the environmentally sensitive areas'. *Proceedings: Symposium*, Lancaster, September 1997.

Sports Turf Research Institute, *Turfgrass Seed*, STRI (annual).

20

Establishing and improving grassland

DOI: 10.1533/9781782423928.4.483

Abstract: This chapter sets out the various ways in which grassland can be established, or improved by renovation. It deals with the terminology normally used in the establishment of leys, with appropriate crop protection measures and with early management. The importance of good drainage is stressed. Detailed information is also given concerning the use of agricultural lime and of nitrogen, phosphorus and potassium fertilisers, with particular reference to the Defra publication RB 209. Information is also given concerning the use of organic manures, the environmental implications of fertiliser use and the current regulations concerning Nitrate Vulnerable Zones.

Key words: reseeding, renovation, drainage, lime, fertilisers.

20.1 Establishing leys

Newly established leys, containing almost 100% of improved grass and legume varieties, constitute the most productive form of grassland and so it is appropriate to commence this chapter with a section on their establishment.

20.1.1 Terminology

The following refer to the various practices associated with sowing grass seeds:

- *Direct sowing.* Sowing grass seeds in spring or autumn without a cover crop.
- *Undersowing.* Sowing grass seeds with a cover crop, e.g. spring cereals or peas, usually at a reduced seed rate, which may be taken for grain harvest or wholecrop silage, leaving the undersown ley established in the stubble. Sometimes a light cover crop of cereals or forage rape may be grazed.
- *Direct reseeding.* Sowing grass seeds again in a field which has previously contained grass with no arable break.

- *Sod seeding.* A form of direct drilling often associated with the improvement of an old ley or permanent pasture by the introduction of new species without cultivation.
- *Oversowing.* The introduction of new species by surface broadcasting, in some cases associated with raking (for example using the 'Vertikator' or 'Einbock' machines).

20.1.2   Spring sowing

If direct sowing without a cover crop, the maiden seeds can give valuable production in the summer. Establishment can be enhanced by stock being able to graze the developing sward within a few weeks of sowing. This is not possible to the same extent with autumn sowing. A limiting factor with spring sowing may be moisture, and in the drier districts the seeds should be sown in March if possible. The plant should thus establish itself sufficiently well to withstand a possible dry period in early summer. Undersowing in a cereal crop in the spring means that the greatest possible use is being made of the field, although with the slower growing grasses establishment may not be so good, especially in a dry year. Weed problems in a ley are often reduced by undersowing.

20.1.3   Late summer/early autumn sowing

Leys are normally sown at this time of year and provided satisfactory establishment is achieved it ensures top yields in the following full harvest year. There is usually some rain at this time, heavy dews have started again, and the soil is warm. When clovers are included in the seed mixture, sowing in August is preferred before the onset of frosts. Whether earlier sowing is possible depends on when the preceding crop in the field is harvested. Winter barley is the ideal crop to follow. Sowing later than September is not advised and can lead to poor establishment.

20.1.4   Direct sowing

Reference has already been made (Section 8.6.2) to the cultivation necessary for preparing the right type of seedbed for grass seeds. It must be re-emphasised that grass and clover seeds are small, and should be sown shallow, and that a fine, firm seedbed is necessary. With ample moisture the seed can be broadcast. This should be on to a ribbed-rolled surface, so that the seeds fall into the small furrows made by the roller. Most fertiliser distributors can be used for broadcasting seeds, and seeds and fertiliser are sometimes sown together. Seeds may then be covered with a light harrow or chain harrow or simply rolled in.

In the drier areas, and on light soils, drilling to about 2 cm is preferred. The seed is then in much closer contact with the soil. The 10 cm coulter spacing of an ordinary cereal seed drill should give a satisfactory cover of seeds, but for an improved result, cross-drilling could be considered. After broadcasting or drilling, a thorough rolling will complete the whole operation. For those with the time to spare and seeking the best possible result, drilling the grass seeds and subsequently broadcasting the clover seeds should give the best chance of good establishment to both. At soil index 2 about 50 kg/ha each of phosphate and potash can be broadcast (up to 120 kg/ha at index 0) and worked in during the final seedbed preparation. For spring sowing, 60 kg/ha of nitrogen is also advised. Some farmers also use up to 40 kg/ha N on seeds sown in August but this may not always be necessary and may adversely affect the establishment of clovers.

20.1.5   Direct drilling

Grassland can be direct reseeded by direct drilling. Although it is no cheaper than conventional reseeding, the practice does allow a much quicker turn-round from the old grass to the new sward. It also reduces poaching. Glyphosate is a desiccant herbicide which can be used to kill off or suppress the existing vegetation prior to the introduction of the new seeds. Cropping with spring-sown direct drilled kale, turnips or forage rape prior to direct drilling with grass seeds in late summer or autumn, ensures the best possible establishment from this technique. Another option, particularly if the old ley is thick and matted, is to spray in the autumn and allow the old sward to die off during the winter months, prior to direct drilling in the spring. Rolling immediately following drilling is particularly beneficial and control of slugs, leatherjackets and frit fly is almost always necessary. A substantial area of leys is direct drilled into cereal stubbles in most autumns with very satisfactory results.

'Sod seeding' sometimes also referred to as 'strip seeding' usually refers to the direct drilling of new seeds without the complete kill of the old sward. Specialist drills have been evolved for this purpose which in some instances incorporate a 'band' spraying facility sufficient to kill off or suppress the old sward adjacent to the strip where the new seeds have been sown. Sod seeding is a useful way of introducing white clover into an established perennial ryegrass pasture.

20.1.6   Undersowing

Good establishment can be achieved when undersowing spring cereals. The cereal is sown first, usually at about three-quarters of the normal seed rate and followed immediately by the seed mixture drilled or broadcast, followed by light harrowing or harrow combing and rolling. This is the best sequence of operations for slower establishing long term leys. With more vigorous species such as Italian ryegrass and red clover, it is often preferable to broadcast the seed after the cereal is established and at about the three leaf stage. This helps to preclude excessive growth of the undersown species which, particularly in a wet season, may seriously interfere with the harvesting of the cereal.

Another useful technique is to take the cover crop of cereals for wholecrop cereal silage several weeks before the crop is ripe for combining. An alternative which has been used in recent years is undersowing in peas (grain or forage varieties are both suitable) which may also be taken for silage. Weed control in undersown crops is sometimes difficult, especially if clovers have been included in the mixture, and it is essential to use a legume-safe herbicide. If peas are being used as the cover crop then a stale seedbed prior to sowing is the preferred option. Undersowing in autumn sown cereals is not normally advised due to the competitive nature of the cover crop. However, in an organic cropping situation with a less densely tillered cereal crop, undersowing in spring has been shown to be successful in conjunction with operations such as harrow combing for weed control which can also rake in and cover the grass seeds.

20.1.7   Weed, pest and disease control in establishing leys

Seedling diseases such as 'damping off' (*Pythium*) may adversely affect the establishment of some grass varieties. Some seed merchants offer seed treated with a fungicidal dressing which may be well worthwhile, particularly when establishing grasses late in the season or under less than ideal conditions.

Particularly important pests are slugs, which can cause damage at any time of year, but particularly in wet conditions, and almost always when sowing in the autumn or direct drilling. Seeds can be supplied with slug pellets already mixed in or slug pellets can be applied to the surface of affected fields after test baiting. Leatherjackets (larvae of the crane fly or 'daddy longlegs') can cause massive damage and complete crop failure especially when direct reseeding (grass after grass) is being practised in the spring. Frit fly larvae can also be present in old turf in enormous numbers and can seriously damage grass seeds sown at any time of year. Both of these pests can also be present in significant numbers in arable situations, particularly in grassy stubbles. Control of both with chlorpyrifos spray is currently approved.

Control of annual broad-leaved weeds in new grass leys is relatively straightforward Table 20.1. The most important point, of course, is to remember to use a legume-safe herbicide when spraying grass/clover mixtures. A mixture of linuron, 2,4-DB and MCPA gives good results against a wide variety of weeds. The preferred material for lucerne is straight 2,4-DB but this is sometimes difficult to obtain. The most important weed to control is chickweed, particularly in autumn sown leys, as this can grow extremely vigorously if left unchecked and causes significant loss of plant through competition and smothering.

20.1.8   Early management

Leys can often be grazed about eight weeks after sowing if conditions are suitable. Grazing an August sown ley with sheep in October can be very beneficial as it encourages the grasses to tiller. Grazing a spring sown ley in June or July with cattle should be possible if conditions are dry enough. Grazing in wet conditions should be avoided, however, as it will inevitably lead to poaching, a loss of sown species and the early ingress of weeds. When weed growth is not severe a light topping will often control annual weeds adequately and precludes the need to use a herbicide.

Table 20.1 Some common weeds and their control in grassland

Weed	Control
Bracken	This weed is a very serious problem which is increasing, especially in hill areas. The fronds (leaves) should be cut or crushed twice a year when they are almost fully open. If possible, the field should be ploughed deep and then cropped with potatoes, rape or kale before reseeding. Rotavating to about 25 cm deep chops up and destroys the rhizomes. Glyphosate can be used as a non-selective spray before reseeding or, selectively, with a rope-wick machine.
Buttercups	Spray with MCPA. The bulbous buttercup is the most resistant type. Improved drainage and soil fertility will lessen this problem.
Chickweed	Several chemicals will control chickweed including mecoprop-P and fluroxypyr (note they kill clover).
Docks	Grazing can be helpful. Triclopyr, fluroxypyr, dicamba mixtures, mecoprop-P, thifensulfuron-methyl and MCPA alone or in mixtures will give good control, although treatment may need repeating. If reseeding, the old sward should be sprayed with glyphosate before ploughing.
Horsetails	If possible, drainage should be improved. Spraying with MCPA or 2,4-D will kill aerial parts only and regrowth will occur. However, if it is done two to three weeks before cutting, any hay crop should be safe for feeding.
Nettles	Spray with triclopyr alone or in mixtures, or aminopyralid mixtures; glyphosate can also be used through a rope-wick applicator.
Ragwort	Cut before the buds develop to prevent seeding. Spray with 2,4-D or MCPA in early May or in the autumn. Grazing with sheep in winter can be helpful. Cut or sprayed plants are palatable to livestock and can cause poisioning.
Rushes	An increasing problem in many areas. If possible, drainage should be improved. Spray with MCPA or 2,4-D. The hard and jointed rush should be cut several times a year or treated with triclopyr. Apply glyphosate with a rope-wick when the rushes are growing well. Grasses and clovers should be encouraged by good management.
Sorrel	Lime should be applied if necessary. Spray with MCPA or 2,4-D.
Thistles	Spray with clopyralid or MCPA. Creeping thistle should be sprayed at the early flower bud stage. Over-grazing should be avoided. Cutting in July is very effective for control of the biennial spear thistle.
Tussock grass	Drainage should be improved. The 'tussocks' should be cut off with a flail harvester or topper. A hay or silage crop could be taken. Glyphosate can be applied with a rope-wick when the tussock grass is growing well.

20.2   Grassland improvement and renovation

20.2.1   Problem identification

This section will deal mainly with the improvement and renovation of permanent pastures which have deteriorated, although this obviously does not preclude similar operations designed to improve the composition of an old ley. Direct reseeding, as described above, is an expensive operation and should only be carried out when absolutely necessary. Evidence of deterioration should be sought from the species composition and from the condition of the soil. The problems may be obvious (e.g. infestation with such weeds as docks, thistles, nettles and ragwort) or slightly more difficult to discern (e.g. a general lack of productivity) in comparison with other pastures.

The first step should be to undertake a detailed botanical analysis of the species present in the field so far as is possible. The obvious points to note are the presence or absence of significant quantities of perennial ryegrass. If it is possible to identify, say 25% + of the grass tillers in the sward as perennial ryegrass then it should be possible to improve the sward by changing the management. If, on the other hand, there is hardly any perennial ryegrass present, then some form of complete sward renewal by direct reseeding or, at least, a renovation operation, may be necessary. However, before any operations are undertaken it is necessary to check on a few of the important 'basics'.

20.2.2   Drainage

Evidence of poor drainage should be fairly easy to find. Obviously standing water and rushy patches are important indicators; also blocked ditches and drainage outfalls which are failing to run. Digging a shallow pit for soil examination may be worthwhile. The grey and rusty red mottling in the topsoil known as 'gleying' will indicate seasonal waterlogging and a probable need for improved drainage although it may not be necessary to lay a complete new pipe system. Frequently there are drains in a field already which just need unblocking or renewing. In other cases, where the soil is clayey, it may be possible to set out a 'mole' system (Section 8.2.3), leading the mole channels into main drains covered with permeable backfill. Improving the drainage of a wet grass field has been shown to be capable of improving dry matter yields by up to 30% and can increase the length of the grazing season by up to a month.

20.2.3   Soil pH

Acidity can also have a major effect on the productivity of grassland. Defra recommend a target pH of 6 for long term grassland. However, in the 'Park Grass' classical experiment at the Institute for Arable Crops Research (IACR) Rothamstead it has been shown that maintaining a soil pH of between 6.5 and 7.5 increases grass dry matter yields by up to 28% when compared to a pH of 5. Furthermore, in the most acid areas the species composition has deteriorated to almost 100% Yorkshire fog. Correcting the soil pH by the application of agricultural lime is very straightforward and cost-effective.

20.2.4   Phosphorus and potassium

Routine soil samples should be taken to check on soil phosphorus and potassium as well as pH. Visual evidence may include grass with very poor stunted growth and purple discoloration in the leaves. This is indicative of a deficiency of phosphate. Potassium or 'potash' deficiency is often well demonstrated by the deep green colour and vigorous growth of grass growing in dung and urine patches compared to an otherwise poor sward with a pale green colour. Correction of low (index 0 or 1) phosphorus and potassium levels should be carried out as recommended by Defra (2010) in their *Fertiliser Manual* (RB 209).

20.2.5   Control of perennial weeds

The control of perennial weeds such as docks, thistles, nettles and ragwort by the use of an effective herbicide can have a markedly beneficial effect on the productivity of the sward (see also section 5.3). It can also (in theory) be enforced under the provisions of the Weeds Act (1959). The information contained in Table 20.1 indicates the appropriate herbicides for use on established grassland. Particular care should be taken where clovers or other perennial forage legumes are a valued sward component, to select a suitable legume-safe herbicide. Regular topping to prevent weeds setting seed is also important (a mature dock plant can set up to 60 000 viable seeds in a year). In the case of ragwort there is often no alternative to pulling by hand, removing and burning, since it is so poisonous and becomes very attractive to animals when it has been sprayed or mown. Sheep grazing during the autumn and winter periods can also help to control ragwort plants in their first winter 'rosette' stage.

20.3   Improving a sward by changing the management

In the case of a sward which contains significant quantities (25%+) of perennial ryegrass it should be possible to improve the species composition by management. The first step is to ensure that the basic good management practices, as set out above, are in train. Then increase the annual application of nitrogen fertiliser to near optimum with a corresponding increase in the intensity of utilisation, either by intensive grazing or by cutting early for silage or, preferably, both. Avoid cutting late for hay as this enables the poorer grasses to come to maturity and to set seed. Autumn grazing by sheep, although it may to some extent reduce the earliness of the spring flush in the following year, will certainly reduce the incidence of the poorer bents and meadow grasses. It is important to avoid both undergrazing, overgrazing and excessive poaching. Regular defoliation on a rotational basis to a sward height of 6–8 cm, plus judicious inputs of nitrogen fertiliser, should substantially favour the perennial ryegrass component of the sward over time and lead to a gradual improvement.

20.4   Improving a sward by renovation

In the case of a sward which contains few species of value it will be necessary to introduce new seeds, specifically of perennial ryegrass and white clover, and to encourage them to establish and thrive. The advantage of this technique is that it is cheaper than complete sward renewal through a direct reseeding operation. Also it is possible to retain much of the 'thatch' supplied by the old turf which is so valuable in physically carrying stock and reducing poaching. Such operations are usually most successful in August. September sowing can lead to disappointing establishments of both grass and clover.

There are two basic approaches. One is to introduce the new seeds by way of one of the sod-seeding machines described above under the section on direct drilling. Examples of these machines currently in use are the Hunter rotary strip seeder, the Vreedo slit seeder and the Opico seeding rake. A degree of sward suppression may be achieved by an application of glyphosate, or the sward may be severely grazed prior to the operation. The application of a suitable compound fertiliser, rolling, and the application of pesticides to control slugs, leatherjackets and frit fly larvae as previously described should complete the operation.

The other approach involves a very severe grazing and/or vigorous surface cultivation to expose soil and obtain a tilth. This can most conveniently be undertaken with disc harrows. Broadcasting seeds and fertilisers then takes place and the operation is concluded by rolling and pesticide application as previously described.

20.5   Fertilisers for grassland

Grass, like all other crops, needs mineral nutrients for its establishment, maintenance and production; nitrogen, phosphorus and potassium are the most important. However, before discussing these major nutrients it is essential to remind the reader of the importance of lime.

20.5.1   Liming grassland

The importance of liming has already been referred to in the section on grassland improvement and renovation. Soil acidity is probably the biggest single factor adversely affecting the productivity of grassland and it is one of the most simple to correct. The ideal pH for permanent pastures is around 6–6.5, whilst for establishing new seeds the soil should be limed to a pH of 6.5. The ideal pH for lucerne is 6.5–7. Agricultural limes are relatively cheap and extremely cost-effective. The usual dressing is about 5 t/ha, and the materials that are used on arable land are equally suited to grassland. Magnesian or dolomitic lime can, in addition, have a useful effect on the soil magnesium level if it is deficient.

Other materials sometimes favoured for use on grassland include calcified seaweed and granular lime. Like any other liming materials they should be evaluated solely on the basis of their neutralising value (NV). Such an evaluation may make them appear expensive; however, availability in small bags or the ability to apply through a farm fertiliser spreader may make them attractive for very small areas such as horse paddocks. When seeding a new ley it is always advisable to have the soil tested for lime and, should it be necessary, the best advice is to lime on the ploughed furrow and work it into the seedbed before the seeds are sown. When direct drilling into an old sward it is advisable to check the pH of the top 2–3 cm of soil as this is frequently more acid than the rest of the soil profile.

20.5.2   Nitrogen

Optimum nitrogen
Nitrogen is the most important nutrient required by grasses. When applied in the form of inorganic fertilisers, nitrogen has an immediate and very positive effect on grass growth provided there are no other limiting factors, as seen in Fig. 20.1. Organic manures contain less readily available nitrogen but still have a substantial effect on yield. It is often possible to observe very strong correlations between the application of nitrogen fertilisers, farm stocking rates and ultimate profitability. Nitrogen fertilisers have been the central pivot to the development of intensive grass-based livestock systems and have also been responsible for the creation of some environmental problems.

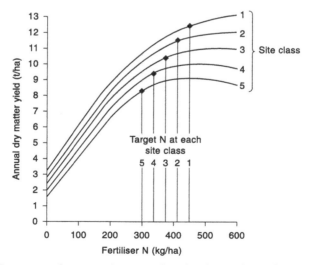

Fig. 20.1   Responses of grass to nitrogen at five site classes (from Thomas *et al.*, 1991).

A great many trials have been carried out to observe the extent to which grasses respond to nitrogen. In the absence of any other limiting factors such as poor drainage, soil acidity or lack of one of the other important nutrients, it is probably true to say that summer rainfall and soil type have the major impact on this response. In that most readable (still relevant, but, sadly, out of print) book *Milk from Grass* (Thomas *et al.*, 1991) the authors set out various hypotheses based on the 'GM 20' series of trials. One of the main suggestions was the subdivision of the UK into five 'site classes' according to their theoretical ability to grow grass. Nitrogen recommendations were then formulated for each site class (Fig. 20.1). In the most recent (2010) edition of the Defra *Fertiliser Manual* (Reference Book 209) this concept has been continued, but in simplified form, setting out 'grass growth classes' dependent on soil type and average (April–September) rainfall. These are reproduced in Table 20.2.

The other major factor likely to impact upon farmers' decisions about nitrogen application is the soil nitrogen supply (SNS) status. This depends mainly upon the levels of use of nitrogen fertilisers and organic manures over recent years as well as the grassland management system. So, for example, fields which have been intensively grazed for several years will be likely to have high levels of soil nitrogen because of accumulations due to the excreta of grazing animals and the use of organic manures; furthermore, nitrogen losses from leaching will be low due to the perennial plant cover. Conversely, fields which have recently been sown to a ley after several years in arable cropping will be likely to have low levels of soil nitrogen. The direct measurement of soil mineral nitrogen (SMN) levels is quite simple, however their interpretation, bearing in mind the complexity of the range of factors influencing the availability of nitrogen from soil sources, means that accurate predictions become very difficult and the general high/average/low classification suggested by Defra is probably good enough. The SNS status classes suggested are set out in Table 20.3.

Table 20.2   Grass growth classes (from Defra, 2010)

Soil available water	Soil types	Rainfall (April–September inclusive)		
		Up to 300 mm	300–400 mm	Over 400 mm
Low	Light sand soils and shallow soils (not over chalk)	Very poor	Poor	Average
Medium	Medium soils, deep clay soils and shallow soils over chalk	Poor	Average	Good
High	Deep silty soils, peaty soils and soils with groundwater (e.g. river meadows)	Average	Good	Very Good

Table 20.3 Soil Nitrogen Supply (SNS) status classes in grassland (from Defra, 2010)

Class	Previous grass management	Previous nitrogen use (kg/ha)
High	Long-term grass, high input. Includes: • Grass reseeded after grass or after one year arable • Grass ley in second or later year	Over 250
Moderate	First year ley after two or more years arable (last crop potatoes, oilseed rape, peas or beans, NOT on light sand soil)	All
	Long-term grass, moderate input. Includes: • Grass reseeded after grass or after one year arable • Grass ley in second or later year	100–250 or Substantial clover content
Low	First year ley after two or more years arable (last crop cereal, sugar beet, linseed or any crop on a light sand soil)	All
	Long-term grass, low input. Includes: • Grass reseeded after grass or after one year arable • Grass ley in second or later year	Up to 100

It is, of course, true to say that there are many more factors impinging on the farmers' decisions about the optimum rate of nitrogen fertiliser in particular circumstances. Quite a lot of them concern the actual system of production (dairy, beef or sheep), the levels of purchased feedingstuffs employed and the socking rate in terms of livestock units (LUs or GLUs) per hectare. In the latest (2010) edition of the Defra *Fertiliser Manual* (RB 209) a very complex series of annual recommendations is proposed for beef, sheep and dairy enterprises of varying-intensities and situated in different grass growth classes, to which the reader is referred. The specific requirements of a range of agri-environmenal schemes as well as the restrictions imposed in nitrate vulnerable zones (NVZs) also need to be borne in mind and, of course, the Defra 'Code of Good Agricultural Practice', *Protecting our Water, Soil and Air*.

Clover

In the absence of fertiliser nitrogen, clovers will make a useful contribution to the requirements of the grass crop. The *Rhizobium* bacteria in the root nodules fix atmospheric nitrogen and, as they die and break down, some of this nitrogen is made available for the companion grass plants. The amount of nitrogen fixed will be proportional to the amount of clover in the sward, as well as to pH, soil temperature and rainfall. With about 30% of a mixed perennial ryegrass based sward present as white clover, a reasonable expectation is for a contribution of up to 180 kg N/ha. Higher levels (up to 300 kg N/ha) are thought to be possible with a greater clover percentage, but it is almost impossible to be precise. If nitrogen fertiliser is applied, then the clover contribution may well decrease although it is common for farmers with clover rich swards to apply some nitrogen (say 50 kg/ha) early in the season, to encourage grass growth before the growth of clover commences, and a similar quantity in August. The vigorous large- and very large-leaved white clover varieties such as *Alice* and *Aran* will survive well in a nitrogen fertilised sward and continue to supply the nutritional benefits for which they were also sown.

Organic manures and slurry
On most farms, with grass-based livestock enterprises, it is necessary from time to time to apply organic manures or slurry to the grass area. These are best applied to the conservation (hay or silage) areas where their contents of phosphate and potash will be most valuable in replenishing that removed by cutting. They may also be applied to fields intended for forage maize, kale, reseeding, etc. It is important, from an environmental as well as an economic point of view, to take account of the nutrient content of these materials when calculating the optimum inputs of inorganic fertilisers. Organic manures and slurry are some of the most important sources of nitrate contamination of surface and ground water. Within Nitrate Vulnerable Zones (NVZs) there are specific regulations governing the amounts of nitrogen in organic form which can be applied to agricultural land and the times of year when they may be applied. Important references for this situation are contained in the Defra (2008) publication, *Guidance for Farmers in Nitrate-Vulnerable Zones.*

Nitrogen losses from organic sources can be minimised by application in the spring and summer rather than the autumn and winter, and such practices as soil injection will reduce volatilisation losses. Nitrification inhibitors have been investigated but their effects are often variable, and their widespread use is precluded by high costs. Dilution, acidification and separation are also techniques which have been investigated and which show some promise.

Timing of nitrogen applications
Grass growth is very seasonal with the main period of rapid growth in the UK falling between April and June. It is sensible therefore for the main applications of nitrogen fertiliser to be made during this period. Two-thirds of the target nitrogen application before the end of May is a good 'rule of thumb'. The usual principle is for fairly large or frequent applications early in the season followed by smaller applications after each cut or grazing. Some further suggestions concerning the timing of applications for grazing and cutting situations is provided in Table 20.4 as a supplement the Defra RB 209 annual recommendations.

The timing of the first nitrogen application is always an area open for discussion. Monitoring soil temperatures is quite straightforward with a simple soil thermomter, and to introduce the first nitrogen application when the soil temperature at 10 cm reaches 5 °C is a fairly logical approach, provided ground conditions are suitable.

The latest date recommended for the application of nitrogen fertiliser is mid-August. In most seasons further applications are unnecessary since warm soils and autumn rainfall frequently result in nitrogen being mineralised from the

Table 20.4   Fertiliser nitrogen timing recommendations for grazing swards and for swards cut for silage

Date	Grazing swards	Swards cut for silage
February/March or when soil temperature at 10 cm = 5 °C and when soil conditions are suitable.	Apply 60 kg N/ha; only apply in February where March grazing is possible; if grazing is carried out in March apply a further 60 kg N/ha.	Apply 40–60 kg N/ha; if an 'early bite' grazing is taken follow this with a further 40 kg N/ha.
April/May	Continue to apply nitrogen monthly at 60 kg N/ha or at up to 75 kg N/ha where SNS levels are low; apply after each grazing on a rotational system; where set stocking is practised either apply monthly to the whole area or weekly to 25% of the area.	Apply the balance of N to bring the total up to 120 kg N/ha (150 kg N/ha where SNS levels are low); this application *must* be carried out more than six weeks before the planned date of cutting; as soon as the first cut is cleared apply a further 100 kg N/ha for the second cut; if the field is to be grazed continue with grazing recommendations.
June/July/August	Extra grazing is normally introduced at this time after first cut silage; continue to apply 40–60 kg N/ha monthly until the target N level is reached; in very dry conditions (typical of poor or very poor grass growth classes in some years) nitrogen application should be discontinued; it is not recommended to apply nitrogen fertiliser after mid-August.	If a third or even fourth cut of silage is to be taken apply 80 kg N/ha prior to each cut; otherwise continue as for the grazing recommendations.

Notes:
1. The recommendations above are most appropriately applied to intensively managed dairy and beef enterprises. In these situations the Defra (2010) *Fertiliser Manual* recommends annual applications of between 200 and 340 kg N/ha for cutting and between 150 and 340 kg N/ha for grazing, in 'average' SNS and grass growth class conditions, depending also on stocking rates and the levels of purchased feedingstuffs used. For more extensive beef and sheep systems the use of N on grazing land especially is much less and in some cases none is recommended to be used at all.
2. Urea is an acceptable fertiliser material to use in the period February–May. After this, ammonium nitrate usually gives better results.
3. For hay crops nitrogen applications of 60–75 kg N/ha are usually sufficient. Higher levels will give very heavy cuts which will be difficult to dry prior to baling.
4. Slurry is most appropriately applied early to the cutting fields, but care is needed to avoid contamination which may adversely affect silage fermentation. Always take account of the likely contribution of nutrients from this source when deciding on individual field applications.
5. For ley establishment between April and mid-August apply up to 60 kg N/ha. For autumn establishment nitrogen is not usually necessary except when direct drilling.
6. Where the clover content of a sward is high then applications of nitrogen should be restricted in order to maintain it to no more than 50 kg N/ha in March followed by a further 50 kg N/ha in August if extra autumn grazing is desired.

large organic reservoirs within many soils. A later application by farmers using mainly clover swards is sometimes advocated, to improve the supply of autumn grazing.

20.5.3   Phosphorus and potassium

Phosphorus and potassium fertilising of grassland is quite well understood. In a grazing situation there is frequently no need to apply any because of the frequent returns of dung and urine. Only in deficient soils (soil indices 0 and 1) will any response be observed. In the case of potassium there is an important connection with the incidence of hypomagnesaemia or grass staggers. High levels of potassium in the herbage impedes the uptake of magnesium. This is a very serious condition affecting all classes of stock and which can cause sudden death or substantial loss of production. The main periods of risk are in the spring, especially during cold, wet spells after turnout. Suckler cows are particularly at risk in the autumn. The best advice is to avoid any applications of potash fertilisers in the spring months except where the soil is seriously deficient (index 0). Where needed, potash can usually be applied safely in the summer. The other remedy (apart from the attentions of the vet) is to feed a suitable magnesium supplement throughout the risk period. This can be variously given as a feed supplement, lick, added to the water supply or as an ingested bolus. Another option is to dust the pasture regularly with feed-grade calcined magnesite, but this can be readily washed away by rainfall. A magnesium based pasture spray is another popular option.

On cut swards phosphorus and potassium are both required in large amounts, especially potassium, where optimum annual dressings for frequently cut swards may total 200–300 kg/ha K_2O. Table 20.5 sets out the phosphorus and potassium, recommendations.

20.5.4   Other nutrients

Magnesium (Mg)
It is possible to obtain a response to up to 85 kg/ha of MgO applied as calcined magnesite where the soil index level for magnesium is 0. Another avenue for the application of magnesium is the use of magnesian or dolomitic lime for correcting soil pH.

Sodium (Na)
Sodium (in compound form) is sometimes applied to grazed swards in an attempt to improve palatability and intake. Sodium application can also reduce the risk from grass staggers in some situations.

Sulphur (S)
ADAS trials in the 1990s have shown yield increases from cut grass of 20–30% where sulphur deficiency has occurred. This is mainly as a result of a marked

Table 20.5 Phosphorus and potassium fertiliser recommendations for grassland (kg/ha P_2O_5 or K_2O)

Management system		Phosphorus soil index level					Potassium soil index level				
		0	1	2	3	3+	0	1	2	3	3+
Ley establishment (seedbed)		120	80	50	30	0	120	80	50	0	0
Grazing (annual application)		60	40	20	0	0	60	30	0	0	0
Cutting for silage	1st cut	90	65	40	20	0	140	110	70	30	0
	2nd cut	25	25	25	0	0	120	100	75	40	0
	3rd cut	15	15	15	0	0	80	60	60	20	0
	4th cut	10	10	10	0	0	70	70	55	20	0

Notes:
1. N, P, and K can be applied as 'straights' or as compound or blended fertilisers of a suitable analysis.
2. Excessive potash applications on grazing fields can give rise to *hypomagnesaemia* (grass staggers). Except in the case of serious potash deficiency, applications to grazing areas should be made in the June/July period.
3. Where individual potash applications in excess of 100 kg/ha are required application in the previous autumn is advised.
4. On fields with inherently low potash levels (index 0–2) an additional 60 kg/ha K_2O should be applied in the autumn after the last cut.
5. Where applications of FYM or slurry have taken place account should always be taken of the nutrient contribution from these sources.
6. Sulphur deficiency can be a problem in second and subsequent silage cuts. Compound fertilisers should be used to apply about 40 kg/ha SO_3 if deficiency is confirmed by herbage analysis.
Based on Defra RB209 (2010).

decline in the level of industrial pollution in recent years. Deficiencies tend to occur in areas removed from the main centres of population. Grass growing on free draining soils is vulnerable and situations where winter rainfall is unusually low have also shown this deficiency. Farms where little or no FYM or slurry is applied to grass, may also be vulnerable. Sulphur, at up to 40 kg/ha of SO_3, should be applied in deficient situations before each cut. Second and third cut silage crops in fairly dry situations have shown the most frequent deficiencies of sulphur. This deficiency can best be assessed by an analysis of herbage.

20.6   Irrigation of grassland

In dry areas with low summer rainfall (i.e. low grass growth class) it is the summer months of June to August which give rise to the poorest rates of grass growth. Soil moisture deficits (SMDs) during this period can frequently reach 100 mm or more in the south and east of England. Grass growth is usually severely impaired when the SMD reaches 50 mm. Irrigation, if it is available at a reasonable cost (e.g. £5/ha mm), can be a way of overcoming such severe summer droughts and (in effect) transforming the farm from grass growth class low to medium, for example. Only a relatively few farms have installed irrigation equipment sufficient to make a worthwhile impact on grass growth. Most frequently, when grass is irrigated, it is through the marginal use of equipment which is already being justified by its use on a high value arable crop such as potatoes, salads or vegetables. A 20 year study carried out in the south-east of England and reported in the 1980s, referred to an increase in grass dry matter yields of 25%, as the result of the sustained use of irrigation.

20.7   Sources of further information and advice

Defra (2008), *Guidance for Farmers in Nitrate Vulnerable Zones* (PB12736 a-i).
Defra (2008), *Manure Planning in Nitrate Vulnerable Zones* (PB3577).
Defra (2009), *Protecting Our Water, Soil and Air: A Code of Good Agricultural Practice for farmers, growers and land managers.*
Frame J (2000), *Improved Grassland Management*, Farming Press.
Hopkins A (ed.) (2000), *Grass: its Production and Utilization*, 3rd edn, Blackwell Science.
Institute of Arable Crops Research (1991) *Guide to the Classical Field Experiments*, Rothamsted Experimental Station, AFRC.
Monsanto (1998), *Renewed Grass for Improved Milk and Meat Production.*
Simpson N A (1993), *Survey Review of Information on the Autecology and Control of Six Grassland Weed Species*, English Nature Research Report no 44.
Thomas C, Reeve A and Fisher G E J (1991), *Milk from Grass*, The British Grassland Society.
Younie D (ed.) (1996), *Legumes in Sustainable Farming Systems*, BGS Occasional Symposium no 30, British Grassland Society.

21

Grazing management

DOI: 10.1533/9781782423928.4.499

Abstract: The use of grass and grass/legume mixtures for grazing can often prove the most economical means of production for agricultural livestock and, for equine enterprises, can provide for a wide range of requirements, from exercise areas to full nutrient provision. The reader is introduced to the principles of grazing management and the matching of grass and animal potential, stocking rates, the importance of a suitable sward height and density and the assessment of dry matter yields. An account is given of the most common systems used for grazing as well as strategies to minimise parasitism in agricultural livestock and equines.

Key words: grazing, stocking rate, sward, paddock, set stocking.

21.1   Introduction

The utilisation of grass by grazing is, for dairy cattle, beef cattle and sheep, undoubtedly the most economical means of production. Reference has already been made to the low cost of energy as grazing compared with conserved grass. United Kingdom Levy Board research has shown that grazed forage costs between 30% and 50% of the costs of conserved feeds and even less when compared with the costs of energy as bought-in feed. There is extensive evidence available from EBLEX, DairyCo, and Hybu Cig Cymru that well managed grazed grass can provide adequate energy and protein for production, although this will vary substantially from farm to farm. Most are agreed, however, that maximising the contribution from grazing, whatever the system of production, is a sensible approach to reducing costs and maintaining an acceptable degree of profitability, even in (or perhaps especially in) a time of depressed prices. The approaches to grassland and grazing for equine enterprises vary even more than those for agricultural livestock, encompassing the whole range from exercise areas to full nutrient provision.

21.2   Stocking rate or density

By convention, stocking rates are usually expressed as 'Grazing Livestock Units' (GLU) per hectare; a unit being a 500 kg Friesian/Holstein cow. Stocking rates should be accurately calculated from the following information, and based on a monthly reconciliation of actual stock numbers in the various categories on a farm and the total forage hectares. There is often a strong correlation between stocking rate, the application of nitrogen fertilisers and farm profitability. High stocking rates (e.g. > 3 GLU/ha) would normally only be associated with the very best grass growing conditions (Table 20.2) and with near optimum applications of nitrogen fertiliser. Lower rates (1.5–2.0 GLU/ha) might be associated with organic farms or with the poorer grass growing conditions. Stocking rate calculations are an integral part of the applications for EU Single Farm Payment Stewardship Schemes, details of which are supplied by Natural England. Horses may be considered to be roughly equivalent to cattle of the same weight.

The Grazing Livestock Unit figures shown in Table 21.1 are those currently in use for farm management calculations:

Table 21.1   Grazing Livestock Unit (GLU) figures currently in use for farm management

Cattle	GLU	Sheep	GLU
Dairy cows (Friesian/Holstein)	1.00	Lowland ewes	0.11
Beef cows (excluding calves)	0.75	Upland ewes	0.08
In-calf heifers	0.8	Hill ewes	0.06
Bulls	0.65	Breeding ewe hogs	0.06
Other cattle 0–1 years old	0.34	Other sheep over 1 year old	0.08
Other cattle 1–2 years old	0.65	Store lambs under 1 year old	0.04
Other cattle over 2 years old	0.8	Rams	0.08

Source: Nix, 2012, as advised by Defra.

21.3   Principles of grazing management

Grazing management necessitates balancing the requirements of stock against the potential of grass and requires a good knowledge of both. It is affected by the situation of the farm and the actual livestock production system. In fact, the choice of production system (especially the dates of calving or lambing) is often strongly influenced by the grass production characteristics of a particular farm. Such considerations would include 'earliness' of grass growth and the potential date of turnout, the likelihood of a mid-season drought and the potential for extended grazing into the autumn and early winter. Consequently, the synchronisation of lambing or calving with the expected growth of grass so as to turn out stock to grass ready to maximise utilisation from grazing seems a good approach to minimising production costs. There are, of course other considerations, such as price seasonality and continuity of product supply such as milk and finished livestock, throughout the year.

21.3.1   Animal potential

So far as milk production is concerned, calving in late winter/early spring should ensure that cows can be turned out to grass at peak lactation and in a position to maximise the contribution of grazed grass to their production. Autumn calvers, too, can achieve good levels of production from grazing, both from autumn grass prior to housing, and again in the spring towards the latter half of the lactation. The susceptibility of a farm to summer drought should influence the decisions about calving date, so the grass-growing site class influences choice of system. The potential is for grazing to provide sufficient energy and protein for a daily milk yield of up to 30 litres from animals capable of producing it. This figure would fall to below 20 litres in late summer, although few animals would have the ability to intake adequate fresh weight to achieve high yields from fresh grass alone (see Section 21.3.4).

In the case of growing stock (young cattle or lambs) the potential is for grazing to provide sufficient nutrients for individual daily live weight gains of up to 1.25 kg for beef cattle and up to 300 g for lambs. Again the need to balance increasing animal liveweight with decreasing grass productivity requires careful management. In all breeding/rearing enterprises the combined grazing of adults and juveniles prior to weaning leads to the inevitable build-up of parasitic worm populations. Grazing management must address the need to integrate parasite control measures together with the provision of sufficient quantities of nutritious grass at the correct stage of growth throughout the grazing season.

21.3.2   Grass potential

The factors which affect grass yields such as sward species and varieties, site class and fertiliser (especially nitrogen) application, have been dealt with at length in Chapters 19 and 20. Seeds mixtures suitable for grazing in various situations have been described in detail in Section 19.8.2 and in Tables 21.2, 19.4, 19.5 and 19.6. Other site factors such as aspect, and soil factors such as texture and drainage, affect the suitability for grazing, and, in particular, the suitability of a field for New Zealand style extended early spring and autumn grazing. On some farms specific crops such as rye, rye plus Italian or hybrid ryegrass or triticale are sown for early grazing. These crops are described in more detail in Section 18.2.3. Winter wheat and oats, sown as grain crops, may also be grazed in early spring so long as soil conditions are suitable and the crop has not reached the 'stem extension' growth stage. Although there may be some suppression of yield the strategic gain in terms of cost saving to the livestock enterprise may well outweigh this. Summer drought may require special provision to be made to feed stock during times of shortage of grazing. This might include sowing an area of quick-growing turnips for summer grazing, or the inclusion of a field of lucerne with its potential for zero grazing (cutting and carting) during the summer months. Most dairy enterprises make available a stock of silage specifically for feeding during the summer in times of grass shortage, and silage or haylage bales are commonly fed to beef and sheep to supplement grazing.

Table 21.2   Suitability of seed mixtures

Purpose	Types of forage plant that should be used
For early grazing, i.e. early bite	Mainly Italian ryegrass and hybrid ryegrass
For optimum production throughout the season	Ryegrass (supported by either non-ryegrass or permanent pasture)
For winter-grazing cattle – foggage (provided conditions allow)	Non-ryegrass
For the grazing block	Herbage plants that produce a closely knit sward
For the cutting block	Herbage plants that produce a relatively tall habit of growth

Note: The heading dates will generally coincide if the above points are borne in mind when deciding on a seeds mixture. As far as possible, varieties should correspond because, with only a few exceptions, the digestibility of a plant starts to decline when the ear emerges. A grass crop cannot be so valuable if its various plant components head at different times. If, for example, the crop is cut or grazed at the average date for the plants making up the sward, the earliest heading varieties will have declined in quality, whereas the latest will not have produced any sort of yield. Compatibility is important.

21.3.3   Sward height

Utilisation is the key to successful grassland farms, reducing wastage by grazing fields at the right time with grass at the right height. If plants are prevented from getting to the fourth leaf stage then maximum grass production occurs at about 5–12 cm average sward height. The most common way of expressing grass growth is in terms of kilogrammes of dry matter per hectare per day, so that coverage of 5–12 cm equates to 1500–2500 kg DM/ha. Estimates of sward height are important in order to maintain good performance at grass from all classes of stock. As a sward is grazed closer to the ground herbage intake and performance will be progressively reduced. The ideals are represented in Table 21.3 which gives information about set stocked and rotationally grazed pastures. Estimates of sward height are best made with a sward stick such as the one shown in Fig. 21.1, but in the absence of anything so sophisticated, a ruler or even a mark on the side of a wellington boot are more than adequate substitutes! To get a reliable estimate of average sward height, at least 20 measurements should be taken in a field at random from both grazed and rejected areas. Sward height estimates are likely to be of most use to farmers grazing cattle or sheep on a set stocked basis. Horses have been found to be very selective in species choice and sward height, usually grazing the shortest forage, although there is great variation in reported trial results.

21.3.4   Sward density and dry matter yield estimates

The water present in grass can vary between 85% in spring, to 75% in summer, giving DM of 15–25%, affecting the nutrient value of the grazing. Wet weather can reduce DM to as low as 10%, which will dramatically affect animals' intake

Table 21.3 Suggested target range of sward heights for set stocked animals or post-grazing sward heights for animals kept on rotational grazing systems

	Target sward height
Sheep	
In spring and summer	
Ewes and lambs (medium growth rates)	4–5 cm
Ewes and lambs (high growth rate)	5–6 cm
Dry ewes	3–4 cm
In autumn	
Flushing ewes	6–8 cm
Finishing lambs	6–8 cm
Store lambs	4–6 cm
Cattle	
Dairy cows	7–10 cm
Finishing cattle	7–9 cm
Beef cows and calves	7–9 cm
Store cattle	6–8 cm
Dairy replacements	6–8 cm
Dry cows	6–8 cm

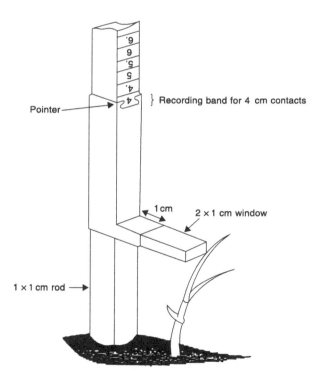

Fig. 21.1   The sward stick designed by the Hill Farming Research Organisation.

and production. When monitoring grass growth and availability, climatic conditions and grazing time (especially for dairy cows) must be borne in mind. The most accurate method is the use of a rising plate meter (RPM), which measures the sward height and density. RPMs can be mechanical or electronic, the latter calculating the kgDM/ha automatically, and have become popular in recent years providing reasonably reliable estimates of sward dry matter yields. Again, it is important to walk the paddocks in a random pattern and make 30 compressions. The readings from an electronic RPM are recorded as an estimated yield of dry matter per hectare and some directly download the information into computer programs. Measurements taken before and after grazing will give a good indication of the average level of dry matter intake per head. This technique is likely to be mainly of use to dairy farmers grazing on a rotational paddock system and gives rise to a 'grazing wedge' diagram, showing grass availability over all the paddocks set against grazing requirements. DairyCo currently provide an excellent resource on grassland management, and suggest that for dairy cows grass should not be allowed to grow beyond 2200–2800 kg DM/ha before grazing, while EBLEX state that when using set stocking for beef maximum cover should be 2000–2500 kg, and for sheep, 1800–2000 kg DM/ha.

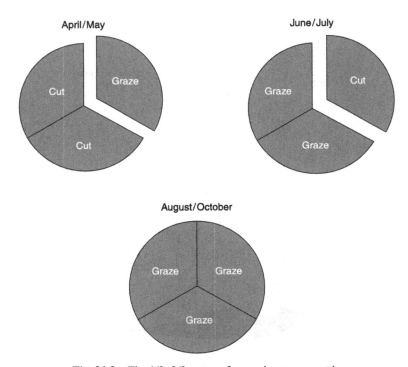

Fig. 21.2   The 1/3 : 2/3 system for grazing young cattle.

21.3.5   Stocking rates at grass

The output of particular grazing systems is strongly correlated with the stocking rate sustained during the grazing season. The rate of grass growth in May can be greater than 40 kg DM/ha/day, but declines to 20–30 kg DM/ha/day as the season progresses. The area devoted to grazing may expand as a result of the progressive introduction of silage or hay aftermath areas into the grazing block. On intensively managed paddock systems, dairy cows at spring turnout would be allocated about 5 kg DM/head. With an initial grazing cover of 2400 kg DM/ha and residual of 1500 kg DM/ha such a herd would require 0.55 ha/100cows/day. Ewes with lambs would normally start the grazing season at about 18–24 ewes with lambs/ha. However, if spring grass growth is poor, it is normal to allocate increased grazing to ewes with lambs at the expense of the conservation area in the expectation that a hay or silage cut may become available later in the season when grass growth improves. With young cattle or dairy replacements at grass, the stocking rate is usually expressed in terms of the liveweight of animals per hectare since individual weights can vary so much. On well-fertilised grassland, about 2.5 tonnes of liveweight/ha would be a good target in the early season falling subsequently to 1–2 t/ha later in the season. Figure 21.2 describes the way grass areas could be allocated throughout the season.

21.4   Grazing systems

21.4.1   Strip grazing

Strip grazing (Fig. 21.3) (also known sometimes as 'flexible block' grazing) is particularly suitable if there is a flush of grass growth (>2800 kg DM/ha) or for the smaller dairy herd. The system allows the animals access to a limited flexible area of fresh grass either twice daily, daily, or for longer intervals, controlled by electric fencing. A back fence should be used, whereby the area just grazed is almost immediately fenced off. This is to protect the recovering sward from constantly being regrazed, slowing the recovery rate. Strip grazing is time-consuming and requires a daily decision as to how much grass is needed for the cows, and a feeding face long enough to allow all the herd simultaneous access. Wet soil conditions can lead to serious poaching along the line of the fence. It is always important to maintain good access to water. Strip grazing is also a suitable system for utilising crops such as forage rye for early bite, or turnips, rape and kale later in the season. Beef cattle can be strip grazed on direct drilled turnips after barley leaving round-baled straw at the edge of the fields to add dietary roughage. Ewes or store lambs grazing rape and turnips during the autumn and early winter are usually rationed in this way, with blocks of forage being introduced for periods of 3–4 days with double strand electric fencing. Welfare regulations require a dry (grass) run-back to allow animals lying areas off wet soil.

21.4.2   Rotational or paddock grazing

The principle of this system is rotational grazing alternating with rest periods (Fig. 21.4). The grazing area is divided into equal-sized paddocks. Good access

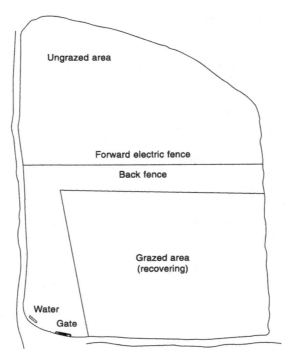

Ungrazed area

Forward electric fence

Back fence

Grazed area
(recovering)

Water

Gate

Fig. 21.3   Strip grazing.

tracks and a reliable water supply to each paddock are essential features of the
system. On a dairy farm the grazing paddocks will normally occupy fields close
to the milking parlour. Fertiliser can be applied immediately after grazing without
any effect on palatability of the sward.

There is huge variation over the number of paddocks considered to be 'ideal',
but the common feature is a system which allows animals to graze for a limited
period (between half a day and one week) followed by a period of recovery.
Rotational grazing can be used for all forms of farm livestock and is beneficial for
horses as well. The more intensive systems with one-day paddocks for example
are obviously most suited to the dairy herd. At the peak of spring growth it would
be expected that animals would rotate around a paddock system on a 21-day
cycle. It is quite possible that some of the paddock areas may become available
for an early cut of silage. As the season progresses the rate at which grass recovers
after grazing will decline. At this time the grazing obtained from part of the area
which has already been cut can be introduced. Fertilising with nitrogen after each
grazing (Table 20.4) and regular 'topping' of rejected material and weeds will
maintain the quality of the grazing throughout the season. 'Leader/follower'
grazing is in operation on some dairy farms with high yielding cows grazing
ahead of a low yielding group. Such a system can lead to very high levels of

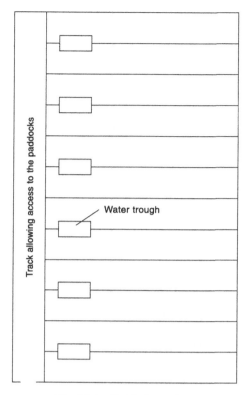

Fig. 21.4   Paddock grazing.

pasture utilisation. The most intensive grass production systems move to new paddocks in the afternoon when sugars (energy) in the sward are at their highest. The grazing wedge system can be used to monitor individual paddock production kgDM/ha.

Young cattle and sheep, when managed on rotational grazing systems, usually have access to fewer paddocks, 4–6 would be normal, but the length of the recovery periods would be roughly the same as for dairy cows. EBLEX recommends efficient grazing strategies such as well managed paddock systems for both beef and lamb production and states that that the financial returns justify the effort. 'Forward creep' grazing, where lambs graze ahead of ewes through a creep gate can also reduce the levels of parasitism. 'Clean' grazing systems for mixed age breeding stock may be applicable to organic production where the use of management strategies to minimise parasite infestations is necessary as routine prophylactic use of anthelmintics is not allowed.

The use of a paddock system for horses, with rotational use and rest phases, mimics the habits observed in semi-feral nomadic equine populations. Groups tend to return to areas in a cycle of 10–21 days which allows for regrowth, and in suitable (warm, dry) weather conditions can reduce the parasite burden by dehydration of the infective larval stage. Permanent wooden or electric tape fencing are the most suitable divisions for equine paddocks.

21.4.3   Set stocking

The majority of beef cattle, sheep and horses in the UK are managed on set stocked grazing systems where animals remain in the same grazing area for extended periods. The obvious appeal of set stocking is the simplicity of management, but it is extremely important to ensure that the swards are of suitable quality (short, dense and leafy) and, most of all, that the grazing pressure, as determined by the stocking rate at grass, is at the correct level to maintain the optimum sward height. Optimum sward height recommendations are set out in Table 21.3. They can have a substantial effect on daily dry matter intakes and animal performance. Sward heights should be checked regularly (e.g. weekly) and adjustments made to the stocking rate by removing animals to other fields or introducing extra grazing (e.g. silage or hay aftermaths) if the need arises. Dairy cows are often managed in set stocking systems, being low in labour requirement, but better output from grass is obtained from rotational or strip grazing. Set stocking is often used for dry cows and heifers, not lactating during the later grazing period.

Nitrogen fertiliser application is controlled by the Nitrate Vulnerable Zone (NVZ) legislation and by the amount of clover in the sward. Fertiliser (solid) application to set stocked areas can follow one of two patterns, either fertiliser is applied to the whole area at set intervals, or to a part of the area (e.g. a quarter) on a regular (e.g. weekly) basis. It has been observed, with sheep in particular, that animals tend to avoid the most recently fertilised areas for several days and this can lead to a kind of rotational grazing without fences. The use of liquid fertiliser is common in the United States and the injection of aqueous or anhydrous ammonia into set stocked pastures was popular in Europe, even to the extent of a whole season's nitrogen fertiliser applied at the start of the grazing season. A cost benefit analysis in the Netherlands found uptake and yield return varied significantly according to seasonal conditions. There have also been concerns about the possible consequences of extra nitrate leaching into ground water.

There are several 'commercial' sources of nutrients advertised specifically for equine grassland such as calcified seaweed or phased release nitrogen, but most single animal horse owners do not use fertiliser to encourage grass growth. Commercial equestrian enterprises such as studs and race-courses follow the recommendations used by other professional grass producers and treat it as a crop. The biggest problem of single species equine grazing areas is the development of lawns and roughs due to horses' using latrine areas to defaecate and subsequently avoiding grazing that area. Mixed grazing with ruminants is recommended to balance equine species and area selectivity, improving the consistency of the sward, and potentially reducing the levels of species specific parasite larvae.

21.4.4   Zero grazing

Cutting and carting green forage to farm animals obviously involves substantial extra costs as well as wear and tear on machinery. On the majority of UK farms it would only be regarded as a technique for green feeding of forage crops such as lucerne, kale or forage maize where either the nature of the crop or conditions underfoot precluded the use of strip or block grazing or the need to cut green forage from outlying fields in times of shortage of grazing. Intensive dairy units sometimes use direct cut grass in the Total Mixed Ration (TMR), instead of allowing cows to walk to field to graze. The utilisation of grass is at its highest when zero grazed, as there is no wastage due to faecal contamination, or from the other factors that lead to rejection.

21.5   Strategies to minimise parasitism at grass

The provision of 'clean' (i.e. relatively parasite free) grazing for young animals is a requirement of good health and welfare management. There is evidence of resistance in ruminant and equine gastrointestinal (GIT) nematodes to most of the common anthelmintic (wormer chemical) groups. This has developed from the continual use of chemical control of infestation of youngstock exposed to high challenge before developing a degree of immunity. Pastures previously or concurrently grazed with older animals present the highest risk. Strategies for minimising this problem include alternating sheep and cattle on different fields for grazing, year on year. The 'leader/follower' and 'forward creep' systems have already been mentioned. A reduction in parasitism can be brought about in this way by reducing the extent to which the young cattle or lambs are forced to graze heavily contaminated material at the base of the sward. The regular introduction of new leys into the system provides clean grazing, and moving young cattle or weaned lambs on to silage or hay aftermaths after worming in mid-season is also extremely effective. The '1/3 : 2/3' system for young cattle, illustrated in Fig. 21.2, incorporates this latter principle. It must be remembered that exposure to infective nematode larvae is necessary for the development of immunity when adult, therefore a low level of challenge should be allowed. Recent wet seasons have increased incidence of liver fluke disease for which mud snails act as an intermediate host. Improving the drainage of liver fluke-infected pastures can reduce the risk, although consideration needs to be given to the cost and to the environment. There is evidence that the inclusion in a sward of herbaceous plants high in tannins, such as legumes, chicory and plantains gives some anthelminthic activity, and improves growth rates and health of grazing young cattle and lambs.

21.6   The energy yield from grass and forage – the Utilised Metabolisable Energy (UME) calculation

Assessment of the real value of the output from grass and forage on livestock farms is difficult for practical reasons. The measurement of the yield of a crop of wheat or potatoes is easy as there is a quantifiable, physical and marketable yield. With grass and forage crops, although the yield of dry matter is obviously present and is sometimes measured directly (e.g. loads of grass or numbers of bales), it has to be utilised by farm animals before any financial or marketable output is created. Where a high proportion of the animal production is obtained from grazing these estimates are particularly difficult.

21.6.1   The Utilised Metabolisable Energy (UME) system

This system takes account of the whole of the theoretical energy requirements of forage-based farm livestock and estimates the proportion of that energy which has been produced on the farm by the forage areas. Energy estimates are made as megajoules (MJ), or gigajoules (GJ), where 1 GJ = 1000 MJ.

The calculations assume that if the total annual ME requirements for the livestock on the farm are known, as well as the annual amount of ME fed to the stock from bought-in feed and concentrates, then the remaining ME must come from home-produced forage. This would include all the forage area on the farm, grazing, silage, hay, and any other forage crops.

Example

To record UME production from grass for the dairy herd on a dairy farm, the following information would be needed:

1. The annual energy (ME) requirement per cow.
    For maintenance this would be about 25 000 MJ and for the production of average quality milk, 5.3 MJ/litre for (say) 6471 litres. (For extra accuracy an allowance might also be made for pregnancy, different breeds and actual animal liveweights and milk of varying quality – however, the above will suffice for a rough calculation.)
2. The annual ME purchased as:
    (a) Concentrates – say 1.6 tonne per cow with a dry matter of 85% and an ME value 13 MJ/kg DM.
    (b) Other purchased feeds – say 100 kg of hay per cow with a dry matter of 84% and an ME of 9 MJ/kg DM plus 1 tonne of wet brewers grains per cow with a dry matter of 20% and an ME of 11 MJ/kg DM.
    The total ME purchased per cow would then be the total of a + b above.
3. UME obtained from grass and forage per cow = the annual ME requirement per cow minus the total ME purchased per cow (a + b).
4. The stocking rate in GLUs per forage hectare – say 1.97.
5. The annual UME per grass and forage hectare for the dairy herd = UME per cow × stocking rate. The complete calculation is shown in Table 21.4.
6. The grass growth site class (see Section 20.5.2). The UME targets suggested in *Milk from Grass* are also given in Table 21.4. The reader can then substitute actual farm figures for this calculation, use the appropriate site class for comparison and assess the efficiency with which forage is being produced and utilised on a particular farm.

Table 21.4   The UME calculation from the example outlined in Section 21.6.1

Maintenance	+ Milk production	= Total annual ME requirements	
25 000	6471 × 5.3 + = 34 296 MJ	= 59 296 MJ	
Supplied by:			MJ of ME
Concentrates	Hay	Brewers grains	purchased/cow
1665 × 85% × 13 = 18 398 MJ	100 × 84% × 9 + = 756 MJ	1000 × 20% × 11 + = 2200 MJ	= 21 354 MJ
Total ME requirements 59 296 MJ	− ME purchased/cow − 21 354 MJ	= UME/cow = 37 942 MJ	
UME/cow	× Stocking rate GLU/ha	= UME/hectare	
37 942 MJ	× 1.97	= 74 745 MJ/ha	= 74.7 GJ/ha

Target UMEs taken from *Milk from Grass*

Site Class (see Appendix 7)	Target GJ/ha
1	126
2	115
3	105
4	93
5	83

The calculation of UME figures for a dairy enterprise is quite simple. The total milk output is always very well documented. The calculation for beef and sheep enterprises is more complex in that the output figures required are total annual liveweight gains which are obviously more difficult to obtain reliably and are subject to a great deal more error.

21.7   Sources of further information and advice

Further reading

British Grassland Society, *Forage Fact Sheets*, http://www.britishgrassland.com/page/fact-sheets.

British Grassland Society (2013), *Proceedings of the BGs 11th Research Conference – Science and Practice for Grass-Based Systems September 2013*.

Crofts A and Jefferson R G (eds), *The Lowland Grassland Management Handbook*, English Nature, 1999.

DairyCo (2012), *Grassland Management Improvement Programme*, http://www.dairyco.org.uk/technical-information/grassland-management/grassplus/.

Defra and Natural England (2013), *Grazing and Pasture: Sustainable Management Schemes*, https://www.gov.uk/grazing-and-pasture-sustainable-management-schemes

EBLEX (2012), *Better Returns Programme*, http://www.eblex.org.uk/returns/nutrition-and-forage/.

Frame J and Laidlaw A. S. (2011), *Improved Grassland Management*, Crowood.

Hopkins, A (ed.), *Grass: Its Production and Utilization*, 3rd edn, Blackwell Science, 2000.

Hybu Cig Cymru (2008), *Grassland Management*, www.hccmpw.org.uk.

Ledgard S, Schils R, Eriksen J. and Luo J. (2009), Environmental impacts of grazed clover/grass pastures, *Irish Journal of Agricultural and Food Research*, 48: 209–226.

Newton J (1999), *Organic Grassland and Sheep Husbandry: 15 Questions Answered*, Newton Publishing.

Nix J, *Farm Management Pocketbook*, AgroBusiness consultants (annual).

O'Beirne-Ranelagh (2005) *Managing Grass for Horses* J.A. Allen.

Pearson C J and Ison R L (1997), *Agronomy of Grassland Systems*, 2nd edn, Cambridge University Press.

Teagasc (2010), *Grassland for Horses*, Agriculture Food Development Authority.

Thomas C, Reeve A and Fisher GEJ (1991). *Milk from Grass*, The British Grassland Society.

Wright I (2005), Future prospects for meat and milk for grass based systems. In S J Reynolds and J Frame (eds), *Grasslands: Developments, Opportunities, Perspectives*, FAO.

Younie D (2012), *Grassland Management for Organic Farmers*, Crowood.

Websites

www.britishgrassland.com
www.dairyco.org.uk
www.eblex.org.uk
www.europeangrassland.org
www.gov.uk
www.grasslanddev.co.uk
www.hccmpw.org.uk
www.naturalengland.org.uk
www.teagasc.ie

22

Conservation of grass and forage crops

DOI: 10.1533/9781782423928.4.513

Abstract: The reasons for the swing away from haymaking and towards silage during the twentieth century form the introduction to this chapter. The main crops used for ensiling are grass and forage maize, but forage legumes, too, are becoming increasingly important. The process of ensiling and the important factors affecting fermentation – such as minimising contamination, wilting, chopping consolidation and the exclusion of air to achieve anaerobic conditions – are all emphasised. The value of silage additives, too, is discussed. There is also a description of the important aspects of good haymaking, barn drying, green crop drying and the making of high-dry-matter baled 'haylage' for enquines.

Key words: silage, hay, haylage, fermentation, silage additive.

22.1 Introduction

22.1.1 Reasons for the change from hay to silage

Grass and a range of other forage crops are conserved as winter feed, mainly as silage in the UK. Hay was the traditional method for conserving grass in Western Europe. Hay, 'cured' in the field, needs a prolonged drying period for the reduction of the moisture content to a safe level for storage and feeding, which is seldom available in the climate of the UK in particular. The quantity of hay conserved virtually halved between 1970 and 1994, while the rise in the popularity of silage, clamp and baled, and haylage rose to an estimated 13 million tonnes over the same period. The most recent figures available are for 2006 indicating that 2.5 million hectares of land were then used for silage production, yielding 15 million tonnes of conserved forage.

The change in British farming systems, increased herd and flock sizes and the reliance on mechanised feeding systems rather than labour-intensive handling of bulk feeds, has also driven the change to fermented, rather than dehydrated, winter forage. Grass varieties have been bred that produce very high yielding, highly digestible crops, responsive to nitrogen fertiliser, virtually forcing farmers into silage making, as discussed in Chapter 19. Other influences include increased intensification, and higher stocking rates which, coupled with improvements in ensiling techniques and reliable systems for mechanically handling and feeding silage, have allowed increased output from forage-based systems. Field machinery has also improved, and the advent of the widespread use of contractor teams for silage making has meant that British farmers now have access to some of the most sophisticated and efficient machinery for making silage anywhere in the world. Speed of operation has become very important since declines in the D value of grass are now well appreciated. A self-propelled forage harvester can achieve an output of 250 t/hour. It is therefore possible for a dairy farmer, with a farm of average size for example, to expect an efficient contract team to make an entire cut of first cut silage within two days, a job which in the 1970s with farm machinery, would have taken weeks!

22.1.2   Losses from silage making

However, all these developments have brought with them associated problems. The costs of silage making have rocketed and although it can be argued that growing heavy crops of grass can minimise the costs per tonne of dry matter conserved, the depressed livestock and milk prices at the turn of the millennium brought home to farmers the true costs of conservation and the need to minimise them whilst maximising the use of grazing. There is also a high potential for losses if silage making is not carried out well, and although modern machinery and practice have reduced such problems it is still routinely possible to lose 15–25% of the dry matter ensiled. An ADAS survey carried out in the 1980s indicated that the true level of losses from silage making (field, storage, effluent and feed-out) ranged from 25–45% of crop dry matter. Precision chopping, good consolidation and clamp sealing can all help to minimise losses, but the greatest potential for losses and for environmental pollution comes from silage effluent.

It is possible, from very wet, direct cut material, to produce as much as 200 litres of effluent per tonne. Silage effluent has a very high biochemical oxygen demand (BOD) which makes it about 200 times as polluting as raw domestic sewage. Pollution of groundwater or watercourses by silage effluent has resulted in many prosecutions and some very heavy fines. The relevant legislation is laid out in the Water Resources (Control of Pollution) (Silage, Slurry and Agricultural Fuel Oil) (England) Regulations 2010 and as amended 2013 (SSAFO).

All silos must have perimeter drains and a sealed effluent pit which is emptied regularly, especially during the period immediately after ensiling when effluent production is greatest. Care must be taken to avoid pollution and to adhere to NVZ regulations when disposing of effluent on land.

Silage effluent can be conserved and fed to livestock as a nutritious liquid feedstuff. One system is to add an absorbent material (e.g. dried sugar beet pulp) at ensiling; another is to conserve the effluent and preserve it either with the addition of a material such as formalin or by enclosing it in an airtight container (e.g. the Eff system). However, the simplest expedient is to wilt silage grass (or other materials to be ensiled) in the field to 25–30% dry matter in which case the production of effluent is minimised. In the case of forage maize, harvesting of the standing crop is delayed until the dry matter reaches about 30%.

22.2   Crops for silage making

22.2.1   Grass

All types of grass can be ensiled. Feed value will be very dependent on the stage of growth and digestibility (D value) of the crops. The fermentation quality will depend on a great many other factors discussed later, but the dry matter percentage and the content of water soluble carbohydrates (sugars) are two of the most important. Grass for silage should be mown at a height of 7.5–10 cm in order to minimise soil contamination and to allow good recovery. Fermented forage for equine use is predominantly grass based.

22.2.2   Legumes

Most legume plants are more difficult to make into well fermented silage than is grass. This is because they usually contain relatively low levels of sugar. The need therefore arises for mixing with grasses or cereals, which have much higher levels of sugar or starch, to improve the fermentation characteristics of the silage.

Red and white clovers are frequently made into silage, usually in mixtures with grasses of various types although occasionally red clover may be sown and harvested as a pure stand. Lucerne is grown on farms in the drier parts of southern and eastern England often as a pure stand, but also mixed with red clover or with grass. D values of red clover and lucerne when cut for silage are often lower than for grasses, but protein content is much higher (up to 20–22% crude protein at early flowering). To make acceptable silage, lucerne and clovers should be wilted to at least 30% dry matter or else some sort of acid additive should be used at an effective rate.

Annual legumes such as peas (forage or grain varieties are both suitable) and vetches can also be made into high protein silage. These crops are also often mixed with spring barley or oats as 'arable silage mixtures' in order to improve the fermentation and to increase the starch percentage of the final product. Such mixtures can be harvested at any stage from when the grain is 'milky' up to the 'cheesey' stage and when the peas or vetches have pods partly filled. Recently, interest has developed into using forage varieties of lupin as an arable silage crop, and trials are under way, too, with forage varieties of soya bean. Both crops are fairly late maturing and have potential in their own right or even as 'bi-crops' with forage maize.

22.2.3   Forage maize

Maize as a silage crop has increased substantially in popularity in recent years and detailed information about its agronomy is given in Section 18.3.1. According to the area where it is being grown, a variety of maize should be selected on the basis of its maturity group, which determines its ability to yield ripe cobs which contain starch. At harvest the cob yield should be about 50% of the total and starch should comprise 25–30% of the dry matter. This usually occurs when the dry matter percentage of the whole crop reaches about 30%. If it is just possible to make a dent with a finger or thumbnail in the grains on a number of cobs then the crop is adjudged ready to harvest.

Very fine (1–2 cm) chopping and grain cracking are desirable for making good maize silage, with recent work suggesting that the longer length is better for digestive health and for milk butterfat. The crop must be well consolidated, which can be difficult with a longer chop length, to exclude air in order to avoid over-heating and the growth of moulds. Care must be taken at feed-out to keep a tidy face and to work quickly across the clamp. It is recommended to have narrow clamps for maize silage (or to divide wider ones with temporary barriers) to avoid secondary fermentation and mould growth at the face. Since there is so much starch and some sugar still in maize when it is harvested, additives should not normally be needed and a satisfactory fermentation should be easy to obtain. 'Ground ear maize' (GEM) and 'corn cob mix' (CCM) are the products of cob-only harvesting options using a maize snapper header either with a forage harvester (for GEM) or a combine harvester (for CCM) to make very high energy silages. In the case of corn cob mix, feeding it to pigs has become popular, especially in other parts of Europe.

22.2.4   Other cereals for 'wholecrop' silage

Any type of cereal can be taken for silage, but the best yields and quality are usually obtained from winter wheat. Advantages are that it is an excellent source of cereal starch (although maize is favoured by livestock farmers) and it can substitute for forage maize in the north of England, Scotland and Northern Ireland where maize cob and starch yields may not be so good. Any surplus can be left for combining and harvest takes place in July/August leaving open many more post-harvest options for the fields in question.

There are two basic systems for making wholecrop cereal silage. One is to cut the crop when it is still fairly green at about 35% dry matter in July for normal fermented wholecrop silage. The other is to leave the crop until it is nearly ripe for combining (about 50% dry matter) and to use feed-grade urea as an additive at the clamp. When the clamp is sealed (good sealing is absolutely essential for this technique) the heat of the initial fermentation releases ammonia gas from the urea. This has the effect then of preserving the grain and straw as a moist relatively non-fermented high pH product. The product known as 'alkalage' is made in the same way, but from a crop at about 70% dry matter and requires the use of a forage harvester with a processing mill. Where clamp space is not available, these crops can be stored in large stretchable bags or tubes on suitable hard standing, provided rodent control is undertaken.

Other systems exist for ensiling moist grain after crimping (partial grain cracking through fluted rollers) and with the use of preserving agents such as propionic acid. The use of brewers' grains for ensiling has also become commonplace either layered within grass clamps, on its own or mixed with an absorbent such as dried sugar beet pulp (grainbeet).

22.2.5   Kale for silage 'kaleage'

This is a system pioneered by farmers whereby leafy kale varieties are ensiled after mowing and wilting for 24–48 hours. Sowing at a high seed rate gives a thick crop which can be wilted on a fairly high stubble. Kale can then be harvested for clamping or, more frequently, for round baling and wrapping (six times to give a really good seal). Other options investigated in the past at IGER and elsewhere have involved growing kale as a bi-crop with spring cereals for making into clamp silage. However, these techniques have not been widely adopted in UK agriculture.

22.2.6   Fodder beet

This crop has been popular for ensiling in some European countries such as Denmark, but has recently been eclipsed by the rise in popularity of forage maize. In the UK some pioneering work was done at ADAS Rosemaund. The system involves growing varieties (e.g. *Kyros*) which can be top lifted in reasonably dry field conditions in September to give a moderate yield of fairly clean beet. A special harvester then chops the whole crop, tops and all, and ensiling takes place between alternate layers of chopped straw (sometimes ammonia treated) and an absorbent such as dried sugar beet pulp. An alternative is to clamp whole beet in the normal way and to make silage from the tops. In both cases excessive soil contamination is a potential problem to the quality of the fermentation of the silage.

22.3   The silage-making process

22.3.1   Silage fermentation

Most silage is made from grass. Therefore most of the information in this section concerns the making of grass silage. Silage making or ensilage is a process of anaerobic fermentation (i.e. fermentation without air). The speed with which air can be excluded from the clamp and prevented from subsequently re-entering it will determine the success of the whole operation. During fermentation the carbohydrates and proteins present in the crop are acted upon by bacteria and plant enzymes. Some of the bacteria will be those naturally occurring on the crop before ensiling. Others may have been added in the form of an inoculant additive. The most desirable outcome of this initial fermentation is a rapid fall in the pH of the silage to about 4. This is caused by an accumulation of lactic acid and of other organic acids which are a by-product of fermentation by a group of bacteria of which the genus *Lactobacillus* is the most common, and one of those which has been cultured to produce inoculant additives. In a well fermented silage the concentration of acids reaches a peak after less than a week and the pH will stabilise at about 4. Such a silage will have a light brown colour, a sharp smell and an acid taste (for those who fancy it!).

In a poorly fermented silage, which may result from very wet grass or where contamination with soil or faecal material has taken place, the bacteria which dominate the fermentation may be from another group known as *Clostridia*. In this case the end result is an accumulation of a weaker acid called butyric acid. The pH of such a silage may well be in the region of 5, and it will appear as an olive green colour with a foul smell. One of the unpleasant smells will be that of ammonia which will almost certainly be present in large quantities as a result of the inefficient preservation of proteins and their subsequent breakdown. Apart from the dry matter percentage and the pH, one of the most useful indicators of silage quality by analysis is the percentage of the total nitrogen in the silage which is present as ammonia. This should be as low as possible (3–5%); if it is high (20–30%) this will almost certainly be as a result of a poor fermentation and the predominance of butyric acid.

22.4   Factors affecting silage fermentation

22.4.1   Type of crop
Crops with high levels of fermentable carbohydrates (sugars or starch) are usually the easiest from which to make good silage. So the Italian and tetraploid ryegrasses with their high sugar content, and maize and cereals with high starch content usually ferment well with no need for additive application. Low sugar crops such as short leafy grass and legumes at most stages of growth are more difficult to ferment and need some extra treatment such as wilting, the use of an effective additive, or both.

22.4.2   Fertiliser treatment
This can affect the sugar content of grasses in particular. It is important to stress that the nitrogen recommendations and timings set out in Table 20.4 should be adhered to. Late or excessive applications of fertiliser can obviously lead to the ensiling of grass containing large quantities of nitrogen in the ammoniacal or nitrate form. This in turn can lead to the depression of grass sugar percentage and the presence of excessive levels of ammonia in the final silage. Where fertiliser application to silage crops is delayed because of poor weather or soil conditions, rates should be appropriately reduced.

22.4.3   Weather
Dry, sunny weather is obviously to be preferred for silage making. Wilting conditions will be improved and the possibility of soil contamination is minimised.

Grass sugar levels are also likely to be highest in sunny weather. Grass mown in the afternoon of a sunny day has been shown to have maximum levels of sugars or 'water soluble carbohydrates' (WSC).

22.4.4   Minimising contamination

The control of moles and the rolling of silage fields are important preparatory tasks in the spring. On most farms they are standard practice. Slurry application, although common on silage fields, should be avoided if possible. Where there is no alternative, it should be applied as early as possible relative to the proposed cutting date. The heavy contamination of silage grass with slurry can lead to poor fermentation and subsequent losses in feed value. Other fairly obvious ways in which contamination can be minimised include power washing the clamp prior to ensiling, tipping the crop on a clean concreted area and ensuring that the wheels of the vehicles undertaking the buckraking are as clean as possible. Levels of contamination can most easily be assessed by reference to the percentage of 'ash' in the silage analysis report. A figure of less than 10% is acceptable whereas 15% or more would indicate that significant levels of contamination have taken place.

Cutting height is also important as cutting too low will increase the risk of soil contamination. In dense swards the residual height should be no lower than 5 cm, but 7.5–10 cm is advisable in more open crops. Cutting lower than this also reduces the seasonal yield as recovery of the crop will be slower.

22.4.5   Wilting

In the last 30 years the average dry matter of clamp silage in the UK has increased from 22% to close to 30%. Wilting in the field after cutting has become a standard treatment when making grass or legume crops into silage. Simply leaving the mown swath in the field, however, can be counterproductive and lead to a degree of aerobic fermentation (heating) of the swath, which is very undesirable. Thick swaths such as those from first cut Italian ryegrass, mown by very wide, high performance mowers should be subsequently spread to full ground cover to obtain optimal wilting. Some high performance mower/conditioners mow, condition and spread in one operation, minimising wilting time in the field and soil contamination since all that is required subsequently is rowing up prior to final harvesting. Conditioning the crop involves physically damaging it with nylon brushes, tines or rollers and increases the rate of wilting substantially so that in good conditions it is easily possible to achieve dry matters of 25% or more from a 24 hour wilt. Most farms use a rotary tedder and may further turn the crop prior to rowing up for harvesting with a precision chop harvester. This process will lead to an increase in dry matter percentage, but it can also lead to quite high levels of field losses (up to 15% for very dry material), and also to a degree of contamination with soil. Wilting to much higher dry matter levels of 30–40% is required for ensiling in towers so that the unloading equipment will work well. This normally requires a

48 hour wilt and some additional tedding. A similar, or even higher, DM content is usual for big bale silage especially for that known as 'haylage' and used as equine forage.

A dry matter of at least 30% is very desirable for ensiling lucerne and red clover and these crops benefit from the action of roller/crimpers or mower-conditioners which crack the thick stems of these plants and improve the wilting rate. Maize and wholecrop cereals are 'self wilting' crops since harvesting does not normally take place until they have started the natural dehydration process associated with ripening. Kale for ensiling should also be wilted for about 48 hours. The best technique is to mow the crop with a mower-conditioner and leave it on a fairly high stubble (about 10 cm) prior to baling direct from the swath.

22.4.6   Harvesting

On the majority of farms grass silage is now harvested using high output metered chop machines. Wilted grass is usually chopped to between 20 and 50 mm, whereas maize benefits from even shorter chopping to between 10 and 20 mm. When harvesting maize, the grains should be cracked (to break the hard seed coat) to aid fermentation and subsequent digestion by farm animals fed the silage. Finely chopped material usually consolidates better and it is possible to achieve true anaerobic conditions in the clamp more quickly. Fine chopping also results in the more rapid release of sugars from grass for fermentation.

22.4.7   Baled silage

About 20% of the silage made on UK farms at the turn of the millennium was harvested by big balers. This is a system of production which has substituted for hay on many of the smaller stock farms. The saving in the large capital expense of a clamp silo has obvious attractions for the smaller farmer. Grass is normally wilted up to about 40% dry matter prior to baling in high density round or square bales. It is easier to stack square profile bales, which are available in many lengths and widths, but round bales are easier to feed in self-feed ruminant systems. As soon as the grass has been baled it should be wrapped. Silage bags are now very infrequently used and the advent of very efficient bale wrapping machines, using polythene stretch film, has meant that this system of covering is now almost universal. Four layers of wrap is the minimum, with six being preferred to increase protection of the crop and to ensure anaerobic conditions maintain in the bale. The more wrapping takes place, and the quicker it takes place after baling, the more efficient will be the exclusion of air and the quality of the final product. It is extremely important to avoid puncturing the wrap during handling and stacking, although if using a bale spike, prompt 'patching' with suitable tape will minimise the risk of poor fermentation. Stacks should be on a hard standing, or fenced so that stock cannot damage them, netted to avoid bird damage, and secured against rodents.

22.4.8   Baled 'haylage'

Crops of grass for very high dry matter baled 'haylage' for horses are normally made from quite mature grass (D value 60–65) and are wilted and tedded for up to three days to achieve a dry matter of about 65%. For competition horses, tetraploid Italian or hybrid ryegrasses ensure a high sugar content and a satisfactory fermentation to provide a very nutritious forage that will reduce the need for high concentrate feeding, improve equine gut health and cut costs. On many equestrian enterprises haylage is also preferred to hay as a result of its dust free characteristics, thus promoting good respiratory health. Wrapping with at least six layers of high quality stretch film is essential, as horses are particularly susceptible to botulism, which can be present in soil-contaminated and poorly sealed bales. Commercial firms produce several other types of haylage, including high fibre (late cut ryegrass), species specific, such as Timothy for a lower nutritional value suited to leisure horses and native ponies and even lucerne haylage for horses in hard work.

22.4.9   Filling a clamp silo

Firstly it should be ensured that the clamp is as airtight as possible and side sheets should be used. Filling a clamp silo should then take place as rapidly as possible. Buckraking and subsequent consolidation should aim at making a wedge-shaped clamp in the initial stages. Rolling and consolidation should be a continuous process during clamp filling and special attention should be paid to consolidating the edges of the clamp. If possible they should be built a little higher than the centre of the clamp. A final layer of heavier, direct cut (not wilted) grass will help final consolidation. If the process of filling the clamp is likely to take several days it is advisable (although not always popular) to sheet the clamp each night. This will help to minimise the intake of air into the clamp.

22.4.10   Final sheeting and sealing

Ideally, clamps should not be opened after ensiling until feeding starts. Sealing should take place as soon as filling and consolidation have finished. The majority of clamps will be sealed with two thicknesses of plastic sheet and the 'shoulders' of the clamp may well be covered by a third layer, from the side sheet. Weight should be placed evenly on top of the sheet all over the clamp. The ideal materials for doing this are the ubiquitous used car tyres or, better still, straw bales placed tightly together.

The recent development of cling film and oxygen barrier films have revolutionised sealing of the top of clamps. The usual black plastic silage sheets are permeable to air, as are the original cling films, but oxygen barrier films used with close-weave netting and sand-bag type weights are claimed to be at least 100 times better at maintaining anaerobic fermentation conditions. These techniques can reduce DM losses by 10–20% due to uniform air exclusion across the clamp top and sides with less risk of contamination from mould growth and the associated mycotoxins which can seriously affect animal and human health.

22.4.11 The use of silage additives

Additives were traditionally used when poor weather or other factors were likely to result in a poorly fermented product. Even when conditions are ideal additives will further improve silage quality and result in more rapid pH reduction and a stable silage. Silage makers have used additives for many years with the earliest including dilute hydrochloric and sulphuric acids and molasses. In spite of quite different modes of action both of these approaches can give rise to improved fermentation, either by direct acidification or by increasing the supply of sugar, leading to an increased rate of lactic acid production in the clamp. Similar materials are still in use today and formic acid marketed as 'Add-F' is one of the longest standing and most effective of all additives. Molasses is also still in use and extremely effective in the ensiling of low sugar crops such as lucerne. However, the inconvenience of applying large quantities of liquid molasses, and the extremely unpleasant and corrosive nature of acids or acid/formalin mixtures (particularly unwelcome on expensive contractor-operated machinery) have led to a wide range of alternatives being available.

The most widely used category today is the inoculant additive which works by the application of very large numbers (to be effective about one million per gramme of grass) of live, lactic acid producing bacteria, at the time of ensiling. There are two types of these additives: *homofermentative*, producing mainly lactic acid and a rapid fall in pH, preserving sugar and protein levels in the crop; and *heterofermentative* producing both lactic and acetic acids, resulting in a slower fermentation, but inhibiting yeasts and moulds, especially during feed-out. Although not always as effective as direct acidification the inoculants have achieved a high degree of popularity mainly because of their total safety.

There is a very large array of commercially available additives of all types, but all should have met the EU 1831/2003 Feed Additive Regulations. Additives and inoculants are most frequently applied as liquids via applicators on the forage harvester, although granular or powder applicators have also now been developed. It would be true to say that both forms of application frequently leave a great deal to be desired in terms of accuracy. Additives can cost between 50 pence and £5 per tonne of grass and so can become quite significant items of expenditure on livestock farms. Extensive trials with beef cattle have shown that both intake and daily liveweight gain were significantly improved with additive treated silage. Finishing time was reduced and carcase weights were higher, resulting in a definite financial advantage for using additives. DairyCo states that baled silage should only ever be treated with additives that inhibit all microbial activity or those that promote a rapid fermentation such as the homofermentative. Organic standards limit the use of additives to the inoculant type.

22.4.12 Secondary fermentation

Secondary fermentation or 'aerobic spoilage' can occur when the silage is being fed. This is particularly the case if the rate at which a clamp is being used is fairly slow, or if the size of the face exposed is large. Aerobic bacteria, yeasts and moulds can affect the silage and if the face is left exposed for a long period of time there can be significant deterioration. Bacteria of the *Listeria* family can be particularly damaging in the case of baled silage where the wrapping has been punctured, and in some cases this can lead to a serious brain infection (listeriosis) of stock feeding the silage. Maize silage is particularly susceptible to secondary fermentation at feed out because of its high carbohydrate content. Really good consolidation at ensiling coupled with ensiling in a long narrow clamp are ways in which such deterioration can be minimised. Some silage additives also claim benefits in improving the stability of silage at feed-out. The use of mechanical block cutters when feeding all types of silage minimises the extent to which the face of the clamp is disturbed.

Before any conserved forage is fed, a sufficiently representative sample should be analysed for feed value. In the case of silage and haylage it is obvious that the differences in dry matter will affect nutrition due to the simple dilution factor of, for example, 30% DM versus 50% DM.

22.5   Hay

With the intensification of grassland management has come a marked reduction in the popularity of hay. On many smaller or more traditional farms the more predictable big bale silage system has replaced hay, and as early as 1999, all the Irish farms surveyed produced the majority of their baled forage as big bale silage. For many small equestrian enterprises hay remains the most manageable winter forage, and there is a ready market for well-made small bale hay in the UK, Europe and N. America. Similarly to the crops grown for silage production, hay cropping can be a species monoculture, usually ryegrass 'seeds hay' in the UK (often lucerne in N. America), a mixture of selected species, such as ryegrass and clover, or more traditionally an indigenous hay meadow that has up to 20 species of grasses and herbaceous plants. The high dry matter of hay can result in a higher nutrient intake and also, of interest to animal keepers, a lower urine output than lower dry matter forage and hence dryer bedding!

22.5.1   Traditional hay production
Making hay traditionally is obviously totally weather-dependent and a very risky method of feed conservation in UK conditions. The trend towards wetter summers found during the first 10–12 years of the twenty-first century resulted in a shortage of well-made hay and further influenced the change to silage/haylage production. The target is to reduce the moisture content of mown grass from about 80–85% to 12–15% for field conditioning and barn storage. This can be achieved in four days of hot sunny weather but can frequently mean that the hay is in the field for a week or more. The crop is subjected to prolonged respiratory, physical, and leaching losses that result from frequent tedding, rowing up and the action of dews or rain showers. Such losses from material which may be already of fairly low feed value

(e.g. a D value of about 60 when cut) frequently result in hay with a feeding value which is little better than straw. Further hazards can arise of course as a result of baling and carrying the hay at too high a moisture content. The heating of bales with the consequent growth of bacteria and moulds, at best results in a further deterioration of feeding value and, at worst, in the spontaneous combustion of the stack. Handling and feeding mouldy hay can be hazardous to stock as well as causing substantial hazards to workers. The human medical condition known as 'farmers lung' can be caused in this way.

There are a few 'dos and don'ts', however, which may have a beneficial effect on hay quality. As with silage, if the sward is cut before full heading then the feed value is at its highest, but this is not the traditional time to harvest a 'hay meadow'. Diploid varieties are recommended when selecting grasses for a ley mixture which may be used for hay. The very sappy, sugary tetraploids, whilst ideal for making haylage, are extremely difficult to dry sufficiently well to make good quality field hay. Another important point is not to condition severely, and certainly not to mow with a flail, any crop which is to be made into field-cured hay. Leaching losses in the event of rain showers from crops so treated will be excessive. Additives to restrict microbial activity in hay bales based on propionic acid or propionate salts are available but have not been widely taken up. For most farmers, field-cured hay will remain merely an opportunist method of conservation should the right crop and weather conditions obtain.

Increases in the designations of traditional species-rich hay meadows for conservation purposes may well increase marginally the quantity of this type of material conserved in the future. The management agreements for many of these meadows usually stipulate very late (e.g. July) cutting to facilitate the setting of seeds from the desired species. By this time the feeding value of the hay will be very low indeed, and some potentially hazardous infections with fungi such as ergot may have occurred.

22.5.2   Barn-drying or conditioning

The obvious value of some form of artificial drying of hay in store was recognised in the 1950s and 1960s when barn-drying installations involving electric or diesel driven fans were popular to a limited extent. The obvious disadvantages of high fuel costs, and the very high demand for labour, have subsequently reduced these installations to a very few, most of which are now producing specialist high value products for the equine market. It is possible, by producing low density bales (about $100\,kg/m^3$), to bale and carry hay with up to 40% moisture content and to dry to less than 20% moisture in store. This can also be achieved with loose chopped hay, and various systems exist for handling and drying in store relatively small tonnages of such material. Although most legume hay is imported into the UK, home-produced lucerne or sainfoin hay, destined for the equine market, can also be successfully barn-conditioned and the stemmy nature of both of these enables air movement through the bales and facilitates drying. Hay should be blown continuously and drying can take up to 21 days.

22.6   Green-crop drying

Green-crop drying can no longer be considered as a farm enterprise but as an industrial operation using agricultural crops. Most of the output is milled and cubed as grass and lucerne nuts but in some cases the material is sold long as dried grass or lucerne or chopped and mixed with molasses and chopped straw (e.g. Dengie equine products such as 'Alfa–A', Hi-Fi).

Crops for drying may be drawn from a single farm source or contracted out on to neighbouring farms as a break from arable cropping. Grass and lucerne constitute the main output. Tall fescue is favoured in some cases because of its resilience to mowing every five to six weeks and its long growing season. Occasionally, other crops such as wholecrop maize or cereals have been dried and sold as blends with the main output of grass or lucerne nuts. In most cases the nuts are sold on to feed compounders, but some farmers and horse keepers feed them direct to stock. One of the valuable constituents of grass meal is the range of pigments in it; this is one of the main reasons for its inclusion in poultry rations, for example, to improve the colour of egg yolks.

The system of production is extremely simple and grass is cut and dried if possible on the same day. The driers themselves are usually large rotating drums and the wet grass is introduced into a stream of very hot air (over 1000°C) fuelled by oil, gas or coal. The crop spends about three minutes in the drier, emerging at a temperature of about 120°C and is then milled and cubed at a moisture content of about 10%. Obviously the logistics of such an operation need to be very efficiently organised, and during the main production season the plants are run 24 hours a day and seven days a week.

Crop agronomy and fertiliser inputs will be much the same as for a livestock based unit, except that the absence of any return of organic manures necessitates the substantial use of both nitrogen and potash fertilisers on grass and mainly potash fertiliser on lucerne. In the latter case potash (K_2O) application can be as much as 375 kg/ha in order to replace that removed by the crop.

In the post-BSE era the popularity of dried grass and lucerne has improved. However, in spite of the availability of EU aid for the industry, green-crop drying has never developed in the UK to the same extent as in the rest of Europe where co-operative ventures have been (as ever) very willing to accept the very generous levels of aid available.

22.7   Sources of further information and advice

Further reading

Baillie J, *Assessment of Silage Additives in High Quality Wilted First Cut Grass Silage*, Kingshay Farming Trust, 1997.

British Grassland Society, *Forage Conservation Fact Sheets*.

British Grassland Society, *Big Bale Silage: the Technology of Making Silage in Big Bales and its Place in the General Forage Conservation Scene*. Proceedings: BGS Conference, 1989.

Cooper M M and Morris D W, *Grass Farming*, 5th edn, Farming Press, 1983.

DairyCo, *Silage Additives*, AHDB, 2012.

EBLEX, 'Better Returns Programme' publication – *Making Grass Silage for Better Returns*, AHDB, 2012

Hopkins A (ed.), *Grass: Its Production and Utilization*, 3rd edn, Blackwell Science, 2000.

Nash M J, *Crop Conservation and Storage in Cool Temperate Climates*, Pergamon Press, 1985.

Pagan J, Forages for horses. In *Advances in Equine Nutrition*, J Pagan (ed.), Kentucky Equine Research, 1998.

Park R S and Stronge M D, *Silage Production and Utilisation*, Wageningen Academic Publishing, 2005.

Raymond W F and Waltham R, *Forage Conservation and Feeding*, 5th edn, Farming Press, 1996.

Robinson D, *Grasses of Agricultural Importance*, Read Books, 2011.

Silage Advisory Centre, A selection of on-line resources available, 2013.

Spedding A and Diekmahns T, *Grasses and Legumes in British Agriculture*, CAB, 1972.

The Environment Agency, Fact sheets and forms relating to the SSAFO regulations.

Westaway M, *A Guide to Forage*, Marksway Horsehage, 2013.

Wilkinson J M, *Silage*, Chalcombe Publishing, 2005.

Wilkinson J M, Allen D M and Newman G, *Maize: Producing and Feeding Maize Silage*, Chalcombe Publishing, 1998.

Websites

www.britishgrassland.com
www.dairyco.org.uk
www.eblex.org.uk
www.environment-agency.gov.uk
www.gov.uk
www.horsehage.co.uk
www.kingshay.com
www.silageadvice.com
www.silagedecisions.co.uk
www.ukgriculture.com

Appendices

Appendix 1

Soil texture assessment in the field

The texture of a soil (i.e. the amount of sand, silt, clay and organic matter present) can be measured by mechanical analysis of a representative sample of the soil in the laboratory. But it is important for the farmer and his adviser to be able to assess the texture of the soil in the field, not only as a guide for cultivations and general management, but also because it affects the recommended application rate of soil-acting pesticides (mainly herbicides). Many of these chemicals are adsorbed by the clay and/or organic matter in the soil and so higher dose rates may be required on soils which are rich in these materials. On sandy soils, surface-applied residual herbicides may be washed into the root zone of the crop too easily and so cause damage.

With practice, it is possible to become reasonably skilled at assessing soil texture by feeling the soil in the following way:

Carefully moisten a handful of stone-free soil until the particles cling together (avoid excess water). Work it well in the hand until the structure breaks down; rub a small amount between the thumb and fingers to assess the texture according to how gritty, silky or sticky the sample feels. The handful of moist soil can also be assessed by the amount of polish it will take, and the ease or difficulty of moulding it into a ball and other shapes.

Sands feel gritty, but are not sticky when wet (loose when dry) and do not stain the fingers.

Clays (at the other extreme of particle size) take a high polish when rubbed, are very sticky, bind together very firmly, and need some pressure to mould into shapes.

Silty soils have a smooth silky feel and the more obvious this is, the greater is the amount of silt present. The amount of polish the sample takes, and its grittiness, are guides to the amount of clay and sand present.

Loams have a fairly even mixture of sand, clay and silt and, because these tend to balance each other, loam soils are not obviously gritty, silky or sticky and they

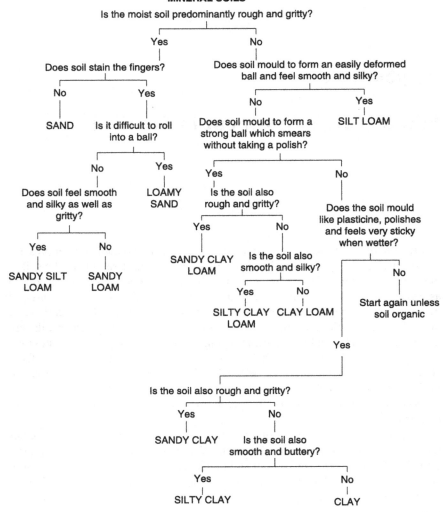

SOIL TEXTURE
MINERAL SOILS

Is the moist soil predominantly rough and gritty?

Based on Defra
RB209 (Crown copyright)

take only a slight polish. A ball of moist loam soil is easily formed, and the particles bind together well.

These are the main texture grades, but a wide range of intermediate grades exist, each having different amounts of sand (of various sizes), silt and clay particles. All this can be complicated by the amount of organic matter present which has a soft silky feel and is usually dark brown or black in colour.

Once the particular qualities described are recognised, it should be possible to use the Texture Key to arrive at a single texture.

The prefix *organic* can be applied to the above classes if organic matter levels are relatively high, i.e. 10–12%; *peat soils* when more than 20% organic matter. The prefix *calcareous* (Calc) can be applied if more then 5% calcium carbonate is present.

The above textural groups are used in advisory work for making recommendations on dose rates for soil-acting herbicides and for assessing available-water capacity, suitability for mole drainage, workability and stability of soils.

Appendix 2

Nomenclature of crops

Crop names	Botanical names
Cereals – wheat	*Triticum aestivum*
durum	*T. durum*
barley	*Hordeum vulgare*
oats	*Avena sativa*
rye	*Secale cereale*
triticale	*Triticosecale*
spelt	*T. spelta*
maize	*Zea mays*
Potato	*Solanum tuberosum*
Sugar beet, mangel, fodder beet	*Beta vulgaris*
Cabbage group	*Brassica oleracea* var. *capitata*
Brussels sprouts	*B. oleracea* var. *gemmifera*
Cauliflower	*B. oleracea* var. *botrytis*
Sprouting broccoli, calabrese	*B. oleracea* var. *italica*
Kohlrabi	*B. oleracea* var. *gongyloides*
Kale	*B. oleracea* var. *acephala*
Turnip group	*B. rapa*
Swede group including oilseed rape (colza)	*B. napus*
Mustard – brown	*B. juncea*
white	*Sinapis alba*
Fodder radish	*Raphanus sativus*
Buckwheat	*Fagopyrum esculentum*
Carrot	*Daucus carota* ssp. *sativa*
Parsnip	*Pastinaca sativa*
Celery	*Apium graveolens*
Onion	*Allium cepa*
Pea	*Pisum sativum*
Beans – field and broad	*Vicia faba*
green, dwarf, French	*Phaseolus vulgaris*
runner	*P. coccineus*
soya	*Glycine max*

Crop names	Botanical names
Vetch (tares)	*Vicia sativa*
Lupins – yellow (white)	*Lupinus luteus (L. albus)*
pearl (blue)	*L. mutabilis (L. augustifolius)*
Lucerne (alfalfa)	*Medicago sativa*
Sainfoin	*Onobrychis viciifolia*
Linseed and flax	*Linum usitatissimum*
Sunflower	*Helianthus annuus*
Grasses: ryegrass – Italian	*Lolium multiflorum*
hybrid	*L. (multiflorum × perenne)*
perennial	*L. perenne*
Grasses: cocks-foot	*Dactylis glomerata*
Timothy	*Phleum pratense*
meadow fescue	*Festuca pratense*
tall	*F. arundinacea*
red	*F. rubra*
Clovers: red	*Trifolium pratense*
white	*T. repens*
alsike	*T. hybridum*
crimson	*T. incarnatum*
Borage	*Borago officinalis*
Evening primrose	*Oenothera biennis*
Fenugreek	*Trigonella fenum-graecum*
Quinoa	*Chenopodium quinoa*
Fibre hemp	*Cannabis sativa*
Elephant grass	*Miscanthus sinensis* and *M. sacchariflorus*
Kenaf	*Hibiscus cannabinus*
Pot marigold	*Calendula officinalis*
Crambe	*Crambe abyssinica*
Vipers bugloss	*Echium plantagineum*
Gold of pleasure	*Camelina sativa*
Opium poppy	*Papaver somniferum*

Appendix 3

Nomenclature of weeds

Common names	Botanical names
Barley – meadow (P)	*Hordeum secalinum*
– wall	*H. murinum*
Bent – black (P)	*Agrostis gigantea*
– common (P)	*A. capillaris*
– creeping (watergrass) (P)	*A. stolonifera*
Bindweed – black (A)	*Fallopia convolvulus*
– field (P)	*Convolvulus arvensis*
Bird's-foot-trefoil – common (P)	*Lotus corniculatus*
Bistort amphibious (P)	*Persicaria amphibia*
Black-grass (A)	*Alopecurus myosuroides*
Borage	*Borage officinalis*
Bracken (P)	*Pteridium aquilinum*
Bristly oxtongue (A) or (B)	*Picris echioides*
Brome – soft (A) or (B)	*B. hordeaceous*
– rye (A) or (B)	*B. secalinus*
– meadow (A) or (B)	*Bromus commutatus*
– barren (sterile) (A) or (B)	*Anisantha sterilis*
– great (A)	*A. diandrus*
Broomrape – common	*Orobanche minor*
Burdock – greater (P)	*Arctium lappa*
Burnet – salad (P)	*Sanguisorba minor* ssp. *minor*
– fodder (P)	*S. minor* ssp. *muricata*
Buttercup – bulbous (P)	*Ranunculus bulbosus*
– corn (A)	*R. arvensis*
– creeping (P)	*R. repens*
– meadow (crowfoot) (P)	*R. acris*
Campion – red (B)	*Silene dioica*
– bladder (P)	*S. vulgaris*
– white (B) or (P)	*S. latifolia*

Common names	Botanical names
Canary grass – awned (bristle-spiked) (A)	*Phalaris paradoxa*
Carrot – wild (A)	*Daucus carota*
Cat's-ear (P)	*Hypochaeris radicata*
Chamomile – corn (A) or (B)	*Anthemis arvensis*
Charlock (yellow) (A)	*Sinapis arvensis*
Chervil – rough (B)	*Chaerophyllum temulum*
Chickweed – common (A)	*Stellaria media*
– common mouse-ear (A)	*Cerastium fontanum*
Cleavers (A)	*Galium aparine*
Colt's-foot (P)	*Tussilago farfara*
Corncockle (A)	*Agrostemma githago*
Cornflower (A)	*Centaurea cyanus*
Corn mint (P)	*Mentha arvensis*
Cornsalad – common (A)	*Valerianella locusta*
Couch – common (P)	*Elytrigia repens*
– onion (false oat grass) (P)	*Arrhenatherum elatius* (var. *bulbosum*)
Cowbane (P)	*Cicuta virosa*
Cow parsley (wild chervil) (P)	*Anthriscus sylvestris*
Cow-wheat – field (A)	*Melampyrum arvense*
Crane's-bill (A)	*Geranium* spp.
Cress – hairy bitter (A)	*Cardamine hirsuta*
– swine (A)	*Coronopus squamatus*
Cuckoo flower (P)	*Cardamine pratensis*
Daisy (P)	*Bellis perennis*
Dandelions (P)	*Taraxacum* agg.
Darnel (A)	*Lolium temulentum*
Dead nettle – red (A)	*Lamium purpureum*
– Henbit (A)	*L. amplexicaule*
Docks – broadleaved (P)	*Rumex obtusifolius*
– curled (P)	*R. crispus*
Duckweed – common	*Lemna minor*
– ivy-leaved	*L. trisulca*
Fat hen (A)	*Chenopodium album*
Fescue – red (P)	*Festuca rubra* agg.
– sheep's (P)	*F. ovina*
Fiddleneck common (A)	*Amsinckia micrantha*
Fleabane – Canadian (A)	*Conyza canadensis*
Flixweed (A)	*Descurainia sophia*
Fool's parsley	*Aethusa cynapium*
Forget-me-not (field) (A)	*Myosotis arvensis*
Foxglove (B) or (P)	*Digitalis purpurea*
Foxtail – meadow (P)	*Alopecurus pratensis*
Fritillary (P)	*Fritillaria meleagris*
Fumitory – common (A)	*Fumaria officinalis*
Gallant soldier (A)	*Galinsoga parviflora*
Garlic – field (P)	*Allium oleraceum*
Gorse (whin, furze) (P)	*Ulex europaeus*
Goosefoot – many seeded (A)	*Chenopodium polyspermum*
Gromwell – field (A)	*Lithospermum arvense*

(Continued)

Continued

Common names	Botanical names
Ground-elder (P)	*Aegopodium podagraria*
Ground-ivy (P)	*Glechoma hederacea*
Groundsel (A)	*Senecio vulgaris*
Hawkbit – autumn (P)	*Leontodon autumnalis*
– rough (P)	*L. hispidus*
Hawk's-beard – rough (B)	*Crepis biennis*
– smooth (A)	*C. capillaris*
Heather (P)	*Calluna vulgaris*
– bell (P)	*Erica cinerea*
Hedge mustard (A)	*Sisymbrium officinale*
Hedge parsley – spreading (A)	*Torilis arvensis*
Hemlock (A) or (B)	*Conium maculatum*
Hemp nettle – common (A)	*Galeopsis tetrahit*
Henbane (A)	*Hyoscyamus niger*
Hogweed (P)	*Heracleum sphondylium*
– giant (P)	*H. mantegazzianum*
Horsetail – field (P)	*Equisetum arvense*
– marsh (P)	*E. palustre*
Knapweed – common (P)	*Centaurea nigra*
Knawel – annual (A) or (B)	*Scleranthus annuus*
Knot-grass (A)	*Polygonum aviculare*
Knotweed – Japanese (P)	*Fallopia japonica*
Loose silky-bent (A)	*Apera spica-venti*
Marigold – corn (A)	*Chrysanthemum segetum*
Mayweed – scented (A)	*Matricaria recutita*
– scentless (A)	*Tripleurospermium inodorum*
– stinking chamomile (A)	*Anthemis cotula*
Meadow-grass – annual (A)	*Poa annua*
– rough (P)	*P. trivialis*
– smooth (P)	*P. pratensis*
Medick – black (A) or (P)	*Medicago lupulina*
Mercury – annual (A)	*Mercurialis annua*
– dog's (P)	*M. perennis*
Mignonette – wild (B) or (P)	*Reseda lutea*
Mugwort (P)	*Artemisia vulgaris*
Mustard – black (A)	*Brassica nigra*
– white (A)	*Sinapis alba*
– treacle (A)	*Erysimum cheiranthoides*
Nettle – common (P)	*Urtica dioica*
– small, annual (A)	*U. urens*
Nightshade – black (A) or (B)	*Solanum nigrum*
– deadly (P)	*Atropa belladonna*
Nipplewort (A)	*Lapsana communis*
Nutsedge – yellow (P)	*Cyperus esculentus*
Oat – bristle (A)	*Avena strigosa*
– spring, wild (A)	*A. fatua*
– winter, wild (A)	*A. sterilis (A. ludoviciana)*

Common names	Botanical names
Oat-grass – downy (P)	*Helictotrichon pubescens*
– false (onion couch) (P)	*Arrhenatherum elatius* (var. *bulbosum*)
Onion – wild (P)	*Allium vineale*
Orache – common (A)	*Atriplex patula*
Pansy – field (A)	*Viola arvensis*
– wild (A)	*V. tricolor*
Parsley – cow (P)	*Anthriscus sylvestris*
– fool's (A)	*Aethusa cynapium*
Parsley-piert (A)	*Aphanes arvensis*
Pearlwort (P) – procumbent	*Sagina procumbens*
Penny-cress – field (A)	*Thlaspi arvense*
Persicaria – pale (A)	*Persicaria lapathifolia*
Pineapple weed (rayless mayweed) (A)	*Matricaria disciodes*
Plantain – greater (P)	*Plantago major*
– ribwort (narrow-leaved) (P)	*P. lanceolata*
Poppy – corn or common (A)	*Papaver rhoeas*
– Californian (A)	*Eschscholzia californica*
Primrose (P)	*Primula vulgaris*
Radish – wild (sea) (A) or (B)	*Raphanus raphanistrum*
Ragwort – common (B) or (P)	*Senecio jacobaea*
– Oxford (P)	*S. squalidus*
Ramsons (P)	*Allium ursinum*
Red bartsia (A)	*Odontites vernus*
Redshank (A)	*Persicaria maculosa*
Reed – common (P)	*Phragmites australis*
Restharrow – common (P)	*Ononis repens*
– spiny (P)	*O. spinosa*
Rush – jointed (P)	*Juncus articulatus*
– soft (common) (P)	*J. effusus*
– hard (P)	*J. inflexus*
– heath (P)	*J. squarrosus*
Saffron – meadow (P)	*Colchicum autumnale*
St John's-wort (P) – perforate	*Hypericum perforatum*
Scabious – field (P)	*Knautia arvensis*
Scarlet pimpernel (A)	*Anagallis arvensis*
Sedges (P)	*Carex* spp.
Selfheal (P)	*Prunella vulgaris*
Shepherd's-needle (A)	*Scandix pecten-veneris*
Shepherd's-purse (A)	*Capsella bursa-pastoris*
Silverweed (P)	*Potentilla anserina*
Soft-grass – creeping (P)	*Holcus mollis*
Sorrel – common (P)	*Rumex acetosa*
– sheep's (P)	*R. acetosella*
Sow-thistle – corn or perennial (P)	*Sonchus arvensis*
– smooth, milk (A)	*S. oleraceus*
Speedwell – common, field (A)	*Veronica persica*
– germander (P)	*V. chamaedrys*
– green field (A)	*V. agrestis*
– ivy-leaved (A)	*V. hederifolia*
– wall	*V. arvensis*

(Continued)

Continued

Common names	Botanical names
Spurge – sun (A)	*Euphorbia helioscopia*
– dwarf (A)	*E. exigua*
Spurrey – corn (A)	*Spergula arvensis*
Stork's-bill – common (A) or (B)	*Erodium cicutarium*
Thistle – creeping (field) (P)	*Cirsium arvense*
– spear (Scotch) (B)	*C. vulgare*
– marsh (P)	*C. palustre*
Toadflax – common (P)	*Linaria vulgaris*
Trefoil hop (A)	*Trifolium campestre*
Tufted hair-grass (P)	*Deschampsia cespitosa*
Venus's-looking-glass (A)	*Legousia hybrida*
Vetch – common (tares) (A) or (B)	*Vicia sativa*
– kidney (P)	*Anthyllis vulneraria*
Viper's-bugloss (B)	*Echium vulgare*
Water-dropwort (hemlock) (P)	*Oenanthe crocata*
Willow-herb (rosebay) (P)	*Chamerion angustifolium*
Wood-rush – field (P)	*Luzula campestris*
Yarrow (P)	*Achillea millefolium*
Yellow iris or flag (P)	*Iris pseudacorus*
Yellow-rattle (A)	*Rhinanthus minor*
Yorkshire fog (P)	*Holcus lanatus*

Useful web site for plant distribution: www.brc.ac.uk/plantatlas/

Appendix 4

Nomenclature of diseases

Crop names	Common names	Botanical names
Cereals		
All cereals	Powdery mildew*	*Blumeria graminis*
Wheat, barley, triticale	Yellow (stripe) rust	*Puccinia striiformis*
Barley	Brown rust	*P. hordei*
Wheat	Brown rust	*P. triticina*
Rye, triticale	Brown rust	*P. recondita*
Oats	Crown rust	*P. coronata*
Barley	Rhynchosporium (leaf scald)	*Rhynchosporium secalis*
Wheat, barley, rye	(Septoria nodorum) leaf + glume blotch	*Phaeosphaeria nodorum* *(Stagonospora nodorum)*
Oats	Dark leaf spot, speckle blotch	*S. avenae 'Leptosphaeria avenaria*
Wheat	Septoria leaf blotch	*Mycosphaerella graminicola* *(Septoria tritici)*
Oats	Leaf spot, seedling blight	*Pyrenophora avenae*
Barley	Net blotch*	*P. teres*
Wheat, barley, rye	Tan spot	*P. tritici-repentis*
All cereals	Black point	*Alternaria* spp.
Barley	Halo spot	*Selenophoma donacis*
Barley	Ramularia leaf-spot	*Ramularia collo-cygni*
All cereals	Ascochyta leaf scorch (spot)	*Didymella exitialis*
All cereals	Pink snow mould	*Monagraphella nivalis* *(Microdochium nivale)*
All cereals	Ergot	*Claviceps purpurea*
Wheat,	Bunt, stinking smut	*Tilletia tritici*
Barley	Bunt, covered, smut	*Ustilago hordei*
Oats	Bunt, covered, smut	*U. segetum*
Oats	Loose smut	*U. avertae*
Wheat, Barley	Loose smut*	*U. nuda*
Rye	Stripe smut	*Urocystis occulta*
Barley	Leaf stripe	*Pyrenophara graminea*

(Continued)

Continued

Crop names	Common names	Botanical names
All cereals	Cephalosporium leaf stripe	*Hymenella cerealis*
All cereals	Eyespot–rye type	*Oculimacula acuformis*
All cereals	Eyespot–wheat type	*O. yallundae*
All cereals	Sharp eyespot	*Cerato basidium cereale* (*Rhizoctonia cerealis*)
Wheat, barley, rye	Take-all	*Gaeumannomyces graminis* var. *tritici*
Oats, wheat, barley, rye	Take-all	*G. graminis* var. *avenae*
All cereals	Brown foot rots and ear blight	*Fusarium* spp. and *Microdochium nivale*
All cereals	Black (sooty) mould	*Cladosporium* spp.
Winter wheat, barley	Snow rot	*Typhula incarnata*
All cereals	Rhizoctonia stunt	*Thanatephorus cucumeris*
Maize	Damping-off	*Pythium* spp.
Wheat, barley	Omphalina patch	*Omphalina pyxidata*
Maize	Bunt, covered, smut	*Usilago maydis*
Maize	Stalk rot	*Fusarium* spp.
Maize	Eyespot	*Kabatiella zeae*
Potatoes		
	Foliage blight and tuber blight	*Phytophthora infestans*
	Pink rot	*P. erythroseptica*
	Early blight	*Alternaria alternata* or *A. solani*
	Common scab	*Streptomyces scabiei*
	Powdery scab	*Spongospora subterranea*
	Gangrene	*Phoma exigua* var. *foveata*
	Watery wound rot	*Pythium* spp.
	Wart disease	*Synchytrium endobioticum*
	Sclerotinia white mould	*Sclerotinia sclerotiorum*
	Blackleg and soft rot	*Pectobacterium* spp. and *Dickeya* spp.
	Skin spot	*Polyscylatum pustulans*
	Black scurf and stem canker	*Rhizoctonia solani* '*Thanatephores cucumeris*'
	Dry rot	*Fusarium* spp.
	Brown rot	*Ralstonia solanacearum*
	Black dot	*Colletotrichum coccodes*
	Silver scurf	*Helminthosporium solani*
	Ring rot	*Clavibacter michiganensis* subsp. *sepedonicus*
	Verticillium wilt	*Verticillium albo-atrum*
Sugar beet		
	Blackleg	*Pleospora bjoerlingii*
	Downy mildew	*Peronospora farinosa*
	Powdery mildew	*Erysiphe betae*
	Ramularia leaf spot	*Ramularia beticola*
	Beet rust	*Uromyces betae*
	Violet root rot	*Helicobasidium purpureum*
	Vector of rhizomania	*Polymyxa betae*
	Cercospera leaf spot	*Cercospera beticola*
Legumes		
Peas, beans	Downy mildew	*Peronospora viciae*
Peas	Leaf and pod spot	*Ascochyta pisi, Mycosphaerella pinodes*
	Ascochyta blight	

Crop names	Common names	Botanical names
Beans	Leaf and pod spot	*A. fabae 'Didymella fabae'*
Beans	Bean rust	*Uromyces viciae-fabae*
Beans, peas	Chocolate spot and grey mould	*Botrytis* spp. '*Botryotinia fuckeliana*'
Peas, beans	White mould, stem rot	*Sclerotinia* spp.
Red clover, lucerne	Clover rot	*S. trifoliorum*
All legumes	Damping-off of seedlings	*Pythium* spp.
Peas	Pea wilt	*Fusarium oxysporium* f.sp. *pisi*
Lucerne	Verticillium wilt	*Verticillium albo-atrum*
Peas	Powdery mildew	*Erysiphe pisi*
Peas	Bacterial blight	*Pseudomonas syringae* pv. *pisi*
Peas, beans	Foot rot	*Fusarium solani* f.sp. *pisi*
Brassicae		
	Club root (finger and toe)	*Plasmodiophora brassicae*
	Powdery mildew	*Erysiphe cruciferarum*
	Downy mildew	*Peronospora parasitica*
	White blister	*Albugo candida*
	Light leaf spot	*Pyrenopeziza brassicae*
	Dark leaf and pod spot	*Alternaria* spp.
	Stem canker and phoma leaf spot	*Leptosphaeria maculans* and *L. biglobosa*
	Ring spot	*Mycosphaerella brassicicola*
	Stem rot	*Sclerotinia sclerotiorum*
	White leaf spot	*Mycosphaerella capsellae*
	Verticillium wilt	*Verticillium longisporum*
	Grey mould	*Botrytis cinerea*
	Black rot	*Xanthomonas campestris* pv. *campestris*
Linseed		
	Alternaria blight	*Alternaria linicola*
	Pasmo	*Mycosphaerella linicola*
	Powdery mildew	*Sphaerotheca lini*
	Grey mould	*Botryotinia fuckeliana*
Onions		
	White rot	*Sclerotium cepivorum*
	Downy mildew	*Peronospora destructor*
	Neck rot	*Botrytis allii*
Grasses		
	Drechslera leaf spot	*Drechslera* spp. '*Pyrenophora* spp.'
	Ergot	*Claviceps purpurea*
	Powdery mildew	*Blumeria graminis*
	Rhynchosporium leaf spot	*Rhynchosporium* spp.
	Crown rust	*Puccinia coronata*
	Brown rust	*P. loliina*
	Stem rust	*P. graminis*

* Different races or forms found on different cereals

Note The sexual (teleomorph) stages for some diseases have only just been identified. These are given as ' '. The anamorph, imperfect or asexual classification is still commonly used.

Appendix 5

Nomenclature of pests

Crop	Common names	Latin names
Cereals		
Wheat, barley, oats	Cereal cyst nematode	*Heterodera avenae*
Oats, rye	Stem nematode	*Ditylenchus dipsaci*
Cereals	Bird cherry aphid	*Rhopalosiphum padi*
Barley, oats	Fescue aphid	*Metopolophium festucae*
Barley, oats, wheat	Rose grain aphid	*Metopolophium dirhodum*
All	Grain aphid	*Sitobion avenae*
All	Leatherjackets	*Tipula* spp. and *Nephrotoma* spp.
All	Wireworms	*Agriotes* spp.
All	Slugs	*Deroceras reticulatum*, *Arion* spp.
Wheat, barley, rye	Wheat bulb fly	*Delia coarctata*
Wheat	Yellow cereal fly	*Opomyza florum*
Barley, wheat	Gout fly	*Chlorops pumilionis*
Oats, barley, wheat, maize	Frit fly	*Oscinella frit*
All	Saddle-gall midge	*Haplodiplosis marginata*
All	Thrips	*Limothrips* spp.
Barley	Cereal ground beetle	*Zabrus tenebrioides*
Wheat	Orange wheat blossom midge	*Sitodiplosis mosellana*
Wheat	Yellow wheat blossom midge	*Contarinia tritici*
All	Grain weevils	*Sitophilus* spp.
All	Saw-toothed grain beetle	*Oryzephilus surinamensis*
All	Mites	*Acarus siro*, *Lepidoglyphus destructor* and *Tyrophagus longior*
All	Swift moth	*Hepialus* spp.
All	Cereal leaf beetle	*Oulema melanopa*

Crop	Common names	Latin names
Potatoes		
	Stubby root nematodes	*Trichodorus* spp. and *Paratrichodorus* spp.
	Golden potato cyst nematode (PCN)	*Globodera rostochiensis*
	White potato cyst nematode	*Globodera pallida*
	PCN	*Myzus persicae, Aulacorthum solani, Macrosiphum euphorbiae*
	Aphids	*Aphis nasturtii* and *A. fabae*
	Colorado beetle	*Leptinotarsa decemlineata*
	Cutworms (noctuid moths), e.g. turnip moth	*Agrotis segetum*
	Large yellow underwing	*Noctula pronuba*
	Garden dart moth	*Euxoa nigricans*
	Slugs, leatherjackets, swift moth, wireworms, see Cereals	
	Chafer grubs	*Melolontha melolontha* and *Phyllopertha horticola*
Sugar beet		
	Beet cyst nematode	*Heterodera schachtii*
	Aphid – peach potato	*Myzus persicae*
	Aphid – black bean (blackfly)	*Aphis fabae*
	Mangold fly	*Pegomya hyoscyami*
	Millipedes	*Blaniulus guttulatus*
	Docking disorder	*Longidorus* and *Trichodorus* spp.
	Cutworms, slugs, wireworms, chafer grubs, leatherjackets, see Potatoes and Cereals	
	Beet flea beetle	*Chaetocnema concinna*
	Silver 'Y' moth	*Autographa gamma*
Peas and beans		
	Pea and bean weevil	*Sitona lineatus*
	Pea cyst nematode	*Heterodera gottingiana*
	Stem nematode	*Ditylenchus dipsaci* and *D. gigas*
	Black bean aphid (blackfly)	*Aphis fabae*
	Bean seed beetle (Bruchid beetle)	*Bruchus rufimanus*
	Pea moth	*Cydia nigricana*
	Pea aphid	*Acyrthosiphon pisum*
	Pea midge	*Contarinia pisi*
	Field thrips	*Thrips angusticeps*
	Pea thrips	*Kakothrips pisivorus*
Oilseed rape		
	Flea beetle	*Phyllotreta* spp.
	Cabbage root fly	*Delia radicum*
	Pollen beetle	*Meligethes aeneus*
	Cabbage seed weevil	*Ceuthorhynchus assimilis*
	Cabbage aphid	*Brevicoryne brassicae*
	Cabbage stem flea beetle	*Psylliodes chrysocephala*
	Brassica pod midge	*Dasineura brassicae*

(Continued)

Continued

Crop	Common names	Latin names
	Turnip sawfly	*Athalia rosae*
	Cabbage stem weevil	*Ceutorhynchus quadridens*
Field brassicae		
	Small white butterflies	*Pieris rapae*
	Diamond-back moth	*Plutella xylostella*
	(Cabbage root fly, flea beetle, and cabbage aphid see oilseed rape)	
Onions and leeks		
	Onion fly	*Delia antiqua*
	Onion thrip	*Thrips tabaci*
	Leek moth	*Acrolepiopsis assectella*
Carrots		
	Carrot fly	*Psila rosae*
	Nematode causing fanging	*Longidorus* spp., *Trichodorus* spp. and *Pratylenchus* spp.
	Carrot willow aphid	*Cavariella aegopodii*
Grasses		
	Frit fly, leatherjacket, swift moth, wireworm, slugs, see Cereals	
	Bibionid flies	*Dilophus febrilis* and *Bibio marci*
Herbage legumes		
	Stem nematode	*Ditylenchus dipsaci*
	Clover weevil	*Sitona* spp.

Appendix 6

Crop seeds

The following are average figures and are only intended as a general guide and for comparisons. Precision drilling of crops requires seed counts per kilogram to be known for the stock of seed being sown and merchants will usually supply these figures.

Crop	1000 seeds weight (g)	Seeds per kilogram (000s)	Seeds per m² for every 10 kg/ha sown	kg/hl
Cereals – wheat	48	21	21	75
barley	37	27	27	68
oats	32	31	31	52
maize	285	3.5	3.5	75
Grasses – ryegrass – Italian	2.2	455	455	29
ryegrass – Italian tetraploid	4	250	250	
ryegrass – hybrid	2.15	465	465	
ryegrass – perennial S 24	2	500	500	35
ryegrass – perennial tetraploid	3.3	303	303	
cocksfoot	1	1000	1000	33
Timothy	0.3	3333	3333	62
meadow fescue	2	500	500	37
tall fescue	2.5	400	400	30
Clovers – red	1.75	571	571	80
white – cultivated wild	0.62	1613	1613	82
	0.58	1724	1724	82
Lucerne (alfalfa)	2.35	425	425	77
Sainfoin – milled	21	48	48	

(*Continued*)

Continued

Crop	1000 seeds weight (g)	Seeds per kilogram (000s)	Seeds per m² for every 10 kg/ha sown	kg/hl
Peas – marrowfats	330	3	3	
large blues and whites	250	4	4	78
small blues	200	5	5	
Beans – broad	980	1	1	
winter and horse types	670	1.5	1.5	
tick	410	2.4	2.4	80
dwarf, green, French	500	2	2	
Vetches (tares)	53	19	19	77
Linseed and flax	10	100	100	68
Carrots	1.5	660	660	
Onions – natural seed	4	246	246	
mini-pellets	18	55	55	
Sugar beet – pelleted	62	16	16	
Cabbages	4.1	240	240	
Kale	4.5	220	220	
Swedes	3.6	280	280	
Turnips	3	330	330	
Oilseed rape	4.5	220	220	
Brussels sprouts	4.7	210	210	
Cauliflower	4.1	240	240	

Appendix 7

Agricultural land classification (ALC) in England and Wales

Defra land classification maps and reports

The 12 million hectares of agricultural land in England and Wales have been classified into five grades by the Department for Environment Food and Rural Affairs (Defra). These are used by planners when considering requests for planning permission which would take land out of agricultural use.

This survey work is published on coloured Ordnance Survey maps on three scales:

(i)   1:625 000 covering the whole of England and Wales.
(ii)   1:250 000, there are six of these, each covering approximately one Defra region.
(iii) 1:63 260 i.e. 1 inch to 1 mile.

The details shown in (i) are more generalised than in (ii) while (iii) shows areas accurate to approx. 50 acres. The classification is based mainly on:

Climate – rainfall, accumulated temperature.
Relief – slope, surface irregularities, flood risk.
Soil – wetness, depth, texture, structure, stoniness and moisture balance.

These characteristics can affect the range of crops which can be grown, the level of yield, the consistency of and the cost of obtaining the yield. Land is graded into five grades under this classification with the third grade divided into two sub-grades.

Grade 1 – Excellent. Land with no, or very minor, limitations to agricultural use. A very wide range of agricultural and horticultural crops can be grown. Yields are high and consistent. This grade occupies 3% of agricultural land in England and Wales and is coloured dark blue on the maps.

Grade 2 – Very good. Land with minor limitations which affect crop yield, cultivations or harvesting. A wide range of crops can be grown, but there may be difficulties with very demanding crops. The level of yield is generally high, but it can be lower or more variable than Grade 1. This grade occupies 14% of agricultural land and is coloured light blue.

Grade 3 – Good to moderate. Land with moderate limitations which affects choice of crop, timing and type of cultivations, harvesting or level of yield. Yields are generally lower or more variable than on land in Grades 1 and 2. Grade 3 is the largest group of soils, occupying 50% of agricultural land and is coloured green.

Sub-grade 3a – Moderate/high yields of a narrow range of crops (especially cereals) or moderate yields of a wide range of crops. Some good grassland.

Sub-grade 3b – Moderate yields of a narrow range of crops and lower yields of a wide range of crops.

Grade 4 – Poor. Land with severe limitations restricting range and/or yield levels. Suited to grass plus some low and variable yields of cereals. It also includes droughty arable land. It occupies 20% of agricultural land and is coloured yellow.

Grade 5 – Very poor. Land with very severe limitations. Permanent pasture or rough grazing. It occupies 14% of agricultural land and is coloured light brown.

Urban areas are coloured red and non-agricultural areas are coloured orange on the maps.

Additionally, Defra is undertaking a physical classification of the hill and upland areas of Grades 4 and 5 which will be divided into two main categories:

The Uplands – The enclosed and wholly or partially improved land; there will be five sub-grades.

The Hills – Unimproved areas of natural vegetation; there will be six sub-grades.

The physical factors to be taken into account will be vegetation, gradient, irregularity and wetness.

The object of this classification is to help the farmer plan possible improvement schemes when conditions permit. It is also intended to be a guide to those concerned with conservation and amenity who are trying to preserve scenery and ecology. Booklets giving full details of all the classifications and grades and maps can be obtained from the Defra.

Soil Series and Soil Survey Maps

In the grading of land for the Land Classification Maps and Reports, use was made of the maps which show the Soil Series Classification of England and Wales. A 'soil series' is a soil formed from the same parent material, and having similar horizons (layers) in their profiles. Each soil series is given a name, usually where it was first recognised and described. The name is used for the soil series, however widespread, throughout the country; it is also used on the maps which are produced on 1:250 000 scale. A few examples of the named soil series are:

Romney series. These soils are deep, very fine, sandy loams found in the silt areas of Romney Marsh, and also in parts of Cambridgeshire, Lincolnshire and Norfolk. They are potentially very fertile soils.

Bromyard series. Red-coloured, silt loam soils (red marls) found in Herefordshire and other parts of the West Midlands and in the south-west. They should not be worked when wet. Suitable crops are cereals, grass, fruit and hops.

Evesham series. Lime-rich soils formed from Lias (or similar) clays, found in parts of Warwickshire, Gloucestershire, Somerset, East and West Midlands. They are normally very heavy soils and are best suited to grass and cereals.

Sherborne series. Shallow (less than 25 cm deep), reddish-brown, loam-textured soils of variable depth and stoniness (and so subject to drought). They are mainly found on the soft oolitic limestones of the Cotswolds, and in parts of Northamptonshire, and the Cliff region of Lincolnshire. The soils of this series are moderately fertile, easy to manage and mainly grow cereals and grass.

Worcester series. Red silt loam (or silty clay loam) formed from Keuper Marl and found in the East and West Midlands and in the south west. They are slow-draining, require subsoiling regularly and are best suited to grass and cereals.

Newport series. Free-draining, deep, easy-working, sandy loams over loamy sands with varying amounts of stones. They are formed from sands or gravels of glacial origin and are found in many areas of the Midlands and north. They are well suited to arable cropping.

Several different soil series may be found on the same farm and sometimes in the same field.

The Land Use Capability Classification (LUCC) in the United Kingdom has been modelled on the United States Department of Agriculture Classification Scheme. It uses the soil series groups in conjunction with limitations imposed by wetness (w), soils (s), gradient and soil pattern (g), erosion (e) and climate (c). The Classification divides all land into seven classes (1–4 classes are very similar to grades 1–4 of the Land Classification Maps prepared by Defra; Classes 6 and 7 are virtually useless for agriculture). Each class has sub-divisions and limitations as indicated by the letters w, s, g, e, or c, after the class number, e.g. Class 2w. This classification was carried out by the Soil Survey and Land Resource Center, Silsoe. Although a more accurate and useful system than ALC the LUCC is hardly ever used.

Land is *Not* graded on suitability for grass only. Wheat and potatoes at scales of 1:250 000 for the whole country, and at 1:63 360 or 1:25 000 for selected areas.

Index

Printed in the United States
By Bookmasters